Lecture Notes in Computer Science

Commenced Publication in 1973
Founding and Former Series Editors:
Gerhard Goos, Juris Hartmanis, and Jan van Leeuwen

Nikos Mamoulis   Thomas Seidl
Torben Bach Pedersen   Kristian Torp
Ira Assent (Eds.)

Advances in Spatial and Temporal Databases

11th International Symposium, SSTD 2009
Aalborg, Denmark, July 8-10, 2009
Proceedings

 Springer

Volume Editors

Nikos Mamoulis
University of Hong Kong
Department of Computer Science
Pokfulam Road, Hong Kong, China
E-mail: nikos@cs.hku.hk

Thomas Seidl
RWTH Aachen University
Department of Computer Science
52056 Aachen, Germany
E-mail: seidl@cs.rwth-aachen.de

Torben Bach Pedersen
Kristian Torp
Ira Assent
Aalborg University
Department of Computer Science
Selma Lagerlöfsvej 300, 9220 Aalborg Ø, Denmark
E-mail: {tbp, torp, ira}@cs.aau.dk

Library of Congress Control Number: 2009929547

CR Subject Classification (1998): H.2.0, H.2.8, H.2-4, I.2.4

LNCS Sublibrary: SL 3 – Information Systems and Application, incl. Internet/Web
and HCI

ISSN	0302-9743
ISBN-10	3-642-02981-7 Springer Berlin Heidelberg New York
ISBN-13	978-3-642-02981-3 Springer Berlin Heidelberg New York

springer.com

© Springer-Verlag Berlin Heidelberg 2009
Printed in Germany

Typesetting: Camera-ready by author, data conversion by Scientific Publishing Services, Chennai, India
Printed on acid-free paper   SPIN: 12716166   06/3180   5 4 3 2 1 0

Preface

SSTD 2009 was the 11th in a series of biannual events that discuss new and exciting research in spatio-temporal data management and related technologies. Previous symposia were successfully held in Santa Barbara (1989), Zurich (1991), Singapore (1993), Portland (1995), Berlin (1997), Hong Kong (1999), Los Angeles (2001), Santorini, Greece (2003), Angra dos Reis, Brazil (2005), and Boston (2007). Before 2001, the series was devoted solely to spatial database management, and called SSD. From 2001, the scope was extended in order to also accommodate temporal database management, in part due to the increasing importance of research that considers spatial and temporal aspects jointly.

SSTD 2009 introduced several innovative aspects compared to previous events. There was a demonstrations track which included ten presentations of systems related to the topics of interest. In addition to that, the event included a poster session with seven presentations of innovative research developed at an early stage. For the first time in the SSTD series, the best paper of the symposium was awarded and a few high-quality papers were selected and the authors were invited to submit extended versions of their work to a special issue of the *Geoinformatica* journal (Springer). Prior to the symposium, there was a two-day advanced seminar, which hosted three half-day tutorials on state-of-the-art topics within spatio-temporal data management, held by distinguished international researchers.

SSTD 2009 received 62 research submissions and 11 demonstration submissions from 20 countries (based on the affiliation of the first author). A thorough review process led to the acceptance of 20 high-quality papers, geographically distributed as follows: USA 6, Greece 3, Germany 2, Belgium 1, Brazil 1, Denmark 1, Hong Kong 1, Ireland 1, Israel 1, Singapore 1, South Korea 1, and United Kingdom 1. The papers are classified in the following categories, each corresponding to a conference session: (1) Spatial and Flow Networks, (2) Integrity and Security, (3) Uncertain Data and New Technologies, (4) Indexing and Monitoring Moving Objects, (5) Advanced Queries, and (6) Models and Languages.

Distinguished members of the community delivered the three keynotes, covering diverse subjects. Mor Naaman (Rutgers University) discussed how the spatio-temporal information tracked by social media services allows for a new type of "spatio-tempo-social" analysis of world facts and human behavior. Sudhakar Menon (ESRI) gave an overview of the design and architecture of GIS Servers for Web-based Information Systems, using ESRI's ArcGIS Server system as a concrete example. Lars Arge described some of the recent advances in the development of worst-case efficient range search indexing structures.

The success of SSTD 2009 was the result of team effort. First, we would like to thank the authors, irrespectively of whether their papers were accepted or not, for their support of the conference series and for sustaining the high quality of the submissions. Second, we are grateful to the members of the Program Committee (and the external reviewers) for their thorough and timely reviews. Third, we thank the invited speakers for their excellent keynotes. Fourth, we are grateful to Christian S. Jensen for his advice and support and to Man Lung Yiu and Hua Lu for maintaining the conference website and publicizing the event. Finally, we would like to thank our sponsors ESRI, Det Obelske Familiefond, and Otto Mønsteds Fond for their generous support. We believe that SSTD 2009 continued the successful tradition of the series, providing an interesting program and lively discussions in a pleasant environment.

May 2009   Nikos Mamoulis
                                                            Thomas Seidl
                                                    Torben Bach Pedersen
                                                           Kristian Torp
                                                             Ira Assent

Organization

Program Chairs

Nikos Mamoulis	University of Hong Kong, Hong Kong, China
Thomas Seidl	RWTH Aachen University, Germany

General Chair

Torben Bach Pedersen	Aalborg University, Denmark

General Co-chair

Kristian Torp	Aalborg University, Denmark

Proceedings Chair

Ira Assent	Aalborg University, Denmark

Program Committee

Walid Aref	Ki-Joune Li	Timos Sellis
Lars Arge	Yannis Manolopoulos	Cyrus Shahabi
Spiridon Bakiras	Mohamed Mokbel	Shashi Shekhar
Claudio Bettini	Kyriakos Mouratidis	Richard Snodgrass
Thomas Brinkhoff	Mirco Nanni	Kian-Lee Tan
Reynold Cheng	Enrico Nardelli	Yufei Tao
Hakan Ferhatosmanoglu	Mario Nascimento	Yannis Theodoridis
Ralf Hartmut Güting	Dimitris Papadias	Agnès Voisard
Marios Hadjieleftheriou	Spiros Papadimitriou	Ouri Wolfson
Erik Hoel	Matthias Renz	Michael Worboys
Panos Kalnis	Philippe Rigaux	Donghui Zhang
George Kollios	Markus Schneider	Baihua Zheng
Peer Kröger	Bernhard Seeger	

External Reviewers

Mohammad Ali Abam	Guadalupe Canahuate	Michel Crucianu
Daniar Achakeyev	Jinchuan Chen	Ugur Demiryurek
Daniel Ayala	Chi-Yin Chow	Tobias Emrich
Joel Booth	Antonio Corral	Mike Evans

Dario Freni
Elias Frentzos
Nikos Giatrakos
Jian Gong
Franz Graf
Herman Haverkort
Ling Hu
James Kang
Leyla Kazemi
Mohamed Khalefa
Ali Khoshgozaran
Onur Kucuktunc
Justin Levandoski
Yimin Lin

Gerasimos Marketos
Sergio Mascetti
Praadeep Mohan
Joe Naps
Stavros Papadopoulos
Linda Pareschi
Kostas Patroumpas
Kyriacos Pavlou
Nikos Pelekis
Michalis Potamias
Philip Prange
Chedy Raissi
Daniele Riboni
Ioannis Roussos

Ahmet Sacan
Houtan Shirani-Mehr
Yannis Stavrakas
Liwen Sun
Panagiotis Symeonidis
Piotr Szczurek
Kostas Tsichlas
Michael Vassilakopoulos
Lixing Wang
Xike Xie
Bo Xu
Yin Yang
Jilian Zhang

Table of Contents

6. Models and Languages

Short Papers

Demonstrations

Spatio-Tempo-Social: Learning from and about Humans with Social Media

Mor Naaman

School of Communication and Information
Rutgers University, New Brunswick NJ 08901, USA
mor@scils.rutgers.edu
http://scils.rutgers.edu/~mor

Social media – online services that encourage content sharing through individual participation – have encouraged and enabled people to share various types of information in social and public settings. Flickr, Twitter, Facebook, YouTube, Blogger, MySpace and their likes have become platforms where millions of participants share nuggets of their life, their knowledge, their creations and their opinions in various manners: from blog posts, to status updates, to multimedia content such as photos and videos.

These social media artifacts often serve as explicit and implicit indicators of attention. Individual participants, driven by their unique set of motivations [1], are uploading, sharing and annotating content on these sites. These resources are often markers of the person's interests and activities. Taken together, the cummulative effect of all these activities is a dynamically created, constantly updated reflection of people and their attention.

Adding spatio-temporal metadata to this content, we may get an idea of what people are paying attention to *in space and time*. Many of the social media services (first and foremost, Flickr) allow the users to annotate any piece of content with location (and time) metadata. In other services, location can be derived from known entities mentioned in the text. Indeed, Flickr alone features over 100,000,000 "geotagged" photos and videos, specifying the location in which media were captured in various level of specificity. On the personal level, these spatio-temporal collections might inform us about the user's spatial breakdown of interests and habits. Taken in aggregate, we can get a spatio-temporal representation of the world's interests, and even attitudes and intentions. In my talk, I will show how these spatio-tempo-social media resources can:

1. Improve the organization and enable new services for a user's media collection [2,3].
2. Allow extraction of place and event semantics and other information that is not easily attainable otherwise [4].
3. Help us understand and explore the world [5,6].

Indeed, using millions of geotagged photos from Flickr, we created an fundamentally new way to view the world. Our system, World Explorer [5], employs simple heuristics and spatial data structures to select and visualize representative tags for every area in the world, in any zoom level (where enough raw information is available), providing new opportunities for map-based exploration. We extended

N. Mamoulis et al. (Eds.): SSTD 2009, LNCS 5644, pp. 1–2, 2009.

the system using image analysis to automatically detect and select representative images for landmarks worldwide [6].

Beyond learning about the world, this type of spatio-tempo-social data may provide opportunities to learn about humans – and humanity. This "social information" can suggest both information created in social settings, as well as information about social actions, relationships and patterns that exist in mass scale. On one hand, social information may mean information that is biased by the properties and identities of the people who contribute. On the other hand, social information is what we can learn from the published information about the very nature of humans, for example, by asking questions using social science methodoligies on the dataset of spatio-temporal social media.

What kind of social information can we derive from spatio-tempo-social data? Firstly, systems can be devised to generate ad-hoc results, for example on-line analysis of social media to detect space-time anomalies from a traffic jam to "mass depression" to a natural disaster, or any other activity that is not expected in that time and place or from a given set of users. Second, we can continue to mine this data for spatial and spatio-temporal trends, providing an ever-improving representation and understanding of our public spaces as reflected in the public eye, or perhaps by different communities. Finally, as mentioned above, the spatio-tempo-social data shared on these online services can arguably be used to extract deep insights about human social behavior in various settings.

To summarize, the increasing amount of user-contributed information on the Web, coupled with the increasing availability of location-aware devices and services, will be making an astonishing amount of spatio-tempo-social data available on the Web. This essentially new type of information may capture and reflect people's relationships with space and time in a way that allows us to construct systems that have a better understanding of space, time, and people.

References

1. Ames, M., Naaman, M.: Why we tag: Motivations for annotation in mobile and on-line media. In: CHI 2007: Proceedings of the SIGCHI conference on Human Factors in computing systems. ACM Press, New York (2007)
2. Naaman, M., Song, Y.J., Paepcke, A., Garcia-Molina, H.: Automatic organization for digital photographs with geographic coordinates. In: JCDL 2004: Proceedings of the Fourth ACM/IEEE-CS Joint Conference on Digital Libraries (2004)
3. Naaman, M., Nair, R.: ZoneTag's collaborative tag suggestions: What is this person doing in my phone? IEEE Multimedia 15(3), 34–40 (2008)
4. Rattenbury, T., Naaman, M.: Methods for extracting place semantics from flickr tags. ACM Trans. Web 3(1), 1–30 (2009)
5. Ahern, S., Naaman, M., Nair, R., Yang, J.H.I.: World explorer: visualizing aggregate data from unstructured text in geo-referenced collections. In: JCDL 2007: Proceedings of the Seventh ACM/IEEE-CS Joint Conference on Digital Libraries, pp. 1–10. ACM, New York (2007)
6. Kennedy, L.S., Naaman, M.: Generating diverse and representative image search results for landmarks. In: WWW 2008: Proceeding of the 17th international conference on World Wide Web, pp. 297–306. ACM, New York (2008)

Recent Advances in Worst-Case Efficient Range Search Indexing

(Invited Talk)

Lars Arge*

MADALGO**, Dept. of Computer Science
Aarhus University, Aarhus, Denmark

`large@madalgo.au.dk`

Abstract. Range search indexing is the problem of storing a set of data points on disk such that the points in a axis-parallel (hyper-) query rectangle can be found efficiently (with as few disk accesses - or I/Os - as possible). The problem is arguably one of the most fundamental problems in spatial databases. Many indexes have been proposed for the problem and its variants.[1] The R-tree for example can be used to solve the more general version of the problem where the data is rectangles.

We describe some of the recent advances in the development of worst-case efficient range search indexing structures, that is, structures where a query is guaranteed to be answered within a certain (asymptotic) number of I/Os. We first discuss the well-known and optimal structure for the one-dimensional version of the problem, the B-tree [8,10], along with its variants weight-balanced B-trees [7], multi-version (or persistent) B-trees [4,9,17] and buffer-trees [3]. Then we discuss structures for the two-dimensional version of the problem, as well as its variants, most notably the external priority search tree [6], the external range tree [6,16], the kdB-tree [14,15] and the O-tree [13]. We also discuss lower bounds techniques that can be used to prove that both the range tree and kdB-tree/O-tree are optimal among query efficient and linear space structures, respectively [1,6,12,13]. We end by discussing the recent worst-case query optimal R-tree variant called the PR-tree [5].

* Supported in part by the Danish National Research Foundation, the Danish Strategic Research Council, and by the US Army Research office.
** Center for Massive Data Algorithms—a center of the Danish National Research Foundation.
[1] Comprehensive surveys of efficient index structures can e.g. be found in [2,11,18].

N. Mamoulis et al. (Eds.): SSTD 2009, LNCS 5644, pp. 3–4, 2009.
© Springer-Verlag Berlin Heidelberg 2009

References

1. Agarwal, P.K., de Berg, M., Gudmundsson, J., Hammer, M., Haverkort, H.J.: Box-trees and R-trees with near-optimal query time. In: Proc. ACM Symposium on Computational Geometry, pp. 124–133 (2001)
2. Arge, L.: External memory data structures. In: Abello, J., Pardalos, P.M., Resende, M.G.C. (eds.) Handbook of Massive Data Sets, pp. 313–358. Kluwer Academic Publishers, Dordrecht (2002)
3. Arge, L.: The buffer tree: A technique for designing batched external data structures. Algorithmica 37(1), 1–24 (2003)
4. Arge, L., Danner, A., Teh, S.-H.: I/O-efficient point location using persistent B-trees. In: Proc. Workshop on Algorithm Engineering and Experimentation (2003)
5. Arge, L., de Berg, M., Haverkort, H.J., Yi, K.: The priority R-tree: A practically efficient and worst-case optimal R-tree. In: Proc. SIGMOD International Conference on Management of Data, pp. 347–358 (2004)
6. Arge, L., Samoladas, V., Vitter, J.S.: On two-dimensional indexability and optimal range search indexing. In: Proc. ACM Symposium on Principles of Database Systems, pp. 346–357 (1999)
7. Arge, L., Vitter, J.S.: Optimal external memory interval management. SIAM Journal on Computing 32(6), 1488–1508 (2003)
8. Bayer, R., McCreight, E.: Organization and maintenance of large ordered indexes. Acta Informatica 1, 173–189 (1972)
9. Becker, B., Gschwind, S., Ohler, T., Seeger, B., Widmayer, P.: An asymptotically optimal multiversion B-tree. VLDB Journal 5(4), 264–275 (1996)
10. Comer, D.: The ubiquitous B-tree. ACM Computing Surveys 11(2), 121–137 (1979)
11. Gaede, V., Günther, O.: Multidimensional access methods. ACM Computing Surveys 30(2), 170–231 (1998)
12. Hellerstein, J., Koutsoupias, E., Miranker, D., Papadimitriou, C., Samoladas, V.: On a model of indexability and its bounds for range queries. Journal of ACM 49(1) (2002)
13. Kanth, K.V.R., Singh, A.K.: Optimal dynamic range searching in non-replicating index structures. In: Beeri, C., Bruneman, P. (eds.) ICDT 1999. LNCS, vol. 1540, pp. 257–276. Springer, Heidelberg (1998)
14. Procopiuc, O., Agarwal, P.K., Arge, L., Vitter, J.S.: Bkd-tree: A dynamic scalable kd-tree. In: Hadzilacos, T., Manolopoulos, Y., Roddick, J., Theodoridis, Y. (eds.) SSTD 2003. LNCS, vol. 2750. Springer, Heidelberg (2003)
15. Robinson, J.: The K-D-B tree: A search structure for large multidimensional dynamic indexes. In: Proc. SIGMOD International Conference on Management of Data, pp. 10–18 (1981)
16. Subramanian, S., Ramaswamy, S.: The P-range tree: A new data structure for range searching in secondary memory. In: Proc. ACM-SIAM Symposium on Discrete Algorithms, pp. 378–387 (1995)
17. Varman, P.J., Verma, R.M.: An efficient multiversion access structure. IEEE Transactions on Knowledge and Data Engineering 9(3), 391–409 (1997)
18. Vitter, J.S.: External memory algorithms and data structures: Dealing with MASSIVE data. ACM Computing Surveys 33(2), 209–271 (2001)

Design and Architecture of GIS Servers for Web Based Information Systems – The ArcGIS Server System

Sudhakar Menon

ESRI
Redlands, CA, USA
smenon@esri.com

This talk provides an overview of the design and architecture of GIS Servers for web based information systems, using ESRI's ArcGIS Server system as a concrete example. A GIS Server allows customers to compile, manage and disseminate geographic information. Information is compiled into and stored within object relational spatial databases using a geodatabase information model that supports the key types needed by applications including features, relationships, networks, imagery, terrains, maps and layers. Information is managed using both short and long transaction models that include support for versioning, archiving and replication. The GIS Server allows administrators to selectively publish this information to clients using stateless web services based on REST and SOAP as well as via OGC interfaces. These geospatial web services support visualization, analysis, data access and replication. Key GIS services include Mapping, Query, Location, Network Analysis, Editing, Geoprocessing and Imaging. These services run on clusters, can access data from local caches, and can be scaled out by growing the cluster. These web services allow information to be exchanged using optimized representations based on JSON and XML as well as client specific optimized protocols. The GIS Server supports a role based authorization model that allows access to services to be controlled, leverages standard web based authentication mechanisms, and allows information to be securely exchanged by leveraging standard transport level security. The GIS Server supports a variety of client architectures including Rich Internet clients based on Flex, Silverlight and Javacript, occasionally connected Mobile Devices that are used for field work, as well as Desktop clients built using .Net and Java. The key to success is simple and elegant Web APIs and online SDKs that allow customers to easily exploit the value of the information managed by their servers. This talk will highlight specific areas of interest to researchers working on spatial and temporal databases as it spans the above aspects of GIS server technology.

N. Mamoulis et al. (Eds.): SSTD 2009, LNCS 5644, p. 5, 2009.
© Springer-Verlag Berlin Heidelberg 2009

Versioning of Network Models in a Multiuser Environment

Petko Bakalov[1], Erik Hoel[1], Sudhakar Menon[1], and Vassilis J. Tsotras[2]

[1] Environmental Systems Research Institute, Redlands, CA 92373, USA
{pbakalov,ehoel,menon}@esri.com
[2] University of California, Riverside, CA 92507, USA
tsotras@cs.ucr.edu

Abstract. The standard database mechanisms for concurrency control, which include transactions and locking protocols, do not provide the support needed for updating complex geographic data in a multiuser environment. The preferred method to resolve conflicts in GIS systems is to encapsulate the modifications generated by the end users through the use of multiple versions. Multiuser (or versioned) geographic databases allow users to operate as though they have full access to the entire dataset. Instead of relying upon row locking, versioned databases allow multiple users to simultaneously edit the same row. They implement a model for conflict detection and resolution where the first to commit the change wins by default (though clients can manually intervene and select the latter change as the winner).

Network models are frequently used as a mechanism to describe the connectivity information between spatial features in many emerging GIS applications. Supporting networks within the context of a versioned database imposes additional requirements – the complex network model must retain integrity irrespective of the sequence of simultaneous edits by various clients. In this paper, we review our network model and discuss the enhancements necessary to maintaining topological network integrity in this complex environment. Our solution is based on the notion of dirty areas and dirty objects (i.e., regions or elements that contain edits that have not been reflected in the network connectivity index). The dirty areas and objects are identified and marked during editing of the network feature data. They are then subsequently cleaned as a byproduct of the incremental update of the connectivity network.

Keywords: Versioning, Network Models, Transportation Networks.

1 Introduction

Network data models have a long history as an efficient way to describe the topological connectivity information among spatial features in geographic information systems [11], [14], [17], [18]. At an abstract level, the network model can be viewed as a graph whose elements explicitly represent the connectivity information about the features in the database. The presence of an edge in the graph depicts the information that the two features represented by the junctions are connected and vice versa. Different versions of the

N. Mamoulis et al. (Eds.): SSTD 2009, LNCS 5644, pp. 6–24, 2009.
© Springer-Verlag Berlin Heidelberg 2009

network model have been implemented in existing operational systems such as ARC/INFO [21] and TransCAD [5]. Because of the large volume of data frequently found in these networks, the model is typically persisted inside a centralized database server. Using connectivity information, those systems can then be utilized to solve a wide range of problems, typical for the transportation or utility network domains (e.g., finding the shortest path between points of interest, finding optimal resource allocation, determining the maximal flow of a resource, and other graph theoretic operations).

A typical requirement for the network models (or to the GIS in general), is that they must provide support for many users simultaneously creating and updating large amounts of geographic information. In scenarios where those users are required to edit the same data at the same time, the system must provide an editing environment that supports multiuser concurrent modifications without creating multiple instances of the data. In contrast to traditional DBMSs, this editing environment must also support edit sessions that typically span a number of days or weeks (e.g., large engineering projects requiring significant interactive editing and revision), the facility to undo or redo changes made to the data, and the ability to develop models and alternative application designs without affecting the published database.

Concurrency control in traditional database management systems is addressed through the use of transactions and the two-phase locking protocol. This is efficient for short-lived edit operations that are typically completed in few seconds. It is not well suited however for the type of editing tasks required when updating geographic data. For a GIS multiuser environment, the row-locking mechanisms adopted by many DBMSs would be prohibitively restrictive for many common workflows.

To deal with long-lasting transactions, a solution based on the use of multiple versions has been proposed [16], [26]. A version can be logically viewed as an alternative, independent, persistent view of the database that does not involve creating a copy of the actual data. Since there is an independent view for every user, a versioned database can support multiple concurrent editors.

In addition, versioning is useful in many other GIS scenarios such as:

- **Modeling "what if" scenarios.** The versioning mechanism allows end users to exploit different alternatives (versions) during a design phase.
- **Workflow management.** Typically the design process goes through multiple steps organized in a workflow process where the output of one step is an input for another. The versioning scheme allows users to save intermediate results during the design process.
- **Historical queries.** The versioning scheme allows the preservation of different states of the data which later can be re-visited and re-examined if necessary.

Existing database versioning approaches cannot easily manage the specifics of the geographical data like topological network relations, the presence of connectivity among the stored elements, and traversability. Such information among spatial features is represented in a GIS by a network model.

Recently, we have proposed an incremental connectivity rebuilding algorithm for network models [1]. In this algorithm, the users are allowed to rebuild portions of the network model using the notions of dirty areas and dirty objects. Changes over portions of the network data are effectively captured and the incremental algorithm is utilized to clean such dirty areas/objects and re-establish the associated portions of the network connectivity index. The connectivity rebuilding algorithm has been implemented in ArcGIS and provides an effective solution to maintain dynamic network models in an incremental manner.

In this paper, we propose a new versioning scheme for network models that utilizes the dirty areas/objects of the connectivity rebuild algorithm (a similar mechanism has also been applied to our topological data model [13]). Versioning of network models is different from version control over simple spatial data ("simple" meaning data that is geometrically unrelated to other data – i.e., no topological structuring). While the same basic principles are still in operation, resolving conflicts between features that are related to other features, as with network models, is different. This is because of the specific internal behavior of the network and the requirement that the connectivity information (or index) in the model should be kept consistent all the time.

The rest of the paper is organized as follows: Section 2 provides a brief description of the network model including logical structure and physical design and provides description of the algorithms used for connectivity establishment. Section 3 provides in depth discussion of versioning spatial databases. Section 4 addresses our proposed extensions of these techniques to the support of versioned network models. Section 5 discusses our implementation experiences, and Section 6 concludes the paper.

2   The Network Model

A *network model* is described as a graph (named *connectivity graph*) that maintains the connectivity information about spatial features with line or point geometry. The basic elements of a network model are (*edges*, *junctions* and *turns*). Features with point geometry are represented with junction elements inside the graph, while lines are represented as one or more edge elements between pairs of junction elements. Figure 1 depicts the network model that we employ [14]. It is composed of spatial features and network elements. Similar designs have been used in many research or commercial implementations [7], [12], [15], [22]. In the network models we are considering, network elements are used only to describe the connectivity information for the spatial features they are representing; they do not carry any geometrical properties.

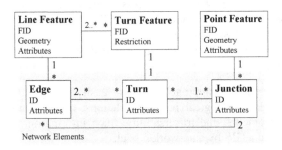

Fig. 1. Network model – features and network elements (which represent connectivity)

Most of the systems that utilize network models have client-server architectures. Because of their very large data size (e.g., many tens of millions of features for some nationwide or continent-wide transportation networks), the network models are usually located in a centralized server, persisted either in a RDBMS tables or in a file system. Typically the process of analysis is done within a GIS server (that acts as a client to the database) or within a thick client [4], [23], [27].

2.1 Traversability

While the connectivity elements (edges and junctions) allow the user to express connections, they are not sufficient for expressing specific restrictions from the real world (for example, no left turn, or, no u-turn allowed at an intersection) [3], [29]. Turn restrictions are used for this purpose. Turn restrictions present a problem to most network models. The presence of turns can greatly impact the movement (or traversability) through a network [20]. A common way to model turns within a network is with a turn table [30]. A turn table represents each explicitly specified turn restriction (or penalty) as a row with references to the associated two edges. Turn tables may be augmented with an impedance attribute if the turns may also represent delays or impedances. When traversing the network, the turn table is queried as necessary. An alternative approach is to employ a transition matrix that represents possible transitions at an intersection [10].

A maneuver is a turn that spans three or more edges. Maneuvers are used to model turning movements at complex street intersections within transportation networks. Consider the following intersection formed by a dual carriageway (i.e., a street where each travel direction is represented as a separate line feature) and a two-way street in Fig. 2.

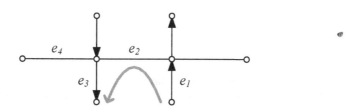

Fig. 2. Example of a three part maneuver e_1-e_2-e_3 at an intersection with a dual carriageway

To restrict the u-turn from edge e_1 to edge e_3, we need a maneuver composed of the edges e_1, e_2 and e_3 in sequence. The maneuver cannot be synthesized from the two overlapping turns e_1-e_2 and e_2-e_3, since restricting the e_1-e_2 turn also incorrectly restricts the left turn specified by the sequence e_1-e_2-e_4.

To introduce the turn restriction in addition to the edge and junction elements, a network model can also have a special network elements called *turns* (see Figure 1). Similar to the edges which are defined as a relation between junctions turns are defined as a relation between edges. A turn element is anchored to a specific junction (the junction where the turn starts) and controls the movement between sequence edges expressed as pairs (firstEdgeId, lastEdgeId).

2.2 Physical Implementation

In our network implementation the connectivity information is maintained as a set of *adjacency pairs* of the form *<edgeId, junctionId>*, stored inside the "junction table" (see Figure 3). This approach is designed to answer the most common type of adjacency queries during the network analysis process. The junction table uses fixed-length records for direct access purposes; this implies a fixed number (four in our implementation) of adjacency pairs per record (see Figure 3). If the junction has more than four connected edges an overflow mechanism is applied.

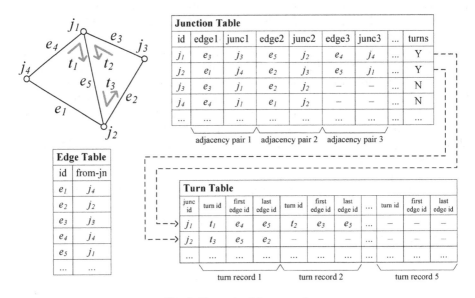

Fig. 3. Network tables example

In a similar way, in the traversal process, it is required that at each junction we know all the turns anchored at this junction. This has influenced the way we implement the turn storage scheme. Information about turns is stored in the "turn table", in the form of *turn triplets <turnId, firstEdgeId, lastEdgeId>*. If there are any turns anchored at a junction j_i, the turn table will have a record with primary key j_i which also contains all the turns anchored on j_i. This storage scheme can be easily optimized for the most commonly used client access patterns [28].

2.3 Maintaining Network Connectivity

Maintaining network connectivity can be viewed as a two phase process [1]:

- Initial establishment of connectivity when the network model is first defined, with the connectivity information being derived from the features participating in the network.

- Incremental rebuilding the connectivity index on a periodic basis after edits occur on the spatial features in the network.

Having an incremental solution is of significant practical value - the amortized cost of maintaining an incrementally rebuildable network is far less than an ordinary network that must be periodically rebuilt in its entirety (e.g., editing a subdivision and only rebuilding that portion of the nationwide network versus rebuilding the whole nation-wide network). In order to keep track of the modifications to the features that occur since the last full or partial rebuilding of the connectivity index, the network model employs the concept of *dirty areas*. Similarly, to track changes to elements without geometrical properties (e.g., turns), we use the concept of *dirty objects*.

Definition 1. *A **dirty area** corresponds to the regions within the feature space where features participating in the network have been modified (added, deleted, or updated) but whose connectivity has not been re-established.*

To simplify its computation and storage, a dirty area in our implementation is defined as a union of envelopes (e.g., bounding boxes) around the features that have been modified. It is possible however to use other shapes - the convex hull of the feature for example. In order to ensure that the network is correct, only the portion of the network encompassed in the dirty areas will need to be rebuilt.

Both the initial establishment of connectivity and the incremental rebuild algorithms follow the same four steps:

- **Geometrical extraction.** Extract the geometry information for all features in the area of interest (the whole area in the case of initial establishment or the dirty area in the case of subsequent rebuild) and analyze the vertices in those geometries. The extracted vertex coordinates and their corresponding feature identifiers are stored in a temporary table, called the "vertex table".
- **Connectivity analysis.** The content of the vertex table is sorted by coordinate values. As a result the coincident vertices from different features are grouped together. The algorithm scans the vertex table sequentially and picks groups of coincident vertices. Every single group is examined to determine if the vertices satisfy the connectivity model specified for the network.
- **Junction creation.** For each group which satisfies the connectivity model a new junction element is created in the network model. The junction id of this newly created junction element is added to all the vertices participating in this connectivity group.
- **Edge creation.** The content of the vertex table is then resorted using the feature identifier as the sorting key. As a result, the vertices for each line feature are again grouped together. The vertex table is scanned sequentially once more and for each pair of adjacent vertices which belong to the same line feature a new edge is created.

The difference between the incremental rebuild and the full (re)build algorithms, is that the incremental rebuild process adds to the vertex table those vertices that are outside of the rebuild region but belong to features which intersect the rebuild region.

These vertices are saved and later reused as connection points through which the rebuild portion of the network is "stitched" together with the rest of the model.

Rebuilding turn features in the network requires additional processing. The complexity comes from the fact that the turn features are defined as a relation between two or more line features and typically do not have geometrical properties. As depicted in Figures 1 and 3, a record in the turn table consists of a turn identifier and a list of the line feature identifiers that participate in the turn. In order to cover network elements without geometrical properties, we extend our dirty area concept with the notion of *dirty objects*.

Definition 2. *A **dirty object** is an object without geometrical properties (like turn features) whose modifications have not yet resulted in the incremental rebuilding of the network connectivity index.*

During the rebuild process, we restore all dirty objects to their clean state. An object is kept as dirty until it is successfully cleaned. Turn features are marked as dirty objects when:

- The turn feature is directly modified (Insert, Update, Delete), or
- The associated line features are modified (Update, Delete), or
- The associated network turn element is deleted (this may happen during the rebuild process).

Using the dirty areas and dirty objects, we can capture the dynamic behavior of network maintenance. It is this dynamic behavior that complicates and thus requires extra attention during the versioning process.

3 Versioned Spatial Databases

Spatial databases have dramatically evolved in their capability to handle multiple simultaneous editors. Some solutions have required organizations to alter their workflow so as to ensure that no two editors are editing the same geographic region within the spatial dataset. Supporting such a constrained workflow can become problematic once the need for supporting long transactions (e.g., design alternatives) is considered. In order to address this problem where design alternatives on the same geographic area are necessary (as well as very long transactions spanning weeks or months are required), versioned geographic data management technologies were developed [6], [8], [9], [19], [30]. Versioning does not prevent editing conflicts from occurring, rather, it provides an infrastructure for the detection and resolution of such conflicts.

Definition 3. *A **version** is a logical entity that represents a unique, seamless view of the database that is distinguished from other versions by the particular set of edits made to the version since it was created.*

Definition 4. *A **state** represents a discrete snapshot of the database whenever a change is made. Every edit operation creates a new database state.*

In versioned databases, there are two fundamental abstractions – *versions* and *states*. Versions are organized into a tree that is used to model the hierarchical relationships between versions (e.g., projects or design alternatives). A version is associated with a current state. A state is used to represent an instance of the database that is associated with a particular version. When a child state is created, it will initially have the same set of rows in each table as its parent state. However, as the state is edited, rows will either be added, deleted, or updated. Changes made in a child state are not visible in the parent state. Updated rows in the child will take precedence over the corresponding row in the parent when materializing the version associated with the child state.

Version model   Version tree

Fig. 4. Model depicting the relationship between versions and states is on the left, while a simple example version tree is shown on the right

Similar to versions, states are also organized into trees. A version will commonly be associated with numerous states over its lifetime; however, it will only be associated with a single state at any given moment in time. A given state may or may not be associated with one or more versions (as shown on the left side of Figure 4).

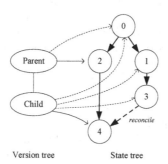

Version tree   State tree

Fig. 5. Example version tree and state tree

In Figure 5, we highlight a simple example where there are two versions, labeled parent and child, and an associated state tree. In the example, the parent version initially is associated with state 0. When a child version is created (as a child of the parent), it will also point to state 0. Following an edit to the child version, the child will then point to state 1. Assuming that the next edit is to the parent version, the parent will then point to state 2. The child is then edited one more time (causing the child version to point to state 3) prior to reconciling (making the changes made in the parent visible to the child – see Section 4.1 for additional details) with the parent version. The reconcile will cause the changes that have been made in the parent version (i.e., the differences between states 0 and 2) to be visible in the new state that the child will point to following the reconcile (i.e., state 4). This sequence of edits and a reconcile leaves the parent version pointing to state 2, while the child version points to state 4.

Versioned databases are useful in supporting a number of database usage patterns and workflows [16]; this includes:

- Direct multiuser editing of the main database,
- Two-level project organizations – work-order processing systems,
- Multi-level project organizations – hierarchical design parts and alternatives,
- Cyclical workflows (multiple stages of approval), and
- Historical states (temporal snapshots).

Some organizations will require the versioned database to support several of these workflows simultaneously; for example, a utility company may organize itself into a two-level project organization for maintaining its 'as built' status, while additionally requiring the maintenance of historical states (temporal snapshots). The key point is that a versioned database must be able to support each of these usage patterns (oftentimes simultaneously).

3.1 Operations on Versioned Databases

There are two fundamental operations that can be performed on versioned databases that are required in order to support versioning. These two operations (note – in the following discussion, we will employ the general terms 'child' version and 'parent' version; child version will refer to a version of interest, while parent version will generically refer to any ancestor version of the child within the version tree). Reconciling is logically the process of taking a child version and merging all the changes that have been made in its parent version (effectively making changes made to the parent version visible in the child). These changes may be either inserted, updated, or deleted features. This results in the creation of a new state that is then associated with the child version (e.g., state 4 in Figure 5). Note that it is possible that conflicts may be detected during reconciliation if a given feature has been modified in both the child version as well as the parent version. Additionally, if a feature is updated in one version and deleted in another, this is also a conflict (an update-delete conflict). When conflicts occur, the changes that are made in the parent version will take precedence by default (note that it is equally reasonable to implement a reconcile process where the child version takes precedence by default). Thus, human intervention is oftentimes necessary in order to resolve the difference if any of the changes made in the child version (that are in conflict with the parent) are to take precedence. In sum, reconciling is the process of making all the changes that were made to a parent version visible in a child version.

Posting is conceptually the converse operation to a reconcile. Posting involves taking a child version that has been reconciled with its parent version, and making all the changes made in the child visible to the parent version. Conceptually, changes in the child are pushed up into the parent. Once two versions have been reconciled and posted (with one version assuming the role of descendent, and the other as the ancestor in both operations), the parent and child versions will represent the same instance of data within the versioned database (at least until another edit is made to either version).

Version reconciliation (and conflict detection) may be implemented using queries against the underlying relational database that allow all inserts, updates, and deletes that occur between two states in the state tree to be detected. We term these queries 'difference queries' (detect the differences between two states). Note that for a conflict to occur between a feature in a child and parent version, the difference queries between the two states associated with the child and parent version relative to their common ancestor state (e.g., state 0 in Figure 5) must show that either the feature was either updated in both, or updated in one and deleted in the other state.

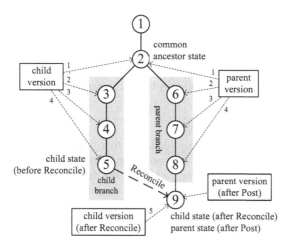

Fig. 6. Example state tree showing the interaction between child and parent versions

Figure 6 depicts a simple example highlighting the interaction between states, versions, and a reconcile. In the example assume that the parent version corresponds to state 2 (as indicated by the dashed arrow labeled "1" between the parent version and the circle labeled "2" (note that states correspond to labeled circles in the diagram). If a child version is now created, it will also reference state 2 (also depicted by a dashed arrow labeled "1" between the child version and state 2). State 2 also becomes what is termed the common ancestor state between the parent and child version. Assume that the child version is then edited three times. Each edit operation (an atomic set of edits) results in a new state; in this instance, states 3 through 5. At the end of the three edit operations, the child version will be referencing state 5. Following the edits to the child, assume that the parent version also has three edits made to it. This results in the creation of states 6, 7, and 8, with the parent version referencing state 8 following the edits. Now assume that the child version is reconciled with the parent version. The reconcile will require that the edits made in the parent version (essentially, the edits represented by states 6 – 8 in what is termed the parent branch) are made visible to the child version. This is accomplished by creating a new state (state 9) off of state 8, and pushing all the changes that have occurred in the child branch (states 3 – 5) into state

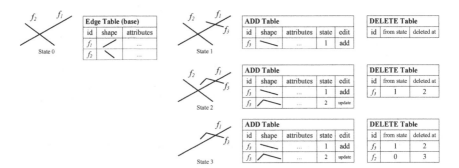

Fig. 7. Simple edit scenario highlighting the ADDs, DELETEs, and the base table

9, and making the child version reference state 9. Finally, the application of the Post operation following the reconcile results in the parent version also referencing state 9, making the changes made to the child visible in the parent.

3.2 Implementation Details

Versions are associated with a state identifier that corresponds to each update that occurs in the view. The state identifiers are unique and map to a set of updates corresponding to a single logical edit. For each state, the database keeps information about the modification type (either an insert, update, or delete). The ADDs table contains information related to inserts and updates, while the DELETEs table maintains the deletes (Figure 7). These two tables are collectively referred to as *delta tables*. One set of delta tables is associated with each base table in the versioned database. Thus, if a data model contained two tables, one representing parcels, and the second representing owners, there would be four additional tables necessary to represent the two sets of delta tables. A versioned dataset, therefore, consists of the original table (referred to as the *base table,* which corresponds to State 0), plus the two delta tables. The versioned database keeps track of which version the user is connected to. In addition, when modifications are made to the data, the versioning system populates the delta tables as appropriate. When a user queries a dataset in a versioned environment, the system assembles the relevant rows from the base table and the delta tables to present the correct view of the data for that particular version.

4   Versioned Network Models

Network models, with their associated network connectivity indexes, dirty areas, and dirty objects, introduce complexities into the standard reconcile and post processes within a versioned database (as described in Section 3). The primary cause of this complexity is the fact that inconsistent network indexes may occur when an edited child and parent version are reconciled. This is irrespective of whether or not each version has its full extent rebuilt (i.e., no dirty areas or objects).

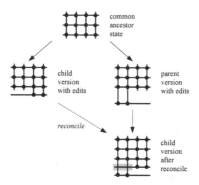

Fig. 8. Example highlighting a reconcile that results in an inconsistent network index (the inconsistent index is depicted by the shaded region at the bottom of the figure)

Consider the situation shown in Figure 8 (an annotated state tree is depicted – the common ancestor state refers to the state that the parent version was pointing to when the child version was originally created). In this example, assume that the network is clean; no dirty areas or objects exist with the features and the network index being in a consistent state. Edits are then made to both the parent and child versions. In the child version, the network is augmented in the southeast direction, while in the parent version the network is augmented toward the southwest. Assume that the network has been incrementally rebuilt following all edits in each version (i.e., no dirty areas or objects exist). In the figure, connectivity between line features is represented by the small black circles. As can be observed, both the parent and child versions have planar connectivity.

If the child version is then reconciled against the parent version, new edits made in the parent version are made visible in the child version. This is depicted in the southeast corner of Figure 8. Making these new features visible in the child version results in an inconsistency between the features and the network connectivity index as depicted in the area enclosed by the gray area. Thus, we observe a simple situation where two versions that are completely rebuilt can have a network connectivity index inconsistency following reconciliation. For this reason, the version reconcile process must be augmented to handle networks correctly.

4.1 Dirty Area and Object Management during Reconciliation

As has been discussed, versioning of network models requires additional functionality on top of the versioning scheme for simple feature classes. This is due to the fact that the model includes both: (i) a feature space with features modeling real world objects, and (ii) a logical network where connectivity information about these features is stored. The connectivity information has to be kept consistent with the state of the feature space during the process of reconciliation when new features have been introduced or existing ones have been updated or deleted in the child version as a result of the reconciliation. All these modifications introduce changes in the connectivity inside the feature space of the network model, which have to be reflected in the logical network.

There are two general approaches to solve this problem. The first one employs the concept of reactive behavior which is applied to the network and has been used in the ArcGIS geometric network model [2]. The reactive behavior refers to the logical connectivity network reacting automatically to the changes in the feature space. Thus, the process of reconciliation will require the maintenance of the connectivity information. This entails both logical networks (in the child and parent versions) being analyzed concurrently during the reconciliation process and merged together in the resultant child version. The main disadvantage to this approach is the complexity of the problem (analyzing and merging graphs) which itself can deteriorate the performance of reconcile.

To avoid this disadvantage when reconciling a Network model, we choose to employ another strategy which we call the *lazy* approach (it is termed lazy because we are deferring the actual rebuilding of the network connectivity to a later, more convenient time). Instead of analyzing and restitching the connectivity information during reconcile, we instead utilize the incremental network rebuild algorithm discussed in [1]. We relax the requirement that the connectivity network must always reflect the state of the feature space. From a connectivity perspective, the logical network is allowed to be in an incorrect state; the regions of inconsistency are marked as dirty areas (or dirty objects in the case of turn features).

Dirty area (and object) management becomes a key concept in the versioned network model. In order to ensure that the incremental rebuilding of the network index is properly handled, we rely upon a strategy where dirty areas or objects are generated for the areas where spatial features or turn features are modified (created, updated, or deleted). The user may then choose to rebuild the network over the portions of the network where these dirty areas are introduced as a byproduct of reconcile at a time of their choosing. More specifically, we may summarize a complete set of rules related to the handling of dirty areas and objects as follows:

- **Rule 1:** All dirty areas and objects that are present in the child or parent that do not exist in the common ancestor state (i.e., before the child and parent were edited) remain in the result state of the reconcile (corresponding to the child version after reconcile). This is depicted in the left side of Figure 9.

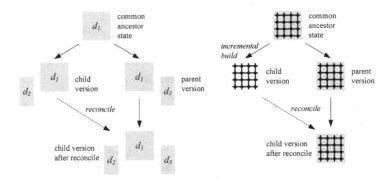

Fig. 9. Example of dirty area management Rules 1 and 2. In this figure (and Figures 10 and 11), dirty areas are represented by shaded areas; dirty objects are not depicted as they have no spatial representation. The arrows represent the sequence of events; the dirty areas are labeled in the left side of Figure 9.

- **Rule 2:** All dirty areas and objects that exist in the common ancestor state but do not in the child (i.e., an incremental network rebuild in the child) will still exist in the child following the reconcile (depicted in the right side of Figure 9).
- **Rule 3:** All dirty areas and objects that exist in the common ancestor state but do not in the parent version (they were validated) will not exist in the child version following the reconcile.
- **Rule 4:** All dirty areas and objects created in the child version, irrespective of whether or not they exist at the time of reconciliation, will exist following the reconcile.
- **Rule 5:** All dirty areas and objects created in the parent version will only exist in the child version following the reconciliation if they exist at the time of reconcile. This situation is shown in Figure 11.

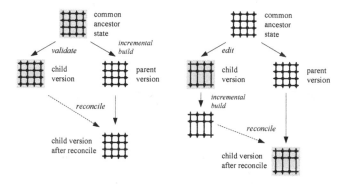

Fig. 10. Example of dirty area management Rules 3 and 4

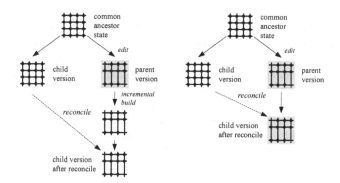

Fig. 11. Example of dirty area management Rule 5. The left side depicts how a dirty area that no longer exists in the parent at the time of reconcile will not exist in the child following the reconcile. The right side of the figure depicts the opposite situation where the dirty area on the parent version exists at the time of reconcile.

4.2 Detailed Example

We illustrate the behavior of versioning networks using the example shown in Figures 12 and 13. In the common ancestor state (see Figure 12a), two new line features l_2 and l_6 have been added to the previously built feature space. Within the common ancestor state, the reactive behavior of the network creates two new dirty areas around the new line features in order to keep track of the modifications in the feature space. Since the area has not been rebuild with the incremental rebuild algorithm, the two new line features l_2 and l_6 have not been reflected in the network connectivity index.

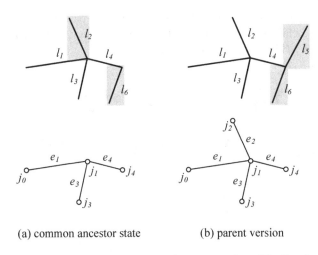

(a) common ancestor state   (b) parent version

Fig. 12. Version example – common ancestor and parent versions. The line features are represented on the top half, and the corresponding connectivity network on the bottom half. Dirty areas are represented by shaded rectangles.

In the next step, a child version is created. In the parent version, additional edits are made. Within the parent, the incremental build is run over the area encompassing the dirty area surrounding feature l_2. This results in new a new edge e_2 and junction j_2 being created in the connectivity index. Finally, an additional line feature is created in the parent version. The result of all edits to the parent version is shown in Figure 12b.

In the child version, a different set of edits are made. In the child, two new line features are created, l_7 and l_8 (this will result in new dirty areas being created). Finally, an incremental build is run over the areas encompassing the dirty areas surrounding line features l_6 and l_8. This results in an update of the connectivity index where edges e_6 and e_8 and junctions j_6 and j_8 are created. Figure 13a represents the results of these modifications.

The next operation performed on the child version (Figure 13a) is to reconcile it with the parent version (Figure 12b). Recall that the reconcile makes the edits made in the parent version visible in the child version. Applying the rules for reconciling networks as described in Section 4.1, the child version will be modified (with the result shown in Figure 13b). The following modifications are of note; first, the application of Rule 1 results in line l_7 being associated with a dirty area (i.e., new dirty areas present in the child or parent remain after reconcile). The application of Rule 2 causes line l_6 to also be associated with a dirty area (a dirty area in the common ancestor state and the parent version, but not the child version prior to reconcile). Rule 3 results in line l_2 being clean and reflected in the connectivity network following reconcile (a dirty area in the common ancestor state and child version, but clean in the parent version). Rule 4 results in line l_8 becoming dirty following the reconcile. Finally, the application of Rule 5 causes line l_5 that was created but not rebuilt in the parent being marked as dirty following the reconcile. Following the reconcile, if the incremental rebuild is applied to all dirty areas in the child version, the result will be a clean and up to date connectivity index as depicted in Figure 13c.

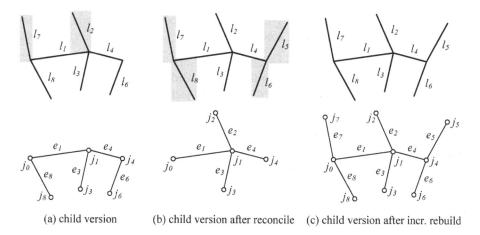

(a) child version   (b) child version after reconcile   (c) child version after incr. rebuild

Fig. 13. Version example – the child and reconcile versions

5 Implementation Experiences

The proposed versioning scheme for network models is currently being implemented in the ESRI ArcGIS system and will be made available to customers in the next release. It has been tested in a multiuser editing environment for large continental wide network models, including a model derived from the set of features representing the full street network within the entire continental United States (35.9 million line

Table 1. Reconcile times for different network datasets and number of child version edits

Dataset	Size (M of features)	Edited features	Ave. reconcile time
Southern California	1.3	100	≈ 2 sec.
Southern California	1.3	500	≈ 3 sec.
Southern California	1.3	1,000	≈ 5 sec.
SW United States	10.6	1,000	≈ 6 sec.
SW United States	10.6	5,000	≈ 7 sec.
SW United States	10.6	10,000	≈ 10 sec.

features). Similarly sized networks were constructed for all of Europe. Table 1 provides examples of the reconcile times (wall clock) for different real world transportation network datasets and differing numbers of edits to the child version.

6 Conclusion

In this paper, we explored the difficulties of managing large network models in a multiuser environment and presented solutions to address these problems using a flexible versioning scheme. Taking into account the dynamic nature of the source data associated with a network model (i.e., the data is regularly edited), we presented innovative versioning schemes that facilitate the notions of dirty areas and dirty objects (used already for maintaining the dynamic network model). The following summarizes key features of the versioning scheme:

- The flexible reconciling rules allow the definition of a resolving mechanism between conflicting edits according to user needs.
- In addition, the utilization of dirty areas/objects minimizes the overhead of tracking editing history.

We have implemented the versioning scheme presented in this paper within the well-established ArcGIS development framework. The proposed ideas have proven to be efficient methods in handling concurrency control of large network datasets.

References

1. Bakalov, P., Hoel, E., Heng, W.L., Tsotras, V.: Maintaining Connectivity in Dynamic Multimodal Network Models. In: Proceedings of the International Conference on Data Engineering (ICDE 2008), Cancun, Mexico, April 2008, pp. 1267–1276 (2008)
2. Borchert, R.: Geometric Network: What Is It and How to Make It? In: Proceedings of the 23rd Annual ESRI User Conference, San Diego (July 2003)
3. Caldwell, T.: On Finding Minimum Routes in a Network with Turn Penalties. Communications of the ACM 4(2), 107–108 (1961)
4. Cho, H.-J., Chung, C.-W.: An Efficient and Scalable Approach to CNN Queries in a Road Network. In: Proceedings of the 31st International Conference on Very Large Data Bases (VLDB 2005), Trondheim, Norway, August 2005, pp. 865–876 (2005)
5. Caliper Corporation: TransCAD Transportation GIS Software Reference Manual. Caliper Corporation (1996)
6. Dittrich, K., Lorie, R.: Version Support for Engineering Database Systems. IEEE Transactions on Software Engineering 14(4) (April 1988)
7. Dueker, K., Butler, A.: GIS-T Enterprise Data Model with Suggested Implementation Choices. Journal of the Urban and Regional Information Systems 10(1), 12–36 (1998)
8. Easterfield, M., Newell, R., Theriault, G.: Version management in GIS - applications and techniques. In: Proc. of the European Conference on Geographical Information Systems (EGIS 1990), Amsterdam, April 1990, pp. 1–8 (1990)
9. ESRI: Building a Geodatabase. Prepared by Environmental Systems Research Institute. ESRI Press, Redlands (2002)

10. Evans, J., Minieka, E.: Optimization Algorithms for Networks and Graphs. Dekker, Marcel Incorporated (1992)
11. Goodchild, M.: Geographic Information Systems and Disaggregate Transportation Modeling. Geographical Systems 5(1-2), 19–44 (1998)
12. Hage, C., Jensen, C., Pedersen, T., Speicys, L., Timko, I.: Integrated Data Management for Mobile Services in the Real World. In: Proceedings of the 29th Intl. Conf. on Very Large Data Bases (VLDB 2003), Berlin, September 2003, pp. 1019–1030 (2003)
13. Hoel, E., Menon, S., Morehouse, S.: Building a Robust Relational Implementation of Topology. In: Hadzilacos, T., Manolopoulos, Y., Roddick, J., Theodoridis, Y. (eds.) SSTD 2003. LNCS, vol. 2750, pp. 508–524. Springer, Heidelberg (2003)
14. Hoel, E., Heng, W.L., Honeycutt, D.: High Performance Multimodal Networks. In: Bauzer Medeiros, C., Egenhofer, M.J., Bertino, E. (eds.) SSTD 2005. LNCS, vol. 3633, pp. 308–327. Springer, Heidelberg (2005)
15. Jensen, C., Pedersen, T., Speicys, L., Timko, I.: Data Modeling for Mobile Services in the Real World. In: Hadzilacos, T., Manolopoulos, Y., Roddick, J., Theodoridis, Y. (eds.) SSTD 2003. LNCS, vol. 2750, pp. 1–9. Springer, Heidelberg (2003)
16. Katz, R.: Toward a Unified Framework for Version Modeling in Engineering Databases. ACM Computing Surveys 22(4) (1990)
17. Longley, P., Goodchild, M., Maguire, D., Rhind, D.: Geographical Information Systems, Principles, Techniques, Applications and Management. Wiley, Chichester (1999)
18. Mainguenaud, M.: Modeling of the Geographical Information System Network Component. International Journal of Geographical Information Systems 9(6), 575–593 (1995)
19. Menon, S., Aronson, P., Brown, T., Muller, M., Ryden, K., Morehouse, S.: Requirements and Design Considerations for Versioned Geographic Data Management. Unpublished manuscript, ESRI, Redlands (July 2000)
20. Miller, H., Shaw, S.-L.: Geographic Information Systems for Transportation. Oxford University Press, Oxford (2001)
21. Morehouse, S.: ARC/INFO: A Geo-relational Model for Spatial Information. In: Proceedings of AUTOCARTO 7, Washington, DC, March 1985, pp. 388–397 (1985)
22. Oracle Corp: Oracle Database 10g: Oracle Spatial Network Data Model: technical white paper (May 2005)
23. Papadias, D., Zhang, J., Mamoulis, N., Tao, Y.: Query Processing in Spatial Network Databases. In: Proceedings of the 29th International Conference on Very Large Data Bases (VLDB 2003), Berlin, September 2003, pp. 802–813 (2003)
24. Peuquet, D., Duan, N.: An Event-based Spatiotemporal Data Model (ESTDM) for Temporal Analysis of Geographic Data. International Journal of Geographical Information Science 9(1) (1995)
25. Ralston, B.: GIS and its Traffic Assignment: Issues in Dynamic User-optimal Assignments. Geoinformatica 4(2), 231–243 (2000)
26. Sciore, E.: Versioning and Configuration Management in an Object-oriented Data Model. International Journal on Very Large Data Bases 3(1) (1994)
27. Shahabi, C., Kolahdouzan, M., Sharifzadeh, M.: A Road Network Embedding Technique for k-nearest Neighbor Search in Moving Object Databases. In: Proceedings of the 10th ACM International Symposium on Advances in Geographic Information Systems (ACMGIS 2002), McLean Virginia, November 2002, pp. 94–100 (2002)
28. Shekhar, S., Liu, D.-R.: Ccam: A Connectivity-clustered Access Method for Networks and Network Computations. IEEE Transactions on Knowledge and Data Engineering 9(1), 102–119 (1997)

29. Speicys, L., Jensen, C., Kligys, A.: Computational Data Modeling for Network-constrained Moving Objects. In: Proceedings of the 11th ACM Intl. Symp. on Advances in Geographic Information Systems (ACMGIS 2003), New Orleans, November 2003, pp. 118–125 (2003)
30. Stokes, A., Balasubramanian, S., Harrison, S.: Building Versioning Applications with the Oracle Internet File System. Oracle Technical Brief, Oracle Corporation (2000)
31. Winter, S.: Modeling Costs of Turns in Route Planning. GeoInformatica 6(4), 345–361 (2002)
32. Worboys, M., Hearnshaw, H., Maguire, D.: Object-oriented Data Modeling for Spatial Databases. International Journal of Geographical Information Systems 4(4), 369–383 (1990)

Efficient Continuous Nearest Neighbor Query in Spatial Networks Using Euclidean Restriction*

Ugur Demiryurek, Farnoush Banaei-Kashani, and Cyrus Shahabi

University of Southern California
Department of Computer Science
Los Angeles, CA 90089-0781
{demiryur,banaeika,shahabi}@usc.edu

Abstract. In this paper, we propose an efficient method to answer continuous k nearest neighbor (CkNN) queries in spatial networks. Assuming a moving query object and a set of data objects that make frequent and arbitrary moves on a spatial network with dynamically changing edge weights, CkNN continuously monitors the nearest (in network distance) neighboring objects to the query. Previous CkNN methods are inefficient and, hence, fail to scale in large networks with numerous data objects because: 1) they heavily rely on Dijkstra-based *blind expansion* for network distance computation that incurs excessively redundant cost particularly in large networks, and 2) they *blindly map* all object location updates to the network disregarding whether the updates are relevant to the CkNN query result. With our method, termed ER-CkNN (short for *Euclidian Restriction* based CkNN), we utilize ER to address both of these shortcomings. Specifically, with ER we enable 1) *guided search* (rather than blind expansion) for efficient network distance calculation, and 2) *localized mapping* (rather than blind mapping) to avoid the intolerable cost of redundant object location mapping. We demonstrate the efficiency of ER-CkNN via extensive experimental evaluations with real world datasets consisting of a variety of large spatial networks with numerous moving objects.

1 Introduction

The latest developments in wireless technologies as well as the widespread use of GPS-enabled mobile devices have led to the recent prevalence of location-based services. Many of the location-based services rely on a family of spatial queries, termed nearest neighbor (NN) queries. In particular, a *Continuous k-NN* query (CkNN for short) continuously monitors the k data objects that are nearest (in

---

* This research has been funded in part by NSF grants IIS-0238560 (PECASE), IIS-0534761,IIS-0742811 and CNS-0831505 (CyberTrust), and in part from CENS and METRANS Transportation Center, under grants from USDOT and Caltrans.Any opinions, findings, and conclusions or recommendations expressed in this material are those of the author(s) and do not necessarily reflect the views of the National Science Foundation.

N. Mamoulis et al. (Eds.): SSTD 2009, LNCS 5644, pp. 25–43, 2009.

network distance) to a given query object, while the data objects and/or the query object arbitrarily move on a spatial network. With CkNN, for example, a driver can use the automotive navigation system of her vehicle to continuously locate the three nearest restaurants as the vehicle is moving along a path, or a pedestrian can use her GPS-enabled mobile device (cell phone, PDA, etc.) to locate the nearest transportation vehicles (e.g., taxis, buses, trams).

Currently, incremental monitoring (IMA) and its extension group monitoring algorithm (GMA) [8] is the only known method for answering CkNN queries with arbitrarily moving data and query objects. GMA extends IMA with shared execution paradigm by grouping the queries in the same sequence and monitoring them as a group (rather than individually). We refer to these algorithms as IMA/GMA in the rest of the paper. IMA/GMA is based on the incremental network expansion (INE) method [9] to support CkNN queries on dynamic/moving objects. However, the performance of IMA/GMA degrades in real-world scenarios where the spatial network is large and the data objects moving on the network are numerous. IMA/GMA is inefficient due to two main reasons. Firstly, in order to identify the k nearest neighbors, IMA/GMA uses the computationally complex Dijkstra based algorithm that relies on *blind network expansion*. With network expansion, starting from q all network nodes reachable from q in every direction are visited in order of their proximity to q until all k nearest data objects are located (see Figure 1). The overhead of executing network expansion is prohibitively high particularly in large networks with a sparse (but perhaps large) set of moving data objects, because such a blind search approach has to redundantly visit many network nodes which are away from the shortest paths to the nearest data objects. For example, Figure 1 depicts a real spatial network (namely, the road network of San Joaquin, CA) and illustrates the set of nodes that network expansion would have to visit (marked by the shaded area) to locate the first nearest data object (1-NN) for the query object q. In this case, 47.2% of the entire set of network nodes (8620 nodes out of total 18262) must be visited to find 1-NN.

Secondly, with IMA/GMA the cost of mapping the object location updates (e.g., current coordinates such as longitude-latitude) to the network is also prohibitively high. While the location updates are continuously received, they must

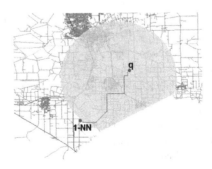

Fig. 1. Blind network expansion

be mapped to the network to locate the current edge of the moving object. However, with IMA/GMA *all* location updates are blindly and redundantly mapped to the network when they are received regardless whether they can possibly affect the CkNN query result, whereas most of the updates are irrelevant and can be ignored. Considering the cost of mapping each object location update (i.e., $O(logN)$) as well as high frequency of location updates in large spatial networks with numerous objects, the overhead incurred due to blind object location mapping with IMA/GMA becomes intolerable with real-world applications.

In this paper, we propose ER-CkNN, a *Euclidean Restriction* (ER) based method for efficient CkNN query answering. ER-CkNN addresses the two shortcomings of IMA/GMA by leveraging ER to enable *guided search* and *localized mapping*, respectively. Firstly, to identify the nearest neighbors of the query point q, ER-CkNN uses a filtering mechanism to rapidly find a set of candidate data objects based on their *Euclidean distance* from q (i.e., filtering by ER), which is then refined by computing their *network distance* from q to identify the exact nearest neighbors. The benefit of this filter-and-refine approach versus the blind network expansion is that once the candidate data objects are identified at the filter step, at the refine step ER-CkNN can use a one-to-one *guided search* algorithm such as A* [10] to perform the costly network distance computation with minimum redundant network traversal. With ER-CkNN, we use an EBE (Edge Bitmap Encoding)-based A*, with a search complexity proportional to the size of the shortest path. Secondly, to avoid the high cost of blind object location mapping, ER-CkNN only maps a location update to the network if it is relevant to the result of the CkNN query; otherwise, the location update is ignored. To determine whether a location update is relevant, ER-CkNN uses ER to rapidly (in $O(1)$) identify whether the location update is within certain Euclidean locality that can potentially change the result of the CkNN query. If the update is within the q locality, ER-CkNN maps the update to the network (i.e., *localized mapping*) which subsequently initiates the query update.

While ER is previously used for kNN query answering [9] assuming *static* objects, to the best of our knowledge ER-CkNN is the first ER-based method proposed to answer CkNN queries on *dynamic/moving* objects. ER-CkNN is fundamentally different from previous ER-based approaches as they unanimously index the objects to apply ER, whereas with moving objects maintenance of such an index is unaffordable. Instead, ER-CkNN indexes the spatial network which is static, and uses a grid file (with $O(1)$ update cost) for efficient access to the objects in order to apply ER (see Section 4). Our experiments with real-world datasets show that ER-CkNN outperforms GMA with at least three times improved response time (see Section 6).

The remainder of this paper is organized as follows. In Section 2, we review the related work about kNN queries on moving objects. In Section 3, we formally define the CkNN query in spatial networks. We mention the theoretical foundation of our algorithms as well as our data structure and indexing schemes in Section 4. In Section 5, we discuss the factors that affect the performance of

ER-CkNN. In Section 6, we present the results of our experiments with a variety of parameters. Finally, in Section 7 we conclude and discuss our future work.

2   Related Work

The research on kNN query processing can be grouped into two main areas, namely, query processing in Euclidean space and query processing in spatial networks.

2.1 kNN Queries in Euclidean Space

In the past, numerous algorithms [19,18,12,14,13] have been proposed to solve kNN problem in Euclidean space. Most of these algorithms, assuming the data objects are static, used tree-based (e.g., R-Tree) structures (or their extensions) to enable efficient query processing. Although the tree-based data structures are efficient in handling stationary spatial data, they suffer from the node reconstruction overhead due to frequent location updates with moving objects. Therefore, some researchers have exclusively used the simple but efficient space-based (i.e., grid) structures to index and query the moving objects [3,17,7]. All of these approaches are applicable to the spaces where the distance between objects is only a function of their spatial attributes (e.g., Euclidean distance). In real-world scenarios, however, the queries move in spatial networks, where the distance between a pair of data objects is defined as the length of the shortest path connecting them. We proceed to mention early proposals for kNN processing in spatial networks below.

2.2 kNN Queries in Spatial Networks

In [9], Papadias et al. introduced INE (discussed in Section 1) and IER. IER exploits the Euclidean restriction principle in spatial networks for achieving better performance. The data and query objects are assumed to be static in this work. Kolahdouzan and Shahabi utilized the first degree *network Voronoi diagrams* [4,5] to partition the spatial network to network Voronoi polygons (NVP), one for each data object. They indexed the NVPs with a spatial access method to reduce the problem to a point location problem in Euclidean space and minimize the on-line network distance computation by precomputing the NVPs. Cho et al. [1] presented a system UNICONS where the main idea is to integrate the precomputed k nearest neighbors into the Dijkstra algorithm. Huang et al. addressed the same problem using *Island* approach [16] where each vertex is associated (and network distance precomputed) to all the data points that are centers of given radius r (so called islands) covering the vertex. With their approach, they utilized a restricted network expansion from the query point while using the precomputed islands. Aside from their specific drawbacks, these algorithms rely on *data object dependent* precomputations (i.e., the distance to the data objects are precomputed) and subdivide the spatial network based on the location of the data objects. Therefore, they assume that data objects are static and/or trajectory of query objects is known. This assumption is undesirable in applications where the query and data objects change their positions frequently.

Recently, Huang et al. [15] and Samet et al. [11] proposed two different algorithms that address the drawbacks of data object dependent precomputation. Huang et al. introduced *S-GRID* where they partition (using grid) the spatial network to disjoint sub-networks and precompute the shortest path for each pair of connected border points. To find the k nearest neighbors, they first perform a network expansion within the sub-networks and then proceed to outer expansion between the border points by utilizing the precomputed information. Samet et al. proposed a method where they associate a label to each edge that represents all nodes to which a shortest path starts with this particular edge. They use these labels to traverse *shortest path quadtrees* that enables geometric pruning to find the network distance between the objects. With these studies, the network edge weights are assumed to be static therefore the precomputations are invalidated with dynamically changing edge weights. This dependence is unrealistic for most of the real-world applications.

Therefore, unlike the previous approaches, we make the fundamental assumption that *both* the query and the data objects make frequent and arbitrary moves on a spatial network with dynamically changing edge weights. Our assumption yields a much more realistic scenario and versatile approach. To the best of our knowledge, the only comprehensive study proposed to this problem is IMA/GMA [8]. We discussed the shortcomings of IMA/GMA in Section 1.

3   Problem Definition

In this section, we formally define CkNN queries in spatial networks. Consider a spatial network (e.g., the Los Angeles road network) with a set of data objects and a query object. We assume the query object and the data objects either reside or move on the network edges. The position of a moving object p at time t is defined as $loc_t(p) = (x_p, y_p)$, where x_p and y_p are the cartesian coordinates of p in the space at time t. We assume all the relevant information about the moving objects and the spatial network is maintained at a central server. Whenever an object moves to a new location and/or the cost of an edge changes, the central server is updated with the new location and weight information, respectively. We formally define the spatial network and CkNN queries as follows:

Definition 1. *A spatial network is a directional weighted graph $G(N, E)$, where N is a set of nodes representing intersections and terminal points, and E ($E \subseteq N \times N$) is a set of edges representing the network edges each connecting two nodes. Each edge e is denoted as $e(n_i, n_j)$ where n_i and n_j are starting and ending nodes, respectively. The network distance d_N between a given source $s \in N$ and a destination $t \in N$ is the length of the shortest path connecting s and t in G.*

Definition 2. *A Continuous k nearest neighbor (CkNN) query in spatial networks continuously monitors the k data objects that are nearest (in network distance) to a* *given query object, while the data objects and/or the query object arbitrarily move on network edges. Considering a set of n objects $S = \{p_1, p_2, ...p_n\}$, the k nearest neighbors of a query object q constitute a set $S' \subseteq S$ of k objects such that for any data object $p' \in S'$ and $p \in S - S'$, $d_N(p', q) \leq d_N(p, q)$.*

4   ER-C*k*NN

Arguably, the main challenges with answering C*k*NN queries in spatial networks are 1) efficient network distance computation, and 2) effective maintenance of the query results given frequent changes of the moving object locations. With ER-C*k*NN, we employ a network partitioning approach that allows us to address the above challenges by enabling 1) guided shortest path algorithm that minimizes redundant network traversal, and 2) localized mapping that allows for effective maintenance of the query results.

ER-C*k*NN involves two phases: an off-line grid partitioning phase and an on-line query processing phase. During the off-line phase, the spatial network is partitioned into grid cells and each edge in the network is encoded whether it is a part of a shortest path to any node in a given grid cell (*edge-bitmap-encoding*). In addition, an *edge-cell-mapping* is computed between the edges of the spatial network and the cells of an overlaid grid. These precomputations are used to expedite the on-line query processing. During the on-line phase, a Euclidean Restriction (ER) based filter-and-refine method is adopted to identify the k nearest neighbors at the time of query arrival. At the filter step, ER-C*k*NN performs a grid expansion to rapidly identify a set of candidate nearest neighbors in the Euclidean proximity. At the refine step, the candidate set is refined (if necessary) by fast guided network distance computation exploiting the edge-bitmap-encoding information. However, considering the often large number of moving objects and their frequent location updates, effective monitoring of this query result remains the main challenge. To address this challenge, we leverage the edge-cell-mapping information to rapidly identify the relevant location updates (without traversing the spatial network index) and ignore the updates that will not affect the query result. Below, we explain the two-phase ER-C*k*NN algorithm.

4.1   Off-Line Grid Partitioning

With the off-line phase, we partition the spatial span of the network with regular grid as illustrated in Figure 2(a). Each grid cell is a square of size $\alpha \times \alpha$ (in Section 5.1 we explain how we choose the optimal size) denoted by its row and column indices $c(i, j)$, where the reference cell $c(0, 0)$ is the bottom-left cell of the grid. The resulting grid partitioning, yields following two main advantages.

Firstly, such network partitioning enables ER-C*k*NN to expedite on-line network distance computations using precomputed information. Specifically, ER-C*k*NN, for each edge, maintains a vector $\overrightarrow{v}_{EBE}$ (proposed by Lauther in [6]) that contains encoded values indicating whether the edge is a part of a shortest path to a given grid cell. ER-C*k*NN utilizes $\overrightarrow{v}_{EBE}$ to avoid exploring unnecessary paths (hence pruning the search space) during an on-line shortest path

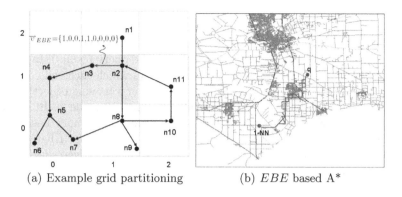

(a) Example grid partitioning   (b) EBE based A*

Fig. 2. EBE based shortest path computation

computation. For example, Figure 2(a) illustrates a simple road network (partitioned to nine regions) where the $\overrightarrow{v}_{EBE}$ of edge $e(n_2, n_3)$ only contains three 1 entries which correspond to $c(0,0)$, $c(1,0)$, and $c(1,1)$ cells (marked by the shaded area). This means that edge $e(n_2, n_3)$ can be a part of a shortest path to any node in those regions. Considering a shortest path search from n_1 with target nodes (e.g., n_{10}) in unmarked cells, the search ignores edge $e(n_2, n_3)$ during the on-line computation.

In order to determine the encoded values (i.e., 1 or 0) contained in $\overrightarrow{v}_{EBE}$ of an edge $e(n_s, n_t)$, we compute a one-to-all shortest path from the head node n_s to all other nodes in the network. If *any* node n_u is reached in a grid cell $c(i,j)$, we set the encoding information to 1 (i.e., true) for the region containing node n_u. We refer to this operation *edge-bitmap-encoding* (EBE) and repeat it for each edge. The integration of $\overrightarrow{v}_{EBE}$ to any shortest path algorithm is very easy. Specifically, any shortest path algorithm can be modified to check the encoded value of the corresponding grid cell (that contains the target node) ever time before traversing an edge (If true, the edge is traversed). With ER-CkNN, we integrate $\overrightarrow{v}_{EBE}$ to A* algorithm (referred as EBE-based A*). This integration further improves the performance of A* algorithm thus minimizing the redundant network traversal. Recall that A* is already much faster than Dijkstra for point-to-point shortest path computations. We refer readers to [10] for the details of A* algorithm and the comparison of it to Dijkstra. Continuing with our example presented in Figure 1, Figure 2(b) shows the set of edges (highlighted around the actual shortest path) that ER-CkNN would traverse to locate the first nearest data object using EBE-based A*. As shown, EBE-based A* algorithm visits significantly less number of network nodes.

The storage requirement of $\overrightarrow{v}_{EBE}$ is extremely low as its size is bounded by the number of grid cells. The space complexity is $O(RE)$ for a network with R regions and E edges. To imagine, the space required to maintain the EBE information of Los Angeles spatial network (with 304,162 edges) divided to 128X128 grid cells is around 20 mega bytes. Note that only one bit value is stored for each region. In addition, the EBE precomputation is not affected from the dynamic edge weights as it is based on the topology of the network. If the network topology changes (less likely), the $\vec{v}_{EBE}$ of the edges should be updated.

Secondly, grid partitioning enables ER-CkNN to efficiently manage the object location updates (hence continuous monitoring of the query results). In particular, ER-CkNN maintains an in-memory grid index in which each cell contains a list of the objects currently residing in the cell. Given a moving object location $loc_t(p) = (x_p, y_p)$, $c(\lfloor \frac{x_p}{\alpha} \rfloor, \lfloor \frac{y_p}{\alpha} \rfloor)$ is the grid cell containing the p. In order to relate the grid index with the network edges (indexed by memory based PMR QuadTree [2]) thus enabling spatial network query processing, ER-CkNN associates each network edge with the overlapping grid cells (*edge-cell-mapping*). This information is stored in an hash table. For example, the edge $e(n_8, n_{10})$ (in Figure 2(a)) is mapped to the cells $\{c(0,1), c(0,2)\}$. We will describe the use of edge-cell-mapping more in Section 4.4.

4.2   On-Line Query Processing

In this section, we first explain how ER-CkNN leverages ER on the grid-partitioned network and employs a filter-and-refine process to generate the initial query results. Next, we discuss how ER-CkNN continuously maintains the query results as new location updates and/or network edge weight changes are received.

4.3   Generating Initial Query Result

ER-CkNN computes the initial result of a query using a filter-and-refine approach. With the filter step, first ER-CkNN performs a grid search to quickly find a set of candidate nearest neighbors based on their Euclidean distance from q. Next, by exploiting the fact that Euclidean distance is a lower-bound for network distance, ER-CkNN uses ER to extend the original candidate to a *super-set* which contains actual nearest neighbors of q. Toward that end, ER-CkNN 1) computes the maximum network distance NDT (short for *Network Distance Threshold*) from q to any of the objects in the original candidate set, and 2) performs a range query with radius NDT to identify all objects (and corresponding edges) that comprise the super set. The super-set contains actual k nearest neighbors for q on the spatial network; hence, no *false misses*. At the refine step, the super set is further refined by removing possible *false hits*, and ER-CkNN returns the top k objects with minimum network distance from q.

4.3.1   Filter Step

When a query object initiates a kNN search, the first step is to perform a grid expansion to identify the k nearest neighbors in the Euclidean proximity. Figure 3 illustrates an example where the goal is to find $k = 2$ nearest neighbor for the query object q. Referring to $q.cell$, we first check the grid cell in which the q resides. Since there is not any potential neighbors in this cell (see Figure 3(a)), the search moves to the next level as illustrated in Figure 3(b). Here we find the two nearest neighbors, namely p_1 and p_2 and, hence, the grid search is stopped. Note that with the grid search we only retrieve the object list from each grid cell without traversing the underlying spatial network. Having found the candidate set, next we move

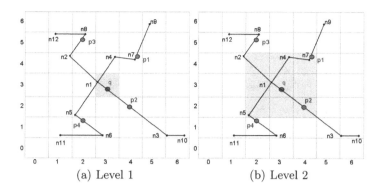

(a) Level 1   (b) Level 2

Fig. 3. Grid search for 2NN query

on to compute the super set that contains actual nearest neighbors. Toward that end, we first compute the respective network distances (using EBE- based A*) of the objects in the candidate set $d(q, p1) = 10$, $d(q, p2) = 6$, and correspondingly update $NDT=10$ (Algorithm 1 Line 3). Next, ER-CkNN performs a range query on the spatial network (using PMR quadtree) with q as the center and NDT as the radius (Line 4-5). With this operation, ER-CkNN retrieves the *active-edges* $q.activeEdges$ (the edges that are within the shaded area in Figure 4(a)) as well as m ($\{p_1, p_2, p_3, p_4\}$) objects that comprise the super set. The crucial observation in this step that If $m = k$ (i.e., there are no false hits), the exact set of k nearest neighbors for q are found (Line 7). Our experiments show that in 68% of the cases $m = k$. This implies that ER-CkNN finds kNN with only a simple grid search and the least number of network distance computations thus incurring fast response time and low computation costs.

At this point, it is important to clarify that ER-CkNN, even with the large values of k where the shortest path executed k times and some edges may be traversed more than once, visits less nodes than the network expansion methods (see Section 6.2.3 for experimental results). This is due to the following reasons. First, ER-CkNN performs EBE-based A* algorithm that enables extensive pruning (to almost the linear function of the shortest path) hence minimizing the invocation of the costly network data access. Second, ER-CkNN computes the shortest path for only feasible data objects. If the corresponding value of a grid cell that contains one or more data objects is false in the $\overrightarrow{v}_{EBE}$ of the query object's edge, ER-CkNN does not attempt to compute the network distance to those objects at all. Finally, it is possible to reuse some network distance computations as the shortest path from multiple queries to some target data objects might overlap. Consider a query point q looking for data object p which was previously found by q'. If an on-line shortest path computation, from q to p, reaches q' during the scan, then there is no need to continue the computation as ER-CkNN already knows the shortest path from q' to p.

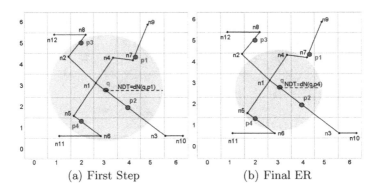

(a) First Step   (b) Final ER

Fig. 4. Using ER (Euclidean Restriction) with ER-CkNN

Algorithm 1. ER-CkNN Algorithm

1: /* k:number of NNs, q:moving query object, p_n: network distance */
2: $[p_k]=searchGrid[q.cell, k]$ /* returns k objects from the grid search*/
3: $q.NDT = max(pnd_1, ..pnd_k)$
4: $edgeList= euclideanRangeQuery(q.loc, q.NDT)$
5: $candidateNNs = retreiveObjects(edgeList)$
6: $m = candidateNNs.length$
7: **if** $(m = k)$ **then** $q.resultSet = [p_k]$; **break;**
8: **else**
9: $j = 1$
10: repeat
11:     $p_j = nextEuclideanNN(q)$ /* find closest data object to q */
12:     **if** $d_N(q, p_j) < NDT$ **then**
13:         $updateResultSet(p_j)$ /*insert p_i and remove kth NN from $q.resultSet$*/
14:         $NDT = d_N(q, p_j)$
15:         $edgeList= euclideanRangeQuery(q.loc, q.NDT)$
16:         $candidateNNs = retreiveObjects(edgeList)$
17:         $m = candidateNNs.length$
18: **end if**
19: $j++$
20: until m=k
21: $updateActiveEdges(edgeList)$
22: **end if**

4.3.2   Refine Step

Once at the refine step, we have $m > k$. We denote the set of $m - k$ objects that were added to the candidate set as P' (in Figure 4(a), $P' = \{p_3, p_4\}$). To find the actual nearest neighbors, ER-CkNN applies ER until $m = k$. Specifically, ER-CkNN performs consecutive range queries with $NDT = d_N(p_j, q)$, where p_j is the nearest Euclidean neighbor of q among P' objects, until $m = k$ (Line 11-20). To illustrate, we continue with our running example from Section 4.3.1. Since the active-edges obtained at this step contain $m = 4$ $(m > k)$ data objects,

ER-CkNN proceeds to refine step, where it computes the network distance to p_4 in P' and compares $d_N(p_4, q) = 8$ with NDT. As $d_N(p_4, q) < NDT$, the new object p_4 becomes the current k-th nearest neighbor and NDT is updated to $d_N(p_4) = 8$. Later, ER-CkNN performs the second range query (with the new NDT) which excludes p_1 and p_3 (see Figure 4(b)). At this point, since there are only two remaining objects in the candidate set (i.e., p_2 and p_4), the refine step terminates by returning p_2 and p_4 as the final result and updating $e.activeEdges$ with the final set of active-edges which are within the shaded area in Figure 4(b) (i.e., $\{e(n_1, n_2), e(n_1, n_3), e(n_1, n_4), e(n_4, n_7), e(n_1, n_5), e(n_5, n_6)\}$). As shown, p_3 is automatically excluded after the second ER. Note that finding objects on the active-edges is in linear time. After finding the active- edges, ER-CkNN first determines the corresponding set of overlapping grid cells using edge-cell-mapping hash table, and then retrieves the objects (in $O(1)$) from the grid file.

4.4 Continuous Maintenance of Query Result

In this section, we explain how ER-CkNN continuously maintains the query result while data and query objects move and the weights of the network edges change. As noted before, ER-CkNN benefits from the grid partitioning to determine the relevancy of the incoming location updates. Specifically, ER-CkNN identifies the active-cells (that overlaps with the active-edges found in the refine step) around the q. If a location update is mapped to one of these active-cells, it is considered as relevant update and processed; otherwise it is ignored. Below, we first classify the location updates and explain each of the classes in isolation; thereafter, we discuss how ER-CkNN implements all cases simultaneously.

4.4.1 Data and Query Object Location Updates
With ER-CkNN, we classify each object location update into following three categories based on the type of the movement: *resident updates*, *inbound updates*, and *outbound updates*. The resident updates are those which indicate an object has been resident among the active-edges of the query since last update, either staying still or moving from one active-edge to another. The inbound updates report movement of the objects that used to reside on an inactive edge but have moved to an active-edge since then. Finally, the outbound updates report movement of the objects that have moved from an active-edge to an inactive edge.

When a data object p sends an update containing its old and new location to the central server, ER-CkNN first updates the grid index with the new location of p by identifying the new grid cell. Next, ER-CkNN checks if the location update falls in to grid cells (*active-cells*) that overlap with the active-edges of a query object. Recall that the active-edges (thus active-cells) of the query objects are already found in the refinement step. The crucial observation here is that only the relevant location updates that are mapped to the active-cells can affect the result set of the queries (*localized mapping*); hence the irrelevant updates are efficiently identified and ignored. The cost to identify whether the location update is relevant or not is in $O(1)$ as ER-CkNN only checks the flag (indicating active or not) of the grid cell that the new location update falls in. For example, assume that p_3 sends a location

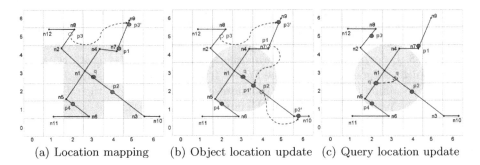

(a) Location mapping   (b) Object location update   (c) Query location update

Fig. 5. Data and query object location update

update which falls into cell $c(6,5)$ in Figure 5(a). Since neither the new nor the old location of p_3 are mapped to the active-cells (marked cells in the Figure 5(a)) which overlap with the active-edges, the movement of p_3 is ignored.

If relevant updates are received, ER-CkNN updates the results of the queries by considering the update categories. Specifically, ER-CkNN considers two cases. The first case is when the number of outbound updates is less than or equal to the number of inbound updates. In this case there still exist at least k objects on the active-edges. Therefore, ER-CkNN first removes the outbound nearest neighbors from the result set and then merges the objects that are in resident and inbound update categories by returning the k nearest objects among them. Note that if the NDT is decreased after this operation, ER-CkNN updates the NDT and $q.activeEdges$ accordingly. For instance, assume that p_1, p_2, and p_3 send location updates, and p_4 stays still as shown in Figure 5(b). In this example, the location updates of p_1, p_2, and p_4 are categorized as inbound, outbound, and resident, respectively (the movement of p_3 is ignored as described before). Since the number of outbound updates is equal to the number of inbound updates, ER-CkNN removes p_2 from the result set and adds p_1 to it. Finally, ER-CkNN returns p_1 and p_4 as the final result set. The second case is the number of outbound updates is more than the number of inbound updates. In this case ER-CkNN needs to expand the search space to retrieve the new results since there are less than k objects on the active-edges. To avoid recomputation, the grid expansion, instead of starting from the cell where q resides, starts from the level of the k th data object that is furthest to q in Euclidean distance.

Similar to data objects the query objects move on or out of the active-edges. If q moves on any of the active-edges, we save some computation. To illustrate, consider Figure 5(c) where q moves to a new location q'. Since the shortest path from p_4 to q was originally computed, we can easily compute network distance from q' to p_4. To retrieve the new results of the q, ER-CkNN continues from the Line 11 of the Algorithm 1. In the case q moves to an in-active edge, ER-CkNN computes the new k nearest neighbors and the active-edges from the beginning.

When an edge weight change is received, ER-CkNN first checks if the edge update corresponds to an active-edge. In case the new edge weight has increased, implying that the network distance between q and the data object p_i is increased, there may exist shorter alternative paths to reach p_i in the network. Therefore, ER-CkNN employs the filter step and continues from the Line 12 of the Algorithm 1. If the new edge weight has decreased and the weight change affects the current NDT, ER-CkNN updates the NDT and the $q.activeEdges$ accordingly.

When ER-CkNN receives object location and edge weight updates simultaneously, it first checks if the q moves to an inactive edge. If so, ER-CkNN recomputes the kNN from the beginning by ignoring the data object and edge updates. Otherwise, ER-CkNN first processes the edge updates since handling the query update before considering the edge update may result in retrieving wrong active-edges. After finishing the edge updates, ER-CkNN processes the query movement. Finally, it handles data object updates based on the finalized active-edges from the previous two updates.

5   Discussion

In this section we discuss the two main factors that affect the performance of ER-CkNN, namely the grid granularity and the network topology. We describe the details of these two factors and explain how we optimize the grid granularity.

5.1   Grid Granularity

As grid size grows (having less cells), the total number of cell retrievals decreases. However, large cells result in retrieving more excessive (unneeded) objects. On the other hand, with small grid cells, the cost of grid cell retrieval, and the memory requirements of EBE increases. We derive the optimal value of the grid size with the following theorem. Similar analysis have been given in [17].

Lemma 1. *Let the moving objects be represented as points distributed on the unit square* $[0, 1] \times [0, 1]$ *partitioned by the cell (which contains P objects) size* $\alpha \times \alpha$. *The number of cells in a region* $(L \times L)$ *(L is the edge length of the square after grid expansion) is given by* $\frac{\lfloor (L+\alpha) \rfloor^2}{\alpha^2}$ *and the number of objects in this region is approximately* $(\lfloor (L + \alpha) \rfloor^2)P$ *under uniform distribution.*

Theorem 1. *For a gird of size* $\alpha \times \alpha$ *which contains P objects,* $\alpha \approx \frac{1}{\sqrt{P}}$ *is the optimal grid size to minimize the query time under uniform distribution.*

Proof. The overall query processing time T is dominated by the grid expansion (i.e., time required to query several grid cells which contain k objects) and processing the objects located in the grid cells (i.e., time required to compute network distance). Therefore $T = t_g n_g + t_o n_o$ where n_g and n_o represents number of grid cells and number of objects, respectively (t_g and t_o are constants). Computing the region for a kNN query involves finding the circle $C(o; r)$ ($r = d_{NDT}$ of kth object in the worst case) which includes the kth nearest object to the query point. Therefore, using the lemma from above, the number of objects (i.e., k) in this region is $k \approx \pi r^2 * P$ so $r \approx \sqrt{\frac{k}{\pi P}}$ and the number of grid cells contained in the area bounding this circle $n_g = \frac{(2r+\alpha)^2}{\alpha^2}$ and $n_o \approx (2r + \alpha)^2 P$. Replacing n_g and n_o in $T = t_g n_g + t_o n_o$ with the above values and setting $\frac{\partial T}{\partial \alpha} = 0$ gives $\alpha^3 = \frac{t_g r}{t_o P}$ or $\alpha = \sqrt[3]{\frac{t_g}{t_o}} \sqrt{\frac{k}{\pi}} \frac{1}{\sqrt{P}}$. For $k \ll n$, the α can be simplified to $\alpha \approx \frac{1}{\sqrt{P}}$.

5.2   Network Topology

In general, the topology of a network may affect the performance of spatial network query processing methods. One topology concern when processing ER based queries in spatial networks is network and Euclidean distance correlation (referred as NEC). Because, ER based methods rely on lower bound restriction that yields better results when NEC is high (i.e., the network distance is close to the Euclidean distance between two points). The experimental evaluations show that the response time of ER-CkNN is not severely affected from NEC. The average response time loss of ER-CkNN is approximately %12 when the NEC between the query and data objects is low. We used two different datasets (i.e., Los Angeles and San Joaquin networks) to show that the behavior of ER-CkNN does not change significantly with the different network topologies. The NEC is high with Los Angeles network where as it is low with San Joaquin network.

On the other hand, ER-CkNN provides a very effective way to handle directional queries. For example, if a query asks for the nearest gas stations towards North (i.e., the driving direction), ER-CkNN only expands the grid search towards that direction thus pruning the search space to half. Directional queries are commonly used in car navigational systems.

6   Experimental Evaluation

6.1   Experimental Setup

We conducted experiments with different spatial networks and various parameters (see Table 1) to evaluate the performance of ER-CkNN and compare it with GMA [8] (as GMA yields better performance results than IMA). With our experiments, we measured the impact of the data object and query cardinality, data object and query distribution, network size, gird size, k and data

Table 1. Experimental parameters

Parameters	Default	Range
Number of objects	15 (K)	1,5,10,15,20,25 (K)
Number of queries	3 (K)	1,2,3,4,5 (K)
Number of k	20	1,10,20,30,40,50
Object Distribution	Uniform	Uniform, Gaussian
Query Distribution	Uniform	Uniform, Gaussian
Object Agility a	10 %	0, 5, 10, 15, 20,25 (%)
Object Speed v	60 kmph	20, 40, 60, 80, 100 (kmph)

object agility and speed. As our dataset, we used Los Angeles (LA) and San Joaquin County (SJ) road network data with 304,162 and 24,123 road segments, respectively. Both of these datasets fit in the memory of a typical machine with 4GB of memory space. We obtained these datasets from TIGER/Line (http://www.census.gov/geo/www/). Since the experimental results with these two networks differ insignificantly (approximately %12) as noted in Section 5.2 and due to space limitations, we only present the results from the LA dataset.

We generated the parameters represented in Table 1 using a simple simulator prototype developed in Java. We conducted our experiments on a workstation with 2.7 GHz Pentium Core Duo processor and 8GB RAM memory. We continuously monitored each query for 50 timestamps. For each set of experiments, we only vary one parameter and fix the remaining to the default values in Table 1.

6.2   Results

6.2.1   Impact of Object and Query Cardinality

First, we compare the performance of two algorithms by varying the cardinality of the data objects (P) from 1K to 25K while using default settings in Table 1 for all other parameters. Figure 6(a) illustrates the impact of data object cardinality on response time. The results indicate that the response time linearly increases with the number of data objects in both methods where ER-CkNN outperforms GMA with all numbers of data objects. From P=1K to 5K, the performance gap is more significant where ER-CkNN outperforms GMA by factor of four. Because, when P is less, the data objects are distributed sparsely on the network which causes GMA to incrementally expand the search area by visiting unnecessary edges and nodes. When P is large, GMA, with each location update, traverses the spatial index to identify the edge of the moving objects. However, ER-CkNN needs to identify the edge of the moving object only if the location update of it falls in to localized grid cells. Figure 6(b) shows the impact of the query objects (Q) ranging from 1K to 5K on response time. As shown, ER-CkNN scales better with large number of Q and the performance gap between the approaches increases as Q grows. Because, in addition to factors mentioned above, GMA maintains an expansion tree (for monitoring) which is invalidated frequently with the continuous location updates of the q thus yielding very high maintenance costs.

6.2.2   Impact of Object/Query Distribution and Network Size

With this experiment set, we study the impact of object, query distribution and network size on ER-CkNN. Figure 7(a) shows the response time of both algorithms where the objects and queries follow either uniform or Gaussian distributions. As illustrated, ER-CkNN outperforms GMA significantly in all cases. ER-CkNN yields better performance for queries with Gaussian distribution. Because, as queries are clustered in the spatial network with Gaussian distribution, their active-cells would overlap hence allowing ER-CkNN to monitor relatively less active-cells. Furthermore, since the shortest paths from clustered queries to some data objects would overlap, ER-CkNN benefits from reusing numerous

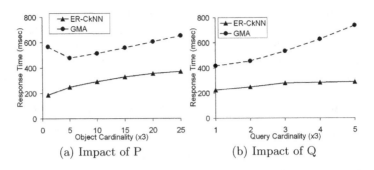

(a) Impact of P   (b) Impact of Q

Fig. 6. Response time versus P and Q

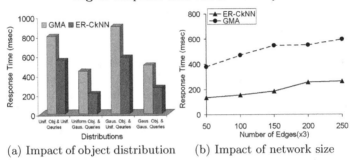

(a) Impact of object distribution   (b) Impact of network size

Fig. 7. Response time versus N and Q

network distance computations. In addition, with these experiments, we also measured our success rate with finding k nearest neighbors by only performing filter step. In average 68% of the cases ER-CkNN was able to find the k nearest neighbors without executing the refinement step as described in Section 4.3.1.

In order to evaluate the impact of network size, we conducted experiments with the sub-networks of LA dataset ranging from 50K to 250K segments. Figure 7(b) illustrates the response time of both algorithms with different network sizes. In general, with the default parameters in the Table 1, the response time increases for both algorithms as the network size increases.

6.2.3 Impact of k

With another experiment, we compare the performance of the two algorithms with regard to k. Figure 8(a) plots the average query efficiency versus k ranging from 1 to 50. The results indicate that ER-CkNN outperforms GMA with all values of k and scales better with both small and the large values of k. Because, when k is small ER-CkNN benefits from the directional search. As k increases the search space of GMA incrementally expands in all directions hence incurring redundant node traversal and the expansion tree (see [8] for the expansion tree) grows exponentially by holding more unnecessary network data hence yielding extensive maintenance and access cost. As illustrated, ER-CkNN outperforms GMA by at least a factor of three for $k{\geq}10$. In addition, we compared the

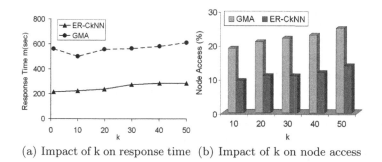

(a) Impact of k on response time   (b) Impact of k on node access

Fig. 8. Response time and node access versus k

average number of node access with both algorithms. As shown in Figure 8(b), the number of nodes accessed by ER-CkNN is less than GMA with all k values. This is because ER-CkNN, rather than expanding the search blindly, utilizes the point-to-point EBE-based A* that minimizes the node access.

6.2.4 Impact of Object Agility and Speed

With this set of experiments, we evaluate the performance of both algorithms with respect to object movements. We use two parameters to measure the affect of object movements namely object agility a and object speed v. Object agility indicates the percentage of objects that move per timestamp (0% represents static data) while the object speed indicates object's speed measured by kilometer per hour. Figure 9(a) illustrate the impact of object agility ranging from 0% to 25%. As the object agility grows, both algorithms query processing time increases slightly due to the frequent updates in the number of inbound and outbound objects. The superior performance of ER-CkNN approach is due to the usage of localized mapping that avoids extensive invalidations with the expansion tree of GMA and unnecessary node and edge access. As Figure 9(b) indicates, both algorithms are unaffected by the object speed because the focus of both algorithms only concern if there are object updates that may invalidate the existing results in the monitoring areas rather than how fast the objects move in or out of the monitoring areas.

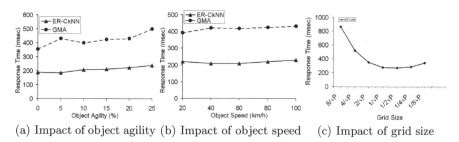

(a) Impact of object agility (b) Impact of object speed   (c) Impact of grid size

Fig. 9. Response time versus object agility, speed and grid size

6.2.5   Impact of Grid Size

In order to compare the theocratical results from the analysis in Section 5.1 and evaluate the the impact of the grid size on ER-CkNN, we run several experiments with different grid sizes and the default values in Table 1. Figure 9(c) illustrates the response time needed with grid sizes ranging from $\frac{8}{\sqrt{P}}$ to $\frac{1}{8\sqrt{P}}$. As illustrated, decreasing the cell size has the effect of reducing the response time. There is a substantial increase in performance as we move from $\frac{8}{\sqrt{P}}$ to $\frac{1}{\sqrt{P}}$, but later the response time starts increasing at finer granularity. This validates the analytical results. Note that memory requirements of ER-CkNN is slightly more than IMA/GMA due to additional grid indexing.

7   Conclusion

With this paper, we proposed an Euclidean restriction based algorithm which avoids the blind network expansion and blind object location mapping shortcomings of the network expansion methods. The key benefit of ER-CkNN is that it enables guided shortest path search that minimizes the redundant node access and localizes the network that facilitates the continuous kNN monitoring. ER-CkNN does not make any simplifying assumption (e.g., static data objects, known trajectories) about the moving objects and edge weights. Therefore, it can easily be applied to real-world road network kNN applications. In the future, we intend to extend this study to handle different monitoring queries such as range and reverse kNN in spatial networks.

References

1. Cho, H.-J., Chung, C.-W.: An efficient and scalable approach to cnn queries in a road network. In: VLDB (2005)
2. Hoel, E.G., Samet, H.: Efficient processing of spatial queries in line segment databases. In: Günther, O., Schek, H.-J. (eds.) SSD 1991. LNCS, vol. 525. Springer, Heidelberg (1991)
3. Kalashnikov, D.V., Prabhakar, S., Hambrusch, S.E.: Main memory evaluation of monitoring queries over moving objects. In: DPDB (2004)
4. Kolahdouzan, M., Shahabi, C.: Voronoi-based k-nearest neighbor search for spatial network databases. In: VLDB (2004)
5. Kolahdouzan, M.R., Shahabi, C.: Continuous k-nearest neighbor queries in spatial network databases. In: STDBM (2004)
6. Lauther, U.: An extremely fast, exact algorithm for finding shortest paths in static networks with geographical background. In: Geoinformation and Mobilitat (2004)
7. Mokbel, M.F., Xiong, X., Aref, W.G.: Sina: scalable incremental processing of continuous queries in spatio-temporal databases. In: SIGMOD (2004)
8. Mouratidis, K., Yiu, M.L., Papadias, D., Mamoulis, N.: Continuous nearest neighbor monitoring in road networks. In: VLDB (2006)
9. Papadias, D., Zhang, J., Mamoulis, N., Tao, Y.: Query processing in spatial network databases. In: VLDB (2003)
10. Russell, S.J., Norvig, P.: Artificial intelligence: A modern approach. Prentice-Hall, Inc., Englewood Cliffs (1995)

11. Samet, H., Sankaranarayanan, J., Alborzi, H.: Scalable network distance browsing in spatial databases. In: SIGMOD (2008)
12. Song, Z., Roussopoulos, N.: K-nearest neighbor search for moving query point. In: Jensen, C.S., Schneider, M., Seeger, B., Tsotras, V.J. (eds.) SSTD 2001. LNCS, vol. 2121, p. 79. Springer, Heidelberg (2001)
13. Tao, Y., Papadias, D.: Time-parameterized queries in spatio-temporal databases. In: SIGMOD (2002)
14. Tao, Y., Papadias, D., Shen, Q.: Continuous nearest neighbor search. In: VLDB (2002)
15. Huang, X., Jensen, C.S., Hua, L., Saltenis, S.: S-GRID: A versatile approach to efficient query processing in spatial networks. In: Papadias, D., Zhang, D., Kollios, G. (eds.) SSTD 2007. LNCS, vol. 4605, pp. 93–111. Springer, Heidelberg (2007)
16. Huang, X., Jensen, C.S., Šaltenis, S.: The island approach to nearest neighbor querying in spatial networks. In: Bauzer Medeiros, C., Egenhofer, M.J., Bertino, E. (eds.) SSTD 2005. LNCS, vol. 3633, pp. 73–90. Springer, Heidelberg (2005)
17. Yu, X., Pu, K.Q., Koudas, N.: Monitoring k-nearest neighbor queries over moving objects. In: ICDE (2005)
18. Zhang, J., Zhu, M., Papadias, D., Tao, Y., Lee, D.L.: Location-based spatial queries. In: SIGMOD (2003)
19. Zheng, B., Lee, D.L.: Semantic caching in location-dependent query processing. In: Jensen, C.S., Schneider, M., Seeger, B., Tsotras, V.J. (eds.) SSTD 2001. LNCS, vol. 2121, p. 97. Springer, Heidelberg (2001)

Discovering Teleconnected Flow Anomalies: A Relationship Analysis of Dynamic Neighborhoods (RAD) Approach

James M. Kang[1], Shashi Shekhar[1],
Michael Henjum[2], Paige J. Novak[2], and William A. Arnold[2]

[1] Department of Computer Science, University of Minnesota, MN, USA
{jkang,shekhar}@cs.umn.edu
[2] Department of Civil Engineering, University of Minnesota, MN, USA
{henj0016,novak010,arnol032}@umn.edu

Abstract. Given a collection of sensors monitoring a flow network, the problem of discovering teleconnected flow anomalies aims to identify strongly connected pairs of events (e.g., introduction of a contaminant and its removal from a river). The ability to mine teleconnected flow anomalies is important for applications related to environmental science, video surveillance, and transportation systems. However, this problem is computationally hard because of the large number of time instants of measurement, sensors, and locations. This paper characterizes the computational structure in terms of three critical tasks, (1) detection of flow anomaly events, (2) identification of candidate pairs of events, and (3) evaluation of candidate pairs for possible teleconnection. The first task was addressed in our recent work. In this paper, we propose a RAD (Relationship Analysis of spatio-temporal Dynamic neighborhoods) approach for steps 2 and 3 to discover teleconnected flow anomalies. Computational overhead is brought down significantly by utilizing our proposed spatio-temporal dynamic neighborhood model as an index and a pruning strategy. We prove correctness and completeness for the proposed approaches. We also experimentally show the efficacy of our proposed methods using both synthetic and real datasets.

1 Introduction

This section first presents the application domain, followed by the problem statement, challenges, related work, contributions, and the scope and outline of this paper.

Application Domain. A teleconnection represents a strong interaction between paired events that are spatially distant from each other. A well-known example of teleconnected event pair involves the warming of the eastern pacific region (i.e. El Niño) and unusual weather patterns throughout the world [1]. In the United States, teleconnections often occur in air travel when a local weather disruption of a single airport (e.g., Chicago) causes other major airports (e.g., New York City, Atlanta, etc.) to delay or cancel flights. Indeed, many events in everyday life display patterns related to other events occurring a distance away. One type of teleconnected event of special interest to scientists occur in environmental systems when a contaminant enters a river (e.g.,

N. Mamoulis et al. (Eds.): SSTD 2009, LNCS 5644, pp. 44–61, 2009.

Fig. 1. Dead Zone, Gulf of Mexico [5] (Best Viewed in Color)

an oil spill) and then vanishes (e.g., the removal of the oil via natural or man-made) downstream. Identifying these teleconnections in environmental systems is important to maintain high water quality, one of the major global challenges facing humanity according the the United Nations [2]. When contaminants enter river networks, they create problems for drinking water sources and point to the need to identify when and where the contaminant entered and exited the river network [3,4].

For the past several years, environmental engineers and scientists have been actively studying contaminants in water by placing advanced sensors along streams or rivers [6]. One of the greatest challenges in this field, however, is to understand how contaminants *emerge* (i.e., when and where a contaminant may enter) and how they *vanish* (i.e., when and where a contaminant is removed). Pairs of *emerging* and *vanishing* events may be teleconnected. A single contaminant may *emerge* as a result of rain fall and then *vanish* downstream in natural catchments (e.g., Dead Zone in the Gulf of Mexico in Figure 1). For example, nitrate (a component of fertilizer) may emerge from storm water runoff (i.e., process of nitrification) and vanish downstream as a result of biological transformation (i.e., process of denitrification). Although there exist several known locations for vanishing events, studies using mass-balance methods show that only a fraction of the entering contaminants are "caught" [7]. Determining when and where all of these contaminants *vanish* in the river is an open area of study in environmental science with many potential benefits. For example, such research is possible to reduce economic costs and the environmental impact of contamination by limiting the location of man-made remedies [7] to areas where natural processes are shown to be inadequate for removing contamination. Thus several environmental scientists (e.g. our collaborators Novak and Arnold) have expressed the need for an efficient and robust method to discover teleconnections between these *emerging* and *vanishing* events.

There are other important and interesting applications for the discovery of teleconnected events outside the realm of environmental science as well. In transportation systems, identifying the time and the location o teleconnected congestion may be important for commuters when choosing the best route to take. In video surveillance, authorities want to be able to determine the time and source of unusual events such as unattended bags being left (i.e, *emerge*) or picked up (i.e., *vanishing*) at an airport terminal. Monitoring thousands of surveillance video streams may result in expensive manual investigations to identify these events. Thus, there is a need to efficiently detect these teleconnected relationships.

Problem Statement. Given a collection of sensors where each sensor has a time series of measured variables, the teleconnected flow anomaly discovery problem identifies strongly connected pairs of events. We are mostly interested in flow anomaly events and pairs of *emerging* and *vanishing* events. We define this notion informally here and formally in Section 2. Flow anomalies represent time-periods with a (user-defined) high fraction of time-instants having significantly different readings across pairs of adjacent sensors. For example, if no pollution events exist within a river, then all the observations seen at each sensor along the river will be similar. If a pollution event occurs between a pair of sensors at a single time instant, then a transient flow anomaly has been found. A persistent flow anomaly may consist of several transient flow anomalies and several observations that appears to be normal. A persistent flow anomaly found between sensors is considered dominant if it is not a subset of any other flow anomaly event occurring at this location. An *emerging* flow anomaly may be found upstream (e.g., at an industrial outfall) whereas a *vanishing* flow anomaly event may be found downstream (e.g., as a result of degradation).

Mining teleconnected flow anomaly events is computationally challenging for many reasons. First, a single flow anomaly event may consist of subsets that may not be anomalous, but are important for the event itself. This makes it difficult to use the dynamic programming principle for designing an algorithm. Second, the temporal length of each flow anomaly may vary. This makes fixed window-based paradigms unnatural. Third, there may be a large number of possible locations for *emerging* and *vanishing* flow anomaly events across all node paths and time paths in the network and all paths and time-instant paths must be searched to identify teleconnected relationships. In addition, teleconnected flow anomalies may consist of one-to-one, one-to-many, or many-to-many relationships between *emerging* and *vanishing* flow anomaly events. Identifying the relationships between flow anomaly events creates a large number of combinations across the entire network. Finally, the length of time series may be very large due to the potentially infinite nature of time.

Related Work. To the best of the authors' knowledge, no techniques have been reported in the literature to find flow anomalies across an entire network and then identify the relationship between these events. The most related technique, called SWEET, is our preliminary work [8] that introduced the problem of discovering flow anomalies for a pair of adjacent sensors addressing the first critical task identified in the abstract for the overall problem of discovering teleconnected flow anomalies. Computation time for SWEET was reduced significantly by introducing the concepts of a smart counter and a pruning strategy. Briefly, the smart counter allowed SWEET to scan the time series once to identify the transient flow anomalies and the pruning strategy reduced the number of candidates (i.e., time periods) to be analyzed. These algorithmic innovations reduced computation time costs by orders of magnitude. For example, for a long time series, SWEET reduced the execution time from hours to seconds. However, SWEET is limited to finding flow anomalies between only two sensors and cannot identify the teleconnected relationship between multiple flow anomalies occurring at different locations and time periods.

In order to make this paper complete, the related work on flow anomalies presented in our previous work [8] is also presented here. Related literature to flow anomalies may appear to occur in string matching, time series analysis, data stream correlations, clustering, and outlier detection. In string matching, Amir et al. uses an inverse string matching method that maximizes and minimizes the number of mismatches [9], and Lee et al. proposes a similar method using wild cards [10]. However, these techniques use an exact matching technique whereas flow anomalies are found using a statistical measure because an exact match may not occur in our problem domain. In time series analysis, several methods assume that the basic property of dynamic programming of sub-optimal substructure exists in their problem domain (e.g. [11]). However, a persistent flow anomaly may have subsets that may not be anomalous which violates this basic principle of dynamic programming. In data stream correlations, relationships between streams are identified using a correlation measure and a fixed sliding window. Chan et al. found local correlations between multiple data streams using a sliding window [12]. Global relationships between data streams were also found using a sliding window to summarize the entire data stream [13]. Multiple pre-defined sliding windows were used to find correlations based on a query [14]. Rarity and similarity of data streams were found using a fixed sliding window [15]. However, use of a fixed window presupposes that the domain specialist knows the duration of the unexpected event (e.g. Rain Events). Also, there may be multiple events occurring between multiple data streams having anomalous events of variable sizes. In clustering, methods that focus on moving clusters (e.g. [16]) or cluster transitions (e.g. [17]) often require the need of spatially dense datasets to identify each cluster. However, flow anomalies may exist in spatially sparse datasets, limiting the ability of these clustering techniques to discover each event. Basic outlier detection techniques (e.g. t-test [18]) may detect transient flow anomalies and persistent flow anomalies (at 100% missmatched time instants) if flow is considered (e.g. [19,20]). However, these techniques are limited in finding all persistant flow anomalies since they may miss several patterns when the mismatched time instants is less than 100%.

Identifying relationships across multiple sensors presents several challenges such as identifying whether a pair of flow anomaly events that may be spatially distant are in fact related based on their spatio-temporal neighborhood. Existing approaches have modeled these relationships as a spatial neighborhood using concepts such as modeling vector fields (e.g. [21]). However, teleconnected relationships cannot be found using these models to find pairs of *emerging* and *vanishing* flow anomaly events because neighborhoods are only defined by their spatial proximity. Spatio-temporal relationships have been discovered while assuming that the temporal dimension is fixed [22]. Whereas in the teleconnected flow anomaly problem, there may exist spatio-temporal relationships having variable temporal lengths.

Contributions. In this paper, we propose a Relationship Analysis of spatio-temporal Dynamic neighborhoods (RAD) approach for steps 2 and 3 (identified in abstract) of the overall problem of that utilizes several inherent properties of the problem to efficiently identify teleconnected flow anomaly events across an entire network. In summary, this paper makes the following contributions:

1. We define new key concepts that utilize our proposed spatio-temporal dynamic neighborhood model.
2. We propose a new interest measure to discover teleconnected flow anomalies
3. We propose a novel RAD method to discover teleconnected flow anomalies.
4. We propose several design alternatives: "On the Fly", spatio-temporal Dynamic Neighborhood, and a pruning strategy.
5. We prove the correctness and completeness of all proposed approaches.
6. We experimentally evaluate our proposed methods using synthetic and real datasets.

Scope and Organization. The following issues are beyond the scope of this paper: (i) inferring the travel time from the dataset, that is, the travel time is given as part of the input for the teleconnected flow anomaly problem, (ii) sensor placement within the network (e.g. [23]), (iii) non-point source flow anomalies (1:M and M:N) are not discovered, that is, flow anomalies only occur between adjacent sensors, (iv) complex networks, that is, only a tree network is examined in this paper, (v) only singleton neighborhoods are explored, (vi) anomalies occurring beyond the set of known sensors in the network, and (vii) arbitrary event relationships, that is, only emerging and vanishing event types are explored.

The rest of the paper is organized as follows. Section 2 presents the basic concepts and the problem statement of discovering teleconnected flow anomalies. Section 3 presents our proposed RAD method, its design decisions, and theoretical analysis. Section 4 gives the experimental evaluation and Section 5 concludes the paper and discusses future work.

2   Key Concepts and Problem Statement

In this section, we first introduce key concepts for modeling the spatio-temporal dynamic neighborhood relationship and then, introduce definitions to characterize teleconnected flow anomalies. Finally, we give a formal description of the problem statement. Figure 2 illustrates the spatio-temporal dynamic neighborhood model with six spatio-temporal locations where the distance between each spatial neighbor is one unit length. Figure 3 depicts the discovery of *Emerging* and *Vanishing* Flow Anomalies respectively. In this example, the input and output is simplified for illustration by using a unit length of 1 between each sensor and assuming the travel time at each instant is given.

2.1   Key Concepts

A spatio-temporal set ST is denoted as $ST = \{st_1, st_2, \ldots, st_m\}$, where $st_i = \{s_i, t_i\}$ and s_i represents a spatial location and t_i represents a time instant. Figure 2 gives an example of six locations, $\{s_1, s_2, \ldots, s_6\}$. A sensor observation, $f(st_i)$, may be associated with (s_i, t_i).

A vector (e.g. velocity) field, $V(st)$ or $V(s,t)$, is also associated with ST where s is the spatial location of the sensor and t is a time instant that maps each $\{s_i, t_i\}$ to a velocity vector.

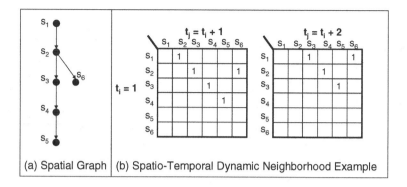

Fig. 2. Spatio-Temporal Dynamic Neighborhood Example

Definition 1. *A spatio-temporal dynamic neighborhood relationship is the association between two ST locations.*

Definition 1 can be formally expressed by the directed $NB(st_i, st_j, t_k)$ where $st_i = (s_i, t_i)$ is a neighbor of $st_j = (s_j, t_j)$ if and only if a particle at s_i in time instant t_i will be propelled by the velocity field $V(s, t)$ to reach location s_j on time instant t_j.

Figure 2 illustrates an example of a spatio-temporal dynamic neighborhood. For illustrative purposes only, suppose the velocity field in this example is a constant function valued 1, i.e., it is a uniformly flowing field with a unit speed downstream of 1 and the unit length between each spatially adjacent neighbor of 1. At $t_i = 1$, suppose we drop a particle at each spatial location and wait one second (i.e., $t_j = 2$). Based on the velocity field, the particle at each spatial location will travel one unit in length downstream and reach its adjacent neighbor. For example, at time instant $t_i = 1$ and $t_j = 2$, the neighbor for s_1 is s_2, i.e., $NB((s_1, 1), (s_2, 2))$. Likewise, the neighbor for s_2 is s_3 and the neighbor of s_6, s_3 is s_4, and the neighbor of s_4 is s_5. Suppose we wait an additional second ($t_j = 3$) after we initially drop the particle at $t_i = 1$ at each sensor. Then, the spatio-temporal neighborhood changes and the particle will travel an additional unit length. Thus, as shown in Figure 2b, the neighbor when $t_i = 1$ and $t_j = 3$ for s_1 is s_3 and s_6, s_2 is s_4, and s_3 is s_5 where the total distance traveled is 2 units in length. This simple example illustrates that a spatio-temporal neighborhood can change over time due to the flow within the network.

Neighbors $N(st_i, t_k)$ of a ST location based on a spatio-temporal dynamic neighborhood relationship can be formally characterized as $\{st_j | st_j \in ST, NB(st_i, st_j, t_k) = True\}$, where t_k represents the travel time from s_i to s_j. Figure 2 gives an example of where the neighbor of s_1 is s_2 when the velocity starting at $t_i = 1$ (i.e., $V(s, t) = 1$) and we wait one second ($t_j = 2$). $N(st_i, t_k)$ is considered a singleton neighborhood if it has only one element. Identifying neighborhoods for all paths and time-instant paths may be very challenging because a directed acyclic graph may merge and disperse, creating an exponential number of paths and time-instant paths due to flow.

A spatial neighborhood gives the relationship of adjacent locations s_i and s_j, whereas a spatio-temporal dynamic neighborhood gives the relationship of a pair of locations s_i and s_j at different travel times. For example, the spatial neighbors of s_1 in Figure 3a is

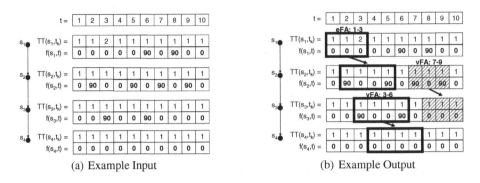

(a) Example Input   (b) Example Output

Fig. 3. Discovering *Emerging* and *Vanishing* Flow Anomalies Example (Best Viewed in Color), $TT(s_i, t_j)$ represents $1/|V(s_i, t_j)|$, i.e. travel time to downstream sensor at unit distance

s_2, whereas the spatio-temporal neighbor of s_1 having a travel time of 2 (i.e., $t_j - t_i$) are s_3 and s_6.

Definition 2. *A transient Flow Anomaly (tFA) is a triple (st_i, t_k, Θ_e) where the difference between corresponding observations (i.e., accounting for the velocity field) from a sensor and its neighboring sensors is larger than the given error threshold, Θ_e.*

Definition 2 can be formally expressed in Equation 1.

$$tFA(st_i, t_k, \Theta_e) \iff \{f(st_i) - AVG(f(st_j)|st_j \in N(st_i, t_k)) > \Theta_e\} \quad (1)$$

There are two types of transient flow anomalies, namely, *emerging* and *vanishing*. An *emerging* tFA (etFA) is defined by $tFA(st_i, t_k) < -\Theta_e$ whereas a *vanishing* tFA (vtFA) is defined by $tFA(st_i, t_k) > \Theta_e$. For simplicity, Figure 3b gives examples of an *emerging* and *vanishing* tFAs for singleton neighborhoods. As can be seen, an *emerging* tFA occurs at time instant 1 between s_1 and s_3 having a value of -90 when the error threshold is 10 and a *vanishing* tFA occurs at time instant 3 between s_3 and s_4.

Definition 3. *A persistent Flow Anomaly (pFA) is a 6-tuple $(s_i, t_k, t_s, t_e, \Theta_e, \Theta_p)$ if and only if (s_i, t_s, Θ_e) and (s_i, t_e, Θ_e) are transient flow anomalies, and at Θ_p fraction of time instants t in time-interval $[t_s, t_e]$ are associated with transient flow anomalies $(<s_i, t>, t_k, \Theta_e)$.*

Definition 3 can be formally expressed in Equation 2.

$$pFA[s_i, t_k, t_s, t_e, \Theta_e, \Theta_p] \iff (tFA((s_i, t_s), t_k)) \ \& \ (tFA((s_i, t_s), t_k)) \ \&$$

$$(\frac{\sum_{t=t_s}^{t_e} tFA((s_i, t), t_k)}{time\ interval\ length(t_e - t_s)} \geq \Theta_p) \quad (2)$$

Persistent flow anomalies are classified as either *emerging* when its tFAs are all etFAs, *vanishing* when its tFAs are all vtFAs; otherwise, they are neither. Figure 3b gives an example of an epFA for the time interval from 1 to 3 between s_1 and s_2 having three etFAs and no vtFAs when the $\Theta_p = 0.5$.

Definition 4. *A dominant persistent Flow Anomaly (dpFA) is a pFA that is not a subset of any other dpFA.*

A dpFA may be characterized as either an *emerging* dpFA (denoted as eFA) or a *vanishing* dpFA (denoted as vFA) based on the type of its pFA. Figure 3 gives an example of an *emerging* dpFA during time instants 1 to 3 between ST locations s_1 and s_2. According to the persistent flow anomaly definition, time instants 1 and 3 each satisfy the persistent threshold and its definition. However, time instants 1 and 3 are not a dpFA because they are a subset of a larger dpFA for period 1 to 3.

Definition 5. *A teleconnected Flow Anomaly (telFA) is an eFA and a vFA pair that are related via a velocity field.*

Intuitively, a telFA may represent a contmination (an eFA) cleaned up later (vFA) by a natural or man-made process. Definition 5 can be formally expressed in Equation 3.

$$telFA(eFA(s_i^1, t_k^1, t_s^1, t_e^1, \Theta_e^1, \Theta_p^1), vFA(s_i^2, t_k^2, t_s^2, t_e^2, \Theta_e^2, \Theta_p^2)) \forall (s_i^1, t_i^1), \iff$$
$$\exists (s_i^2, t_i^2) \; s.t. \; \{t_i^1 \in [t_s^1, t_e^1]\} \; AND \; \{t_i^2 \in [t_s^2, t_e^2]\}$$
$$AND \; \{NB(< s_i^1, t_i^1 >, < s_i^2, t_i^2 >)\} \quad (3)$$

where s_i^1 and s_i^2 is the starting location in the eFA and the vFA respectively for the time period of t_s to t_e.

Figure 3b gives an example of one telFA consisting of one eFA (period 1-3, between s_1 and s_2) and one vFA (period 3-6, between s_3 and s_4). For simplicity, suppose that in this example, the unit length between each immediate neighbor is 1 and the velocity field is 1. When $t_1 = 1$ and the travel time $t_2 = 2$, the neighbor of s_1 is s_3. Likewise, at $t_1 = 2$ and $t_1 = 3$, the neighbor of st_1 is again st_3. A teleconnected flow anomaly may be statistically interpreted to identify emerging and vanishing events. Those events that do not satisfy the criteria for a emerging or a vanishing anomaly are not considered to be a telFA.

2.2   Problem Statement

The teleconnected flow anomaly discovery problem can be defined as follows:

Given. (1) A directed acyclic network consisting of ST locations; (2) A set of observations at each ST location for $t = 1 \ldots n$, where n is the length of the time series; (3) The relevant aspects of the velocity field are represented by the travel time information from each sensor to its neighboring sensor at different start time instants; (4) An error threshold Θ_e; (5) A persistent threshold Θ_p; and (6) A spatial neighborhood (W-Matrix [24]) which maps the spatial locations to a boolean value.
Find. All Teleconnected Flow Anomaly relationships.

Objective. Minimize the computational costs.

Constraints. The directed acyclic network has a tree structure.

Example. Figure 3a gives an example of an input time series for four sensors where the travel time is the temporal length when one observation is expected to be made between each spatially neighborhood sensor. Figure 3b gives an example output of a teleconnected flow anomaly consisting of one eFA and one vFA when the error threshold is zero and the persistent threshold is 0.5. The eFA between ST locations s_1 and s_2 occurs for time period 1 to 3 and satisfies the persistent threshold, is dominant, and emerging. There are two *vanishing* flow anomalies. The first vFA occurs between s_2 and s_3 for the time period 7 to 9 and the second occurs between s_3 and s_4 for the time period of 3-6. These events are *vanishing* because the degree of change is negative. Also, they both satisfy the persistent threshold and are dominant. Based on the spatio-temporal dynamic neighborhood model, when $t_1 = 1$ and the travel time $t_2 = 2$, the neighbor of s_1 is s_3. Likewise, at $t_1 = 2$ and $t_1 = 3$, the neighbor of s_1 is again s_3. The vFA observed in period 7-9 between s_2 and s_3 is not linked to the eFA found between s_1 and s_2 because the travel time (t_2) between s_1 and s_2 is not part of the neighborhood at 7-9 when the travel time t_1 is between period 1-3.

3   Mining Teleconnected Flow Anomaly Events

In this section, we first introduce our proposed RAD (Relationship Analysis of spatio-temporal Dynamic neighborhoods) approach. We then explain key design decisions in the approach and provide its theoretical analysis.

3.1   RAD Approach

This section presents the RAD (Relationship Analysis of spatio-temporal Dynamic Neighborhoods) approach to discover teleconnected flow anomalies among *emerging* and *vanishing* flow anomalies. The RAD method has three phases, namely, *identify flow anomalies*, *identify candidate pairs of flow anomaly events*, and *identify teleconnected flow anomalies* (Figure 4).

Phase I: Identify Flow Anomalies. This phase is concerned with identifying all the flow anomaly patterns across the entire network that satisfy the dpFA definition (Definition 4). Each pair of ST locations is analyzed based on its spatial neighborhood as defined by the W-matrix. For each pair of neighboring sensors, flow anomalies are retrieved using the SWEET[1] method.

Phase II: Identify Candidate Pairs of Flow Anomaly Events. This phase is concerned with identifying pairs of *emerging* and *vanishing* flow anomalies that can be validated in the third phase. Candidate pairs are formed by the cross product of eFAs and vFAs. A **pruning strategy** is introduced to reduce the number of candidates.

Phase III: Identify Teleconnected Flow Anomalies. This phase is concerned with identifying all the teleconnected flow anomalies (Definition 5) based on the dpFAs found in the first phase. For a pair of *emerging* and *vanishing* flow anomalies respectively, their expected and actual travel times are found. The expected travel time is found based on the pair of time instants t_i and t_j at the time periods for the *emerging*

---

[1] To keep this paper self-contained, key ideas of SWEET are discussed in the Related Work (Section 1). Due to space limitations, readers interested are encouraged to see [8] for details.

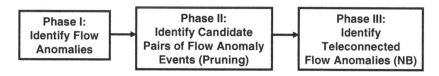

Fig. 4. RAD Approach

and *vanishing* flow anomalies respectively. The actual travel time can be found **"On the Fly"** or using our proposed spatio-temporal **Dynamic Neighborhood**. If the expected travel time is the same as the actual travel time for all time instants in the *emerging* flow anomaly, then a teleconnection is found.

The composition of the phases may be executed sequentially or in a pipeline manner. A sequential approach executes Phase I until completion, followed by the second phase and then the third phase. By contrast, in the pipelined approach, Phase II is executed after a few eFAs and vFAs are determined in Phase I.

The rest of the section describes the design decisions applied in Phase II and Phase III (Due to space limitations, we omit significant design decisions made for Phase I; these are detailed in our previous work [8]). We begin with Phase III because it is easier to describe our pruning strategy for Phase II after defining our "On the Fly" and spatio-temporal Dynamic Neighborhood methods.

On The Fly. The "On the Fly" design decision identifies the travel time between the spatio-temporal locations between an *emerging* FA and a *vanishing* FA by traversing the path between the two locations. For example, Figure 3b gives three examples of dpFAs found between ST locations: (1) Between s_1 and s_2 for period 1 to 3, (2) Between s_2 and s_3 for period 7-9, and (3) Between s_3 and s_4 for period 3-6. After all the dpFAs have been found, the teleconnected flow anomalies are discovered.

In this example, there is one emerging flow anomaly (eFA) between s_1 and s_2 and the other two are vanishing flow anomalies (vFAs). There are two possible pairs of dpFAs that may be teleconnected and will need to be analyzed. First, the eFA found between s_1 and s_2 and the vFA found between s_2 and s_3 are analyzed by checking their pairs of time instants. In the eFA, the expected travel time from time instant 1 (t=1) and time instant 7 (t=7) in vFA is found by taking its difference, which is 6. The actual travel time is found by traversing the path from s_1 to s_2, shown in the spatial graph in Figure 3, which is a subset of the spatial graph in Figure 2a, starting at (t=1) which is one. This on-the-fly computation may be based on a general path computation algorithm such as Dijkstra's [25] or A^* [26], or a custom algorithm for trees. We used a custom algorithm for trees which has a linear (i.e., number of nodes and edges) complexity. For this eFA and a vFA pair, not every time instant in the eFA is a neighbor of the vFA and there is no teleconnected relationship. Then, the next eFA (between s_1 and s_2) and vFA (between s_3 and s_4) pair is evaluated. Here, a check of every time instant in the eFA with the vFA reveals a teleconnection relationship. For example, the expected travel time between the eFA and the vFA at their respective first time instant is $3 - 1 = 2$. Also, the travel time from s_1 to s_3 starting at 1 is also 2. Thus, each neighbor in eFA is a neighbor of at least one time instant of the vFA resulting in a teleconnection.

Algorithm 1. Generation of the Spatio-Temporal Dynamic Neighborhood (DN)

Inputs:
  – The travel time at each ST location, $TT[M][N]$

Outputs:
  – Spatio-Temporal Dynamic Neighborhood (DN)

Algorithm
  1: DN[N][N][M] $\leftarrow$ 0
  2: **for** each pair of ST locations, s_i and s_j where i, j=1 to M and a directed path exists **do**
  3:    **for** each time instant, t_k = 1 to N **do**
  4:       actualTT = 0
  5:       **for** each ST location s_k from s_i to s_j at t_k **do**
  6:          actualTT += TT[s_k][t_k+actualTT]
  7:       **end for**
  8:       DN[s_i][s_k][t_k] = actualTT
  9:    **end for**
 10: **end for**
 11: **return** *DN*

Dynamic Neighborhood. The Dynamic Neighborhood based design decision uses a pre-computed spatio-temporal Dynamic Neighborhood to identify the actual travel times between the *emerging* and *vanishing* flow anomalies (denoted as RAD-index). Unlike the RAD-fly approach, the actual travel time can be determined using the Dynamic Neighborhood index for the RAD-index approach. If the expected and actual travel times are equal for all time instants in the *emerging* flow anomaly, then a teleconnection has been found.

Algorithm 1. gives the pseudocode for the construction of the spatio-temporal dynamic neighborhood (stDN). The stDN approach has one input consisting of the travel time (TT) required between each spatial neighbor of each node at every time instant. The travel time is generated based on the velocity field within the network. The output for Algorithm 1. is the spatio-temporal dynamic neighborhood itself.

The spatio-temporal dynamic neighborhood (DN) consists of three dimensions: (1) the starting ST location, (2) the ending ST locations that a particle may arrive at, and (3) the starting time instant. Initially, each element in the DN matrix is set to zero (Line 1 of Algorithm 1.). Each pair of ST locations (s_i and s_j) is analyzed where a directed path exists between these two locations (Line 2 of Algorithm 1.). At each time instant t_k for the entire time series, the path between s_i and s_j is traversed to calculate the total travel time (Line 3-6 of Algorithm 1.). The total travel time between s_i and s_j at time instant t_k can then be stored in the DN matrix (Line 8 of Algorithm 1.). The process continues until all time instants are examined for each pair of ST locations and the DN is returned (Line 11 of Algorithm 1.).

Table 1 gives the execution trace of the construction of the spatio-temporal Dynamic Neighborhood from the example in Figure 3. The first three rows in the table give the input travel times for each edge, s_1 to s_2, s_2 to s_3, and s_3 to s_4. First, the pair s_1 and s_3 is analyzed to get the total travel times starting at st_1 and arriving at s_3. The travel times are obtained at each edge from the start to its destination. For example, time instant 1,

Table 1. Execution Trace for the Construction of the spatio-temporal Dynamic Neighborhood

	Time Instants									
Edge	**1**	**2**	**3**	**4**	**5**	**6**	**7**	**8**	**9**	**10**
$s_1 \to s_2$	1	1	2	1	1	1	1	1	1	1
$s_2 \to s_3$	1	1	1	1	1	1	1	1	1	1
$s_3 \to s_4$	1	1	1	1	1	1	1	1	1	1
$s_1 \to s_3$	2	2	3	2	2	2	2	2	2	-
$s_1 \to s_4$	3	3	4	3	3	3	3	3	-	-
$s_2 \to s_4$	2	2	2	2	2	2	2	2	2	-

starting at s_1 has a travel time of 1 to s_2. Then, at time instant 2 of s_2, the travel time is again 1. Thus, the total travel time starting at time instant 1 from s_1 to s_3 is 2. The travel times may vary across the times series and at multiple edges. For example, the travel time from s_1 to s_2 at time instant 3 is 2. The travel time from s_2 to s_3 at time instant 5 is 1. Thus, the travel time from s_1 to s_3 starting at time instant 3 has a total travel time of 3. The dashes in this table represent unknown information because the travel time is not available during part of the path. This process is continued for all node pairs and all time instant pairs until all the travel times are found as shown in Table 1.

We acknowledge that the storage cost may be an issue when the number of time instants grows, there is no periodicity, and travel time fluctuates greatly over time. We plan to address this in more detail in future work. Our current source of real data, a sensor setup at Shingle Creek, MN, does not require modeling of a large number of possibilities for travel time between adjacent sensor pair s due to periodicity, low variation in elevation, rainfall amount, and snow melt-rates.

Pruning. A key pruning design decision can be applied to the second phase when the candidate pairs are identified. In this phase, we can prune any *vanishing* flow anomalies (vFA) where each vFA is not a neighbor of the first time instant (sTime) of an *emerging* flow anomaly. The pair of ST locations are analyzed for a single path in a tree network starting at the root node. As dpFAs are found, for any two *emerging* and *vanishing* flow anomalies, the expected travel time can be determined based on their time periods and the actual travel time can be found "On the Fly" or using the spatio-temporal dynamic neighborhood. If there exists at least one *emerging* flow anomaly whose first time instant is a neighbor to a *vanishing*, then this vFA is added to the dpFAs. All other *emerging* flow anomalies are also placed in the dpFAs. In future, we plan to explore other pruning methods such as those found on spatial relationships (e.g., ancestor-descendant) among sensors.

For example, Figure 3 gives the input and output used in this example and Table 1 contains the travel times at all node and time instant pairs. In phase 1, the first pair of ST locations (s_1 and s_2) is analyzed for dpFAs. The SWEET technique discovers one emerging dpFA during the period of 1 to 3. The second pair of ST locations (s_2 and s_3) is analyzed and a vanishing dpFA is discovered. In the second phase, this vFA is checked for a teleconnection with the first time instant of any eFA. Examining the eFA found previously within this vFA reveals that the total travel time from s_1 to s_2 starting at time instant 1 is 1 as also shown in Table 1. This vFA cannot be a valid telFA with any eFAs found so far and nor can any other eFA found in the dataset be linked to this vFA. Thus, this vFA is not added as a dpFA. Finally, the last vanishing dpFA found from period 3-6 is analyzed. If we examine this vFA with the original eFA discovered earlier we find that the total travel time from s_1 to s_3 at time instant 1 is 3 and that the expected travel time between the eFA and vFA pair is also 3 at the first time instant. Thus, this vFA is a possible telFA and is considered for evaluation.

Lemma 1. *The pruning based on the first time instants is a true filter, i.e., it does not eliminate any teleconnected flow anomalies.*

Proof. In the second phase of RAD, all dpFAs are initially found using SWEET [27] and then the vanishing flow anomalies are pruned if the first time instant of an emerging flow anomaly is not its neighbor and violates the telFA definition (Definition 5). Thus, no telFA patterns will be missed in the second phase for both approaches. In the third phase of RAD, only the pairs of emerging and vanishing flow anomalies that satisfy the telFA definition will be found. Thus, no telFA patterns will be missed in the third phase.

3.2 Theoretical Analysis

In this section, we present the theoretical analysis of the RAD-fly and RAD-index methods and prove that: (1) both are correct, i.e., each pattern found is teleconnected and satisfies the telFA definition, and (2) both are complete, i.e., all patterns satisfying the telFA definition are found.

Theorem 1. *The design decisions "On the Fly" and DN-based are correct, i.e. each pattern $< p, q >$ found by RAD satisfies the telFA definition.*

Proof. The pair p and q is a teleconnected flow anomaly if both satisfy the following conditions: each satisfies the dpFA definition (Definition 4) and the relationship between p and q satisfies the telFA definition (Definition 5). The dpFA patterns p and q are found in the first phase using SWEET, a method previously proved correct in [27]. The pattern is then identified as either emerging or vanishing (Phase II). Both p and q are neighbors if every time instant in p is a neighbor of at least one time instant in q. For each pair of time instants, the actual travel time is found either "On the Fly" (by traversing the path between p and q) or by using the spatio-temporal DN model to identify all the travel times for all paths in the network. A teleconected flow anomaly is identified in the final phase when each time instant in p is found to be a neighbor of q.

Theorem 2. *The design decisions "On the Fly" and DN-based are complete, i.e. all teleconnected FA patterns are found by RAD.*

Proof. In the first phase of both methods, all dpFAs are found using the SWEET approach [27]. In the second phase, all *emerging* and *vanishing* flow anomalies are found. In the third phase for both methods, only the pair of emerging and vanishing flow anomalies that satisfy the telFA definition will be found. Thus, no telFA patterns will be missed in the second phase for both approaches.

4   Experimental Evaluation

In this section, we present our experimental evaluations of our proposed approaches and the workload parameters for our proposed design decisions. We performed our experiments based on the number of nodes in the network and time instants in the series.

Experimental Setup. We evaluated the RAD approach using the "On-the-Fly" design decision with no pruning (RAD-fly), the DN-based design decision with no pruning (RAD-index), and the DN-based design decision with pruning (RAD-index(p)). Figure 5 shows the experimental setup. The synthetic generator takes five inputs: (1) the number of ST locations, (2) the length of the time series, (3) the travel time at each node, (4) the percent of tFAs in each time series, and (5) the error threshold to create the synthetic datasets (see Section 4.1). RAD-fly, RAD-index and RAD-index with pruning were analyzed using a generated dataset and a real dataset (measurement of Turbidity). All approaches were compared in terms of execution time and the number of dpFAs found. Execution time was measured based on the system time call in Java before the first phase was executed till the after the third phase was completed. Number of dpFAs was based on the number of flow anomaly patterns found after the second phase of the RAD method. All experiments were performed on an Intel P4 2 GHz 1.2 GB RAM.

4.1   Experiments Using Synthetic Data

The synthetic dataset was generated based on the following parameters: (1) the number of ST locations in the network, (2) the size of the time series for each ST location, (3) the percent number of transient flow anomalies across the entire network, (4) the travel time for each ST location, and (5) the error threshold, Θ_e. Based on these parameters, the generator created a single time series of equal length that was randomly generated and used for each station. The observations in a downstream ST location location was shifted by its specified travel time based on their upstream neighbor. The location of each tFA was chosen randomly and ensured that there will be exactly the percent number of anomalies specified in the input. For experiments to measure the effect on the number of ST locations, the parameters were set as follows: (1) the size of ST locations from 20 to 100, (2) a length of 1000 time instants, (3) TT=10, (4) 30% tFAs, and $\Theta_e =$ 10. For the experiments to measuure the effect on the size of the time series, the parameters were set as follows: (1) 5 ST locations, (2) a length of 6000 to 30000 time instants,

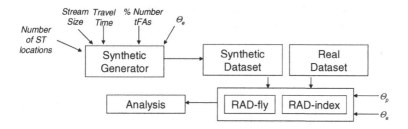

Fig. 5. Experimental Setup

(3) TT=10, (4) 10% tFAs, and $\Theta_e = 10$. The parameters used in this experiment were intended to overlap with those of the real dataset experiments.

Comparison of Phase II and III design decisions over the ST locations. Figure 6 gives the results for all three methods; RAD-fly, RAD-index, and RAD-index(p) in terms of the execution time and the number of dpFAs generated after Phase I as the number of ST locations increases. Figure 6a gives the execution time of all three methods. RAD-fly performs more poorly than RAD-index due to the need to compute the travel time between each time instant in the eFAs and vFAs. By contrast, RAD-index uses the spatio-temporal Dynamic Neighborhood (stDN) model to identify the neighborhoods efficiently. The RAD-index(p) method results in further reduction in execution time by removing the *vanishing* flow anomalies that do not have an *emerging* pattern, resulting in fewer dpFAs to analyze in the second phase.

Figure 6b gives the number of dpFAs found after the second phase of each method. RAD-fly and RAD-index give the highest number of dpFAs because there are no filters in the first phase, causing an increase in the number of dpFAs as the number of ST locations increase. RAD-index(p) show a significant reduction in the number of dpFAs after the first phase. This is due to the removal of invalid *vanishing* flow anomalies whose time instants are not neighbors of the first time instant of *emerging* flow anomalies found previously.

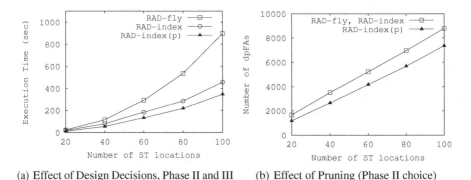

(a) Effect of Design Decisions, Phase II and III   (b) Effect of Pruning (Phase II choice)

Fig. 6. Phase II and III design decisions over the number of ST locations using synthetic data

Comparison of Phase II and III design decisions over the time instants. Figure 7 gives the results for RAD-fly, RAD-index, and RAD-index(p) in terms of the execution time and the number of dpFAs generated after Phase II as the number of time instants increases for each ST location. Figure 7a shows that RAD-fly and RAD-index perform very similarly because there are fewer ST locations in the dataset. However, RAD-index(p) outperforms both methods because the pruned *vanishing* flow anomalies result in fewer combinations to compare against in the second phase. Figure 7b gives the number of dpFAs found after Phase I as the number of time instants increases. RAD-index(p) shows fewer dpFAs than the approaches without pruning.

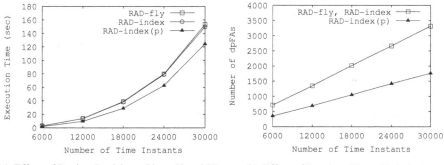

(a) Effect of Design Decisions, Phase II and III   (b) Effect of Pruning (Phase II choice)

Fig. 7. Phase II and III design decisions over the number of time instants using synthetic data

4.2   Experiments Using Real Data

The real datasets were obtained from a study site in Shingle Creek, MN where three sensors were placed along a river. The measurement used was turbidity (approximately 30,000 time instants at each sensor). All errors due to sensing problems were removed from the data and the travel time was given. Since the real dataset has only 3 sensors, experiments were performed based on the number of time instants.

Comparison of Phase II and III design decisions the time instants. Figure 8 gives the execution time and the number of dpFAs for the real dataset using turbidity for RAD-fly, RAD-index, and RAD-index(p) as the number of time instants increase. Figure 8a gives the execution time for all three proposed methods. As shown in Figure 7a, RAD-index with pruning again performs the fastest. RAD-fly and RAD-index exhibit little difference in execution time, presumably because there were only three ST locations. Figure 7b shows that RAD-index with pruning produces far fewer dpFAs than the methods without pruning. For both sets of results, RAD-index(p) is more efficient because of the pruning property to eliminate the *vanishing* flow anomalies that do not

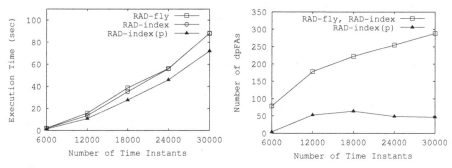

(a) Effect of Design Decisions, Phase II and III   (b) Effect of Pruning (Phase II choice)

Fig. 8. Phase II and III design decisions over the number of time instants using real data

have an *emerging* neighbor based on the spatio-temporal dynamic neighborhood. It is important to note that the up and down pattern exhibited by RAD-index(p) in Figure 8b is the result of the small flow anomalies collapsing into larger flow anomalies as the time series increase, resulting in various numbers of flow anomalies being pruned.

5   Conclusion and Future Work

Conclusion. We introduced a novel problem of discovering teleconnected flow anomalies. This problem a number of important applications for environmental monitoring, video surveillance, and transportation systems. Several new concepts and interest measures were introduced. A RAD approach was proposed that uses novel design decisions of "On the Fly", spatio-temporal Dynamic neighborhoods based, and a pruning strategy. The proof of correctness and completeness for each proposed method was shown. Experimental evaluation was performed on both synthetic and real datasets.

Future Work. The teleconnected flow anomaly problem faces further challenges when the network allows for multiple islands. For example, simply adding one island to a tree network may create two additional paths, two islands may create different paths between nodes, and so, leading a possible exponential number of time paths between nodes. Thus, future work will investigate the discovery of teleconnected flow anomalies within more complex networks. Further investigation will also be needed to explore alternatives for managing the storage of the spatio-temporal dynamic neighborhoods. Finally, a generalized model will be explored to handle relationships between arbitrary events.

Acknowledgments. This work supported by NSF IGERT, USDOD, NSF (EAR 0607138) and USGS/ National Institutes for Water Resources. We would like to thank Kim Koffolt for her comments.

References

1. Pastor, R.: El niño climate pattern forms in pacific ocean (2006),
   http://www.usatoday.com/weather/climate/
   2006-09-13-el-nino_x.htm
2. WFUNA: Millenium project: Global challenges facing humanity (2007)
3. Mason, M.: World's highest drug levels entering india stream, USA today (2009),
   http://www.usatoday.com/tech/science/environment/
   2009-01-26-drug-india-stream_n.htm
4. Saulny, S.: Fish-killing virus spreading in the great lakes, New York Times (2007)
5. Bruckner, M.: The gulf of mexico dead zone, montana state university (2008),
   http://serc.carleton.edu/microbelife/topics/deadzone/
6. Matthews, D.A., Effler, S.W., Driscoll, C.T., O'Donnell, S.M., Matthews, C.M.: Electron budgets for the hypolimnion of a recovering urban lake, 1989-2004. Limnology and Oceanography, American Society of Limnology and Oceanography 53(2), 743–759 (2008)
7. Hyer, K.E., Hornberger, G.M., Herman, J.S.: Processes controlling the episodic streamwater transport of atrazine and other agrichemicals in the agricultural watershed. Journal of Hydrology 254, 47–66 (2001)
8. Kang, J.M., Shekhar, S., Wennen, C., Novak, P.: Discovering Flow Anomalies: A SWEET Approach. In: IEEE International Conference on Data Mining, pp. 851–856 (2008)

9. Amir, A., Apostolico, A., Lewenstein, M.: Inverse pattern matching. Journal of Algorithms 24(2), 325–339 (1997)

10. Lee, H., Ng, R.T., Shim, K.: Estimating Rarity and Similarity over Data Stream Windows. In: VLDB, pp. 195–206 (2007)

11. Berndt, D.J., Clifford, J.: Using Dynamic Time Warping to Find Patterns in Time Series. In: KDD 1994: AAAI Workshop on Knowledge Discovery in Databases, pp. 359–370 (1994)

12. Chen, A., Tang, C., Yuan, C.-a., Peng, J., Hu, J.: Mining Correlations Between Multi-streams Based on Haar Wavelet. In: Grumbach, S., Sui, L., Vianu, V. (eds.) ASIAN 2005, vol. 3818, pp. 270–271. Springer, Heidelberg (2005)

13. Sayal, M.: Detecting time correlations in time-series data streams. Technical Report HPL-2004-103, Hewlett-Packard Company (2004)

14. Bulut, A., Singh, A.K.: A unified framework for monitoring data streams in real time. In: IEEE ICDE, pp. 44–75 (2005)

15. Datar, M., Muthukrishnan, S.: Estimating Rarity and Similarity over Data Stream Windows. In: Möhring, R.H., Raman, R. (eds.) ESA 2002. LNCS, vol. 2461, pp. 323–334. Springer, Heidelberg (2002)

16. Kalnis, P., Mamoulis, N., Bakiras, S.: On discovering moving clusters in spatio-temporal data. In: Bauzer Medeiros, C., Egenhofer, M.J., Bertino, E. (eds.) SSTD 2005, vol. 3633, pp. 364–381. Springer, Heidelberg (2005)

17. Spiliopoulou, M., Ntoutsi, I., Theodoridis, Y., Schult, R.: MONIC - Modeling and Monitoring Cluster Transitions. In: ACM SIGKDD (2006)

18. DeGroot, M., Scheverish, M.J.: Probability and Statistics, 3rd edn. Addison Wesley, Reading (2002)

19. Knorr, E., Ng, R.: A Unified Notion of Outliers. In: ACM KDD (1997)

20. Shekhar, S., Lu, C.T., Zhang, P.: A unified approach to spatial outliers detection. GeoInformatica 7(2), 139–166 (2003)

21. Li, X., Hodgson, M.E.: Vector field data model and operations. GIScience and Remote Sensing 41(1), 1–24 (2004)

22. Zhang, P., Huang, Y., Shekhar, S., Kumar, V.: Exploiting Spatial Autocorrelation to Efficiently Process Correlation-Based Similarity Queries. In: Hadzilacos, T., Manolopoulos, Y., Roddick, J., Theodoridis, Y. (eds.) SSTD 2003. LNCS, vol. 2750, pp. 25–27. Springer, Heidelberg (2003)

23. Leskovec, J., Krause, A., Guestrin, C., Faloutsos, C., VanBriesen, J., Glance, N.: Cost-effective Outbreak Detection in Networks. In: ACM SIGKDD (2007)

24. Shekhar, S., Chawla, S.: Spatial Databases: A Tour. Prentice Hall, Englewood Cliffs (2002)

25. Dijkstra, E.W.: A note on two problems in connexion with graphs. Numerische Mathematik 41 (1959),
www2.informatik.hu--berlin.de/alkox/lehre/lvws0809/verkehr/
dijkstra.pdf

26. Hart, P.E., Nilsson, N.J., Raphael, B.: A formal basis for the heuristic determination of minimum cost paths. IEEE Transactions on Systems Science and Cybernetics 4(2) (1968)

27. Kang, J.M., Shekhar, S., Wennen, C., Novak, P.: Discovering Flow Anomalies: A SWEET Approach. University of Minnesota, MN, Technical Report, 09-006 (2009)

Continuous Spatial Authentication

Stavros Papadopoulos[1], Yin Yang[1], Spiridon Bakiras[2], and Dimitris Papadias[1]

[1] Dept. of Computer Science and Engineering,
Hong Kong University of Science and Technology
{stavros,yini,dimitris}@cse.ust.hk
[2] Dept. of Mathematics and Computer Science,
John Jay College, City University of New York
sbakiras@jjay.cuny.edu

Abstract. Recent advances in wireless communications and positioning devices have generated a tremendous amount of interest in the continuous monitoring of spatial queries. However, such applications can incur a heavy burden on the *data owner* (DO), due to very frequent location updates. Database outsourcing is a viable solution, whereby the DO delegates its database functionality to a *service provider* (SP) that has the infrastructure and resources to handle the high workload. In this framework, *authenticated query processing* enables the clients to verify the correctness of the query results that are returned by the SP. In addition to correctness, the dynamic nature of the monitored data requires the provision for *temporal completeness*, i.e., the clients must be able to verify that there are no missing results in between data updates. This paper constitutes the first work that deals with the authentication of continuous spatial queries, focusing on ranges. We first introduce a baseline solution (BSL) that achieves correctness and temporal completeness, but incurs false transmissions; that is, the SP has to notify clients whenever there is a data update, even if it does not affect their results. Then, we propose CSA, a mechanism that minimizes the processing and transmission overhead through an elaborate indexing scheme and a virtual caching mechanism. Finally, we derive analytical models to optimize the performance of our methods, and evaluate their effectiveness through extensive experiments.

1 Introduction

In *database outsourcing* [9], a *data owner* (DO) delegates its DBMS tasks to a *service provider* (SP) that has the necessary resources to perform advanced query processing. The SP is then responsible for processing client queries on behalf of the DO. *Authenticated query processing* allows the SP to prove to the client that (i) the results are *authentic* (i.e., originated from the DO), (ii) *sound* (i.e., no result object is fictitious or modified), and (iii) *complete* (i.e., all objects satisfying the query are present). We refer to these three terms collectively as *correctness*. Figure 1 illustrates the general framework, commonly used in the outsourcing literature. Initially, the DO obtains, through a trusted key distribution center, a *private* and a *public* key [20]. The private key is known only to the DO, while the public key is accessible by all the clients. The DO signs the data with its private key, generating one (or more) signatures. Then, it

N. Mamoulis et al. (Eds.): SSTD 2009, LNCS 5644, pp. 62–79, 2009.

sends the signature(s) and the data to the SP, which constructs an *authenticated data structure* (ADS) for efficient query processing. The ADS is essentially an index that contains additional authentication information (typically, hash digests and signatures). When the SP receives a query from a client, it generates a *verification object* (*VO*) by accessing the ADS. The *VO* contains the result set along with the necessary authentication information. The SP sends the *VO* to the client, which can verify the results by matching the *VO* against the public key of the DO.

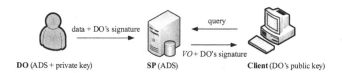

Fig. 1. Database outsourcing framework

While there is extensive literature on authenticated processing in conventional databases, there is very limited work on outsourced data in the presence of frequent updates, especially for spatio-temporal databases. In this paper, we focus on authenticated processing of continuous spatial ranges, motivated by advances in wireless communications and GPS-enabled devices (e.g., RFID chips, sensor networks, navigation systems, etc.). Consider, for instance, a SP that receives locations of shipments around the globe (using RFID technology). A company (i.e., a client) that wishes to track its products through the SP registers long-running queries that monitor certain locations of interest. Whenever an update (arrival or departure) influences a query, the corresponding client is immediately informed. In addition to the timely delivery of query results, it is crucial for the subscribers of such a system to be able to verify their correctness.

The dynamic nature of the data in the above scenario, and the potentially large number of long-running queries, pose several technical challenges. First, a system for continuous authentication on dynamic data must accommodate very fast updates and also support efficient query processing. Second, it must provide effective mechanisms for minimizing the communication cost with the clients, and their verification effort. Third, in addition to correctness, the clients must be able to verify the *temporal completeness* of their results, i.e., confirm that they receive all the updates that affect their queries.

This paper constitutes the first work on continuous authentication of dynamic spatial data. We first introduce a *baseline solution*, called BSL, that achieves correctness and temporal completeness, but incurs false transmissions; that is, the SP has to notify clients whenever there is a data update, even if it does not affect their results. Then, we propose CSA (for *continuous spatial authentication*), a mechanism that minimizes the processing and transmission overhead through an intricate indexing scheme and a virtual caching mechanism. Third, we derive accurate models for estimating the size of the *VO*, which is the most important factor that determines the performance of an outsourcing system. We apply these models to optimize CSA. Finally, we empirically show that CSA outperforms BSL significantly in all aspects and is, therefore, applicable in highly dynamic environments. The remainder of the paper is organized as follows.

Section 2 reviews existing ADSs for database outsourcing. Section 3 describes BSL, while Section 4 proposes the CSA technique. Section 5 presents our experimental results and Section 6 concludes the paper.

2 Related Work

The *Merkle Hash Tree* (MH-Tree) [12] is a main-memory binary tree that provides efficient authentication of equality queries on single-dimensional data. It assumes that the database is sorted on the query attribute and, at the leaf level, every node stores the digest of the binary representation of the record. The digests are computed with a *one-way, collision-resistant hash function* (e.g., SHA1 [15]). The tree is built bottom-up and internal nodes store the digest of the concatenation of the digests of their children. After the tree is constructed, the DO signs the digest stored in the root of the tree and sends it, along with the data, to the SP. During query processing, the SP traverses the tree and, apart from the requested record, it inserts into the *VO* the digest stored in the sibling of every visited node. Having the *VO*, the DO's signature and the DO's public key, the client can verify the authenticity of the result by re-constructing the digest of the root. Devanbu et al. [5] modify the query processing mechanism of the MH-Tree for answering one-dimensional range queries, while satisfying soundness and completeness. They also extend their methods to multiple dimensions, combining the MH-Tree with the *Range Search Tree* [2].

The *Verifiable B-Tree* (VB-Tree) [18] is the first *signature-based* approach that augments a standard B^+-Tree with signed digests. However, this method only guarantees the correctness of the results, but not the completeness. To address this problem, Pang et al. [17] and Narasimha and Tsudik [16] introduce a technique called *signature chaining*. They assume that the dataset is sorted on one attribute, and every record is associated with one signature. This signature combines hashed information about the record, and both its immediate successor and predecessor in the sorted order. In addition, the DO inserts two special (boundary) records at the two ends of the sorted order. To assure integrity for a range query, the constructed *VO* contains (i) the result set, (ii) the signature for each record in the result set, and (iii) the boundary records.

The *Merkle B-Tree* (MB-Tree) [10] extends the MH-Tree to external memory (the node fanout depends on the disk page size). Every node has a digest, which is computed by applying the hash function to the concatenation of all its children's digests. The DO then signs the hash of the concatenation of the digests of the entries contained in the root node of the tree. Range query processing is performed by two top-down traversals (one for each boundary record). At each visited node, the digests of the nodes that do not overlap the query range are inserted into the *VO* (along with the result set and the signed root).

In the context of multi-dimensional databases, which is closely related to this work, there have been very few ADSs proposed in the literature. First, Cheng et al. [3] introduce two authenticated structures, namely the *Verifiable KD-Tree* (VKD-Tree) and the *Verifiable R-Tree* (VR-Tree). Both structures are modified versions of the standard KD-Tree and R-Tree, respectively. Specifically, in every node of the tree, the points and/or MBRs (Minimum Bounding Rectangles) contained therein are sorted according to their x-coordinate, and a signature chain is generated. Range queries are processed by following the signature chain at every node that overlaps the query range. However, these structures incur large space and query processing overhead for the SP, high initial construction cost for the DO, and considerable verification burden for the clients. Furthermore, they lack algorithms for insertions and deletions (updates are not discussed in [3]), which render them inapplicable to dynamic environments.

Currently, the most efficient ADS for multi-dimensional databases is the *Merkle R-Tree* (MR-Tree) [21], which combines the idea of the MB-Tree with the structure of the R*-tree. In particular, every leaf node is associated with a digest that is computed on the concatenation of the binary representation of all objects in the node. Internal nodes are assigned a digest that summarizes the child nodes' MBRs and digests. Each node digest is stored at the corresponding parent entry. The root digest is signed by the DO. Range queries are handled by a depth-first traversal of the tree. The resulting *VO* contains (i) all the points in every leaf node visited, (ii) the MBRs and digests of all the pruned nodes, and (iii) the DO's signature. Nevertheless, the MR-Tree cannot support very fast updates and is, thus, not suitable for our problem.

Also related to our work are two recent solutions that authenticate continuous range queries on one-dimensional data streams. First, Li et al. [11] deal with authentication in sliding window streams, i.e., a tuple expires w time units after its arrival. Their method segments the time into slots, and builds a separate MH-Tree on the tuples that arrived in each slot. Its goal is to reduce the communication cost at the expense of delayed result updates. Papadopoulos et al. [19] introduce CADS, which also deals with streaming environments, but focuses on real-time reporting. CADS combines space (i.e., domain) partitioning with MH-Trees for effective indexing.

Our work extends the general methodology of [19] for continuous spatial ranges. Specifically, we integrate space partitioning, MH-Trees and Hilbert curves for indexing highly dynamic spatial data. In addition to data structures, we develop a comprehensive set of algorithms for the initial computation and the continuous monitoring of the results. Finally, we propose accurate models for determining the best space partitioning granularity, a factor that significantly affects the scheme effectiveness in our setting.

3   Baseline Solution

Since there is no spatial ADS that can handle frequent location updates, in this section we devise a baseline solution (BSL). Each point[1] p is represented by a tuple of the form $<p.id, p.x, p.y>$, where $p.id$ is a unique identifier and $(p.x, p.y)$ are p's coordinates. BSL maps all the 2D points into the 1D domain utilizing a space-filling curve. We employ the Hilbert curve because it preserves spatial locality and leads to low query processing cost [13]. Let $D:[L_x, L_y, U_x, U_y]$ be a square dataspace, where (L_x, L_y) and (U_x, U_y) are the lower left and upper right corners. D is partitioned in $2^{2 \cdot o}$ regular cells, where o is an integer called the *order* (or *resolution*). Figure 2a depicts a Hilbert curve with $o=3$. The cell at the lower left corner has Hilbert value 0, and the values of the remaining cells follow the Hilbert curve (for simplicity, we only include the values of selected cells). Each point p is associated with the Hilbert value $p.hv$ of the cell that covers it, e.g., $p_1.hv=2$, $p_2.hv=7$, $p_3.hv=8$, etc.

---

[1] We use terms *point* and *object* interchangeably.

The DO indexes the points with a 2-3 MB-Tree using their Hilbert values as search keys. The 2-3 MB-Tree is similar to a main memory MB-Tree, where each node may have either 2 or 3 entries. An insertion in a full node causes it to split in two nodes, each containing 2 entries. On the other hand, a deletion from a node with 2 entries leads to an underflow. Similarly to B⁺-Trees, the node first tries to *borrow* an entry from a full sibling node. If this is not possible, the node is *merged* with a sibling. To support multiple updates at the same timestamp, we do not alter any digest, but temporarily mark the visited paths. Then, the marked paths are revisited and the digests are computed bottom-up. In this way, the (expensive) hash computations are performed only once.

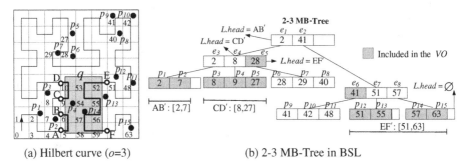

(a) Hilbert curve (o=3)   (b) 2-3 MB-Tree in BSL

Fig. 2. Indexing and query processing in BSL

Each leaf entry p in the 2-3 MB-Tree has the form $<p.id, p.x, p.y, p.hv>$. An intermediate entry e is a triplet $(e.h, e.k, e.ptr)$, where $e.k$ is the Hilbert value of the first point in the subtree of e, $e.ptr$ is a pointer to the corresponding child node and $e.h$ is a digest computed on the concatenation of the digests of the entries in $e.ptr$. Figure 2b contains the tree for the points of Figure 2a, showing only the Hilbert value of each entry. The DO computes a signature on H_{root}, D and o, i.e., it performs $sign(h(H_{root} \mid L_x \mid U_x \mid L_y \mid U_y \mid o))$, where h is the hash function (in our work we employ SHA1 [15]) and 'I' denotes concatenation. As shown later, D and o are necessary during the verification process. Then, it sends the tree, D, o and the signature to the SP. Upon receiving a 2D window query q, the SP finds the parts of the Hilbert curve corresponding to cells that intersect with q. Given the shaded query in Figure 2, (i) poly-line AB corresponds to cells 5 and 6, (ii) CD to 9 and 10, and (iii) EF to cells 52-59. The union of points in these cells constitutes the result of q; i.e., the result is $\{\} \cup \{p_4\} \cup \{p_{13}, p_{14}\}$ for points in AB, CD, EF, respectively. Note that the result may contain some false positives, e.g., p_{13}, that fall out of the query window but reside in an intersecting cell. Such points are filtered out by the client.

Each poly-line corresponds to a 1D range in the 2-3 MB-Tree. One solution would be for the SP to process these ranges one by one. This involves an expansion of each range to include boundary records. For instance, AB is extended to AB':[2,7] so that it covers p_1 and p_2. Similarly, CD and EF are extended to CD':[8,27] and EF':[51,63] to include boundary tuples p_3, p_5 and p_{12}, p_{15}, respectively. Finally, the SP should construct a separate *VO* for each of the expanded ranges. However, executing the 1D ranges individually and generating separate *VO*s would be inefficient, because (i) tree nodes may be visited multiple times, and (ii) *VO* components (i.e., digests and/or boundary points) may either appear in several *VO*s, or they may not be necessary as they can be reconstructed from information contained in other *VO*s. To avoid these problems, we integrate the execution of all sub-queries in one traversal that produces a single *VO*. The generated *VO* has no redundancy and can be verified efficiently by a linear scan. The detailed algorithm, called *MultiRangeMB*, is shown in Figure 3. Note that the algorithm utilizes special tokens [and] that indicate the scope of a node.

MultiRangeMB (*MBNode n, List L*)

1.   Append [to the *VO*
2.   For each entry *e* in *n*
3.   If *n* is an intermediate node
4.   if *e* intersects *L.head* // *e* may contain results
5.   *MultiRangeMB(e.ptr, L)* // *e.ptr* points to a child node
6.   Else append *e.h* to the *VO*
7.   Else // *n* is a leaf node and *e* is a point
8.   Append <*e.id, e.x, e.y*> to the *VO*
9.   If *e* is the last entry of *n* AND *e.hv* is ≥ the upper bound of *L.head*
10.   Evict *L.head* from *L*
11. Append] to the *VO*

Fig. 3. Algorithm *MultiRangeMB* of BSL

MultiRangeMB takes as arguments the root of the 2-3 MB-Tree, and a list L that stores the 1D (Hilbert) sub-ranges of query q sorted in ascending order. The algorithm traverses the tree in a depth-first fashion, and checks all the entries contained in each visited node. For an intermediate node, if an entry e overlaps with the head of L (*L.head*), the traversal continues in e's subtree (lines 4-5). Otherwise, the digest of e is inserted into the *VO* (line 6). Line 6 also captures the case where L is empty, so that the digests of all unexamined entries along the path from the last leaf visited up to the root are appended to the *VO*. When the algorithm reaches a leaf node, it first appends all point entries to the *VO* (line 8). If the Hilbert value of the last point contained in the leaf is greater than or equal to the upper bound of *L.head* (i.e., the boundary point entry for *L.head* is already inserted in the *VO*), the latter is evicted from L.

We illustrate this multi-range traversal using the example of Figure 2. The SP sorts the expanded ranges AB′, CD′ and EF′ in ascending order of their lower boundary value and inserts them in the ordered list L. Then, it starts by processing range AB′:[2,7] at the head of L. Every point (p_1, p_2) satisfying AB is appended to the *VO* (such entries are shown in grey). After the leaf accommodating the last point (p_2) is visited, AB′ is evicted from L. Subsequently, the algorithm continues at entry e_4, where it starts processing CD′:[8,27] and includes p_3, p_4, p_5 in the *VO*. CD′ is evicted from L, EF′:[51,63] commences at e_5 and $e_5.h$, $e_6.h$, p_{12}, p_{13}, p_{14}, p_{15} are appended to the *VO*. Note that it is not necessary to include *p.hv* in the *VO* because, along with the *VO* and the signature, the SP sends D and o. Having this information available, the client can compute the Hilbert values of the points locally.

In general, the *VO* contains a sequence of point entries for each processed 1D interval, and (possibly) digests interleaved between pairs of point sequences. Upon receiving the *VO*, the client first decomposes q into the same set of 1D intervals as the

SP (before their expansion) using D and o, sorts them on their lower boundary value and inserts them in a list L. Then, it utilizes an algorithm to reconstruct the digest of the root (H_{root}). This algorithm is similar to the evaluation of parenthesized arithmetic expressions, where the tokens play the role of the parentheses. When the algorithm encounters a token], it has all the information (digests or records) to compute the digest of the node that started at the corresponding [. The digests and records are appended to a buffer B, which after termination is used to derive $H_{root}=h(B)$. Further-more, for each encountered point sequence, the algorithm computes the Hilbert values of the points, checks whether the boundary points for $L.head$ exist, and evicts the latter from L. Also it reports the points that satisfy q during this process. At the end of the algorithm, L must be empty and the reconstructed H_{root} combined with D and o must verify the signature.

The proof of soundness is straightforward, since if the SP modifies any VO com-ponent, the signature will not be verified (due to the collision-resistance of the hash function). Furthermore, recall that the client receives D and o intact (otherwise the signature verification fails) and, therefore, it can determine the exact 1D queries proc-essed by the SP. It can then ensure completeness by simply checking the existence of the boundary objects for *each* sub-query.

The above discussion captures the initial result computation in BSL; next we de-scribe the continuous monitoring component. Whenever there is a data modification, the DO alters its tree and forwards the update(s) to the SP, according to the *positive-negative* model. An object insertion is denoted as ($+<p.id, p.x, p.y, p.hv>$), and a dele-tion as ($-<p.id, p.hv>$). An object movement is handled by a deletion followed by an insertion. In addition to the actual data, each transmission contains a DO signature and two timestamps: LT is the current time and ST is the time of the previous trans-mission. The signature incorporates the new H_{root}, LT and ST. The two timestamps are necessary so that the clients can detect temporal attacks, i.e., situations where the SP avoids reporting some result updates. Specifically, an authentication scheme satisfies *temporal completeness*, if it is impossible for the SP to omit sending a result change to the client, without the latter detecting it [19]. The DO periodically sends updates to the SP, along with the new signature, LT and ST. The SP updates the tree structure, the timestamps and the signature accordingly, and generates a new VO for *every* monitored query. It then sends the new VO, LT, ST and the signature to the corre-sponding clients (D and o do not need to be re-sent).

Proof of temporal completeness (sketch): Suppose that at time t the SP avoids sending the VO for an update that affects the client's result. At a later time t' the SP transmits a new VO to the client. Note that multiple omissions may have occurred between t and t'. The client will detect the attack by noticing that the time of the previous update (included in the new VO) is $ST \geq t$, at which it did not receive any VO. Note, however, that temporal completeness cannot be guaranteed if the client does not receive any VO for a long time, in which case it cannot be sure whether the last results are still up-to-date. This problem can be solved using the concept of *query freshness* [10], according to which the DO revokes old signatures at periodic time intervals.   □

The efficient query processing and update operations of the 2-3 MB-Tree render BSL suitable for dynamic environments. However, BSL incurs *false transmissions* of VOs for queries whose result is not affected by the latest data updates. This imposes significant CPU cost to the SP (for computing the *VOs*) and to the clients (for verifying them). Furthermore, it leads to excessive network overhead. The next method aleviates these problems by integrating sophisticated indexing schemes and query processing algorithms.

4 Continuous Spatial Authentication – CSA

Section 4.1 describes the indexing scheme of CSA and Section 4.2 explains the query processing algorithms. Section 4.3 presents the analytical models used to optimize the performance of CSA.

4.1 Indexing Scheme

CSA subdivides the dataspace into partitions, in order to reduce the area affected by an update and limit the number of false transmissions. Let $D:[L_x, L_y, U_x, U_y]$ be a square dataspace. We build an $m \cdot m$ regular grid over D, by decomposing each axis into m equal segments. Let l_P be the extent of each partition along the two axes. A point p with co-ordinates $(p.x, p.y)$ can be located in constant time in partition P_{ij}, where $i = \lfloor p.x / l_P \rfloor$ and $j = \lfloor p.y / l_P \rfloor$. In order to capture skewed point distributions, we embed a *Temporal Merkle Hash-Tree* (TMH-Tree) in each partition P. The TMH-Tree is a modified 2-3 MB-Tree that incorporates temporal information used by a virtual caching mechanism. Specifically, every entry e in an intermediate node contains a timestamp $e.t$ that signifies the latest (i) record insertion/deletion/update that occurred in the subtree of e, or (ii) movement of e to another node due to a split/merge operation. Figure 4 summarizes the index structures in CSA.

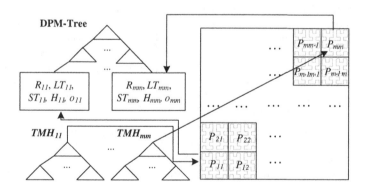

Fig. 4. Indexing structures in CSA

For each partition, we construct a Hilbert curve of order $P.o$ in P and compute the Hilbert values of the residing points. The TMH-Tree then indexes the points using their Hilbert values as search keys. Each leaf entry p has the form $<p.id, p.x, p.y, p.hv>$ and an intermediate entry e is a quadruplet $(e.h, e.k, e.ptr, e.t)$, where the semantics are the same as in BSL, except that $p.hv$ is computed locally within each partition (instead of the entire dataspace). To avoid confusion, the term *cell* is used only for the Hilbert grid. The term *partition* is used for the grid constructed over D. Note that the value of o may be different for each partition. Similar to m, the choice of o may have a significant effect on performance. Section 4.3 contains models for choosing appropriate values of m and o.

All partitions are indexed by a *Domain Partition Merkle-Tree* (DPM-Tree). The DPM-Tree is a binary tree that organizes digests in a way similar to the MH-Tree. A leaf node of the DPM-Tree corresponding to partition P contains a pointer $P.R$ to the root of the TMH-Tree embedded in P, timestamps $P.LT$ and $P.ST$, order $P.o$, and a digest $P.H$. $P.LT$ ($P.ST$) is the timestamp of the last (second last) update that occurred in P ($P.LT \geq P.ST$). $P.H$ is computed on the concatenation of the digests contained in the root of the TMH-Tree along with $P.ST$ and $P.o$. The information in intermediate nodes is inserted bottom-up. An intermediate node N contains value $N.H$, which is the digest of the concatenation of the digests of its children, and timestamp $N.T$ which is the maximum between the timestamps of its children. In order to establish a neighborhood relationship among the nodes of the DPM-Tree, we consider that the root corresponds to the entire dataspace. Its two children are generated by splitting the space *vertically* into two equal subspaces. Subsequently, each child generates two new children by dividing its subspace *horizontally* into two new equal subspaces. This process is repeated recursively (selecting the splitting axis in a round-robin fashion) until the final subspaces are single partitions (leaf level 0).

Let H_{DPM} (T_{DPM}) be the digest (timestamp) in the root of the DPM-Tree. The DO computes $h(H_{DPM} \mid T_{DPM} \mid L_x \mid L_y \mid U_x \mid U_y)$, signs it and sends it to the SP along with the dataset. CSA supports multiple updates at the same timestamp as follows. The TMH-Trees are first modified, as discussed in Section 3, without altering any hash or timestamp value, and the visited paths are marked. When an entry is deleted from a full *intermediate node* (i.e., there is no underflow), it is replaced with a *dummy* value, so that the order of the remaining entries in the node remains the same. Then, the marked paths are revisited and the digests and timestamps are computed bottom-up, only once. Finally, a single depth-first traversal of the DPM-Tree locates the leaf nodes that correspond to the affected partitions and computes the appropriate digests and timestamps bottom-up. However, if at some instant the number of points in a partition P changes significantly, the DO may decide to change $P.o$ in order to improve performance. In this case, it notifies the SP that computes the new Hilbert values for the residing points and re-builds the embedded TMH-Tree.

Finally, the SP maintains some book-keeping structures for query monitoring. Specifically, each partition P is associated with an influence list $P.IL$ that stores the identifiers of the queries that overlap with P. A hash table QT on $q.id$ maintains a tuple $<q.id, q.rg, q.t>$ for every running query q, where $q.id$ is a unique identifier, $q.t$ is the timestamp of q's last *VO* update., and $q.rg$ is the spatial range $[q_{Lx}, q_{Ly}, q_{Ux}, q_{Uy}]$ of q.

4.2 Query Processing

First we discuss snapshot query processing and verification. Upon receiving a spatial range q, the SP starts its execution by traversing the DPM-Tree. At every visited node, the SP obtains the subspace covered by its subtree. This is performed by recursively breaking D into two equal spatial subspaces, either horizontally or vertically, in a round-robin fashion. If q overlaps with the corresponding subspace of a node, the algorithm continues traversing its subtree. Otherwise the node's digest is appended to the *VO*. Upon reaching the leaf level of the DPM-Tree, if q does not overlap with a partition P, $P.H$ is inserted into the *VO*. Otherwise, the algorithm appends $P.ST$ and $P.o$ into the *VO*, and decomposes q into a set of 1D intervals, by determining its intersections with the embedded Hilbert curve. Then, it expands the intervals to include boundary records, sorts them on their lower boundary and stores them in a list L. The expanded sub-queries are issued to the embedded TMH-Tree, using the multi-range algorithm of BSL (Figure 3).

Given the *VO* and D, the client can verify its correctness, by computing the digest H_{DPM} at the root of the DPM-Tree. The process is similar to the one described in Section 3, except that intervals are used to determine the extents of each partition on-the-fly. After the client computes H_{DPM}, it hashes it with T_{DPM} and D, and matches it against the signature of the DO. The actual results are extracted during the verification process.

Proof of soundness (sketch): Soundness is ensured by the hierarchical organization of the hashes in the two trees and the collision-resistance of the hash function. If an adversary alters or deletes a point from the dataset, or inserts a bogus one, the change will propagate until H_{DPM}. Thus, the client will reconstruct an H_{DPM} that does not match the DO's signature. □

Proof of completeness (sketch): Completeness is satisfied for two reasons. (i) The client has D and, therefore, while reconstructing H_{DPM}, it can verify that the SP returns a partial *VO* for every partition overlapping the query. (ii) For each such partition it also has $P.o$ and, therefore, it can establish completeness in the way that we discussed in Section 3. Finally, note that, for every P, $P.o$ is incorporated in $P.H$ and D is in the signature. Consequently, the SP must send them intact in order for the signature to be certified. □

When the SP receives updates from the DO, it alters the indices and determines the affected queries whose range overlaps with partitions where at least one update has occurred. Finally, it generates a new *VO* for each such query. Motivated by the observation that an updated *VO* shares common components with the previous one, we propose a *virtual caching mechanism* (*VCM*) to further reduce the communication cost. The term *virtual* is due to the fact that the SP does not store the *VO* for any query. Instead, each client keeps in its own cache the *VO* that was received last. *VCM* works as follows: whenever a node N of the DPM-Tree (or node entry e of the TMH-Tree) is visited during processing query q, its corresponding timestamp is checked against $q.t$. If $q.t$ is equal or larger, then token *Hit* is inserted into the *VO* and the traversal for the N's (e's) subtree stops. This token instructs the client to retrieve the partial *VO* corresponding to N's (e's) subtree from the cached *VO*. Upon receiving a new *VO*, the client merges the components contained in the updated *VO* with the ones in the cache. Eventually, it reconstructs H_{DPM} (as described above) and matches it against the signature.

Proof of temporal completeness (sketch). Suppose that the initial computation of a query q occurs at a time t and the *VO* is sent to the client. The client successfully verifies its correctness and stores it as *cachedVO*. Now assume that at later time t' ($>t$)

one (or more) update(s) takes place in some partition P that overlaps with q, but the SP does not send a new *VO* to the client. Subsequently, another update occurs that affects q. This time the SP generates *newVO* and sends it to the client (along with new signature and T_{DPM}). We distinguish two cases: (i) the *newVO* contains a partial *VO* corresponding to P, thus also *P.ST*. The client compares *P.ST* with the cached T_{DPM} (=t). Since *P.ST* > t, at least a potential result update (at *P.ST*) was omitted and the client is alarmed. (ii) *newVO* contains a *Hit* token that corresponds to P. Since the actual *P.ST* is different than the one included in *cachedVO*, the client reconstructs a false *P.H* value and soundness is violated. ☐

4.3 Computing the Grid Granularity

The granularity m of the dataspace partitioning greatly affects the efficiency of CSA. If m is too coarse (i.e., there are very few partitions), the ability of CSA to reduce false transmissions is subdued. On the other hand, a large number of partitions leads to a tall DPM-Tree and numerous TMH-Trees, which also adversely affects performance. Moreover, for skewed distributions, many of the partitions may contain few or no records at all. Since manually tuning m at the DO is both costly and error-prone, in the sequel we first establish cost models and then clarify the selection of m based on these models.

Our analysis focuses on the expected *VO* (*EVO*) size, for two reasons. First, the *VO* must be transmitted from the SP to the client through the network, which is usually the bottleneck of the entire system. This is especially true for mobile clients (e.g., PDAs), where battery consumption is a major concern (wireless transmissions consume significantly more power than offline computations [6]). The second reason is that other performance goals, such as minimizing the computation at the SP and the client, are strongly correlated with *EVO*. Intuitively, the larger the *EVO*, the more nodes are visited during query processing, and subsequently processed by the client to reconstruct the root digest.

Without loss of generality, we normalize the values along each axis of the dataspace to [0, 1]. In order to keep the analysis tractable, we make the following simplifying assumptions: (i) all partitions have the same length. (ii) The updates follow the distribution of the initial dataset, i.e., the cardinality of each partition does not change significantly over time. When this assumption does not hold, the DO and SP can periodically re-compute m and rebuild the structures of CSA accordingly. (iii) Each query q has expected length $l_q \in$ (0, 1] along each axis. Furthermore, its lower boundary (along each axis) is uniformly distributed in [0, 1− l_q]. (iv) The virtual caching mechanism is disabled as its effects are not significantly influenced by the partitioning granularity m.

We use symbol $P_{i,j}$ ($1 \le i, j \le m$) to denote the partition covering the region $[(i-1)/m,$ $i/m] \cdot [(j-1)/m, j/m]$, which contains a known number $|P_{i,j}|$ of points. Query q takes the shape of a square with length l_q along each axis. CSA involves an initial *VO* computation for a query q, as well as the construction of a new *VO* whenever q is affected by updates. Let $EVO_{init}(q)$ be the expected size of the initial *VO* of q, and $EVO_{upd}(q)$ the expected size of the *VO* generated due to an update. For a given random query sample *QS* (e.g., drawn from a past query log) with cardinality $|QS|$, and a number of timestamps *NU* that updates occur, *EVO* is computed by

$$EVO = \frac{\sum_{q \in QS} EVO_{init}(q) + NU \cdot EVO_{upd}(q)}{|QS| \cdot (NU + 1)} \tag{1}$$

Regarding EVO_{init}, CSA includes in the VO five different types of information: (i) the result set of q, (ii) two boundary records for proving completeness, (iii) time-stamps of each partition overlapping q, which collectively prove temporal complete-ness, (iv) the digests inserted during the traversal of the DPM- and the TMH-Trees that is used by the client to verify correctness and, finally, (v) the signature of the DO. We do not consider the tokens since their sizes are negligible. Let S_r be the length of a record and $|q|$ be the average number of tuples in the query result set. Types (i) and (ii) consume $S_r \cdot (|q|+2)$. Given the query extent l_q, $|q|$ can be calculated using standard selectivity estimation techniques (e.g., sampling [1], histograms [7], probabilistic models [8]). If qp is the number of partitions intersecting q, and S_t is the size (in bytes) of a timestamp representation, the size of (iii) is $qp \cdot S_t$. Since each partition has extent $1/m$ on each axis, the expected value for qp is $\lfloor 2 \cdot m \cdot l_q \rfloor + 1$. Regarding (iv), we use symbols EVO_D and EVO_T to denote the total size of the digests appended to the VO when traversing the DPM-Tree and all the TMH-Trees, respectively. Finally, (v) equals the size of one signature (let S_s). Summarizing, EVO_{init} is given by:

$$EVO_{init}(q) = (|q|+2) \cdot S_r + qp \cdot S_t + EVO_D + EVO_T(q) + S_s \tag{2}$$

We next focus on EVO_D. Note that EVO_D consists of the digests of all *pruned* nodes of the DPM-Tree. A node is pruned, if and only if (i) it does not overlap the query, and (ii) none of its ancestors is pruned (otherwise it is not visited at all). Let $OVN(i)$ be the number of nodes at depth i (the root being at depth 0) that overlap with q. The number of nodes at depth i satisfying condition (i) is $2^i - OVN(i)$. Among these nodes, some are descendents of pruned nodes at higher levels and, thus, violate condi-tion (ii). Given that the DPM-Tree is binary, a node at depth j ($j < i$) has 2^{i-j} descen-dents at depth i. Therefore, assuming that $PN(i)$ is the number of pruned nodes at depth i, Equation 3 gives both $PN(i)$ and EVO_D (S_h is the size of a digest). Note that since m is the partitioning granularity for each of the two axes, the height of the DPM-Tree is $\lfloor \lg m^2 \rfloor = 2 \cdot \lfloor \lg m \rfloor$.

$$EVO_D = \sum_{i=0}^{2 \cdot \lfloor \lg m \rfloor} PN(i) \cdot S_h \tag{3}$$

$$\text{where } PN(i) = 2^i - OVN(i) - \sum_{j=0}^{i-1} PN(j) \cdot 2^{i-j}$$

We next determine OVN. Figure 5 depicts the spaces covered by the subtrees of the DPM nodes at an odd and even depth (the root being at depth 0). Specifically, Figure 5a (5b) shows the 5th (6th) depth. In general, each node at depth i covers $1 / 2^{\lfloor i/2 \rfloor}$ extent on the x-axis and $1 / 2^{\lceil i/2 \rceil}$ on the y-axis. The number of cells overlapping q is thus calculated by:

$$OVN(i) = \left(\lfloor 2^{\lfloor i/2 \rfloor} \cdot l_q \rfloor + 1 \right) \cdot \left(\lfloor 2^{\lceil i/2 \rceil} \cdot l_q \rfloor + 1 \right) \tag{4}$$

(a) Odd depth   (b) Even depth

Fig. 5. DPM-Tree nodes and query q

The derivation of EVO_T depends on the order of the Hilbert curve used in each partition. Let $P.o$ be the order in partition P. An overly coarse (fine) $P.o$ leads to limited indexing effectiveness (many empty cells) and consequently sub-optimal performance. We determine $P.o$ based on the following observation: when records are uniformly distributed in partition P, ideally each Hilbert cell should contain exactly one record. Note that according to our experiments, the optimal m is usually large enough for this local uniform distribution assumption to hold. Therefore, an appropriate value for $P.o$ is $\lg|P|/2$.

For partitions completely contained in q, all data records (and no digests) are inserted into the VO. For partitions that partially overlap with q (i.e., those on the boundary of q as shown in Figure 5), the digests of the pruned sibling nodes during the TMH-Tree multi-range traversal are added to the VO (in addition to the records satisfying q). However, the exact analysis of this traversal is very complicated because (i) it involves calculating the number of ranges q is broken into, which itself is a difficult task [14], and (ii) the ranges have different sizes, meaning that the common ancestors are at different levels. Instead, we employ the following approximation for the multi-range traversal: for each TMH-Tree in a partition that partially overlaps q, we count two complete root-to-leaf paths, adding to EVO the digest of the siblings of all visited nodes. This method overestimates the traversal path by counting the root-to-split-node part twice, but on the other hand also underestimates it by not taking into account the small up-and-down paths inside the envelop of the two root-to-leaf paths. These two contradicting factors are expected to partially cancel each other out. Moreover, with reasonably fine partitioning granularity m, (i) the number of partitions on the boundary of q is much smaller than those within q, and (ii) each TMH-Tree is expected to have a small height, both of which render the approximation error insignificant. Summarizing, the formula for EVO_T is (f is the expected fanout of the TMH-Tree):

$$EVO_T(q) = \sum_{P \in PP} 2 \cdot S_h \cdot (f-1) \cdot \left\lfloor \log_f |P| + 1 \right\rfloor \tag{5}$$

where $PP = \{P \mid P \text{ partially overlaps } q\}$

Combining Equations 2, 3 and 5 yields the complete model for EVO_{init}. We next derive EVO_{upd}. Recall that in CSA, the SP sends a new VO only when at least one update happens in a partition intersecting with q. According to the assumption that the updates follow the same distribution as the initial dataset, the probability that an update falls in any one of QP_1, QP_2, ..., QP_{qp} is $\Sigma_i|QP_i|/(\Sigma_j\Sigma_k|P_{j,k}|)$, $1\leq i\leq qp$, $1\leq j,k\leq m$.

Therefore, for a batch of $|U|$ independent update operations (i.e., insertions or deletions) occurring at a timestamp, the probability $Prob_{VO}(q)$ that the SP transmits a new VO (i.e., q is affected by any one of these updates) is:

$$Prob_{VO}(q) = 1 - \left(1 - \left(\sum_{i=1}^{qp}|QP_i| \Big/ \sum_{j=1}^{m}\sum_{k=1}^{m}|P_{j,k}|\right)\right)^{|U|} \tag{6}$$

Similar to the case of $EVO_T(q)$, $Prob_{VO}(q)$ is a function of the values of $|QP_i|$ ($1 \leq i \leq qp$), which, in turn, depends on the position of q. Let $EVO_{init}(q)$ be the expected VO size for a particular query q, which is obtained by combining Equations 2, 3 and 5. Then $EVO_{upd}(q) = EVO_{init}(q) \cdot Prob_{VO}(q)$.

Equipped with the above models, we present a simple and effective algorithm *Bestm* to compute an appropriate value for the partitioning granularity m. Initially, *Bestm* sets m to a maximum value m_{max}. A good choice for m_{max} is half of the first power of 2 that is larger than the data set cardinality. It then scans the dataset once to compute the cardinality for each partition, and utilizes this information to compute EVO using the cost models. After that, it decreases m to $m_{max}/2$, and computes the corresponding EVO. Observe that at this stage it is unnecessary to scan the dataset again to compute the cardinality of the partitions, since these can be obtained by aggregating the corresponding partitions in the previous step. At subsequent steps, m is reduced by half each time, and EVO is estimated, until $m = 1$. Among all considered values for m, the one achieving the minimum EVO is chosen as the partitioning granularity for CSA.

5 Experimental Evaluation

We implemented our methods using the Crypto++ library [4], and deployed them on a Core 2 Duo 2.2GHz CPU with 2GBytes of RAM. Each record r consumes 100 bytes and contains two search keys $r.x$ and $r.y$. The values of $r.x$ and $r.y$ are obtained from the real dataset CAR (California Roads, available at *www.rtreeportal.org*), and are normalized to [0, 1]. At every timestamp, updates arrive at a rate of AR. An update involves a deletion of a random tuple and an insertion of a new one with the same id but different keys. Consequently, the number of update operations $|U| = 2 \cdot AR$, and dataset cardinality DC is constant at all times. The new key values follow their initial distribution. We monitor QC running queries, which are uniformly distributed in the dataspace and cover approximately 0.1% of the data domain.

First, we determine the optimal partitioning granularity m for CSA, using the models of Section 4.3. We set $DC = 100K$, $QC = 1K$ and $AR = 100$, and we disable the VCM. Figure 6a depicts the estimated total VO size generated for all 1K queries with respect to m, as well as its actual size computed in our experiments. Figure 6b zooms into the part of Figure 6a, where $2^3 \leq m \leq 2^7$. The error of our estimation is 5-17%. Our cost models successfully determine the best granularity, which in this case is $m=2^5$. In the sequel, we set m to the best granularity as estimated by our models.

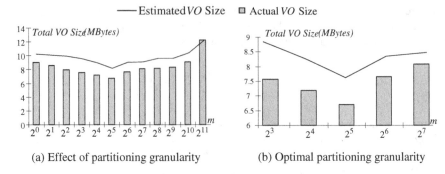

(a) Effect of partitioning granularity   (b) Optimal partitioning granularity

Fig. 6. Total *VO* size vs. partitioning granularity *m*

Figure 7 assesses the effect of the dataset cardinality (*DC*), when *QC* = 1K, *AR* = 100 and the *VCM* is switched on. Figure 7a shows the total query processing time per timestamp at the SP. BSL incurs a considerable computational overhead since it re-processes all queries at each timestamp. On the other hand, CSA executes only the queries affected by the updates, as well as a small number of queries that correspond to false transmissions. Figure 7b depicts the total *VO* size generated for all queries per timestamp versus *DC*. CSA outperforms BSL, because (i) it executes only the queries affected by the updates, and (ii) the *VCM* enables the SP to omit sending *VO* components corresponding to DPM/TMH subtrees that are not altered by the updates. Figure 7c illustrates the verification cost per timestamp at each client. In BSL the

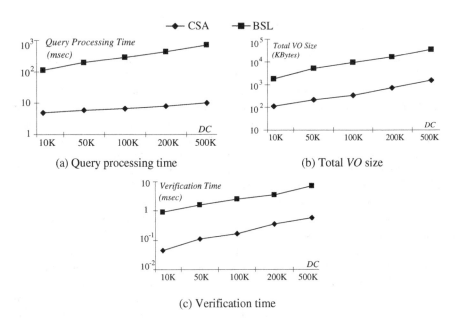

(a) Query processing time   (b) Total *VO* size

(c) Verification time

Fig. 7. Performance vs. dataset cardinality *DC*

client has to verify its query at every timestamp, whereas in CSA it establishes correctness only when its query is affected by an update.

We next investigate the impact of the query cardinality (QC), after setting $DC = 100K$ and $AR = 100$. Figures 8a and 8b plot the query processing cost at the SP and the communication cost at every timestamp, respectively. Both costs grow linearly with QC. In BSL, each query is evaluated at every timestamp. Therefore, the computational effort at the SP as well as the information communicated between the SP and the client increases with the number of running queries. In CSA, more queries are likely to be affected by the updates (in which case a new VO is generated and transmitted to the client) in the presence of a large number of running queries.

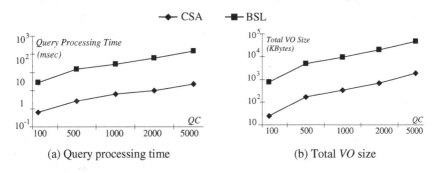

Fig. 8. Performance vs. number of queries QC

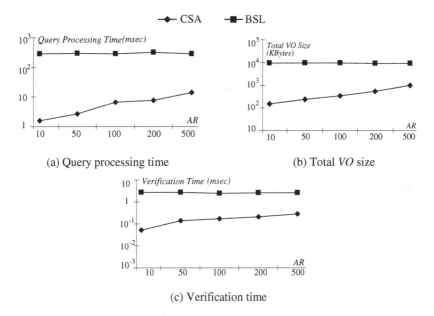

Fig. 9. Performance vs. arrival rate AR

Figures 9a and 9b demonstrate the query processing and the communication cost, respectively, versus the arrival rate AR ($DC = 100K$, $QC = 1K$). BSL is not influenced by AR. The overhead of CSA converges to that of BSL for large values of AR because, as more updates occur, more queries are affected and re-evaluated. For these queries, a new VO must be produced and transmitted. Furthermore, a high AR reduces the effectiveness of VCM because the updates alter a large part of the DPM- and TMH-Trees and, consequently, invalidate many VO components in the clients' cache.

Finally, Figure 9c shows the verification time at the client at every timestamp versus AR. As expected, the verification burden at the client increases for high arrival rates, because its query is affected by an update with higher probability. An interesting observation is that the curve of CSA converges faster to that of BSL, in comparison to the corresponding curves of Figures 9a and 9b because the VCM alleviates the processing and communication costs, but not the verification effort; to establish correctness, the client first has to combine the newly received VO with the one in its cache. Therefore, the client eventually verifies a VO with size equivalent to the one generated when the VCM is disabled.

6   Conclusions

In this paper we address continuous range processing and authentication on highly dynamic spatial databases. We assume a database outsourcing environment, where a service provider (SP) returns to the clients the query results, as well as authentication information necessary to establish their correctness. Due to the dynamic environment, clients must also be able to prove temporal completeness, i.e., that they did not miss any results in-between successive updates. We first propose BSL, a method that achieves these goals at the expense of false transmissions. Next, we introduce CSA, a scheme that utilizes a space partitioning scheme and an efficient caching mechanism to reduce the cost (processing and communication) for both the SP and the client. We optimize the performance of CSA through a detailed analytical study. Finally, we conduct an exhaustive experimental evaluation and show that CSA significantly outperforms BSL for all performance metrics.

Acknowledgments. This work was supported by grant HKUST 6181/08 from Hong Kong RGC.

References

1. Babcock, B., Chaudhuri, S., Das, G.: Dynamic Sample Selection for Approximate Query Processing. In: SIGMOD (2003)
2. de Berg, M., van Kreveld, M., Overmars, M., Schwarzkopf, O.: Computational Geometry: Algorithms and Applications. Springer, Heidelberg (1997)
3. Cheng, W., Pang, H., Tan, K.-L.: Authenticating Multi-dimensional Query Results in Data Publishing. DBSec (2006)
4. Crypto++ Library, www.eskimo.com~weidai/benchmark.html
5. Devanbu, P., Gertz, M., Martel, C., Stubblebine, S.: Authentic Data Publication Over the Internet. Journal of Computer Security 11(3), 291–314 (2003)

6. Datta, V., Vandermeer, D., Celik, A., Kumar, V.: Broadcast Protocols to Support Efficient Retrieval from Databases by Mobile Users. ACM TODS 24(1), 1–79 (1999)
7. Guha, S., Shim, K., Woo, J.: Rehist: Relative Error Histogram Construction Algorithms. In: VLDB (2004)
8. Getoor, L., Taskar, B., Koller, D.: Selectivity Estimation using Probability Models. In: SIGMOD (2001)
9. Hacıgümüş, H., Iyer, B., Mehrotra, S.: Providing Databases as a Service. In: ICDE (2002)
10. Li, F., Hadjieleftheriou, M., Kollios, G., Reyzin, L.: Dynamic Authenticated Index Structures for Outsourced Databases. In: SIGMOD (2006)
11. Li, F., Yi, K., Hadjieleftheriou, M., Kollios, G.: Proof-Infused Streams: Enabling Authentication of Sliding Window Queries on Streams. In: VLDB (2007)
12. Merkle, R.: A Certified Digital Signature. In: Brassard, G. (ed.) CRYPTO 1989. LNCS, vol. 435, pp. 218–238. Springer, Heidelberg (1990)
13. Mokbel, M., Aref, W., Kamel, I.: Analysis of Multi-Dimensional Space-Filling Curves. GeoInformatica 7(3), 179–209 (2003)
14. Moon, B., Jagadish, H.V., Faloutsos, C., Saltz, J.H.: Analysis of the Clustering Properties of the Hilbert Space-Filling Curve. TKDE 13(1), 124–141 (2001)
15. National Institute of Standards and Technology. FIPS PUB 180-1: Secure Hash Standard. National Institute of Standards and Technology (1995)
16. Narasimha, M., Tsudik, G.: Authentication of Outsourced Databases Using Signature Aggregation and Chaining. In: Li Lee, M., Tan, K.-L., Wuwongse, V. (eds.) DASFAA 2006. LNCS, vol. 3882, pp. 420–436. Springer, Heidelberg (2006)
17. Pang, H., Jain, A., Ramamritham, K., Tan, K.-L.: Verifying Completeness of Relational Query Results in Data Publishing. In: SIGMOD (2005)
18. Pang, H., Tan, K.-L.: Authenticating Query Results in Edge Computing. In: ICDE (2004)
19. Papadopoulos, S., Yang, Y., Papadias, D.: CADS: Continuous Authentication on Data Streams. In: VLDB (2007)
20. Rivest, R.L., Shamir, A., Adleman, L.: A method for Obtaining Digital Signatures and Public-key Cryptosystems. Communications of the ACM 21(2), 120–126 (1978)
21. Yang, Y., Papadopoulos, S., Papadias, D., Kollios, G.: Spatial Outsourcing for Location-based Services. In: ICDE (2008)

Query Integrity Assurance of Location-Based Services Accessing Outsourced Spatial Databases

Wei-Shinn Ku[1], Ling Hu[2], Cyrus Shahabi[2], and Haixun Wang[3]

[1] Dept. of Computer Science and Software Engineering, Auburn University, USA
[2] Computer Science Department, University of Southern California, USA
[3] IBM Thomas J. Watson Research Center, USA
weishinn@auburn.edu, {lingh,shahabi}@usc.edu, haixun@us.ibm.com

Abstract. Outsourcing data to third party data providers is becoming a common practice for data owners to avoid the cost of managing and maintaining databases. Meanwhile, due to the popularity of location-based-services (LBS), the need for spatial data (e.g., gazetteers, vector data) is increasing exponentially. Consequently, we are witnessing a new trend of outsourcing spatial datasets by data collectors. Two main challenges with outsourcing datasets is to keep the data private (from the data provider) and ensure the integrity of the query result (for the clients). Unfortunately, most of the techniques proposed for privacy and integrity do not extend to spatial data in a straightforward manner. Hence, recent studies proposed various techniques to support either privacy or integrity (but not both) on spatial datasets. In this paper, for the first time, we propose a technique that can ensure both privacy and integrity for outsourced spatial data. In particular, we first use a one-way spatial transformation method based on Hilbert curves, which encrypts the spatial data before outsourcing and hence ensures its privacy. Next, by probabilistically replicating a portion of the data and encrypting it with a different encryption key, we devise a technique for the client to audit the trustworthiness of the query results. We show the applicability of our approach for both k-nearest-neighbor and spatial range queries, the building blocks of any LBS application. Finally, we evaluate the validity and performance of our algorithms with real-world datasets.

1   Introduction

Due to the rapid advancements in network technology, the cost of transmitting a terabyte of data over long distances has decreased significantly in the past five years. In addition, the total cost of data management is five to ten times higher than the initial acquisition costs and it is likely that computing solution costs will be dominated by people costs in the future [22]. Consequently, there is a growing interest in outsourcing database management tasks to third parties that can provide these tasks for a much lower cost due to the economy of scale. This new outsourcing model has the apparent benefits of reducing the costs for running DBMSs independently and enabling enterprises to concentrate on their main businesses. On the other hand, there are two new concerns with this model. First, the data

N. Mamoulis et al. (Eds.): SSTD 2009, LNCS 5644, pp. 80–97, 2009.

owner may not want to reveal the data to the data provider due to either the sensitivity of the data (e.g., medical records) or the value of the data (e.g., Navteq road vector data). Second, data users need to be confident of the integrity of the data they receive. To illustrate, consider the scenario that Zagat (a data owner) gives its restaurants data to Google (a data provider) to make it available to its customers. First, Zagat does not want to reveal the data to Google as this is its main business and value-add. Second, a user asking for all restaurants with a certain rating wants to be confident that he is indeed receiving every and all the Zagat's restaurants and not some extra ones injected by Google (e.g., paid advertisers) or some missing ones deleted by Google. Several previous studies [1,5] proposed solutions for supporting encrypted queries over encrypted databases to protect data owners' privacy. Another set of studies [14,21,17,15,3] focus on the problem of integrity in outsourced databases by guaranteeing that the results returned by the service provider for a client query are both *correct* and *complete*.

Meanwhile, due to the recent advances in wireless technology, mobile devices (e.g., cell phones, PDAs, laptops) with wireless communication capabilities are increasingly becoming popular. Hence, we are witnessing the emergence of many location-based services (LBS) that allow users to issue spatial queries from their mobile devices in a ubiquitous manner. Obviously, these applications are in desperate need of quality spatial data, resulting in an exponential increase in the customers of spatial data acquirers. Recent mergers between data providers (e.g., TomTom) and data owners (Tele Atlas) are the immediate consequences of this phenomenon. Therefore, the outsourcing of spatial data is becoming an appealing business model for both data owners and data providers. Unfortunately, while the exact same concerns of privacy and integrity exist for outsourcing the spatial data, there has not been much work in addressing these issues for spatial data except for [24,25,18]. To the best of our knowledge none of these studies consider both privacy and integrity at the same time. It is not clear whether the proposed approaches for one problem can easily extend to the other problem or even worse if they conflict with the solutions proposed to the other problem.

In this paper we propose an innovative approach that simultaneously ensures both the privacy and the integrity of outsourced spatial data. This is achieved by using space encryption as the basis of our approach and then devising techniques that enable the data users to audit the integrity of the query result for the most important spatial query types: range queries and k-nearest-neighbor queries (kNN). In particular, we first use a one-way spatial transformation method based on Hilbert curves, which encrypts the spatial data before outsourcing and hence ensures its privacy. Next, by probabilistically replicating a portion of the data and encrypting it with a different encryption key, we devise a technique for the client to audit the trustworthiness of the query results. We evaluated our approaches with both synthetic and real-world datasets. The process of computing Hilbert curves is efficient at client side when the space encryption key is known. Experiment results show that with more than 20% duplication of the original dataset on the server, clients can detect query result deletion attacks with very high confidence.

The remainder of this paper is organized as follows. Section 2 surveys the related work. The system architecture and an overview of our approach is introduced in Section 3. The design of our space encryption based data privacy protection approach is presented in Section 4. In Section 5, we address our spatial query integrity auditing solutions for both range query and k-nearest-neighbor query. The experimental validation of our design is presented in Section 6. Finally, Section 7 concludes the paper with a discussion of future work.

2   Related Work

The outsourcing of databases to a third-party service provider was first introduced by Hacigümüs et al. [6]. Generally, there are two security concerns in database outsourcing: *data privacy* and *query integrity*. We summarize the related researches as follows.

2.1   Data Privacy Protection

Hacigümüs et al. [5] proposed a method to execute SQL queries over encrypted databases. Their strategy is to process as much of a query as possible by the service providers, without having to decrypt the data. Decryption and the remainder of the query processing are performed at the client side. Agrawal et al. [1] proposed an order-preserving encryption scheme for numeric values that allows any comparison operation to be directly applied on encrypted data. Their technique is able to handle updates and new values can be added without requiring changes in the encryption of other values. Generally, existing methods enable direct execution of encrypted queries on encrypted datasets and allow users to ask identity queries over data of different encryptions. The ultimate goal of this research direction is to make queries in encrypted databases as efficient as possible while preventing adversaries from learning any useful knowledge about the data. However, researches in this field did not consider the problem of query integrity.

2.2   Query Integrity Assurance

In addition to data privacy, an important security concern in the database outsourcing paradigm is query integrity. Query integrity examines the trustworthiness of the hosting environment. When a client receives a query result from the service provider, it wants to be assured that the result is both *correct* and *complete*. Correct denotes that the query must be evaluated honestly with the outsourced database to retrieve the result and complete means that the result includes all the records satisfying the query. Devanbu et al. [3] proposed to employ the Merkle hash tree [12] to authenticate data records. The technique computes a signature based on the Merkle hash tree structure and distributes it to clients as a proof of correctness. Mykletun et al. [15] studied and compared several signature methods which can be applied in data authentication. The authors identified the problem of completeness, however they did not propose correspondent solutions. Pang et al. [17] utilized an aggregated signature to sign each record with the information from neighboring records by assuming that all the records are sorted with a certain order. The method assures the completeness of a selection query by checking the aggregated signature. The *challenge token* idea was introduced in [21] for a server with outsourced databases to provide a proof of actual query execution which is then checked at the client side for integrity verification. Compared with [17], the mechanism supports more query types without assuming all the records are sorted. However, all the aforementioned solutions cannot support spatial queries directly.

For auditing spatial queries, Yang et al. [24] proposed the MR-tree which is an authenticated data structure suitable for verifying queries executed on outsourced spatial databases. The authors also designed a caching technique to reduce the information sent to the client for verification purposes. Four spatial transformation mechanisms are presented in [25] for protecting the privacy of outsourced private spatial data. The data owner selects transformation keys which are shared with trusted clients and it is infeasible to reconstruct the exact original data points from the transformed points without the key. Mouratidis et al. [14] proposed the *Partially Materialized Digest scheme* which avoids unnecessary query processing costs and outperforms existing solutions by employing separate indexes for the data and for their associated verification information. However, these researches [24,25,14] did not consider data privacy protection and query integrity auditing jointly in their design. The related work closest to ours is presented by Wang et al. [23] which focuses on numerical data integrity authentication. Nevertheless, the solution cannot be applied in auditing spatial queries because spatial locality information of records is destroyed after encryption.

3   System Overview

In this section, we introduce the architecture of our system and provide an overview of our approach.

3.1   System Architecture

Figure 1 illustrates the architecture of a spatial database outsourcing environment with three main components: mobile user, location-based service provider, and database owner. We consider mobile clients such as cell phones, personal digital assistants, and laptops, which are instrumented with Global Positioning System (GPS) receivers for continuous position information. Moreover, we assume that there are access points distributed in the system environment for mobile devices to communicate with LBS providers. Generally, mobile devices cannot store any significant amount of the outsourced data in local memory for integrity checking. Therefore, a feasible way of obtaining integrity assurance of query results for mobile devices is through asking queries and analyzing results. On the other hand, a LBS provider is able to store and access all the outsourced data for answering spatial queries from clients. However, LBS providers could be malicious (e.g., returning incomplete query results) and they are not trusted

Fig. 1. The system architecture of spatial database outsourcing

by the clients. The third element – the database owner (e.g., possessing point of interest datasets) outsources its data management tasks to service providers (e.g., providing location-based services).

3.2   Overview of Our Approach

We assume that the database owner can embed additional information in the outsourced spatial dataset for query integrity verification. Let $\mathbb{D}$ denote the spatial database to be outsourced. The database owner first replicates a portion of $\mathbb{D}$ with randomly selected objects. Then, $\mathbb{D}$ and the replicated portion are encrypted with different Hilbert curve based encryption keys. Afterward, the two encrypted datasets are combined and stored at the LBS provider. We employ `dataPreprocess()` to denote the replication and encryption process and $\mathbb{D}_E =$ `dataPreprocess`($\mathbb{D}$) to denote the spatial data stored at the service provider. For requesting LBS based on encrypted spatial databases, a mobile user rewrites spatial queries against $\mathbb{D}$ to spatial queries against $\mathbb{D}_E$ by making use of a query rewriting method `queryRewrite()`. In addition, the user also launches auditing queries for verifying a group of previously executed spatial queries. By exploiting the replicated data, the client is able to determine that the results are correct

Table 1. Symbolic notations

Symbol	Meaning
s	Spatial object
s_E	Encrypted spatial object
r	Data replication percentage
$\mathbb{D}$	Spatial database
$\mathbb{R}$	Query result set
$|\mathbb{A}|$	The number of elements in set $\mathbb{A}$
$V_{\mathcal{H}}$	Hilbert value
O	Order of a Hilbert curve
$\mathcal{T}$	One-way function for space encryption
I_d	Dual information
Ψ	Cryptographic signature
S_K	Symmetric key
SEK_P	Primary spatial encryption key
SEK_S	Secondary spatial encryption key
$Dist(p, q)$	The Euclidean distance between two objects p and q

and complete and the confidence is beyond a user-specified level according to the replication ratio. The mechanism to discriminate a replicated data object from an original data object is only shared between the database owner and the users. LBS providers cannot tell the duplicated dataset from other encrypted data in the outsourced spatial database. Table 1 summarizes the set of notations of this paper.

4   Space Encryption Based Privacy Protection

In this section, we first introduce the one-way function based space encryption solution. Next, space-filling curves are introduced and applied as one-way functions in our system for protecting the privacy of outsourced spatial data.

4.1   Space Encryption

In order to protect the privacy of outsourced spatial databases, we exploit the power of one-way functions to preserve privacy by encoding the locations of all spatial objects. A one-way function is easy to compute but difficult to invert, meaning that some algorithms can compute the function in polynomial time while no probabilistic polynomial-time algorithm can compute an inverse image of the function with better than negligible probability. Our space transformation method is capable to map each point from the original space to a point in the encrypted space to prevent the service provider from obtaining the original spatial object locations. Because we focus on managing spatial data, an ideal one-way transformation should respect the spatial proximity of the original space. If the encrypted space is able to maintain the distance properties of the original space, it will enable efficient evaluation of spatial queries. Transforming spatial object locations with such a locality-preserving one-way mapping can be viewed as encrypting the elements of the two-dimensional (2-D) space for securing privacy and facilitating spatial query processing. In this research, we apply the parameters of our space encryption function as the trapdoor [4] which is only provided to users to encode queries and decode the encrypted query results for retrieving the original spatial object positions.

4.2   Space Filling Curves

A space-filling curve is a continuous curve, which passes through every point of a closed space. The formal definition of a space-filling curve is as follows. If a mapping $f : I \rightarrow \mathbf{E}^n$ ($n \geq 2$) is continuous, and $f(I)$, the image of I under f, has positive Jordan content (area for $n = 2$ and volume for $n = 3$), then $f(I)$ is called a space-filling curve. $\mathbf{E}^n$ denotes an n-dimensional Euclidean space. An important property of space-filling curves is that they retain the proximity and neighboring aspects of the indexed data. Because space-filling curves can preserve the locality between objects in the multidimensional space in the transformed linear space, we investigate the applicability of space-filling curves as ciphers for preserving privacy of outsourced spatial databases. Since the main goal of this research is to provide both privacy protection and integrity assurance of location-based services with outsourced spatia databases, we focus on the transformation of a 2-D space which covers the locations of POIs. However, our solution can be easily extended to high dimensional space.

The Hilbert curve [7,2] is a continuous fractal space-filling curve which is broadly used in multidimensional data management. The superior distance preserving properties [11] makes the Hilbert curve an ideal choice as a space cipher. In addition, the Hilbert curve achieves better clustering than the Z curve [16] and the Gray-coded curve [8]. Therefore, we apply the Hilbert curve in our system for encrypting the original space. As Ref [13], we define $\mathcal{H}_O^D$ for $O \geq 1$ and $D \geq 2$, as the Oth order Hilbert curve for a D-dimensional space. Consequently, $\mathcal{H}_O^D$ maps an integer set $[0, 2^{OD} - 1]$ into a D-dimensional integer space $[0, 2^O - 1]^D$. The mapping determines the Hilbert value $V_{\mathcal{H}}$ of each point in the original space based on their coordinates where $V_{\mathcal{H}} \in [0, 2^{OD} - 1]$. Accordingly, we can formulate the relationship in a two-dimensional space as $V_{\mathcal{H}} = \mathcal{T}(x, y)$ where x and y are the coordinate of a point in the original space and $\mathcal{T}$ is the one-way transformation function. Note that it is possible for two or more points to have the same Hilbert value in a given curve. Figure 2 illustrates an example of mapping 2-D space POIs into their Hilbert values. In the illustration, we can retrieve the Hilbert values of the points of interest A, B, C, and D as 0, 2, 8, and 12 respectively with an order two Hilbert curve. Depending on the desired resolution, more fine-grained curves can be recursively generated based on the Hilbert curve production rules.

Based on the aforementioned properties of the Hilbert curve, it can be employed as a one-way function to support space encryption. The curve parameters including the curve's starting point (x_0, y_0), curve order O, and curve orientation θ make up the *Space Encryption Key* (SEK) of the Hilbert curve based one-way function [9]. Consequently, adversaries who do not have the decryption key have to exhaustively check for all possible combinations of curve parameters to decipher the physical locations of interested objects. However, with reasonable curve parameters it is computationally impossible to reverse the transformation and retrieve the physical locations of interested points in polynomial time.

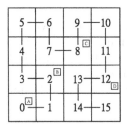

Fig. 2. The Hilbert curve transforms a 2-D space into corresponding Hilbert values

5 Spatial Query Integrity Auditing with Dual Space Encryption Keys

5.1 Dual Space Encryption

In order to audit the integrity of query results retrieved from outsourced spatial databases, we encrypt the original spatial database $\mathbb{D}$ with dual space encryption keys. We first encrypt $\mathbb{D}$ with a primary space encryption key SEK_P. Then, we replicate r percent of $\mathbb{D}$ and encode the duplicate with a secondary space encryption key SEK_S which possesses different curve parameters. Afterward, we combine the two encrypted datasets as $\mathbb{D}_E$ and store $\mathbb{D}_E$ at the service provider. After space encryption, a service provider can only see the Hilbert value of each spatial data object in $\mathbb{D}_E$ instead of their original coordinates. Since $\mathcal{T}$ is a one-way function, for any spatial object s in $\mathbb{D}_E$, a service provider cannot tell whether s was encoded by SEK_P or SEK_S. In addition, the Hilbert values generated by the two space encryption keys may overlap which makes it even more difficult to distinguish if an object is the original or the duplicate.

On the other hand, we need corresponding techniques to enforce query integrity on the client side. For any spatial object s in the query result set, a client should be able to verify whether s is a valid record of $\mathbb{D}$ and if s has a counterpart which is encrypted with another SEK. For supporting object verification, we encrypt the coordinate, non-spatial attributes, and dual information I_d with a symmetric key S_K which is shared by the database owner and all the clients. In addition, we apply cryptographic hash functions [20] to generate a signature Ψ for each spatial object with the coordinate and non-spatial attributes as the input message. The purpose of the dual information field is for clients to tell if a spatial object has a duplicate in the outsourced database. I_d has three values which stand for (i) primary encryption without duplication, (ii) primary encryption with duplication, and (iii) secondary encryption respectively. The structure of an encrypted spatial object stored in $\mathbb{D}_E$ is as follows:

$$s_E = \{V_{\mathcal{H}}, \{x, y, \text{non-spatial attributes}, I_d\}_{S_K}, \Psi\}$$

For each server returned spatial object s_E, a client first decrypts s_E with the symmetric key and executes the cryptographic hash function for verifying the object with its attached signature. Since it is computationally infeasible to forge a cryptographic hash function generated signature, any tampering with the object will be detected. If the spatial object is valid, the client will check its I_d field for determining whether the object is an original or a duplicate. Replicated objects are utilized to audit query result integrity as described in the following two subsections.

5.2 Range Query

With a given range query Q_R, a client first identifies the Hilbert values covered by the range query based on the parameters of SEK_P. Afterward, the client queries the service provider for retrieving the objects covered by the query range. In order to hide the SEK parameters from malicious service providers, the client may

---

Algorithm 1. Query Integrity Assured Range Query (Q_R)

1: Compute Hilbert curve segments covered by Q_R on the curve defined by SEK_P
    and store the segments in $\mathbb{S}$
2: **for** each segment $e \in \mathbb{S}$ **do**
3:    Retrieve the spatial objects covered by e and store them in $\mathbb{R}$
4: **end for**
5: Filter out the spatial objects encrypted with SEK_S in $\mathbb{R}$
6: **for** each spatial object $s \in \mathbb{R}$ **do**
7:    Verify s with its signature
8:    **if** s is a valid object and s has a duplicate **then**
9:      Store s in $\mathbb{C}$
10:   **else**
11:     Report the anomaly to the client and exit
12:   **end if**
13: **end for**
14: Create an auditing query Q_A based on Q_R and SEK_S
15: Compute Hilbert curve segments covered by Q_A on the curve defined by SEK_S
    and store the segments in $\mathbb{S}$
16: **for** each segment $e \in \mathbb{S}$ **do**
17:   Retrieve the spatial objects covered by e and store them in $\mathbb{R}'$
18: **end for**
19: Filter out the spatial objects encrypted with SEK_P in $\mathbb{R}'$
20: **if** $\mathbb{C} \neq \mathbb{R}'$ **then**
21:   Report the anomaly to the client and exit
22: **end if**
23: Return $\mathbb{R}$

---

interleave the Hilbert value segments covered by a group of range queries. After receiving the query result set $\mathbb{R}$, the client first filters out objects encrypted with SEK_S and verifies the validity of all the remaining objects with their attached signatures. If all the objects in $\mathbb{R}$ are valid, the client generates an auditing range query Q_A with the same query range size as Q_R and the parameters of SEK_S. If the service provider carries out queries honestly, the query result set of the auditing query must contain counterparts of all the objects with duplicates in $\mathbb{R}$. In practice, the client can launch a single auditing query for verifying a number of regular queries by combining their query ranges for saving resources.

Figure 3 demonstrates an example of range query integrity auditing. The query window of a range query Q_R covers three Hilbert curve segments, $[17 - 18]$, $[23 - 24]$, and $[27-31]$, based on the primary encryption key as shown in Figure 3(a). After receiving the three curve segments, the service provider retrieves all the spatial objects whose Hilbert values are embraced by the three curve sections and returns the retrieved spatial object set $\mathbb{R}$ as the query result. Then, the client removes records encrypted with SEK_S in $\mathbb{R}$, verifies the remaining records, and identifies the records which have duplicates by checking the I_d filed. Subsequently, the client creates an auditing query Q_A with equal query range as Q_R on the Hilbert curve defined by SEK_S. In this example, Q_A encompasses two Hilbert curve segments,

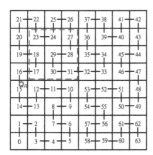

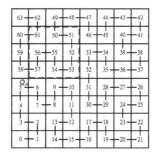

Fig. 3(a) Original range query Q_R Fig. 3(b) Auditing range query Q_A

Fig. 3. A range query Q_R covers three Hilbert curve segments based on SEK_P as illustrated in (a). The auditing query Q_A encloses two Hilbert curve segments based on SEK_S as demonstrated in (b).

[50 − 57] and [61], based on the secondary encryption key. With the I_d field, the client is able to filter out the objects which are encrypted with SEK_P in the result of Q_A. Finally, the client checks if all the duplicates retrieved by Q_A have counterparts in $\mathbb{R}$. If there is any mismatch, the discrepancy proves that the service provider is malicious. The complete procedure of a *Query Integrity Assured Range Query* (QIARQ) is formalized in Algorithm 1.

5.3   *k* Nearest Neighbor Query

We design a *Query Integrity Assured k Nearest Neighbor* (QIAKNN) search algorithm by extending our range query solution in Section 5.2. For a given kNN query point Q located at position (x_Q, y_Q), a client first employs SEK_P to compute $V_{\mathcal{H}} = T(x_Q, y_Q)$ as the query point in the encrypted space. Because there is r percent duplicate data in $\mathbb{D}_E$ which should be filter out from query results, we multiply k by $(1 + r)$ to get k' and apply k' as the query parameter. Thereafter, the client transmits the values of $V_{\mathcal{H}}$ and k' to the service provider for retrieving k' nearest neighbors of Q. The service provider searches $\mathbb{D}_E$ with both directions (ascending and descending) of $V_{\mathcal{H}}$ until k' closest spatial objects are found and then returns the query result set $\mathbb{R}$ to the client. After being receipt of $\mathbb{R}$, the client first removes objects encrypted with SEK_S and checks if there are k objects leftover in $\mathbb{R}$. If $\mathbb{R}$ contains fewer than k objects, the client repeats the aforementioned steps with a multiple of r until obtaining k valid objects. Subsequently, the client retrieves the object s^* which has the longest distance to Q in $\mathbb{R}$. Because of loss of a dimension in the encrypted space, the objects in $\mathbb{R}$ may not precisely match the actual k nearest neighbors of Q. Consequently, the client utilizes the distance between Q and s^* ($Dist(Q, s^*)$) as a *search upper bound* and launches a range query Q_R with $Dist(Q, s^*)$ to decide the query window size. Following acquiring $\mathbb{R}'$ as the result of Q_R, the client audits the range query result as described in Section 5.2. Because $Dist(Q, s^*)$ is the search

Algorithm 2. Query Integrity Assured k Nearest Neighbor Query(Q, k)

1: Compute $V_{\mathcal{H}} = \mathcal{T}(x_Q, y_Q)$ based on SEK_P
2: Set δ = true, $\lambda = 1$, and $\gamma = 0$
3: **while** δ **do**
4:    Set $\mathbb{R} = \emptyset$
5:    $k' = k(1 + \lambda * r)$
6:    $\mathbb{R} \cup$ Retrieve k' objects closest to Hilbert value $V_{\mathcal{H}}$ from the server
7:    Filter out the spatial objects encrypted with SEK_S in $\mathbb{R}$
8:    **if** $|\mathbb{R}| \geq k$ **then**
9:       δ = false
10:    **else**
11:       $\lambda = \lambda + 1$
12:    **end if**
13: **end while**
14: **for** $i = 0$; $i < |\mathbb{R}|$; $i++$ **do**
15:    **if** $Dist(Q, s_i) > \gamma$ **then**
16:       $\gamma = Dist(Q, s_i)$
       /* $s_i \in \mathbb{R}$ */
17:       $s^* = s_i$
18:    **end if**
19: **end for**
20: Compute the edge length of Q_R by $Dist(Q, s^*)$
21: $\mathbb{R}' = \text{QIARQ}(Q_R)$
22: Set $\gamma = \infty$, $\lambda = 0$, and $\mathbb{R} = \emptyset$
23: **for** $j = 0$; $j < |\mathbb{R}'|$; $j++$ **do**
24:    **if** $\lambda < k$ **then**
25:       $\lambda = \lambda + 1$
26:       $\mathbb{R} = \mathbb{R} \cup s_j$
27:    **else**
28:       sort $\mathbb{R}$ in ascending order of distance to Q and retrieve the last element as s_k
29:       **if** $Dist(Q, s_k) > Dist(Q, s_j)$ **then**
30:          Replace s_k with s_j
31:       **end if**
32:    **end if**
33: **end for**

upper bound, the client has to identify the top k objects in $\mathbb{R}'$ based on their distance to Q to acquire the final query result.

We illustrate k-nearest-neighbor query integrity auditing with an example in Figure 4. The client first encodes the location of the query point with SEK_P and computes its $V_{\mathcal{H}} = 30$. Afterward, the client launches a kNN query with the numbers of $V_{\mathcal{H}}$ and k'. The service provider searches $\mathbb{D}_E$ for objects with Hilbert values ≥ 30 and < 30 in parallel until k' objects are found as demonstrated in Figure 4(a). Next, the client interacts with the server until k objects encrypted in SEK_P are retrieved. Among the k valid objects, assume the one which has the longest distance to Q has Hilbert value 32 and then we can obtain the search window edge length as five units. Subsequently, the client launches a

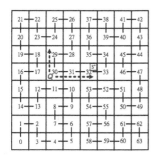 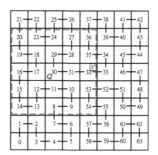

Fig. 4(a) Finding the search upper bound

Fig. 4(b) Auditing range query Q_A

Fig. 4. A query integrity assured k-nearest-neighbor query

query integrity assured range query for searching k nearest objects of Q as the exact query result as showed in Figure 4(b).

5.4 Attack-Aware Auditing Query Composition

The purpose of our dual space encryption design is to allow for sophisticated cross examination. Mobile users carry out cross examination against a single spatial database that has two different encryptions. However, negligent auditing queries launched by clients may reveal critical information to allow malicious LBS providers to detect the correspondence among the data with different encryption keys. For example, assume a client launches an auditing query after every regular query. Then, a malicious service provider can easily learn the relationship between the two queries and remove the query results of both queries to jeopardize future queries without being detected by clients.

In order to defend against the aforementioned attack, we need more advanced solutions for composing auditing queries. Generally, we want to create a checking query Q_A, which will not leak any correspondence information among the data objects in $\mathbb{D}_E$. In addition, Q_A should be hard to differentiate from other regular queries. Consequently, the main principle is to apply a single query to evaluate the integrity of multiple queries. Because spatial queries launched by the same mobile client usually exhibit locality [10], the query range overlap between successive queries from identical user is significant. By evaluating the integrity of multiple queries at a time, we can improve security and decrease integrity auditing overhead to save energy of mobile devices. Based on the memory capacity, a client can decide the threshold to generate a checking query for a group of executed regular queries $\mathcal{Q} = \{q_1, \ldots, q_n\}$ by merging their query regions. Only queries whose results contain replicas should be included in $\mathcal{Q}$.

6 Experimental Validation

We use Hilbert curves as a space encoding approach to encrypt spatial information in outsourced databases. It has been proved in [9] that without knowing

Table 2. The simulation datasets

Name	Number of POIs	Source
Uniform	10, 163	Synthetic
Skewed	10, 163	Synthetic
Los Angeles (LA)	10, 163	NAVTEQ
California (CA)	62, 556	US Census Bureau
North America (NA)	569, 120	US Census Bureau

the space encryption key, a brute force attack will need to exhaustively search all possible key combinations and the complexity of the attack is $O(2^{4b})$ where b is the number of bits for each parameter. Hence, the Hilbert curve based encryption method is employed as a one way function in our design. Table 2 illustrates two synthetic datasets and three real-world datasets utilized in our experiments. The two synthetic datasets of 10K data points each represent uniform and skew distributions, respectively. Los Angeles is a dataset containing around 10K restaurants inside a geographic area measuring 26 miles by 26 miles in the City of Los Angeles, California. The last two datasets consist of points of interest (POI) across California (61K) and North America (556K). Our query integrity assurance algorithms were implemented in Java and the experiments were conducted on a Windows Vista PC with Intel Core 2 Duo 3.16GHz processor and 4GB memory. All simulation results were recorded after the system model reached a steady state.

6.1   Encoded POI Density

We first show the relationship between Hilbert curve orders and the number of POIs encoded by one Hilbert value (POI density). A higher Hilbert curve order increases the security level of the corresponding space encryption key while on the other hand, a higher curve order will incur higher computational complexity. As Figure 5 shows with both synthetic and real-world datasets the number of POIs per Hilbert value decreases rapidly as the curve order increases. The average num-

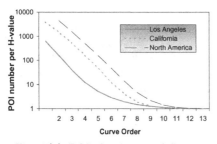
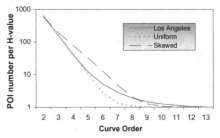

Fig. 5(a) POI density on different size of datasets

Fig. 5(b) POI density on different data distributions

Fig. 5. The relationship between curve order and POI density

ber of POIs per non-empty Hilbert value reaches 1 when the curve order is greater than 12. An empty Hilbert value means that there is no POI associated with it and we discard these empty values during the initialization process. Hence, we use the default curve order of 12 for space encryption unless specified explicitly.

6.2   Spatial Database Outsourcing Initialization

There are three major operations in the initialization process for our spatial database outsourcing approach: (1) computing Hilbert values for all data objects based on their locations; (2) encrypting each data object with the symmetric key; and (3) calculating the cryptographic signature for each object. There are various algorithms for (2) and (3) and the cost of each may vary. We used the Blowfish encryption algorithm [19] and MD5 (Message-Digest algorithm 5) for signature computation. We perform the data initialization process on all the three real-world datasets with 40% duplication rate and curve order of 12. The cost of each operation in Figure 6 shows that computing the Hilbert values is efficient and less expensive than the other two operations.

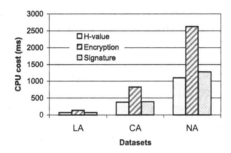

Fig. 6. Initialization cost of the proposed spatial database outsourcing approach

6.3   Query Processing on the Client Side

For range queries, the client performs a transformation from range query window to Hilbert curve segments and sends these segments to the service provider. A kNN query, as described in Algorithm 2, is split into two parts: retrieve k nearest POIs simply based on the Hilbert value of the query point and launch a range query using the distance of the kth point in the previous operation as the search upper bound.

The transformation cost on the client side is analyzed in this experiment. We extended the size of range query windows from 0.01 to 0.05 on the normalized dataset and varied curve orders from 10 to 15. Figure 7 demonstrates the average cost of 50 range queries on the Los Angeles dataset with the aforementioned settings. Based on the results, we can see that it is efficient to compute Hilbert curve segments with a given SEK and the client does not need to store any information other than the SEK. Hilbert curve segments are represented by the start and the end values and are transmitted to the service provider to retrieve spatial objects.

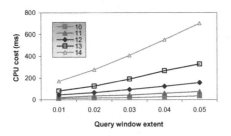

Fig. 7. Query processing cost on the client side

6.4   Integrity Auditing

Clients employ QIARQ and QIAKNN to verify if the spatial query results are both correct and complete when receiving query results. There are three operations in the query result authentication process: (1) decrypt each data object; (2) verify the signature of every retrieved object; and (3) check the counterpart existence of each object with a duplicate. The cost of (1) and (2) are constant per object given the same encryption method. Therefore, we only show the cost of (3) in Figure 8 for range queries at different range extents and data replication percentage. Verifying query results is efficient when the size of the data returned is small, which is the case in general for common kNN queries and range queries with small extents. To better illustrate the authentication cost, we use the California dataset here with query window extent varying from 0.02 to 0.1 and replication percentage changing from 10% to 50% on the normalized data. The returned POI number and CPU cost increase linearly when we enlarge the query window and replication ratio as shown in Figure 8.

6.5   Communication Cost

We study the communication cost by considering the size of data transferred between the client and the service provider. Network delays and packet retransmission due to unstable connections are not the focus of this research and hence they are not considered in this experiment. A round trip of a query-and-answer process between a client and a server can be split into two parts and each of them

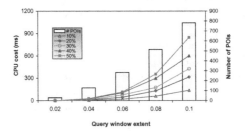

Fig. 8. Cost of authenticating query results

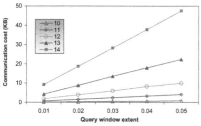
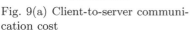
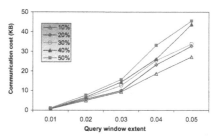

Fig. 9(a) Client-to-server communication cost

Fig. 9(b) Server-to-client communication cost

Fig. 9. The cost of communication between a client and a service provider

are affected by different factors. The query transfer from a client to a server is
composed of multiple Hilbert curve segments and the size of which is related to
the Hilbert curve order in the applied SEK. The result returning back from the
service provider contains all the POIs with respect to the query and the size is
determined by the distribution of POIs, the extent of the query range (or the
number k for a kNN query), and the replication percentage of the outsourced
database. We assume the size of every data object is 1KB. In Figure 9(a), the
communication cost is measured by the number of segments transmitted and
in Figure 9(b), it is measured by the number of objects in the transmission.
The experiments were performed on the Los Angeles dataset and the trend was
similar for other datasets. As it can be observed from the figures, the communication cost increases linearly when we expand the curve order or the replication
percentage with the same query window extent.

6.6   Against Malicious Attacks

Modifying and adding data objects in an outsourced spatial database can be
easily detected by our integrity auditing algorithms, which results in one of the
two cases: unable to perform decryption on the tampered data or inconsistent
cryptographic signatures. Consequently, the attack model studied in this experiment primarily focuses on data object deletion by malicious service providers.
We conducted the simulation on the Los Angeles dataset using randomly generated queries with the extent of 0.04 on the normalized coordinates. With different data replication ratio, the server launches random deletion attacks on
query results. Figure 10 shows the probability that the attacker can escape from
client auditing process versus the total number of data objects deleted from a
query result. As we can see from the figure, with more than 20% replication, the
probability of not detecting a deletion declines rapidly as the service provider
deletes more data objects.

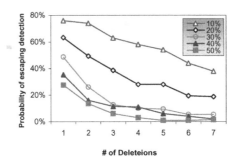

Fig. 10. Probability of escaping detection of deletion attacks

7 Conclusions

Outsourcing of spatial databases for supporting location-based services has become a trend in recent years due to the economy of scale. Existing solutions are designed for data privacy protection or query integrity auditing, respectively, instead of considering both data privacy and query integrity as a whole. We have introduced query integrity assured algorithms for both range query and k-nearest-neighbor query with space encryption techniques to secure data privacy. We have demonstrated through simulation results that our mechanisms have remarkable performance. For future work, we plan to extend our algorithms to support more spatial query types such as spatial join, spatial path queries, etc.

Acknowledgements

This research has been funded in part by the US National Science Foundation (NSF) grants IIS-0238560 (PECASE), IIS-0534761, IIS-0742811, CNS-0831502 (CT), and CNS-0831505 (CT), and in part from the METRANS Transportation Center, under grants from USDOT and Caltrans. Any opinions, findings, and conclusions or recommendations expressed in this material are those of the author(s) and do not necessarily reflect the views of the NSF.

References

1. Agrawal, R., Kiernan, J., Srikant, R., Xu, Y.: Order-Preserving Encryption for Numeric Data. In: SIGMOD Conference, pp. 563–574 (2004)
2. Butz, A.R.: Alternative Algorithm for Hilbert's Space-Filling Curve. IEEE Trans. Comput. 20(4) (1971)
3. Devanbu, P.T., Gertz, M., Martel, C.U., Stubblebine, S.G.: Authentic Third-party Data Publication. In: DBSec, pp. 101–112 (2000)
4. Diffie, W., Hellman, M.E.: New Directions in Cryptography. IEEE Transactions on Information Theory 22(6), 644–654 (1976)

5. Hacigümüs, H., Iyer, B.R., Li, C., Mehrotra, S.: Executing SQL over Encrypted Data in the Database-service-provider Model. In: SIGMOD Conference, pp. 216–227 (2002)
6. Hacigümüs, H., Mehrotra, S., Iyer, B.R.: Providing Database as a Service. In: ICDE, p. 29 (2002)
7. Hilbert, D.: Über die stetige Abbildung einer Linie auf ein Flächenstück. Mathematische Annalen (38), 459–460 (1891)
8. Jagadish, H.V.: Linear Clustering of Objects with Multiple Atributes. In: SIGMOD Conference, pp. 332–342 (1990)
9. Khoshgozaran, A., Shahabi, C.: Blind Evaluation of Nearest Neighbor Queries Using Space Transformation to Preserve Location Privacy. In: Papadias, D., Zhang, D., Kollios, G. (eds.) SSTD 2007. LNCS, vol. 4605, pp. 239–257. Springer, Heidelberg (2007)
10. Ku, W.-S., Zimmermann, R., Wang, H.: Location-Based Spatial Query Processing in Wireless Broadcast Environments. IEEE Trans. Mob. Comput. 7(6), 778–791 (2008)
11. Lawder, J.K., King, P.J.H.: Querying multi-dimensional data indexed using the hilbert space-filling curve. SIGMOD Record 30(1), 19–24 (2001)
12. Merkle, R.C.: A Certified Digital Signature. In: Brassard, G. (ed.) CRYPTO 1989. LNCS, vol. 435, pp. 218–238. Springer, Heidelberg (1990)
13. Moon, B., Jagadish, H.V., Faloutsos, C., Saltz, J.H.: Analysis of the Clustering Properties of the Hilbert Space-Filling Curve. IEEE Trans. Knowl. Data Eng. 13(1), 124–141 (2001)
14. Mouratidis, K., Sacharidis, D., Pang, H.: Partially Materialized Digest Scheme: An Efficient Verification Method for Outsourced Databases. VLDB J. 18(1), 363–381 (2009)
15. Mykletun, E., Narasimha, M., Tsudik, G.: Authentication and Integrity in Outsourced Databases. In: NDSS (2004)
16. Orenstein, J.A.: Spatial Query Processing in an Object-Oriented Database System. In: SIGMOD Conference, pp. 326–336 (1986)
17. Pang, H., Jain, A., Ramamritham, K., Tan, K.-L.: Verifying Completeness of Relational Query Results in Data Publishing. In: SIGMOD Conference, pp. 407–418 (2005)
18. Papadopoulos, S., Papadias, D., Cheng, W., Tan, K.-L.: Separating Authentication from Query Execution in Outsourced Databases. In: ICDE (2009)
19. Schneier, B.: Description of a New Variable-Length Key, 64-bit Block Cipher (Blowfish). In: Anderson, R. (ed.) FSE 1993. LNCS, vol. 809, pp. 191–204. Springer, Heidelberg (1994)
20. Schneier, B.: Applied Cryptography (2nd ed.): Protocols, Algorithms, and Source Code in C. John Wiley & Sons, Inc., New York (1996)
21. Sion, R.: Query Execution Assurance for Outsourced Databases. In: VLDB, pp. 601–612 (2005)
22. Sommerville, I.: Software Engineering, 8th edn. Addison-Wesley, Reading (2006)
23. Wang, H., Yin, J., Perng, C.-S., Yu, P.S.: Dual Encryption for Query Integrity Assurance. In: CIKM, pp. 863–872 (2008)
24. Yang, Y., Papadopoulos, S., Papadias, D., Kollios, G.: Spatial Outsourcing for Location-based Services. In: ICDE, pp. 1082–1091 (2008)
25. Yiu, M.L., Ghinita, G., Jensen, C.S., Kalnis, P.: Outsourcing of Private Spatial Data for Search Services. In: ICDE (2009)

A Hybrid Technique for Private Location-Based Queries with Database Protection*

Gabriel Ghinita[1], Panos Kalnis[2], Murat Kantarcioglu[3], and Elisa Bertino[1]

[1] Purdue University, West Lafayette, IN 47907, USA
{gghinita,bertino}@cs.purdue.edu
[2] King Abdullah University of Science and Technology, Jeddah, Saudi Arabia
panos.kalnis@kaust.edu.sa
[3] University of Texas at Dallas, Richardson, TX 75080, USA
muratk@utdallas.edu

Abstract. Mobile devices with global positioning capabilities allow users to retrieve points of interest (POI) in their proximity. To protect user privacy, it is important not to disclose exact user coordinates to un-trusted entities that provide location-based services. Currently, there are two main approaches to protect the location privacy of users: *(i)* hiding locations inside cloaking regions (CRs) and *(ii)* encrypting location data using private information retrieval (PIR) protocols. Previous work focused on finding good trade-offs between privacy and performance of user protection techniques, but disregarded the important issue of protecting the POI dataset D. For instance, location cloaking requires large-sized CRs, leading to excessive disclosure of POIs ($O(|D|)$ in the worst case). PIR, on the other hand, reduces this bound to $O(\sqrt{|D|})$, but at the expense of high processing and communication overhead.

We propose a hybrid, two-step approach to private location-based queries, which provides protection for both the users and the database. In the first step, user locations are generalized to coarse-grained CRs which provide strong privacy. Next, a PIR protocol is applied with respect to the obtained query CR. To protect excessive disclosure of POI locations, we devise a cryptographic protocol that privately evaluates whether a point is enclosed inside a rectangular region. We also introduce an algorithm to efficiently support PIR on dynamic POI sub-sets. Our method discloses $O(1)$ POI, orders of magnitude fewer than CR- or PIR-based techniques. Experimental results show that the hybrid approach is scalable in practice, and clearly outperforms the pure-PIR approach in terms of computational and communication overhead.

1 Introduction

Mobile devices with positioning capabilities (e.g., GPS) facilitate access to location-based services that provide information relevant to the users' geo-spatial context. Typically, users are interested in finding nearby *points of interest (POI)*, and send

---

* The work reported in this paper has been partially supported by NSF grant 0712846 "IPS: Security Services for Healthcare Applications", and MURI award FA9550-08-1-0265 from the Air Force Office of Scientific Research.

nearest-neighbor (NN) queries to *location servers (LS)* that own databases of POI. However, users are reluctant to disclose their exact locations to the un-trusted LS, since sensitive details about lifestyle, political or religious affiliation, etc., can be revealed by a person's whereabouts.

To address this threat, user locations are perturbed before being reported to the LS. On the other hand, replacing exact locations with coarse regions requires the disclosure of a large number of POIs to the user, such that result correctness is preserved. However, the LS wishes to protect its data against excessive disclosure, since the POI dataset represents a valuable asset to the service provider. For instance, consider that Bob asks the query "find the nearest restaurant to my current location". The LS may reward Bob with certain discounts, in the form of electronic coupons (e.g., digital gift card codes) that are associated with each POI. If the user is billed on a "per-retrieved-POI" basis, then a large number of results will increase the cost of using the service. On the other hand, if the LS offers the service with no charge to the user (e.g., advertisement-generated income), then users could abuse the system by redeeming a large number of coupons. This causes the LS to lose its competitive edge, and to cease providing the service.

Existing solutions for private location queries focus on user protection only, and can be broadly classified into two categories:

1. *Location Cloaking* techniques replace the exact location of a user with a *cloaking region (CR)*, typically of rectangular shape. To ensure result correctness, the CR must enclose the actual user location. Furthermore, CRs must satisfy certain constraints dictated by a privacy paradigm, which expresses the privacy requirements of the user (e.g., *spatial k-anonymity (SKA)* [1,2,3,4] requires each CR to contain at least k distinct users). Regardless of the method used to generate the CR, query processing at the LS side is performed with respect to a rectangular region, as opposed to an exact user location. In consequence, the result returned by the LS is a super-set of the actual query result.

2. *Private Information Retrieval (PIR)* techniques rely on a cryptographic protocol to achieve query privacy [5]. In a pre-processing phase, the LS organizes the POI database into a data structure relevant to the supported type of query[1], and maps it to an ordered array $D[1 \ldots n]$. At runtime, a query is transformed from a context-based (i.e., spatial) query to a query-by-index (i.e., return the ith item), according to the pre-defined data organization which is known by the users. When a user wishes to retrieve $D[i]$, s/he creates an encrypted query object $q(i)$. Using a mathematical transformation, the LS computes privately (i.e., without learning the value of i) the result $r(D, q(i))$ and sends it back to the user. PIR protocols ensure that it is computationally hard for the LS to recover the value i from $q(i)$, but at the same time the user can easily re-construct $D[i]$ from r.

Previous work [3,4,5] evaluates location privacy techniques based on two criteria: *privacy* and *performance*. With respect to privacy, PIR offers strong guarantees for both one-time, as well as repetitive (i.e., continuous) queries. Furthermore, PIR does not require trusted components, such as anonymizer services or other trusted users. On the other hand, CR methods operate under a more restrictive set of trust assumptions, but

---

[1] For instance, to answer NN queries, [5] uses a Voronoi diagram mapped to a regular grid.

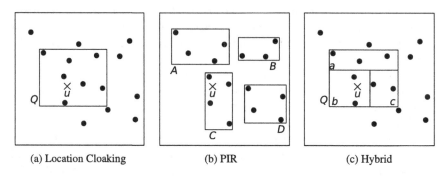

(a) Location Cloaking   (b) PIR   (c) Hybrid

Fig. 1. Benefit of the Hybrid Approach

are considerably more efficient in terms of computational and communication overhead. The cryptographic elements incorporated in PIR require powerful computational resources (e.g., parallel machines), and high-bandwidth communication channels.

However, there is a third, equally-important dimension in evaluating techniques for private location queries: the amount of protection provided to the database. To the best of our knowledge, this aspect has not been addressed before[2]. Nevertheless, as illustrated by the earlier customer-reward example, it is important to control tightly the amount of POI disclosure.

To illustrate the limitations of existing approaches, consider the example of Figure 1, where the location server stores a database D of 15 POI (marked as full dots). User u asks a query for the nearest POI. If location cloaking is used (Figure 1(a)), the user will retrieve all the 7 POI enclosed[3] by query CR Q. As CRs grow large, location cloaking methods may disclose a large fraction of the database (possibly linear to $|D|$). On the other hand, the NN protocol from [5] does not use CRs (a detailed protocol description is given in Section 2). Instead, the dataset is partitioned into rectangular tiles $A \ldots D$, containing at most $\lceil \sqrt{15} \rceil = 4$ POI each (Figure 1(b)). The boundaries of the tiles are sent in plain text to u, who determines that his/her location is enclosed by tile C. Only the POIs in tile C are revealed to u through a PIR request. This method discloses $O(\sqrt{|D|})$ exact POI locations. However, revealing the tile boundaries may result in additional disclosure of POI locations, especially if the tiles have small spatial extent.

We propose a hybrid approach, outlined in (Figure 1(c)). The CR Q is sent to the LS, which determines a set of fine-grained tiles $\{a, b, c\}$ that cover the query area. We impose a constraint that each tile encloses at most a constant number F of POI (a system parameter). The boundaries of the tiles are not sent to the user. Instead, the user and the LS engage in a novel cryptographic protocol that privately determines which one of the tiles encloses the location of u. At the end of the protocol, the LS learns nothing about the user location (except that u is inside Q), whereas the user only learns the identifier of the tile that encloses u (but not the boundaries of any of the tiles). Finally, the user

---

[2] Previous work considered result set size only in the context of communication cost. However, this indirect approach is not effective due to other factors that influence bandwidth consumption (e.g., POI size may be negligible in comparison with other traffic components).

[3] The example considers an approximate query, where candidate NNs outside Q are ignored.

requests through PIR the contents of the enclosing tile[4] (in this case, b). The hybrid approach has two benefits: first, it controls strictly the amount of POI disclosed, which is bounded by a constant. This improvement is clearly superior to location cloaking and pure-PIR approaches, which disclose $O(|D|)$ and $O(\sqrt{|D|})$ POI, respectively. Second, the hybrid approach incurs considerably less overhead than the pure PIR method, since the cryptographic protocol is applied only on a partition of the database.

The contribution of this work is two-fold:

(*i*) We propose a cryptographic protocol which allows private evaluation of point-rectangle enclosure. We use this protocol as a building block in determining the nearest POI to a given user location. This protocol can be easily adapted to other types of spatial queries (e.g., private spatial joins), and represents an interesting finding in itself.

(*ii*) We develop a hybrid approach that efficiently supports PIR processing with respect to a user-generated cloaked region Q. The proposed method can handle CRs with large extents, and controls tightly the amount of disclosed POI. Furthermore, we show experimentally that it is considerably more efficient than its PIR-only counterpart.

The rest of the paper is organized as follows: Section 2 surveys related work. Section 3 outlines the system architecture and the privacy assumptions. Section 4 introduces the proposed protocol for private evaluation of point-rectangle enclosure, whereas Section 5 presents the hybrid technique for processing PIR requests based on dynamic cloaking regions. We present the results of our experimental evaluation in Section 6, and conclude with directions for future research in Section 7.

2   Related Work

Several approaches to private location queries have been proposed. In [6], the querying user sends to the server $k - 1$ fake locations to reduce the likelihood of identifying the actual user position. *SpaceTwist* [7] performs a multiple-round incremental range query protocol, based on a fake *anchor* location that hides the user coordinates. In [8], a random cloaking region that encloses the user is generated. However, neither of these techniques is suitable if an adversary possesses background knowledge about user locations. Most CR-based solutions [2,1,4,3] implement the spatial k-anonymity (SKA) paradigm, and rely on a three-tier architecture: a trusted anonymizer service intermediates all interaction between users and LS, and generates CRs that contain at least k *real* user locations. If the resulting CRs are *reciprocal* [4], SKA guarantees privacy for snapshots of user locations. However, supporting continuous queries [9] requires generating large-sized CRs. In [10,11], the objective is to prevent the association between users and sensitive locations. Users define privacy profiles [11] that specify their sensitivity with respect to certain *feature types* (e.g., hospitals, bars, etc.), and every CRs must cover a diverse set of sensitive and non-sensitive features. A common limitation of CR-based techniques is that they disclose an excessive number of POIs.

---

[4] The indexing scheme we employ (Section 5) guarantees that the retrieved tile is not empty.

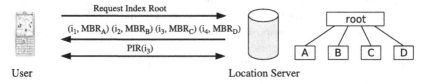

Fig. 2. Approximate NN PIR Protocol from [5]

In [12], the set of POI is first encoded according to a secret transformation by a trusted entity. A Hilbert-curve mapping (with secret parameters) transforms 2-D points to 1-D. Users (who know the transformation key) map their queries to 1D, and the processing is performed in the 1-D space. However, the mapping can decrease the result accuracy, and the transformation may be vulnerable to reverse-engineering.

Private Information Retrieval (PIR) protocols allow users to retrieve an object X_i from a set $X = \{X_1 \ldots X_n\}$ stored by a server, without the server learning the value of i. The PIR concept was first formulated in [13], where it is shown that in the information theoretic setting, any single-server solution requires $\Theta(n)$ communication cost. In practice, this bound can be reduced by employing *Computational* PIR (cPIR), which offers protection against an attacker with polynomially-bounded computational capabilities. The PIR protocol in [14] relies on the *Quadratic Residuosity Assumption (QRA)*, which states that it is computationally hard to find the quadratic residues (in modulo arithmetic) of a large composite number $N = q_1 \cdot q_2$ (q_1, q_2 are large primes). Specifically, given a number $y \in \mathbb{Z}_N^{+1}$ ($\mathbb{Z}_N^{+1}$ is the sub-set of $\mathbb{Z}_N$ for which the Jacobi symbol [15] is $+1$) it is computationally hard (without knowing the factorisation of N) to determine whether y is a quadratic residue (QR) (i.e., $\exists x \in \mathbb{Z}_N | y = x^2 \mod N$) or a non-residue (QNR). Assume that all objects in X are bits. The client sends the server an ordered array of n numbers $Y = [y_1 \cdots y_n]$, such that y_i is QNR, whereas all the others are QR. The server performs a *masked* multiplication of values in Y, i.e., it multiples together only the y_j values for which $X_j = 1$. The client, who knows the factorisation of N, can determine that if the result of the multiplication is QNR, then $X_i = 1$, otherwise $X_i = 0$. The protocol can be applied bit-by-bit to support more complex objects.

The work in [5] extends the above-mentioned protocol for binary data to the LBS domain, and proposes approximate (*ApproxNN*) and exact (*ExactNN*) protocols for nearest-neighbor queries. However, ExactNN is very expensive. Furthermore, ApproxNN achieves very good accuracy in practice. The idea behind [5] is to organize the POI set such that spatial queries (e.g., NN) can be translated to queries "by-index", which are then answered using the QRA-based protocol. Our work proposes a hybrid alternative to answer approximative queries, and since ApproxNN is used as a baseline in our experimental evaluation, we provide an overview of its functionality. In an off-line phase, the server performs a partitioning of the POI set D using an R^*-tree index, which is constrained to have exactly two levels. Therefore, each leaf node holds at most $\sqrt{|D|}$ POI, and the root node contains at most $\sqrt{|D|}$ minimum bounding rectangles (MBR). Figure 2 shows the obtained index for the partitioned dataset in Figure 1(b). At query time, the user u first retrieves the root node in plaintext, and determines which leaf node encloses, or is nearest to, u's location. Next, u retrieves privately the contents of the selected leaf node. There are three limitations of this approach: *(i)* a large number $(O(\sqrt{|D|}))$ of POI are directly disclosed, *(ii)* sending MBRs of leaf nodes to the user can indirectly disclose additional POI locations and *(iii)* the computational complexity of the PIR phase is $O(|D|)$, as all data elements are considered, and bandwidth consumption is high.

Several protocols that support secure multy-party computational geometry have been proposed. For instance, in [16] it is shown how to compute privately point-rectangle inclusion using secure scalar products, whereas [17] introduces a protocol for private point-circle inclusion evaluation. However, these protocols rely on SMC [18] primitives, and as a result they are very expensive and require multiple communication rounds. In contrast, our proposed point-rectangle evaluation protocol uses homomorhpic encryption, and only requires a single communication round.

3   System Architecture and Assumptions

3.1   Privacy Model

Many privacy models that rely on location cloaking have been proposed in literature [1,2,3,4,10,11]. The proposed hybrid approach can be used in conjunction with any of these methods. For instance, CRs can be built according to the spatial k-anonymity paradigm [1,2,3,4], which requires that at least k distinct user locations must be enclosed by the CR. Alternatively, CRs can be determined based on user-specified sensitivity thresholds with respect to a set of sensitive feature types [10,11]. The particular choice of privacy paradigm and CR generation technique is outside the scope of this work. We consider the CR as an input to our method, and we focus on two aspects: *(i)* how to efficiently perform PIR with respect to dynamically-generated CRs, and *(ii)* how to control tightly the amount of disclosed POIs. We do, however, factor in our system design provisions for CRs with *large* spatial extents, suitable to accommodate highly-demanding privacy requirements.

Note that, it has been discussed previously [5] that location cloaking may not be suitable for highly-mobile users issuing continuous queries. However, as shown in [9], cloaked regions can be generated in a manner that accommodates continuous queries. Furthermore, if the CR is large enough to cover an entire user trajectory, private continuous queries can be supported with strong privacy guarantees. To illustrate this claim, consider the example of user Jin, who often visits karaoke lounges. Jin wishes to keep her passion for karaoke secret, so she does not want a malicious attacker to learn that she was in the proximity of such an establishment. On the other hand, Jin may be comfortable with disclosing the fact that she is currently in Koreatown, which is a large area. In addition, while Jin remains within the perimeter of Koreatown, her privacy is protected even if she issues continuous queries. In Section 6, we experimentally evaluate our proposed method using CRs that cover large portions of the dataspace.

3.2   System Overview

The proposed system architecture is shown in Figure 3. The system model is flexible, and can accommodate several distinct solutions for creating input CRs. For instance,

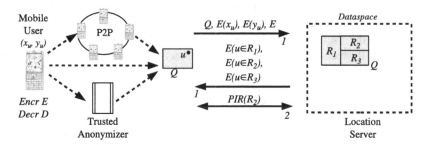

Fig. 3. System Architecture

users can cloak their locations by themselves, as considered in [10,11]. Alternatively, users can send their queries to a trusted anonymizer service which creates the CRs [1,2,3,4]. Or, users can build CRs in a collaborative fashion [19,20,21].

Given the query CR Q, the LS returns the approximate NN POI of the user by executing a two-round protocol, as shown in Figure 3. In the first round (arrows labeled 1), the user[5] generates an encryption (E)/decryption (D) key pair, which are part of a homomorphic encryption family, such as Paillier [22]. The user sends to the LS the query CR Q, together with the encryption (i.e., public) key E and the encrypted user coordinates $E(x_u)$ and $E(y_u)$. The LS processes the query Q, and partitions the result set into disjoint rectangular regions, or tiles. Each tile contains a number of POI bounded by constant F, which is a system parameter. In this case, the set of tiles $\{R_1, R_2, R_3\}$ is obtained. The LS evaluates privately, using the properties of homomorphic encryption[6], the enclosure condition between point (x_u, y_u) and the resulting tiles. The encrypted evaluation outcome is returned to the user, who will decrypt and find which of the given rectangles encloses its location, in this case R_2. The private point-rectangle enclosure evaluation is necessary because the resulting query result tiles can be arbitrarily small. Sending these tiles in plain text to the user (as it is done in [5], with the root of the two-level index) would give away excessive information about the distribution of POI. Finally, in the second round of the protocol, the user issues a private request for the contents of R_2, and determines which of the retrieved POI is closest to his/her location.

4   Private Evaluation of Point-Rectangle Enclosure

In this section, we introduce a two-party protocol between parties A and B, which determines privately whether a given point p owned by A is enclosed in a rectangle R owned by B. The protocol protects the privacy of both parties involved. Specifically, A learns only if the point p is enclosed by R, but does not find any additional information about R. In addition, B does not learn any information about the point p of A.

Our protocol relies on the Paillier public-key homomorphic encryption scheme introduced in [22]. Paillier encryption operates in the message space of integers $\mathbb{Z}_N$, where

---

[5] Alternatively, the trusted anonymizer or a trusted peer can perform the described protocol on behalf of the user.

[6] Details about the private evaluation of point-rectangle enclosure are given in Section 4.

N is a large composite modulus. Similar to the PIR protocol in [14] (described in Section 2), the security of Paillier encryption relies on the QRA assumption with respect to modulus N. Denote by D and E the decryption and encryption functions, respectively. Given the ciphertexts $E(m_1)$ and $E(m_2)$ of plaintexts m_1 and m_2, the ciphertext of the sum $m_1 + m_2$ can be obtained by multiplying individual ciphertexts:

$$D(E(m_1) \cdot E(m_2)) = (m_1 + m_2) \mod N \tag{1}$$

In addition, given ciphertext $E(m)$ and plaintext $r \in \mathbb{Z}_N$, we can obtain the ciphertext of the product $r \cdot m$ by exponentiation with r, as follows:

$$D(E(m)^r) = r \cdot m \mod N \tag{2}$$

Furthermore, Paillier encryption provides semantic security, meaning that encrypting the same plaintext with the same public key E twice will result in distinct ciphertexts. Therefore, the scheme is secure against chosen plaintext attacks.

In our setting, the querying user wishes to find whether his/her location is enclosed inside some rectangular region R stored by the server. This can be achieved by privately evaluating the difference between the user coordinates and the boundary coordinates of rectangle R. Furthermore, to prevent leakage of POI locations, only the sign of the difference should be revealed to the user, and not the absolute value.

We introduce the protocol for private evaluation of point-rectangle enclosure in an incremental fashion. Assume that parties A and B hold two numbers a and b, respectively. In Section 4.1 we show how to privately evaluate $sign(b-a)$. Next, in Section 4.2 we give the complete protocol for point-rectangle inclusion.

4.1 Private Evaluation of $sign(b - a)$

We show how to evaluate privately $sign(b - a)$ in two steps: first, we give an auxiliary protocol that privately evaluates the difference $(b - a)$. Then, we extend the auxiliary protocol to disclose only the sign of the difference, but not its absolute value. Note that, the difference protocol has no practical value by itself, since disclosing the value of $(b - a)$ to one of the parties (say A) automatically discloses the value held by the other party (since A can determine the value of b based on $b - a$ and a). However, the private difference protocol introduces a construction that is later used in the private evaluation of $sign(b - a)$.

Paillier encryption allows the computation of the ciphertext of sums based on the ciphertexts of individual terms. However, only the addition operation is supported, and not subtraction. Furthermore, the message space $\mathbb{Z}_N$ consists of positive integers only, hence the trivial solution of setting $m_1 = (-a)$, $m_2 = b$ and computing $E(m_1) \cdot E(m_2) = E(b-a)$ is not suitable. We overcome this limitation imposed on the message space by simulating complement arithmetic for N-bit integers.

Assume that $a, b \in \mathbb{Z}_{N'}$, where $N' < N$. Party A computes $m_1 = N - a$ and sends $E(m_1)$ to B, who in turn sets $m_2 = b$, and determines

$$E(m_3) = E(m_1) \cdot E(m_2) = E(m_1 + m_2) = E(N + (b - a)) \tag{3}$$

Fig. 4. Determining the value of $b - a$

Party B returns $E(m_3)$ to A who decrypts the message and learns the value of $m_3 = N + (b - a)$. The difference $b - a$ can be computed from m_3 as shown in Figure 4.

Let $I_1 = \{0, 1, \ldots, N'\}$ and $I_2 = \{N - N', \ldots, N - 1\}$. If $(b - a) \geq 0$, then $m_3 \in I_1$, otherwise $m_3 \in I_2$. To correctly interpret the result, it is necessary that $I_1 \cap I_2 = \emptyset$. A sufficient condition to ensure that the two intervals are disjoint is

$$[(N' - 0 + 1)] + [(N - 1) - (N - N') + 1] \leq N \Leftrightarrow N' \leq \left\lfloor \frac{N - 1}{2} \right\rfloor \quad (4)$$

Party A determines that

$$b - a = \begin{cases} m_3, & 0 \leq m_3 \leq N' \\ -(N - m_3), & N - N' \leq m_3 \leq N - 1 \end{cases} \quad (5)$$

The pseudocode in Figure 5 details the protocol for private computation of $(b - a)$. The protocol requires only one round of communication. Note that, A can immediately learn from $(b - a)$ the value of b. Next, we show how to protect against this inference.

Private Evaluation (b-a)
Input: value a held by party A, b held by party B
Output: A learns $(b - a)$, B learns nothing

```
1.    A(Client):   m₁ = N - a
2.                 Send E, E(m₁) to B
3.    B(Server):   m₂ = b
4.                 E(m₃) = E(m₁) · E(m₂)
5.                 Send E(m₃) to A
6.    A(Client):   m₃ = D(E(m₃))
7.                 if (0 ≤ m₃ ≤ N′)
8.                     b - a = m₃
9.                 else
10.                    b - a = -(N - m₃)
```

Fig. 5. Private Evaluation of $(b - a)$

We modify the protocol for evaluating $(b - a)$ to only disclose $sign(b - a)$, without revealing any additional information about b. The main idea is to multiply m_3 in the previous protocol by a random blinding factor[7], such that the absolute value of $(b - a)$

---

[7] Random blinding is a frequently-used operation in cryptographic protocols [23].

can no longer be reconstructed by A. Consider random integer ρ uniformly distributed in the set $\{1, 2, \cdots, M\}$, such that

$$M \leq \left\lfloor \frac{N-1}{2N'} \right\rfloor \quad (6)$$

(we will give the rationale for this condition shortly). Steps $1 - 4$ of the protocol in Figure 5 remain unchanged. However, in step 5, instead of sending $E(m_3)$ back to A, B sends $E(m_4)$ obtained through exponentiation with plaintext ρ:

$$E(m_4) = E(m_3)^\rho = E(\rho \cdot m_3) = E(\rho \cdot (N + b - a)) \quad (7)$$

The value of $sign(b - a)$ can be computed from m_4 as shown in Figure 6. In a similar

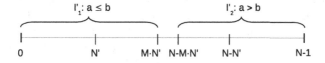

Fig. 6. Private Evaluation of $sign(b - a)$

manner to the protocol for difference, let $I'_1 = \{0, 1, \ldots, M \cdot N'\}$ and $I'_2 = \{N - M \cdot N', \ldots, N - 1\}$. If $(b - a) \geq 0$, then $m_4 \in I'_1$, otherwise $m_4 \in I'_2$. This time, the condition $I'_1 \cap I'_2 = \emptyset$ is equivalent to

$$[(M \cdot N' - 0 + 1)] + [(N - 1) - (N - M \cdot N') + 1] \leq N \Leftrightarrow N' \leq \left\lfloor \frac{N-1}{2M} \right\rfloor \quad (8)$$

hence the requirement in Eq. (6). Party A determines that

$$sign(b - a) = \begin{cases} +1, & 0 \leq m_4 \leq M \cdot N' \\ -1, & N - M \cdot N' \leq m_4 \leq N - 1 \end{cases} \quad (9)$$

The proof of Eq. (9) is immediate: if $(a \leq b)$, then $0 \leq m_3 \leq N'$, and therefore $0 \leq \rho \cdot m_3 \leq M \cdot N'$. On the other hand, if $(a > b)$ we have $N - N' \leq m_3 < N$, therefore

$$M(N - N') \mod N \leq M \cdot m_3 < N \Leftrightarrow (N - M \cdot N') \mod N \leq M \cdot m_3 < N$$

Note that, in practice, the additional constraint imposed on the domain size N' by Eq. (8) does not represent a limitation. For security considerations, the magnitude of modulus N must be at least 768 bits large. Consider values of a and b that can be represented on 64 bits, for instance. Such values are sufficiently large for many applications. In this case, the random blinding factor domain will be bounded by $M = \frac{2^{768}}{2} \cdot \frac{1}{2^{64}}$, which is in the order of 2^{700}, sufficiently large to obtain a strong degree of protection through random blinding.

Security Discussion. The proposed private sign evaluation protocol (and consequently the point-rectangle enclosure evaluation protocol) inherits the security strength provided by the random blinding. Note that, this level of security is weaker than the information-theoretic security features offered by other security primitives, such as secure multi-party computation (SMC) [18], for instance. However, SMC protocols are prohibitively expensive. On the other hand, random blinding offers good security features given that the blinding factors are large.

4.2 Private Evaluation of Point-Rectangle Enclosure

The protocol for private evaluation of point-rectangle enclosure builds upon the sign evaluation protocol of Section 4.1. Denote the user location by coordinates (x_u, y_u), and let the server-stored rectangle R be specified by its lowest-left (LL_x, LL_y) and upper-right (UR_x, UR_y) coordinates. We maintain the notations from the previous sections, i.e., all coordinates $x, y \in \{0, 1, \ldots, N'\}$ and the random blinding factors in the set $\{0, 1, \ldots, M\}$, such that Eq. (8) is satisfied. Consider the example in Figure 7(a): the user location is situated inside the rectangle if and only if the four inequalities hold simultaneously. Conversely, if any of the inequalities does not hold (Figure 7(b)), the user is outside the rectangle (or on the boundary of R).

The enclosure condition can be privately evaluated by running the $sign(b - a)$ protocol for each of the four inequalities, as shown in the pseudocode of Figure 8. The user sends the server (lines 1-2) its public key E, as well as the encryption of messages m_x and m_y that encode the coordinates x_u and y_u as described in Section 4.1. The server will compute the ciphertext of the four subtraction operations (two for each of the x and y axes of coordinates), and blind them with random factors (lines 4-5). Note that, the protocol incurs only one round of communication. Furthermore, if the user wishes to evaluate enclosure with respect to more than one rectangle, the server can repeat the steps 4-5 for all rectangles, but the number of communication rounds does not increase (although the communication cost from the server to the user increases linearly to the number of rectangles).

In practice, spatial coordinates are represented as floating point numbers, either in single (32-bit) or double (64-bit) precision. On the other hand, Paillier encryption requires the use of positive integers alone. Nevertheless, the message space $\mathbb{Z}_N$ is large enough to accommodate even the most demanding application requirements with respect to coordinate precision. During the protocol execution, floating point values are converted to fixed precision. For instance, assume that the spatial data domain is $[0, 10^6]^2$ and 6 decimal points are required. Then, $2 \cdot \lfloor \log(10^6) \rfloor = 34$ bits are sufficient

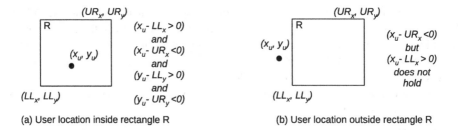

(a) User location inside rectangle R   (b) User location outside rectangle R

Fig. 7. Arithmetic Conditions to Determine Point-Rectangle Enclosure

---

Private Point-Rectangle Enclosure

Input: user location $p = (x_u, y_u)$, server rectangle $R = (LL_x, LL_y, UR_x, UR_y)$

Output: *true* if $p \in R$, *false* otherwise

1.	Client:	$m_x = N - x_u, m_y = N - y_u$
2.		Send $E, E(m_x), E(m_y)$ to the server
3.	Server:	Generate random numbers r'_x, r'_y, r''_x, r''_y
4.		$E(m'_x) = (E(m_x) \cdot E(LL_x))^{r'_x}, E(m''_x) = (E(m_x) \cdot E(UR_x))^{r''_x}$
5.		$E(m'_y) = (E(m_y) \cdot E(LL_y))^{r'_y}, E(m''_y) = (E(m_y) \cdot E(UR_y))^{r''_y}$
6.		Send $E(m'_x), E(m'_y), E(m''_x), E(m''_y)$ to the client
7.	Client:	if ($(0 \leq m''_x \leq M \cdot N')$ **and** $(N - M \cdot N' \leq m'_x \leq N - 1)$ **and**
8.		$(0 \leq m''_y \leq M \cdot N')$ **and** $(N - M \cdot N' \leq m'_y \leq N - 1)$)
9.		**return true**
10.		**else**
11.		**return false**

Fig. 8. Protocol for Private Evaluation of Point-Rectangle Enclosure

for this representation, much lower than the magnitude of N. This leaves a very large domain for the values of the random blinding factors.

5   Hybrid Protocol for Nearest-Neighbor Query Processing

We introduce a technique for processing PIR requests with respect to dynamically-generated query CRs. This method overcomes the drawbacks of [5] (discussed in Section 2), which performs PIR with respect to the entire POI dataset D. In the hybrid approach, the server knows that the user is located inside query CR Q, and therefore it can return a query result which discloses fewer POI and incurs less overhead.

A naive approach to restrict the set of POI included in the PIR protocol would work as follows: first, the server determines the set P_Q of POI that are located inside Q. Next, the points in P_Q are bulk-loaded into a two-level spatial index. Finally, the PIR retrieval is performed as in [5] with respect to the obtained index. There are several drawbacks of this approach: first, the index must be built on-line, which is time consuming. Second, although the number of disclosed POI is reduced from $\sqrt{|D|}$ to $\sqrt{|P_Q|}$, the resulting POI count can still be quite large, and it depends on the query Q (hence, it is not constant). Third, the root node of the index is sent in plain-text to the user. This discloses excessive information about the distribution of POI, since the minimum bounding rectangles (MBRs) of the leaf nodes may be small in size (especially if Q is not very large). The proposed hybrid technique addresses all these limitations.

The requirement of a two-level index restricts the flexibility in determining customized results for dynamic query CRs. We employ a multi-level index structure (computed off-line) that can efficiently find at run-time the leaf nodes that intersect query Q. Furthermore, we choose an index structure that strictly bounds the leaf node cardinality below a threshold F. Another important factor in developing the index structure is the fact that the cryptographic protocol of Section 4 allows private evaluation of point-rectangle inclusion, but not distance evaluation. This is a direct consequence of

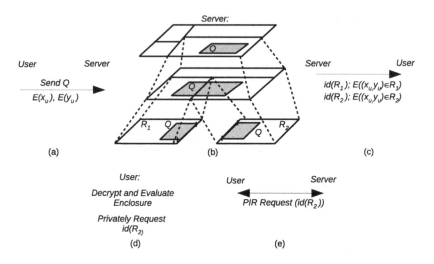

Fig. 9. Hybrid Technique Overview

protecting the location of the POI. In order to ensure query correctness (i.e., that at least one of the leaf nodes includes the user location) we employ a space-partitioning index, rather than a data-partitioning one. We provide more details about the indexing structure used in Section 5.1.

Figure 9 gives an overview of the entire query processing protocol. In step *(a)*, the user sends to the server the CR Q, as well as the encrypted user coordinates $E(x_u)$ and $E(y_u)$. The server processes a range query with parameter Q (step *(b)*) and identifies all leaf nodes (in this case, R_1 and R_2) that intersect Q. The server also executes the private point-rectangle evaluation protocol (Section 4) and sends back to the user (step *(c)*) tuples $(id(R_i); E((x_u, y_u) \in R_i))$, i.e., a rectangle identifier and the encrypted result of enclosure evaluation[8]. Next, in step *(d)*, the user decrypts the enclosure evaluation results and determines the identifier of the leaf node[9] that encloses (x_u, y_u), in this case R_2. Finally, the user and the server engage in a PIR round to retrieve the contents of R_2 (step *(e)*). For clarity of presentation, we have highlighted each step individually. However, there are only two communication rounds, as in the case of [5].

5.1   Indexing Structure

The choice of POI indexing structure is very important to the objectives of minimizing the POI disclosure and reducing query processing overhead. We consider a structure reminiscent of k-d-trees [24], which recursively cuts the space based on the number of data points in each partition. However, as opposed to k-d-trees, we do not require partition cuts to intersect data points. Furthermore, we do not restrict the axis of the cut

---

[8] Note that, if disclosing the number of leaf nodes that intersect Q represents a privacy concern for the database, the server can include randomly generated rectangles (that do not intersect Q) in the enclosure evaluation phase, without affecting correctness.

[9] Due to the non-overlapping indexing of POI, exactly one rectangle will enclose the user.

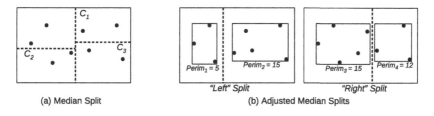

(a) Median Split   (b) Adjusted Median Splits

Fig. 10. Split Heuristic

at each step, and we use a more advanced split heuristic that factors criteria such as the perimeter of resulting partitions.

Consider the example of Figure 10(a), where the data is split according to median cut C_1, resulting in two sub-sets of equal cardinality (four points each). Assume that the node capacity is $F = 3$. Two additional splits are performed according to cuts C_2 and C_3, resulting in four leaf nodes of two points each. The median split has two drawbacks: first, the number of POI retrieved by the user is less than the allowed value 3, which may decrease the result accuracy (recall from Section 2 that we support approximate NN queries). Second, there are a total of four leaf nodes, although the original 8 points could be split into $\lfloor 8/3 \rfloor = 3$ nodes. A larger number of leaf nodes increases the cost of the point-rectangle enclosure evaluation.

We employ a variation of the median split that controls tightly the cardinality of leaf nodes. Given the cardinality c of the current partition, we ensure that at least one of the resulting partitions is a multiple of F. If this requirement is met at each cut, the amount of fragmentation (which is the reason why the median split under-performed) is considerably reduced. Consider Figure 10(b): there are two candidate splits across the x axis, *Left* and *Right*. *Left* places $\lfloor c/2/F \rfloor \cdot F$ points to the left of the cut axis and $c - \lfloor c/2/F \rfloor \cdot F$ to the right, whereas *Right* places $(\lfloor c/2/F \rfloor + 1) \cdot F$ points to the left and $c - (\lfloor c/2/F \rfloor + 1) \cdot F$ to the right. For each of these candidates, a *benefit* metric is evaluated, which measures the sum of perimeters[10] for the minimum bounding rectangles of points in each partition. The candidate that minimizes the sum of perimeters (in the example the *Left* split) is chosen[11]. A similar evaluation of candidate splits is performed for the y axis. Figure 11 shows the pseudocode of the proposed *NodeSplit* technique for data partitioning. *NodeSplit* considers both the x and y axes, and chooses the split with the largest benefit (i.e., minimum sum of perimeters). Data points in the initial node U are sorted with respect to the selected axis (line 1). Next, the costs of the candidate splits $Cost_{left}$ and $Cost_{right}$ are evaluated as the sum of perimeters for the points in each region (lines $2 - 5$). The split position that yields the lowest cost is chosen (lines $6 - 9$). The computational complexity of the index creation is $O(|D| \log |D|)$, where $|D|$ is the dataset cardinality.

---

[10] A similar benefit metric has been used for R-trees [24].

[11] Although the MBRs are used in the benefit evaluation, the resulting partition is not pruned to the MBR, due to the requirement that the index must cover the entire data space.

NodeSplit

Input: Initial Node U, Leaf Cardinality Threshold F

Output: Two children nodes U_1 and U_2

/* $x - axis$ */

1.   sort points in U increasingly according to x coordinate

      /* We use array notation to refer to the ponts in U */

      /* "$Left$" split* /

2. $Count_{left} = \lfloor |U|/2/F \rfloor \cdot F$

3. $Cost_{left} = perimeter(MBR(\{U[1], \ldots, U[Count_{left}]\})) +$
                    $perimeter(MBR(\{U[Count_{left} + 1], \ldots, U[|U|]\}))$

      /* "$Right$" split* /

4. $Count_{right} = (\lfloor |U|/2/F \rfloor + 1) \cdot F$

5. $Cost_{right} = perimeter(MBR(\{U[1], \ldots, U[Count_{right}]\})) +$
                     $perimeter(MBR(\{U[Count_{right} + 1], \ldots, U[|U|]\}))$

6.   **if** $(Cost_{left} < Cost_{right})$

7. $U_1 = U[1 \ldots Count_{left}], U_2 = U[Count_{left} + 1 \ldots |U|]$

8.   **else**

9. $U_1 = U[1 \ldots Count_{right}], U_2 = U[Count_{right} + 1 \ldots |U|]$

/* Repeat steps $1 - 9$ for $y - axis$ and choose the lowest cost*/

Fig. 11. Heuristic for Index Partitioning

6   Experiments

We evaluate experimentally the proposed hybrid method with respect to the effectiveness in controlling the disclosed POI and the incurred computational and communication overhead. We use a real database with points of interest: the Sequoia set[12] with $62,556$ data points (Figure 12). We consider values of F, the threshold for disclosed POI, in the range $20 - 80$, and we randomly generate square-shaped cloaking regions Q with side between 1% and 10% of the dataspace side. Recall that, a larger CR provides stronger privacy for the user. For each experimental run, we randomly generate 1000 user queries. The size of the modulus N used in the cryptographic protocols for PIR retrieval and private enclosure evaluation is 768 bits. The experiments were run on an Intel P4 3.0 GHz machine with 1GB of RAM.

First, we evaluate the amount of protection offered to the database by the hybrid method, in comparison with location cloaking (label *CR-only*) and the pure-PIR technique (label *PIR-only*), for varying CR size. We consider approximate NN queries. For fairness of comparison, only candidate POI inside Q are returned by the CR-only method (this decreases the number of disclosed POI compared to the exact methods in [3,4]). Figure 13 shows that the CR-only technique discloses an excessive amount of POI, especially as the CR size grows larger. Therefore, the privacy of the database is sacrificed for the sake of user privacy. The PIR-only method does not use CRs, and always discloses approximately 250 POI (square root of database cardinality). Note that, the hybrid method controls strictly the number of disclosed POI in the narrow band

---

[12] http://www.rtreeportal.org

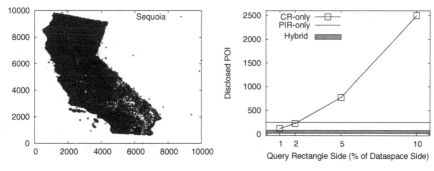

Fig. 12. Sequoia Dataset

Fig. 13. Number of Disclosed POI

$20 - 80$, up to one order of magnitude superior to PIR-only, and up to two orders of magnitude better than the CR-only method. This improvement is obtained for the same level of privacy offered to the user by the CR-only method (i.e., same CR sizes).

For the rest of the experiments, we compare the performance between the hybrid and the PIR-only methods with respect to computational and communication overhead incurred by query processing. We do not include the CR-only method any further in the head-to-head comparison, since it offers virtually no amount of protection for the database. It is, however, well-understood [5] that CR-only techniques are more efficient in terms of overhead, because they do not make use of cryptographic elements. In general, the processing time is expected to take on average around one second [4]. Based on the number of POI returned obtained in the previous experiment (which is the only communication factor in CR-only methods), the communication overhead is expected to be around 20 kilobytes.

Similar to previous work [4,5], we consider that the set of POIs fits in memory, and that the processing time is dominated by CPU time. This is a reasonable assumption, especially since the compared methods use heavily cryptographic transformations, which are not I/O bound. Note that, in [5] optimizations based on parallel processing are proposed to improve execution time. Such optimizations are directly applicable for the hybrid methods as well. In our tests, we run both methods on a single-CPU machine,

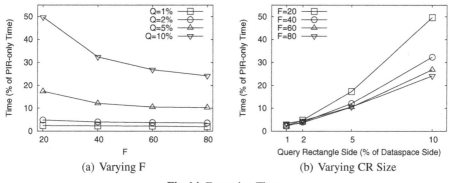

(a) Varying F

(b) Varying CR Size

Fig. 14. Execution Time

and we report the hybrid method execution time as the percentage of the time incurred by the PIR-only method.

Figure 14(a) shows the execution time when varying the POI disclosure bound F. In the worst case, the hybrid method is twice as fast as the PIR-only method. On the other hand, for all CR sizes with less than 10% of the dataspace side, the hybrid method is at least 5 times faster. The decreasing trend with F can be explained as follows: since the size of query Q is fixed, the number of POI included in the PIR step does not vary with F. On the other hand, a smaller F results into more rectangles for which the private point-rectangle enclosure evaluation protocol must be performed, leading to an increase in processing time. In absolute values, the execution time of the hybrid method on a single CPU requires roughly 0.5 sec for queries spanning 2% of the dataspace, and between 1.2 and 1.9 sec for queries spanning 5% of the dataspace. Figure 14(b) shows the variation of execution time with query CR size. A larger query window translates into more leaf nodes being included in the enclosure evaluation protocol. Furthermore, a larger number of data points are considered in the PIR retrieval phase. Hence the increase in processing time.

Figure 15 presents the result of communication overhead, also expressed as a percentage of the overhead incurred by the PIR-only method. In the worst case, the bandwidth consumption of the hybrid method is 30% that of PIR-only, whereas the overall improvement can be as high as 20 times. The cost increases with F (Figure 15(a)) since more POI are retrieved from the server. For varying size of CR Q (Figure 15(b)), the number of retrieved POIs remains unchanged as Q grows, but the number of leaf nodes considered in the point-rectangle enclosure protocol increases, hence the higher communication overhead. In absolute values, the communication cost of the hybrid method is in the range $40 - 140$KB for queries spanning 2% of the dataspace, and $100 - 280$KB for queries spanning 5% of the dataspace.

Finally, Table 1 shows the accuracy of NN results. Since both compared methods are approximative, the closest POI reported to the user may differ from the actual NN POI. Accuracy is measured as the average difference between the user-to-reported-NN distance and the user-to-actual-NN distance. The value is then normalized, and expressed as a percentage of dataspace side. Since the data points that are returned to the user depend only on the leaf node that encloses the user location, the accuracy of the hybrid

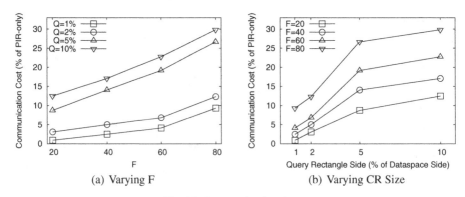

(a) Varying F   (b) Varying CR Size

Fig. 15. Communication Cost

Table 1. Query Result Accuracy

Threshold F	Hybrid Accuracy	PIR-only Accuracy
20	0.014%	
40	0.011%	0.003%
60	0.007%	
80	0.005%	

method is independent of the query size. The only factor that influences accuracy is the POI disclosure threshold F. The accuracy of the PIR-only method is better, since it returns an excessive amount of POI to the user. On the other hand, in absolute values, the hybrid method achieves good precision. For instance, assume a city area of 50×50 kilometers. An approximation error of 0.014% corresponds to a distance of 28 meters. This is a reasonable error, considering that in practice, positioning devices report locations with accuracy of $10 - 20$ meters.

7   Conclusions

This paper proposed a hybrid technique for private location-based queries which provides protection for both the users and the service provider. To our knowledge, this is the first work to consider the protection of the POI database. Furthermore, the proposed technique is efficient in practice. In future work, we plan to study efficient methods for exact nearest-neighbor queries. We also plan to extend our work to support more complex types of queries, e.g., skyline.

References

1. Gruteser, M., Grunwald, D.: Anonymous Usage of Location-Based Services Through Spatial and Temporal Cloaking. In: Proc. of USENIX MobiSys. (2003)
2. Gedik, B., Liu, L.: Location Privacy in Mobile Systems: A Personalized Anonymization Model. In: Proc. of ICDCS, pp. 620–629 (2005)
3. Mokbel, M.F., Chow, C.Y., Aref, W.G.: The New Casper: Query Processing for Location Services without Compromising Privacy. In: Proc. of VLDB (2006)
4. Kalnis, P., Ghinita, G., Mouratidis, K., Papadias, D.: Preserving Location-based Identity Inference in Anonymous Spatial Queries. IEEE TKDE 19(12) (2007)
5. Ghinita, G., Kalnis, P., Khoshgozaran, A., Shahabi, C., Tan, K.L.: Private Queries in Location Based Services: Anonymizers are not Necessary. In: SIGMOD (2008)
6. Kido, H., Yanagisawa, Y., Satoh, T.: An anonymous communication technique using dummies for location-based services. In: International Conference on Pervasive Services (ICPS), pp. 88–97 (2005)
7. Yiu, M.L., Jensen, C., Huang, X., Lu, H.: SpaceTwist: Managing the Trade-Offs Among Location Privacy, Query Performance, and Query Accuracy in Mobile Services. In: International Conference on Data Engineering (ICDE), pp. 366–375 (2008)
8. Cheng, R., Zhang, Y., Bertino, E., Prahbakar, S.: Preserving User Location Privacy in Mobile Data Management Infrastructures. In: Danezis, G., Golle, P. (eds.) PET 2006. LNCS, vol. 4258, pp. 393–412. Springer, Heidelberg (2006)

9. Chow, C.Y., Mokbel, M.F.: Enabling Private Continuous Queries for Revealed User Locations. In: Papadias, D., Zhang, D., Kollios, G. (eds.) SSTD 2007. LNCS, vol. 4605, pp. 258–275. Springer, Heidelberg (2007)

10. Gruteser, M., Liu, X.: Protecting Privacy in Continuous Location-Tracking Applications. IEEE Security and Privacy 2, 28–34 (2004)

11. Damiani, M., Bertino, E., Silvestri, C.: PROBE: an Obfuscation System for the Protection of Sensitive Location Information in LBS. Technical Report 2001-145, CERIAS (2008)

12. Khoshgozaran, A., Shahabi, C.: Blind Evaluation of Nearest Neighbor Queries Using Space Transformation to Preserve Location Privacy. In: Papadias, D., Zhang, D., Kollios, G. (eds.) SSTD 2007. LNCS, vol. 4605, pp. 239–257. Springer, Heidelberg (2007)

13. Chor, B., Goldreich, O., Kushilevitz, E., Sudan, M.: Private information retrieval. In: IEEE Symposium on Foundations of Computer Science (1995)

14. Kushilevitz, E., Ostrovsky, R.: Replication is NOT Needed: SINGLE Database, Computationally-Private Information Retrieval. In: FOCS (1997)

15. Flath, D.E.: Introduction to Number Theory. John Wiley & Sons, Chichester (1988)

16. Atallah, M.J., Du, W.: Secure multi-party computational geometry. In: Dehne, F., Sack, J.-R., Tamassia, R. (eds.) WADS 2001. LNCS, vol. 2125, pp. 165–179. Springer, Heidelberg (2001)

17. Luo, Y., Huang, L., Zhong, H.: Secure two-party point-circle inclusion problem. J. of Computer Science and Technology 22(1), 88–91 (2007)

18. Goldreich, O., Micali, S., Wigderson, A.: How to play any mental game. In: Proceedings of ACM Symposium on Theory of Computing (STOC), pp. 218–229 (1987)

19. Chow, C.Y., Mokbel, M.F., Liu, X.: A Peer-to-peer Spatial Cloaking Algorithm for Anonymous Location-based Service. In: GIS, pp. 171–178 (2006)

20. Ghinita, G., Kalnis, P., Skiadopoulos, S.: PRIVE: Anonymous Location-based Queries in Distributed Mobile Systems. In: WWW (2007)

21. Ghinita, G., Kalnis, P., Skiadopoulos, S.: MobiHide: A Mobile Peer-to-peer System for Anonymous Location-based Queries. In: Papadias, D., Zhang, D., Kollios, G. (eds.) SSTD 2007. LNCS, vol. 4605, pp. 221–238. Springer, Heidelberg (2007)

22. Paillier, P.: Public-key cryptosystems based on composite degree residuosity classes. In: Stern, J. (ed.) EUROCRYPT 1999. LNCS, vol. 1592, pp. 223–238. Springer, Heidelberg (1999)

23. Atallah, M.J.: Algorithms and Theory of Computation Handbook. CRC Press, Boca Raton (1998)

24. de Berg, M., van Kreveld, M., Overmars, M., Schwarzkopf, O.: Computational Geometry: Algorithms and Applications, 2nd edn. Springer, Heidelberg (2000)

Spatial Cloaking Revisited: Distinguishing Information Leakage from Anonymity

Kar Way Tan, Yimin Lin, and Kyriakos Mouratidis

Singapore Management University
School of Information Systems
80 Stamford Road, Singapore 178902
{karway.tan.2007,yimin.lin.2007,kyriakos}@smu.edu.sg

Abstract. Location-based services (LBS) are receiving increasing popularity as they provide convenience to mobile users with on-demand information. The use of these services, however, poses privacy issues as the user locations and queries are exposed to untrusted LBSs. *Spatial cloaking* techniques provide privacy in the form of k-anonymity; i.e., they guarantee that the (location of the) querying user u is indistinguishable from at least k-1 others, where k is a parameter specified by u at query time. To achieve this, they form a group of k users, including u, and forward their minimum bounding rectangle (termed *anonymizing spatial region*, ASR) to the LBS. The rationale behind sending an ASR instead of the distinct k locations is that exact user positions (querying or not) should not be disclosed to the LBS. This results in large ASRs with considerable dead-space, and leads to unnecessary performance degradation. Additionally, there is no guarantee regarding the amount of location information that is actually revealed to the LBS. In this paper, we introduce the concept of *information leakage* in spatial cloaking. We provide measures of this leakage, and show how we can trade it for better performance in a tunable manner. The proposed methodology directly applies to centralized and decentralized cloaking models, and is readily deployable on existing systems.

1   Introduction

The increasing trend of location-aware mobile devices, such as GPS-enabled mobile phones and palm-tops, has lead to a growing market of location-based services (LBS). Users of these services query the LBS to retrieve information about data (points of interest, POI) in their vicinity. The main issue arising in this environment is that the users reveal their locations to the untrusted LBS. In turn, this information may lead to the identity of the users (a process generally termed re-identification) through publicly available information, physical observation, mobile device tracking, etc [1]. The nature of the POIs (e.g., HIV clinics) may disclose sensitive personal information to the LBS, or lead to receipt of unsolicited targeted advertisements (e.g., if the POIs are providers of particular services or products).

N. Mamoulis et al. (Eds.): SSTD 2009, LNCS 5644, pp. 117–134, 2009.

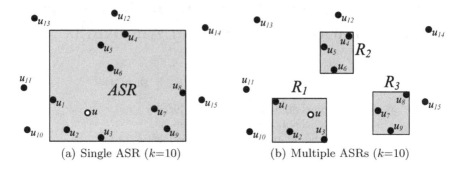

(a) Single ASR (k=10)   (b) Multiple ASRs (k=10)

Fig. 1. Anonymity versus information leakage

To solve the above problem, *spatial cloaking* methods replace the user location with an *anonymizing spatial region* (ASR) prior to sending his/her query to the LBS. The ASR is typically an axis-parallel rectangle[1] that encloses the querying user u and at least $k - 1$ additional users; the set of these k users is called the anonymizing set (AS) of u. Parameter k is specified by u at query time and reflects the degree of anonymity required. Figure 1(a) shows a cloaking example where the query originator u, shown as a hollow point, requests for 10-anonymity (i.e., $k = 10$). The figure shows the computed ASR, assuming that the anonymizing set of u additionally contains $u_1, ..., u_9$.

Depending on the assumed architecture, there exist centralized and decentralized ASR computation methods. In the first case, the ASR is created by a trusted server (the *anonymizer*) who maintains all user locations [2,3]. In the second case, the ASR is computed in a collaborative way by the users themselves (assuming that they are mutually trusted and communicate wirelessly) [4,5]. In either case, the ASR is subsequently sent to the LBS. The latter computes the candidate query answers (i.e., POIs that satisfy the user query) for any possible query position inside the ASR. The set of returned POIs is called *candidate set* CS, and is filtered by the anonymizer or the querying user (in the centralized and the decentralized model, respectively) in order to retrieve the actual query result.

A principle underlying the general approach described above is that user locations should not be disclosed to the LBS. However, if the LBS already knew the locations of all users, there would be no need for an ASR. Anonymity would still be preserved by directly sending the exact user locations in the anonymizing set, e.g., forwarding to the LBS all 10 user locations in Figure 1(a) instead of the ASR. This approach not only would honor anonymity, but it would also lead to a smaller CS and thus a better performance; as we explain later, the main factor affecting the overall performance of the system is the size of the CS (i.e., the number of POIs inside the CS). Furthermore, the cost in dollars may be lower, since in commercial LBSs, the amount paid by the user is proportional to the number of POIs provided by the LBS.

---

[1] Circular ASRs have also been studied, without however resulting in performance benefits over rectangular ASRs [2].

Following the above reasoning, one could argue that even if the LBS did not know the user locations, we could break the ASR into smaller sub-ASRs in order to avoid CS results that correspond to dead-space (i.e., regions of the ASR that contain no user). This technique is exemplified in Figure 1(b), where three sub-ASRs (R_1, R_2, and R_3) are used instead of one, to deal with the same querying user u and the same anonymity requirement $k = 10$. According to the principles and objectives of existing spatial cloaking approaches, this multiple sub-ASR method both (i) preserves the anonymity of the querying user, and (ii) does not disclose any user locations. Furthermore, it leads to a smaller CS and hence to a better performance. However, this approach somehow reveals more location information than a single ASR, because the LBS would acquire more precise knowledge about where users are located. Specifically, in Figure 1(a) the LBS would infer that there are some users inside the ASR.[2] On the other hand, the multiple sub-ASRs in Figure 1(b) would disclose additional and more precise information, because now the LBS infers that each of the three rectangles R_1, R_2, and R_3 contains some users.

In this paper, we define the concept of *information leakage* to capture the location information revealed by spatial cloaking, and provide measures to quantify it. We propose the *Information Leakage-aware Cloaking* (ILC) methodology that incorporates this notion into existing techniques and enables control over the trade-off between performance/cost and information leakage. In particular, we forward multiple sub-ASRs to the LBS, constructed in a way which guarantees that anonymity is preserved and that information leakage does not exceed the amount tolerable by the users/system. Note that this contrasts with existing methods, where there is no control over the amount of location information revealed to the LBS. Our method is readily applicable to existing spatial cloaking systems, and works transparently in both the centralized and the decentralized model.

The rest of the paper is structured as follows. Section 2 surveys related work. Section 3 states out assumptions and objectives, and presents definitions central to our work. Section 4 presents our methodology and shows how it can be incorporated into available spatial cloaking systems. Section 5 experimentally evaluates our approach, studying the trade-offs between performance and information leakage. Finally, Section 6 concludes the paper with directions for future work.

2   Background and Related Work

k-anonymity [6,7] has been used for publishing sensitive data (e.g., medical records) in a way that each record is indistinguishable from at least k-1 others. In the context of location-based services, spatial k-anonymity is achieved by obfuscating the locations of querying users so that they cannot be identified with a probability higher than $1/k$. Location obfuscation is performed by a *cloaking algorithm*. Most systems adopt the *centralized architecture* [2,3]. In this setting, the cloaking algorithm is executed by a trusted third party (anonymizer), which is regularly being updated with the most current user locations. On the other

---

[2] Note that the LBS does not know the value of k, but only sees the query ASR.

hand, in a *decentralized architecture* [4,5], no anonymizer is required. Instead, the users collaboratively construct ASRs communicating via an overlay network (e.g., a peer-to-peer system). Our proposed methodology affects primarily the cloaking process and as such it can be applied to both the centralized and the decentralized architectures. Section 2.1 reviews existing cloaking techniques, and Section 2.2 discusses processing techniques for cloaked queries. Section 2.3 describes alternative location privacy models.

2.1   Cloaking Techniques

Interval Cloak [1] is one of the first cloaking techniques. The anonymizer indexes the users with a Quad-tree [8]. To form an ASR for user u, *Interval Cloak* descends the Quad-tree up to the topmost node that contains at least k users (including u). The extent of this node is returned as the ASR. In Figure 2, if user u_1 issues a query with $k = 2$, *Interval Cloak* will search till quadrant $[(0,0),(1,1)]$ which contains less than 2 users. Then, it will backtrack for one level, and return the parent quadrant $[(0,0),(2,2)]$ as the ASR. The returned quadrant may contain much more than k users, burdening query processing at the LBS.

Casper Cloak [3] is similar to *Interval Cloak*, with two major differences. First, *Casper Cloak* identifies and accesses the leaf level of the Quad-tree directly through the use of a hash table. Second, instead of immediately backtracking to the parent quadrant, it first checks the two neighboring quadrants to see if their combination with the user quadrant contains k (or more) users. In Figure 2, if u_1 issues a query with $k = 2$, *Casper Cloak* first checks the neighboring quadrants $[(0,1),(1,2)]$ and $[(1,0),(2,1)]$. If combination with one of them results in k users, then this composite rectangle is returned as the ASR. In this example, rectangle $[(0,0),(1,2)]$ is returned.

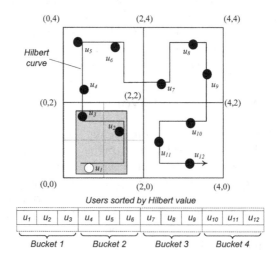

Fig. 2. Cloaking example

[2] shows that there are situations where anonymity is breached with the above methods, and proves that absolute anonymity can be guaranteed if *reciprocity* is honored. Reciprocity is defined as follows:

Definition 1. *Let $AS_k(u)$ be the anonymizing set of u for anonymity degree k. A cloaking algorithm satisfies reciprocity iff (i) $AS_k(u)$ contains at least k users, and (ii) for every user u' in $AS_k(u)$ it holds that $AS_k(u') \equiv AS_k(u)$ (i.e. all users in $AS_k(u)$ have the same AS).*

[2] proposes *Hilbert Cloak*, an algorithm that satisfies this property. The users are sorted according to the Hilbert space-filling curve [9]. The sorted sequence is equally divided into buckets of k consecutive users. AS is formed by the bucket that contains the querying user u. The reported ASR is computed as the minimum bounding rectangle of the AS. In Figure 2, if u_1 issues a query with $k = 3$, then 4 Hilbert buckets are created as shown at the bottom of the figure. User u_1 belongs to the first bucket, and its AS includes u_1, u_2, u_3. The derived ASR is the shaded bounding box. Due to its simplicity, *Hilbert Cloak* has been applied to decentralized systems too [4].

2.2   Query Processing at the LBS

The two most common spatial queries are the *Range Query* and the *Nearest-Neighbor (NN) Query*. Given only an ASR and the query type/parameters, the LBS needs to search for the POIs that satisfy the query for any possible user location within the ASR. Typically, the LBS stores the POIs in secondary storage, indexed by an R-tree [10,11]. If an R-range query is given, the LBS computes CS as the union of all POIs that fall inside the ASR or are within distance R from its boundary. In the example of Figure 3(a), the LBS expands the ASR (shown with a dashed contour) by R, and performs an ordinary range query. The CS contains P_1, P_2, and P_3.

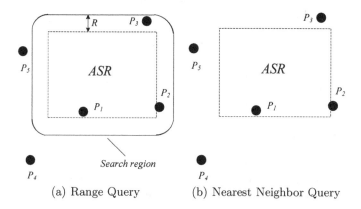

(a) Range Query   (b) Nearest Neighbor Query

Fig. 3. Types of Query Processing at LBS

If a K-NN query is given, the CS contains the union of K nearest POIs[3] to any point within the ASR. To derive the CS for a NN query (i.e., $K = 1$) in Figure 3(b), the LBS needs to retrieve (i) all objects located inside the ASR (i.e., P_1) and (ii) the NN of any location along the boundary of the ASR (i.e., P_2, P_3, P_5). The latter component is processed using the *linear NN* method of [12] for each of the 4 edges of the ASR; the input to this method is one or more line segments, for which NNs are found in a single R-tree traversal.

2.3   Alternative Location Obfuscation Approaches

There exist alternative location privacy approaches. [13] ignores non-querying users, and instead groups only querying users among themselves. [14] proposes a location privacy method specifically for approximate NN processing. In [15,16] the user u forwards to the LBS a set of dummy locations in addition to his/her own. In [17], the user sends only a fake location to the LBS and incrementally retrieves its nearest neighbors. [18] applies *private information retrieval* to process NN queries. The above methods cannot ensure k-anonymity, are limited in the type of queries supported, and/or incur prohibitively high query processing cost.

3   Preliminaries

In this section we state the assumptions underlying our approach (in Section 3.1), our central observation and our design objectives (in Section 3.2). In Table 1 we list frequently used acronyms/symbols, and their interpretation.

Table 1. Description of acronyms and symbols

Term	Description
LBS	Location-Based Service
POI	Point Of Interest
ASR	Anonymizing Spatial Region
AS	Anonymizing Set
CS	Candidate Set
U	The set of users in the system
k	Anonymity parameter
m	Strictness on information leakage parameter
IL	Degree of information leakage

3.1   Assumptions

The *Information Leakage-aware Cloaking* (ILC) methodology applies to (and is orthogonal to the choice between) the centralized and the decentralized cloaking models. However, to avoid confusion and for the sake of tangibility, we assume the

---

[3] Note that parameter K used here is different from the k-anonymity requirement used in the anonymity context.

centralized model in our examples unless otherwise stated. We focus on spatial (i.e., 2-dimensional) user and POI locations. Similar to existing spatial cloaking systems, we consider that the users (forming set U) are mobile, and constantly update the anonymizer with their most recent locations. The set of POIs is static, and it is indexed by a disk-resident R-tree. Note that ILC deals mostly with the cloaking part and, as such, indexing at the LBS side or the mobility of the POIs has little impact on it; alternative contexts can be dealt with in a straightforward manner. We focus on the most common spatial queries, that is, *snapshot*[4] range and nearest neighbor (NN) queries. However, our technique can be directly incorporated into the model of [19] to capture *continuous* queries too; this extension is discussed in Section 4.4. Regarding the communication channel we assume that:

1. The connection between the querying user u and the anonymizer (in the centralized model) or among users (in decentralized systems) is encrypted and secure.
2. The communication channel between the anonymizer (or the users, in a decentralized system) and the LBS needs not be secure.

Point 1 above implies that eavesdropping is not possible for the LBS, and that k is unknown to it. Point 2 practically implies that the LBS is not the only possible adversary, but our method should ensure anonymity and controlled information leakage versus any malicious entity that may intercept the cloaked queries (on their way to the LBS). For simplicity, we consider the LBS as the adversary, but ILC is safe against any of the aforementioned types of entities. Note that ensuring the authenticity of the POIs reported to the users is outside the scope of this paper; result verification methods (e.g., [20,21,22]) could be used in conjunction with ILC to detect any man-in-the-middle tampering with the results.

3.2 Main Observation and Design Objectives

Our motivating observation is that information leakage requirements have always existed in spatial anonymity approaches, but they have never been identified and treated with independently. Specifically, the methods described in Section 2.1 assume two kinds of adversaries:

- **User-aware adversaries:** Adversaries of this type know the user locations. To achieve anonymity against such adversaries, it suffices to send to the LBS the exact positions of all users in the AS. If the AS is formed in a reciprocal way, anonymity is guaranteed. By definition, this would incur the smallest possible CS for the specific AS.
- **User-unaware adversaries:** Adversaries of this type do not know the user locations. Concealing the position of the querying user u from such adversaries is easy and can be done arbitrarily (e.g., by sending to the LBS a

---

[4] Term snapshot refers to queries that are evaluated once and then terminate. It is used to distinguish from continuous evaluation where the queries are standing and request constant updating of their results.

rectangle that encloses u). What is important here is that user-unaware adversaries stay unaware of exact user locations (be them querying or not).

What has been implicitly assumed by previous systems is that there exist both kinds of the above adversaries at the same time (plus possibly adversaries with partial user location knowledge), and that they must be dealt with collectively; the AS was "masked" with a minimum bounding rectangle, so that exact user locations are not revealed to user-unaware adversaries, while anonymity is ensured even against user-aware ones. The sacrifice made in this approach is that the CS contains more POIs (than sending the AS locations directly).

Our observations here is that two different concepts underlie this design principle (i.e., anonymity and information leakage), and that the anonymity-centric approach taken so far ensures anonymity, but fails to control or even to quantify the degree of location information disclosed to user-unaware adversaries. Thus, our first contribution is to provide a meaningful measure of information leakage, and then suggest a methodology (i.e., ILC) to control it; the twofold objective of ILC is to ensure user anonymity *and* guarantee no more than the permissible degree of information leakage. In terms of anonymity, we adopt its strict, reciprocity-based definition (described in Section 2.1); recall that reciprocity is a sufficient (though not necessary) condition to achieve strongly k-anonymous services [2]. We elaborate on the information leakage requirements in Section 4.

Subject to the degree of information leakage tolerable, our second objective is to reduce the CS size; this is the primary factor that (i) determines the dollar cost paid by the user to the LBS, and (ii) determines the end-to-end query response time. Regarding (i), commercial LBSs often charge by the amount of information provided, i.e., the number of POIs returned. In terms of end-to-end response time, in the centralized model, the experiments of [2] and [3] indicate that the major performance factor is the I/O time spent at the LBS and secondarily the communication cost. Both these costs are proportional to the size of the CS.[5] In the decentralized model, the CS (on its way from the LBS to the querying user u) must pass through the overlay user network, and it must subsequently be filtered by u to retrieve the actual query result. The communication and processing costs incurred prolong the end-to-end time, but also consume the (typically scarce) power resources of the user devices. Thus, our aim is to exploit any leeway in terms of information leakage to reduce the CS size.

4   Information Leakage-Aware Cloaking

In this section we define a measure of information leakage and describe the ILC framework.

---

[5] The page accesses performed at the higher levels of the POI R-tree are minimal compared to the POI (leaf) level. That is, the I/O cost is roughly proportional to the CS size.

4.1   Measuring Information Leakage

Our approach is to control information leakage via a parameter m. This parameter is specified by the system (e.g., the anonymizer) as a requirement from the cloaking mechanism. We establish that:

Definition 2. *The strictness on information leakage of a cloaking algorithm is* m *iff any ASR (or sub-ASR) forwarded to the LBS contains at least m users.*

Intuitively, information leakage is inversely proportional to m, hence we quantify IL, the degree of information leakage as $\frac{1}{m}$. It holds that $\frac{1}{|U|} \le IL \le 1$, where $|U|$ is the total number of users in the system. Case $IL = 1$ (maximum information leakage) corresponds to $m = 1$, where exact user locations may be revealed to the LBS. Case $IL = \frac{1}{|U|}$ (minimum information leakage) corresponds to $m = |U|$, where a single ASR enclosing all users is sent to the LBS. In the situation illustrated in Figure 1(b), $m = 3$ and $IL = \frac{1}{3}$, because the smallest number of users contained in any sub-ASR is 3 (regardless of the fact that R_1 contains 4).

Regarding the rationale behind the IL measure, one may wonder why we do not express it in an absolute way as the maximum accuracy that the LBS would get about individual user locations. In other words, this alternative would require that each constructed ASR (or sub-ASR) would not be narrower than some threshold δ_x on the x axis and another threshold δ_y on the y axis. The reason for disqualifying this method is that it fails to capture the user distribution; the distribution can be easily estimated using publicly available information. For example, Figure 4 plots a real dataset of 25,000 locations in North America that could model our users. Illustrated rectangles ASR_1 and ASR_2 have the same extents (say δ_x and δ_y), but the first lies in a very sparse area, while the second includes numerous users (covering, for instance, the highly populated New York city). Clearly, ASR_1 reveals much more information to the LBS than ASR_2 regarding where users lie. Thus, we select a relative IL measure using parameter m.

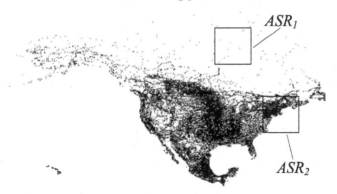

Fig. 4. Shortcomings of an absolute definition based on spatial precision

Another approach is to define IL according to the area of the ASR (or the minimum area of any sub-ASR constructed). Figure 5 demonstrates the weaknesses

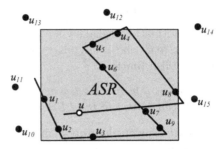

Fig. 5. Shortcomings of an area-based definition

of this approach with a counter-example. Here, after computing the AS of u (say, using *Hilbert Cloak*), we construct an ASR in the form of a poly-line, i.e., a set of connected line segments. The poly-line passes from all users in AS, and its turning points are selected in a randomized way so that they do not coincide with any user location. This ASR has zero area; under an area-based definition, this plain method would incur no information leakage[6]. However, it is obvious that this is not the case, as the location information revealed now is much more precise. Actually, the situation can be even worse if, for instance, the users move on a road network; the LBS could compute the user positions by retrieving the intersections between the ASR poly-line and the edges of the road network. We thus avoid an area-based definition, and adopt the one described in the beginning of the section.

Before presenting the ILC cloaking methodology, we need to clarify a few issues regarding the distinction between anonymity (defined by parameter k) and *IL* (defined by m):

- k and m implement different requirements and control different cloaking functions. However, a higher k implies higher privacy, just like a higher m implies lower information leakage. In this sense, increasing these parameters leads to "safer" cloaking in a general point of view.
- k is specified by each querying user individually, while m is a system-wide parameter (e.g., defined by the anonymizer); m is a global parameter, because it does no longer reflect individual user preferences, but the release of information from the system (as a whole) to the LBS. Note, however, that ILC can be easily adapted to contexts where it makes sense for m to be user-specific.
- Controlling information leakage does not violate anonymity (or reciprocity). In Figure 1(b), for example, if the AS is derived by a reciprocal anonymization algorithm, the LBS still cannot identify the querying user with a probability greater than $\frac{1}{k} = \frac{1}{10}$.
- Typically, m is considerably smaller than k, because the information that user u is at some location *and* asks a particular query (relating to anonymity)

---

[6] Furthermore, this approach would significantly reduce the CS size. Imagine an R-range query. The LBS would only return POIs within distance R from the ASR poly-line, leading to a much smaller CS than a traditional rectangular ASR.

is more sensitive than simply knowing that there is a user at location u (relating to information leakage). In the following, we focus on situations where $m < k$, but we also consider the rare scenario where $m \geq k$.

4.2 The Multiple ASR Approach

The main idea in ILC is to cloak the AS using multiple (sub-)ASRs, none of which contains less than m users. In addition to providing the desired degree of information hiding (i.e., keeping IL lower than its maximum permissible value), we attempt to reduce the CS size by limiting the dead-space within the ASRs. The general idea in this approach is similar to Figure 1(b). Note that ILC does not violate (or interfere with) anonymity, because it does not affect the AS itself.

Although ILC can be applied in conjunction with other cloaking methods, here we choose to incorporate it into *Hilbert Cloak* since it is the current state-of-the-art. The AS for a user u (as output by *Hilbert Cloak*) has the form of a Hilbert-sorted list. ILC splits the AS (Hilbert-sorted) list into m-buckets. Each bucket has exactly m users, except for the last one which contains from m up to $2m - 1$ users. ILC returns a sub-ASR for each m-bucket, by computing its minimum bounding rectangle. The pseudo-code for sub-ASR generation is illustrated in Figure 6. Figure 7 shows an example where $k = 10$ and $m = 3$. The user order on the Hilbert curve is shown at the bottom of the figure. The AS output by *Hilbert Cloak* contains users $u_1, u_2, ..., u_{10}$; *Hilbert Cloak* would return their bounding box as the ASR. Instead, ILC breaks the AS into m-buckets (where $m = 3$), and creates one sub-ASR for each of them. This leads to sub-ASRs SA_1, SA_2, SA_3 shown striped. Observe that the last m-bucket/sub-ASR contains 4 users (i.e., more than $m = 3$).

Algorithm. Creating IL-aware sub-ASRs

```
1. Given a Hilbert-sorted AS
2. Split the AS into m-buckets
3. For each m-bucket do
4.    Create a sub-ASR as the minimum bounding rectangle of users inside
5. Return the list of all sub-ASRs computed in Step 4
```

Fig. 6. Algorithm for deriving sub-ASRs in ILC

Note that, given the AS, the construction of sub-ASRs is not concerned with which of the users was the querying one, and thus no reciprocity requirement underlies the IL-related handling. Therefore, ILC could work with any other sub-ASR creation method, subject to the IL constraint. Our objective is not to propose the best such algorithm, as similar bucketization problems are well-studied in the spatial database and computational geometry literatures. We use the aforementioned Hilbert-based technique merely to provide an example where ILC is readily deployable on an existing system that uses the *Hilbert Cloak* method

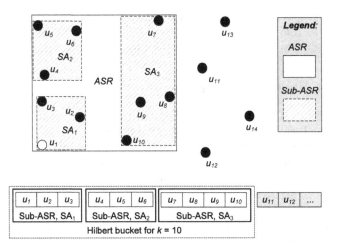

Fig. 7. ILC example

(the system may be centralized or decentralized, with sub-ASR computation performed by the anonymizer or the users, respectively). It is worth mentioning that we did experiment with alternative partitioning techniques, such as STR [23] and the R^+-tree splitting algorithm [24], which however performed similar to the Hilbert-based method.

In the above discussion we assumed that $m < k$. However, in the rare case where $m \geq k$, a single ASR is returned. Moreover, in situations where $m > k$, if the ASR contains fewer than m users[7], we enlarge it in a greedy way so that it encloses m users with the minimum area increase. Greedy enlargement is permissible, because as explained above, reciprocity is not a requirement when information leakage is considered.

4.3 Query Processing at the LBS

At the LBS side, processing multiple ASRs is based on the same primitive functions that handle a single ASR. In both the range and nearest neighbor cases, processing is possible in a single traversal of the POI R-tree, so that multiple reads of the same disk pages are avoided. Specifically, for an R-range query, the LBS descends the R-tree visiting any node with minimum distance smaller than R from any of the sub-ASRs. In the case of K-NN processing, the CS contains the K-NN of any possible query point inside any sub-ASR. Handling is similar to single ASR processing, the difference being that instead of 4 line segments (i.e., single ASR edges), there are 4 segments for each sub-ASR input to the algorithm of [12] (note that this work describes how batch processing is possible in a single R-tree traversal, and that this feature is already being exploited in the standard *Hilbert Cloak* technique).

---

[7] Here containment refers to the spatial domain, and not to the Hilbert range of the AS.

4.4   Different Query Types

So far we considered simple snapshot queries, but ILC can be easily applied to continuous processing too, extending the method of [19]. In particular, sub-ASR computation can be performed in the first evaluation of the query, with users in each m-block forming a group. In subsequent timestamps, the sub-ASRs can be found as the minimum enclosing rectangles of each group, subject to the updated positions of its users. An alternative to that, is to perform sub-ASR computation in every timestamp. This latter approach is expected to lead to smaller ASRs, and thus to smaller CS.

5   Experimental Evaluation

In this section, we empirically evaluate the performance of the ILC approach. Section 5.1 describes our experimental setting, while Section 5.2 presents the results and their interpretation.

5.1   Experimental Setting

We executed the experiments using prototypes written in C++ on an Intel Pentium IV 3GHz machine. We use a real dataset as the POIs available at the LBS; the dataset (denoted as NA) contains 569,120 endpoints of main road segments in North America, and is acquired from www.maproom.psu.edu/dcw. It is normalized into a 10,000 by 10,000 data-space. The set of users U is formed by randomly picking a percentage (1%, 5%, 10%, 15%, or 20%) of the POIs to serve as the users; this choice represents a realistic scenario where the users follow the distribution of the queried facilities (POIs). However, we explore different user and POI distributions towards the end of the section, where the POIs correspond to 1,314,620 locations in Los Angeles (dataset denoted as LA, and acquired from www.rtreeportal.org).

We use a centralized architecture due to its proliferation, and assume that the anonymizer stores the users in main memory, while the LBS indexes the POIs with a disk-resident R*-tree. We use the *Hilbert Cloak* as a basis, and denote its traditional (single ASR) version as HC, and our adaptation as ILC. We quantify the performance benefits of ILC in terms of the size of the candidate set retrieved ($|CS|$); as we demonstrate, the communication cost and the I/O cost (and, as a

Table 2. Experiment parameters and their respective default values

Parameter	Default	Range
Anonymity parameter k	50	5,10,50,100,150
Strictness on Information leakage parameter m	5	2, 3, 5, 7, 10, 15
Number of NNs K	4	1, 2, 4, 8, 16
Range R	10	1, 5, 10,15,20
User-to-POI percentage ρ	10%	1, 5, 10, 15, 20 (%)

result, the total end-to-end response time too) are proportional to $|CS|$. In each experiment, we vary one parameter, while setting the remaining to their default values. The parameter ranges and default values are shown in Table 2.

5.2   Experimental Results

The first set of experiments (in Figure 8) explores the effect of query selectivity, by varying K (in the case of K-NN queries) and R (in the case of R-Range queries). Additionally, it verifies our claim that the I/O time, the communication cost and the end-to-end time are proportional to $|CS|$. Figure 8(a) plots the CS size. HC and ILC exhibit the same pattern of increase as K and R increase. However, ILC has approximately 34%-44% (28%-46%) improvement in $|CS|$ over HC on the average for NN (range) queries. Note that the two lower curves correspond to ILC and the upper ones to HC. This is expected as HC returns numerous candidate POIs corresponding to ASR regions that contain no users. The gap between ILC and HC shrinks for large K or R (to around 34% and 28%, respectively) as for high selectivity, the union of the search regions of the sub-ASRs converges to the search region of a single ASR.

Figure 8(b) plots the total communication cost between the anonymizer and the LBS in terms of Kbytes transferred towards either direction. We assume that each coordinate is 4 bytes, that an ASR (or sub-ASR) is represented by 4 coordinates, and that each POI has an additional 64 byte non-spatial information attached to it. Figure 8(c) shows the I/O cost (of query processing at the LBS) in terms of the number of R-tree node accesses; note that NN queries require more I/Os than ranges because the linear NN algorithm of [12] accesses a considerable amount of POIs that do not belong to the CS. Figure 8(d) sums up all costs, and plots the end-to-end query response time assuming a 10 Mbps connection between the anonymizer and the LBS. From the aforementioned three figures it becomes clear that the individual communication and I/O costs, as well as the overall result delay are proportional to $|CS|$. As such, and in order to avoid cluttering the paper with correlated figures, we use $|CS|$ as the key measurement for ILC's performance in the subsequent experiments.

Figure 9(a) explores the effect of k on $|CS|$, while setting m to 5. In HC, $|CS|$ grows almost linearly to k, because most candidate POIs returned fall inside the (growing) ASR. However, in ILC, we can see that performance degradation with k is slighter. This is because the sub-ASRs prune relatively more dead-space from a large ASR. Thus, ILC scales better than HC with k. To support our previous claim regarding the dead-space pruned, in Figure 9(b) we show the effect of k on the ASR area and the cumulative sub-ASR area per query, for HC and ILC, respectively. In the case of ILC, the gradient of the curve decreases significantly for $k > 100$.

Figures 10(a) and 10(b) illustrate the effect of m on $|CS|$ and on the total ASR/sub-ASR area, respectively. Parameter m does not affect HC since it is only introduced in ILC. As such, all the HC related curves are horizontal lines. Figure 10(a) shows the trade-off between information leakage and performance. As m increases (i.e., IL decreases), the performance improvement of ILC over

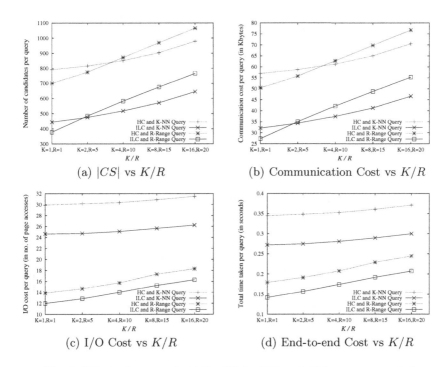

Fig. 8. Effect of query selectivity (K for NN, and R for range queries)

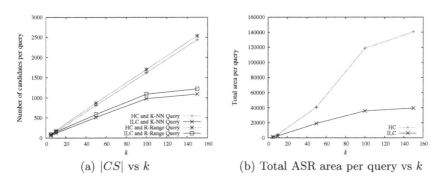

Fig. 9. Effect of varying degree of anonymity, k

HC decreases because there are fewer sub-ASRs and, thus, weaker dead-space pruning. Intuitively, the more information hidden from the LBS, the higher the cost. Figure 10(b) supports our previous claim regarding the dead-space pruned; it shows clearly that the total sub-ASR area per query increases with m.

Figure 11(a) shows the effect of user density (denoted by ρ), e.g., $\rho = 10\%$ implies that there are 56,912 users, which is one tenth of the POI cardinality. ILC has 55% to 63% the CS size of HC. The gains of ILC drop slightly for large ρ, because a dense user set implies an already small amount of dead-space in

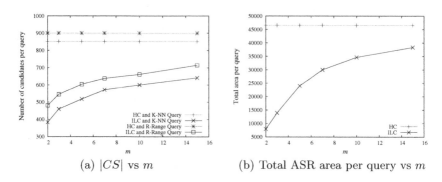

(a) $|CS|$ vs m (b) Total ASR area per query vs m

Fig. 10. Effect of varying strictness on information leakage, m

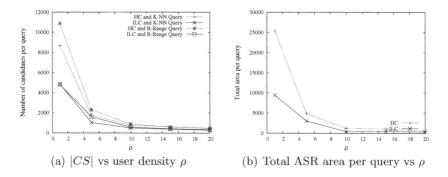

(a) $|CS|$ vs user density ρ (b) Total ASR area per query vs ρ

Fig. 11. Effect of user density, ρ

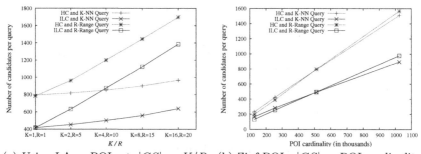

(a) Using LA as POI set: $|CS|$ vs K/R (b) Zipf POIs: $|CS|$ vs POI cardinality

Fig. 12. Effect of different POI datasets

the (single) ASR. Therefore, the benefits of using multiple sub-ASRs become smaller, but ILC still performs significantly better than HC. This trend is also obvious in Figure 11(b), which plots the total ASR/sub-ASR area in the same experiment.

In Figure 12(a), we use LA as the POI dataset, and keep the same users as in previous experiments (i.e., following the NA distribution). The observed trends are similar to the results in Figure 8(a), even though the relative cost of range queries is higher than NN. In Figure 12(b) we generated synthetic POI sets, using a Zipfian distribution (with parameter 0.8) and varying their cardinality from 128K up to 1M POIs. The user set is the same as before. ILC scales better than HC; note that the two lower curves correspond to ILC. The results in Figure 12 show the generality of ILC, and its superiority over the traditional (single ASR) approach, regardless of data skewness and user/POI distribution.

6   Conclusion

In this paper, we define the concept of information leakage in anonymous location-based queries. We describe meaningful leakage measures and propose a methodology to control it. Compared to previous systems, our technique can guarantee that information leakage does not exceed its maximum permissible degree and, moreover, it leads to significant performance benefits. Our method can be easily incorporated into existing cloaking systems and it is applicable to both the centralized and the decentralized cloaking models. A promising direction for future work is to design incremental versions of our methodology suited for continuous query evaluation, in order to improve upon the straightforward extensions of Section 4.4.

References

1. Gruteser, M., Grunwald, D.: Anonymous usage of location-based services through spatial and temporal cloaking. In: MobiSys. (2003)
2. Kalnis, P., Ghinita, G., Mouratidis, K., Papadias, D.: Preventing location-based identity inference in anonymous spatial queries. IEEE Transactions on Knowledge and Data Engineering 19(12), 1719–1733 (2007)
3. Mokbel, M.F., Chow, C.Y., Aref, W.G.: The new casper: Query processing for location services without compromising privacy. In: VLDB, pp. 763–774 (2006)
4. Ghinita, G., Kalnis, P., Skiadopoulos, S.: Prive: anonymous location-based queries in distributed mobile systems. In: WWW 2007: Proceedings of the 16th international conference on World Wide Web, pp. 371–380. ACM, New York (2007)
5. Chow, C.Y., Mokbel, M.F., Liu, X.: A peer-to-peer spatial cloaking algorithm for anonymous location-based service. In: GIS, pp. 171–178 (2006)
6. Samarati, P.: Protecting respondents' identities in microdata release. IEEE Trans. Knowl. Data Eng. 13(6), 1010–1027 (2001)
7. Sweeney, L.: k-anonymity: a model for protecting privacy. Int. J. Uncertain. Fuzziness Knowl.-Based Syst. 10(5), 557–570 (2002)
8. de Berg, M., van Kreveld, M., Overmars, M., Schwarzkopf, O.: Computational Geometry: Algorithms and Applications, 2nd edn. Springer, Heidelberg (2000)
9. Butz, A.R.: Alternative Algorithm for Hilbert's Space-Filling Curve. IEEE Trans. Comput. 20(4), 424–426 (1971)
10. Guttman, A.: R-trees: A dynamic index structure for spatial searching. In: SIGMOD Conference, pp. 47–57 (1984)

11. Beckmann, N., Kriegel, H.P., Schneider, R., Seeger, B.: The r*-tree: An efficient and robust access method for points and rectangles. In: SIGMOD Conference, pp. 322–331 (1990)
12. Tao, Y., Papadias, D.: Spatial queries in dynamic environments. ACM Trans. Database Syst. 28(2) (2003)
13. Gedik, B., Liu, L.: Location privacy in mobile systems: A personalized anonymization model. In: ICDCS, pp. 620–629 (2005)
14. Khoshgozaran, A., Shahabi, C.: Blind Evaluation of Nearest Neighbor Queries Using Space Transformation to Preserve Location Privacy. In: Papadias, D., Zhang, D., Kollios, G. (eds.) SSTD 2007. LNCS, vol. 4605, pp. 239–257. Springer, Heidelberg (2007)
15. Duckham, M., Kulik, L.: A Formal Model of Obfuscation and Negotiation for Location Privacy. In: Gellersen, H.-W., Want, R., Schmidt, A. (eds.) PERVASIVE 2005. LNCS, vol. 3468, pp. 152–170. Springer, Heidelberg (2005)
16. Kido, H., Yanagisawa, Y., Satoh, T.: An Anonymous Communication Technique using Dummies for Location-based Services. In: ICPS (2005)
17. Yiu, M.L., Jensen, C.S., Huang, X., Lu, H.: SpaceTwist: Managing the Trade-Offs Among Location Privacy, Query Performance, and Query Accuracy in Mobile Services. In: ICDE (2008)
18. Ghinita, G., Kalnis, P., Khoshgozaran, A., Shahabi, C., Tan, K.L.: Private Queries in Location Based Services: Anonymizers are not Necessary. In: SIGMOD Conference (2008)
19. Chow, C.Y., Mokbel, M.F.: Enabling private continuous queries for revealed user locations. In: Papadias, D., Zhang, D., Kollios, G. (eds.) SSTD 2007. LNCS, vol. 4605, pp. 258–275. Springer, Heidelberg (2007)
20. Devanbu, P.T., Gertz, M., Martel, C.U., Stubblebine, S.G.: Authentic third-party data publication. In: DBSec., pp. 101–112 (2000)
21. Pang, H., Jain, A., Ramamritham, K., Tan, K.L.: Verifying completeness of relational query results in data publishing. In: SIGMOD Conference, pp. 407–418 (2005)
22. Li, F., Hadjieleftheriou, M., Kollios, G., Reyzin, L.: Dynamic authenticated index structures for outsourced databases. In: SIGMOD Conference, pp. 121–132 (2006)
23. Kanth, K.V.R., Ravada, S., Sharma, J., Banerjee, J.: Indexing medium-dimensionality data in oracle. In: SIGMOD Conference, pp. 521–522 (1999)
24. Sellis, T.K., Roussopoulos, N., Faloutsos, C.: The r+-tree: A dynamic index for multi-dimensional objects. In: VLDB, pp. 507–518 (1987)

Analyzing Trajectories Using Uncertainty and Background Information

Bart Kuijpers, Bart Moelans, Walied Othman, and Alejandro Vaisman

Hasselt University & Transnational University of Limburg, Belgium
{bart.kuijpers,bart.moelans,walied.othman,alejandro.vaisman}@uhasselt.be

Abstract. A key issue in clustering data, regardless the algorithm used, is the definition of a distance function. In the case of trajectory data, different distance functions have been proposed, with different degrees of complexity. All these measures assume that trajectories are error-free, which is essentially not true. Uncertainty is present in trajectory data, which is usually obtained through a series of GPS of GSM observations. Trajectories are then reconstructed, typically using linear interpolation. A well-known model to deal with *uncertainty* in a trajectory sample, uses the notion of *space-time prisms* (also called *beads*), to estimate the positions where the object could have been, given a maximum speed. Thus, we can replace a (reconstructed) trajectory by a necklace (intuitively, a a *chain of prisms*), connecting consecutive trajectory sample points. When it comes to clustering, the notion of uncertainty requires appropriate distance functions. The main contribution of this paper is the definition of a distance function that accounts for uncertainty, together with the proof that this function is also a metric, and therefore it can be used in clustering. We also present an algorithm that computes the distance between the chains of prisms corresponding to two trajectory samples. Finally, we discuss some preliminary results, obtained clustering a set of trajectories of cars in the center of the city of Milan, using the distance function introduced in this paper.

1 Introduction

The study of *Moving Object Databases* (MODs) [7,21] has been increasingly attracting the attention of the GIS (Geographic Information Systems) community. Most of the time, in this field, a moving object's trajectory is obtained from *trajectory samples*, i.e., finite sequences of time-space points. A *trajectory sample database* contains a finite number of labeled trajectory samples. There are various ways of reconstructing trajectories from trajectory samples, of which linear interpolation is the most popular one [7]. An important issue in these databases is the problem of *uncertainty*, arising from various sources (e.g., errors in measurements, interpolation). The uncertainty of the moving object's position in between sample points has been studied using *space-time prisms* or *beads*, where it is assumed that besides the time-stamped locations of the object also some *background* knowledge is known, like, for instance, a speed limit v_{max} at a location (x_i, y_i). Informally, the space-time prism between two consecutive sample

N. Mamoulis et al. (Eds.): SSTD 2009, LNCS 5644, pp. 135–152, 2009.
© Springer-Verlag Berlin Heidelberg 2009

points is defined as the collection of space-time points where the moving objects may have passed, given the speed limitation.

One of the most common data mining techniques is Clustering [9]. This technique partitions the dataset into collections of data objects, such that within each partition the objects are 'similar' to each other and 'different' from the objects contained in other partitions. In the context of moving object data, the clustering technique is aimed at identifying groups of objects that followed similar trajectories. These kinds of data present particular problems for clustering. Clustering moving object trajectories requires, for example, finding out a proper spatial granularity level, and it is not obvious to identify the best clustering algorithm among the wide corpus of work on the subject. In the presence of uncertainty, the problem of representing a trajectory of moving objects and formalizing the notion of trajectory similarity is even more involved. This issue takes us to the problem we address in this paper: studying a distance function that accounts properly for the notion of uncertainty.

1.1 Problem Statement and Contributions

There exist two classic approaches to trajectory clustering: one based on the notion of similarity, typically operationalized through a so-called *distance function* between trajectories. A second approach is denoted trajectory-specific [16], and exploits the characteristics of the data type in the clustering algorithm. In this paper we position ourselves in the first group: we study the impact over clustering of the uncertainty of trajectory samples, by means of the definition of a distance function that accounts for the uncertainty involved in a trajectory. In short, a distance function measures the similarity of two trajectories. Many different distance functions can be defined (we give a formal definition, and a review, later in the paper), ranging from the most simple ones (like, for instance, clustering trajectories with the same origin and/or destination), to very complex mathematical functions. The former are obviously more computationally efficient, probably at the expense of returning less reliable clusters.

A well-known model to deal with *uncertainty* in a trajectory sample, uses the notion of *space-time prisms* (also called *beads*), to estimate the positions where the object could have been, given a maximum speed. We therefore introduce a distance function that accounts for uncertainty, and prove that this function is a metric which can be used to cluster trajectory (and hence, uncertain) data. Given two trajectory samples $T1$ and $T2$, their uncertainty is represented by two *chains of space-time prisms* (also called *lifeline necklaces*), N_1 and N_2, respectively, that connect consecutive sample points of each trajectory. Intuitively, in our proposal, the largest the intersection of the necklaces with respect to their union, the smallest the distance between both trajectories. In other words, the more uncertainty shared by $T1$ and $T2$, the closer they are. On the other hand, if N_1 and N_2 do not intersect, this indicates that these trajectories could not have met, given the speed limit. Then, a clustering algorithm will not group together these two trajectories. We also present an algorithm that, given the *chains of space-time prisms* of two trajectories, computes the distance between them. Finally, we present preliminary experiments, clustering a set of data corresponding to movement of cars in the center of the city of Milan, using the distance function we introduce in this paper.

In Section 2 we review related work on clustering and space-time prisms. Section 3 introduces the concepts of space-time prisms and their relation with uncertainty and background information. In Section 4 we introduce the new distance function, denoted d_u, and we show it is a metric apt for trajectory clustering. Section 5 presents an algorithm to compute d_u. Section 6 presents preliminary experimental results. We conclude in Section 7.

2   Related Work

Space-time prims and uncertainty. In this paper we work with *trajectory samples*, which are well-known in MODs, namely finite sequences of time-space points. A trajectory sample database contains a finite number of labeled trajectory samples. There are various ways of reconstructing trajectories from samples, of which linear interpolation is the most popular in the literature [7]. However, linear interpolation relies on the (rather unrealistic) assumption that between sample points, an object moves at constant minimal speed. It is more realistic to assume that moving objects have some physically determined speed bounds. Given such upper bounds, *an uncertainty model* has been proposed which constructs *space-time prisms* between two consecutive space-time points in a trajectory sample. Basic properties of this model were discussed a few years ago by Egenhofer et al. [3] and Pfoser et al. [17], but space-time prisms were already known in the time-geography of Hägerstrand in the 1970s [8]. In short, a *space-time prism* is the intersection of two cones in the space-time space such that all possible trajectories of the moving object between the two consecutive space-time points, given the speed bound, are located within them. Space-time prisms manage uncertainty more efficiently than other approaches based on cylinders [21]. A chain of space-time prisms connecting consecutive trajectory sample points is called a *lifeline necklace* [3].

Trajectory clustering. In one of the first works on trajectory clustering, Ketterlin [10] presents an structured methodology for discovering patterns in sequences of composite objects. These objects, basically time-series data, can be considered an abstract representation of a trajectory (i.e., a composite object may be described as a sequence of simpler data). The author studies the generalization of sequences of complex objects, and integrate this in a general-purpose clustering algorithm. This generalization-based approach, together with the limitation to time-series, could be considered as shortcomings of this first approach.

Another proposal applies to trajectories some multidimensional scaling technique for non-vectorial data. This is performed in Fastmap [5], where a data space is mapped to an Euclidean space approximately preserving the distances between objects. Then, any standard clustering algorithm for vectorial data can be applied.

Gaffney and Smyth [6] use a different approach, denoted model-based clustering. They work with continuous trajectories, grouping together objects which likely to be generated from a common core trajectory, by adding Gaussian noise. In this way, a cluster contains all objects which can be obtained by a regression function. The problem of this approach is the lack of flexibility for different application contexts.

Typically, the *distance* between two trajectories is computed by fixing two time instants and considering points within this interval. Clustering trajectory data usually produces clusters containing geographically close trajectories. A classical approach for clustering trajectories, is to adapt traditional algorithms like k-means or hierarchical clustering, to the trajectory setting, leading to the notion of *distance-based clustering*.

Nanni [14] defines a distance measure that describes the similarity of trajectories of objects across time, computed by analyzing the way the distance between the objects varies. He considers only pairs of contemporary instantiations of objects, i.e., for each time instant compares the positions of the objects at that moment, aggregating the set of distance values obtained this way. The distance between trajectories is computed as the average distance between objects:

$$D(\tau_1; \tau_2) \mid T = \frac{\int_T d(\tau_1(t); \tau_2(t))dt}{\mid T \mid}$$

where $d()$ is the Euclidean distance over R^2, T is the temporal interval over which trajectories τ_1 and τ_2 exist, and $\tau_i(t)$ is the position of object τ_i at time t. This is the more general expression. However, the author showed that due to the piece-wise linearity of the trajectories, the distance can be computed as a finite sum by means of $O(n1 + n2)$ Euclidean distances, where $n1$ and $n2$ are the number of observations for τ_1 and τ_2.

Other proposals compute the longest common subsequence of two series [19], and the least common sub-sequence of two series, and take these measures as the distance between trajectories [1].

A different line of work is *Density-based clustering*, which can be considered as a combination of the two approaches mentioned above. Initially proposed by Ester *et al.* [4], clusters are populated by objects which can reach each other through densely populated regions, instead of objects close to each other. In other words, these algorithms agglomerate objects in clusters based on the population within a given region. This form of clustering is used in the OPTICS algorithm, proposed by Ankerst *et al.* [2]. It is particularly well-suited to trajectory clustering, given that trajectories of cars in urban traffic tend to agglomerate in non-convex clusters, and that many outlier trajectories should be considered as noise. Nanni and Pedreschi [15] generalize the spatial notion of distance between objects to a spatio-temporal notion of distance between trajectories, leading to a natural extension of the density-based clustering technique to trajectories. More recently, the concept of *progressive clustering* has been introduced, as a process that, using a visual analytics approach, starts from the simpler functions to complex ones, in an incremental way [18], using an iterative approach that filters clusters using simpler but efficient distance functions in the firsts steps.

3   A Model for Moving Object Data with Uncertainty

A well-known model for the management of the uncertainty of the moving object's position in between sample points is the *space-time prism* model, where it is assumed that besides the time-stamped locations of the object, also some background knowledge, in particular a (e.g., physically or law imposed) speed limitation v_i at location (x_i, y_i) is known. The space-time prism between two consecutive sample points is defined as the set of time-space points where the moving objects may have passed, respecting the speed limitation. The chain of space-time prisms connecting consecutive trajectory sample points is called a *lifeline necklace* [3].

In this paper we focus on space-time prisms on *road networks*. Early adaptations of the space-time prism model to road networks were done by Miller [12,13]. We view road networks as a graph embedding in $\mathbf{R}^2$ where all edges are embedded as straight lines between vertices. All edges have a (strictly positive) speed limit as well an associated weight, called their *time span*, which is equal to the time needed to get from one end of the edge to the other when traveling at the speed limit. Also Kuijpers and Othman [11] studied the problem of space-time prisms on road networks, and introduced an algorithm for computing and visualizing space-time prisms. In Section 5 we use this algorithm to compute the surface of a space-time prism.

In the remainder of the paper we work with the speed limits of the road network, simply setting a uniform speed limit, namely v_i, on the network to construct the space-time prism between two sample times t_i and t_{i+1}. We first review basic concepts about trajectories and road networks, that we use throughout the paper.

3.1   Preliminaries: Trajectories and Trajectory Samples

Let $\mathbf{R}$ denote the set of the real numbers and $\mathbf{R}^2$ the 2-dimensional real plane. We consider objects moving in a subset of the two-dimensional (x, y)-space $\mathbf{R}^2$ and describe this movement in the (t, x, y)-space $\mathbf{R} \times \mathbf{R}^2$, where t represents time. Moving objects, which we assume to be points, produce a special kind of curves, denoted *trajectories*. The definitions we present next, necessary to understand the remainder of the paper, are based on [11].

Definition 1. [Trajectory] Let $I \subseteq \mathbf{R}$ be an interval. A *trajectory* T is the graph of a mapping $\alpha : I \to \mathbf{R}^2 : t \mapsto \alpha(t) = (\alpha_x(t), \alpha_y(t))$, i.e., $T = \{(t, \alpha_x(t), \alpha_y(t)) \in \mathbf{R} \times \mathbf{R}^2 \mid t \in I\}$. We call I the *time domain* of T.   □

In practice, trajectories are only known at discrete moments in time. This partial knowledge of trajectories is formalized by the notion of sample.

Definition 2. [Trajectory sample] A *trajectory sample* is a finite set $S = \{(t_0, x_0, y_0), (t_1, x_1, y_1), ..., (t_N, x_N, y_N)\}$ of time-space points. The order on time, $t_0 < t_1 < \cdots < t_N$, induces a natural order on the sample.   □

A trajectory T, which contains a trajectory sample $S = \{(t_0, x_0, y_0), (t_1, x_1,$ $y_1), ..., (t_N, x_N, y_N)\}$, i.e., $(t_i, \alpha_x(t_i), \alpha_y(t_i)) = (t_i, x_i, y_i)$ for $i = 0, ..., N$, is called a *geospatial lifeline* for S [3].

In this paper, we consider movement in $\mathbf{R}^2$ that is constrained to a road network. Thus, we need to formalize the notion of a road network.

Definition 3. [Road network] A *road network* RN is a graph embedding in $\mathbf{R}^2$ of a labeled graph given by a finite set of vertices $V = \{(x_i, y_i) \in \mathbf{R}^2 \mid i = 1, ..., N\}$, and a set of edges $E \subseteq V \times V$ that are labeled by a *speed limit* and an associated *time span*. This graph embedding satisfies the following conditions. Edges are embedded as straight line segments between vertices[1]. If an edge between (x_i, y_i) and (x_j, y_j) is labeled by the speed limit $v_{ij} > 0$, then its time span w_{ij} is $\frac{\sqrt{(x_i-x_j)^2+(y_i-y_j)^2}}{v_{ij}}$, i.e., it is the time needed to get from one side of an edge to another when traveling at the speed limit. □

A trajectory (sample) on a road network RN is then a trajectory (sample) whose spatial projection is in RN. More formally, if T is a trajectory given by the functions α_x and α_y, then it must satisfy $(\alpha_x(t), \alpha_y(t)) \in$ RN for all t in the time domain of T, and for a trajectory sample $S = \{(t_0, x_0, y_0), (t_1, x_1, y_1), ..., (t_N, x_N, y_N)\}$ we must have $(x_i, y_i) \in$ RN for all $i = 0, ..., N$.

3.2 Space-Time Prisms in Road Networks

Given a point in a trajectory sample, if we assume that a speed limit is valid until the next point, we can use this limit to define space-time prisms which model the uncertainty of the object's location in between sample points. In this paper, we consider movement and space-time prisms in road networks in $\mathbf{R}^2$.

In general, suppose p and q are points in some space, and an object is traveling from p to q, leaving p at time t_p and arriving in q at t_q; also assume a speed limit $v_{\max}$ is given. We know that at a time t, $t_p \leq t \leq t_q$, the object's distance to p is at most $v_{\max}(t - t_p)$ and its distance to q is at most $v_{\max}(t_q - t)$. The object is therefore somewhere in the intersection of the sphere with center p and radius $v_{\max}(t - t_p)$ and the sphere with center q and radius $v_{\max}(t_q - t)$. This is illustrated in Figure 1. The geometric location of these points, for $t_p \leq t \leq t_q$, is referred to as a *space-time prism*.

Although we are interested in space-time prisms on a road network, the problem does not simply amount to taking the intersection of a space-time prism representing unconstrained movement and the road network. To see this, consider the projection of the unconstrained space-time prism along the time axis onto the xy-plane. This projection is an ellipse such that its foci are the points of departure and arrival, i.e., p and q. We recall that at a time t, $t_p \leq t \leq t_q$, the object's distance to p is at most $v_{\max}(t - t_p)$ and its distance to q is at most $v_{\max}(t_q - t)$. Adding those distances gives $v_{\max}(t - t_p) + v_{\max}(t_q - t) = v_{\max}(t_q - t_p)$, which

---

[1] These edge embeddings may intersect in non-vertex points. So, we can model bridges and tunnels in our model.

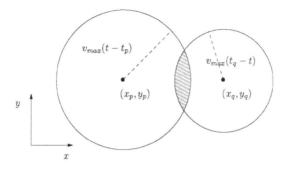

Fig. 1. A moving object's location at time t

is constant. Therefore, all possible points a moving object with speed limit $v_{\max}$ could have visited must lie within this ellipse with foci p and q. Moreover, the sum of their distances to p and q is less or equal to $v_{\max}(t_q - t_p)$. Any trajectory that touches the border of the ellipse and has more than two straight line segments, is longer than $v_{\max}(t_q - t_p)$. This particular trajectory lies in the ellipse and hence in the intersection of the unconstrained space-time prism and the road network, *but it does not lie in the road network space-time prism entirely* because there are points on it which can be reached in time, but from which the destination can not be reached in time and vice versa. Figure 2 depicts such a situation. There is no path on the road network from p that reaches q in the given time interval. The intersection of the space-time prism with the road network is nonempty, whereas the road network space-time prism clearly is. Figure 3 depicts a space-time prism on a road network, and its spatial projection.

To define space-time prisms on a road network, we need to define an appropriate distance function on the network. This distance measure is derived from the *shortest path*-distance used in graph theory [20]. Consider a road network RN, given by a set of vertices V and a set of labeled edges E. Let $p = (x_p, y_p)$ and $q = (x_q, y_q)$ be two points on RN, not necessarily vertices.

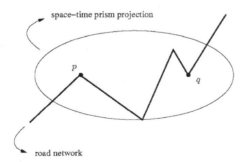

Fig. 2. Road network space-time prisms can not be easily derived from space-time prisms in $\mathbf{R}^2$

Fig. 3. Space-time prism (red) on road networks (green and black) and its spatial projection (green)

Also suppose that p lies on (the embedding of) the edge $((x_{p,0}, y_{p,0}), (x_{p,1}, y_{p,1}))$ and q lies on the edge $((x_{q,0}, y_{q,0}), (x_{q,1}, y_{q,1}))$. We construct a new road network RN_{pq} from RN. Its set of vertices is $\mathsf{V}_{pq} = \mathsf{V} \cup \{p, q\}$ and its set of edges is $\mathsf{E}_{pq} = \mathsf{E} \cup \{((x_{p,0}, y_{p,0}), (x_p, y_p)), ((x_p, y_p), (x_{p,1}, y_{p,1})), ((x_{q,0}, y_{q,0}), (x_q, y_q)), ((x_q, y_q), (x_{q,1}, y_{q,1}))\}$. So, we have split the edges on which p and q are located. The speed limits are the ones of the original edges, and the time spans of the new edges are computed according to Definition 3. It is precisely this construction we need to define the distance along the road network RN and the space-time prism on RN. To do this we need a metric on RN.

Definition 4. [Road network time] Let RN be a road network and let $p, q \in \mathsf{RN}$. The *road network time* between p and q, denoted by $\mathsf{d}_{\mathsf{RN}}(p, q)$, is the shortest-path distance (i.e., the shortest-path distance as usual in graph theory and that can be computed by the Dijkstra's algorithm [20]) between p and q in the graph $(\mathsf{V}_{pq}, \mathsf{E}_{pq})$, with respect to the time-span labeling of the edges.   □

Note that the *road network time* between p and q in the above definition has minimal total weight and returns the earliest possible time you can reach p from q and vice versa. The metric that we describe takes two points from a road network and returns the shortest time needed to get from one to the other when traveling at the allowed maximal speed at each segment. We remark that if all edges in road network have the same speed limit $v_{\max}$, then the metric defined in Definition 4 is the shortest-path metric (up to a scaling factor $v_{\max}$) on the graph embedding RN. If, on the other hand, there are different speed limits per edge, then the metric of Definition 4 is the shortest time-span metric on the temporal projection of the spatio-temporal data. It follows that in the latter case, the shortest paths are not always the fastest paths. Figure 4 depicts a situation where neither one of the two shortest paths in the network is also the

Fig. 4. Road Network space-time prism where the fastest path does not coincide with the shortest

fastest path. Indeed, looking at the evolution of the prism over time, it is clear that the fastest path starts at the leftmost node, goes to the upper, then the lower node, and ends at the rightmost node.

A space-time prism on a road network is the geometric location in $\mathbf{R} \times \mathrm{RN} \subset \mathbf{R} \times \mathbf{R}^2$ of all points a moving object could have visited when traveling, restricted to RN, from an origin p to a destination q with in a time-frame ranging from t_p to t_q, respecting the speed limits on the edges of RN.

4   An Uncertainty-Aware Distance Function

The goal of clustering algorithms is, basically, grouping together the objects which are similar to each other and keep them separated from the objects which are different. A key issue of this technique is the definition of a distance measure. To be useful for clustering, the function must be a metric, which means that it has to verify four well-known conditions stated in the next definition.

Definition 5. *A function $d(\cdot,\cdot)$ is a metric if:*

1. *$\forall\, i:\; d(i,j) = 0$ if and only if $i = j$;*
2. *$\forall\, i,j$ such that $i \neq j : d(i,j) > 0$ (positive definite)*
3. *$\forall\, i,j : d(i,j) = d(j,i)$ (symmetry)*
4. *$\forall\, i,j,k : d(i,j) + d(j,k) \geq d(i,k)$ (triangle inequality).*   □

There are many (mainly Euclidean-based) distance functions used for clustering. In particular, for trajectory clustering, a definition of distance is aimed at considering similar two objects that followed approximately the same spatio-temporal trajectory. The main problem in this scenario is to find out which are the objects that moved together. However, depending on the application at hand, or the analysis a user needs to perform, other forms of distances could be useful. For instance, we may want to cluster together trajectories starting and ending at the same locations [18]. Nevertheless, as far as we are aware of, no distance function has been proposed to account for an intrinsic problem trajectory data have, that is, the uncertainty involved in GPS or GSM observations that originate the trajectory samples. The intuition behind this function is that the temporal projection of the intersection of the space-time prisms of two trajectories, represents the instants when the two trajectories *could have met*. Our claim is that the longer this period, the more similar the trajectories are. This notion is captured by the distance we introduce in Definition 6 below.

Definition 6 (Uncertainty-aware distance). *Let us denote A and B two necklaces corresponding to two trajectory samples τ_1 and τ_2, respectively. Let us define the volume of a 3-dimensional figure C by V_C. Then, the expression*

$$d_u(A, B) = 1 - \frac{V_{A \cap B}}{V_{A \cup B}} \tag{1}$$

is named the Uncertainty-based distance *between τ_1 and τ_2.*   □

Theorem 1 (Metric). *The function $d_u(\cdot, \cdot)$ of Definition 6 is a metric.*

Proof. Let us prove first the identity property. If we replace B by A in (1), we obtain:

$$\begin{aligned}
d_u(A, A) &= 1 - \frac{V_{A \cap A}}{V_{A \cup A}} \\
&= 1 - \frac{V_A}{V_A} \\
&= 0
\end{aligned}$$

The second property is straightforward. Symmetry is proved in a way analogous to the simple proof for the first property:

$$\begin{aligned}
d_u(B, A) &= 1 - \frac{V_{B \cap A}}{V_{B \cup A}} \\
&= d(A, B)
\end{aligned}$$

Now, we prove the triangle inequality in Definition 5. For that, let us consider the diagram of Figure 5, where we have three sets, let us call them A, B, and C, representing three space-time prisms with these names. We replace the terms in Equation 1 with the elements in the partition of Figure 5.

$$d_u(A, B) = 1 - \frac{a_2 + a_5}{a_1 + a_2 + a_3 + a_4 + a_5 + a_6} = \frac{a_1 + a_3 + a_4 + a_6}{a_1 + a_2 + a_3 + a_4 + a_5 + a_6}$$

$$d_u(B, C) = \frac{a_2 + a_3 + a_4 + a_7}{a_2 + a_3 + a_4 + a_5 + a_6 + a_7}$$

$$d_u(A, C) = \frac{a_1 + a_2 + a_6 + a_7}{a_1 + a_2 + a_4 + a_5 + a_6 + a_7}$$

Then,

$$d_u(A,B) + d_u(B,C) = \cfrac{a_1 + a_3 + a_4 + a_6}{a_1 + a_2 + a_3 + a_4 + a_5 + a_6 + a_7} + \cfrac{a_2 + a_3 + a_4 + a_7}{a_1 + a_2 + a_3 + a_4 + a_5 + a_6 + a_7}$$

And, since $a_1, a_7 \geq 0$, we have:

$$d_u(A,B) + d_u(B,C) \geq \cfrac{a_1 + a_3 + a_4 + a_6}{a_1 + a_2 + a_3 + a_4 + a_5 + a_6 + a_7} + \cfrac{a_2 + a_3 + a_4 + a_7}{a_1 + a_2 + a_3 + a_4 + a_5 + a_6 + a_7}$$

$$= \cfrac{a_1 + a_3 + a_4 + a_6 + a_2 + a_3 + a_4 + a_7}{a_1 + a_2 + a_3 + a_4 + a_5 + a_6 + a_7}$$

$$\geq \cfrac{a_1 + a_2 + a_6 + a_7 + a_3}{a_1 + a_2 + a_3 + a_4 + a_5 + a_6 + a_7}$$

And, since "If $0 \leq a \leq b \neq 0$ and $c \geq 0$, then $\frac{a}{b} \leq \frac{a+c}{b+c}$", then we get:

$$\geq \frac{a_1 + a_2 + a_6 + a_7}{a_1 + a_2 + a_4 + a_5 + a_6 + a_7} = d(A,C). \qquad \square$$

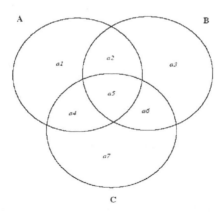

Fig. 5. A schematic description of the sets

5   An Algorithm to Compute d_u

We now describe the algorithm that computes the intersection between two chains of space-time prisms computed for two trajectories whose distance we need to calculate. In order to speed-up the computation, we pre-process information related to the road network. This pre-processed structures are presented next.

Data structure. The basic structure for storing the road network is a graph G, such that for each two nodes n_i, n_j, we store the weight of the edge connecting them, given by $\frac{d(n_i, n_j)}{v_{max}}$, where v_{max} is the speed limit in the road segment. There are also two lists. The first one is denoted `rowNonZeroes`, defined as:

$$\texttt{rowNonZeroes} = \{\dots, \{.., \texttt{n}_\texttt{j}, \dots\}_\texttt{i}, \dots\}$$

We can see that `rowNonZeroes` is a list of lists, such that the list in position i contains the nodes n_j such that $n_i, n_j \in G$. (i.e., the nodes reachable through just one edge from n_i). Analogously, there is another list, denoted `columnNonZeroes`:

$$\texttt{columnNonZeroes} = \{\dots, \{.., \texttt{n}_\texttt{j}, \dots\}_\texttt{i}, \dots\}$$

Thus, `columnNonZeroes` is a list of lists, such that the list in position i contains the nodes n_j such that $n_j, n_i \in G$. (i.e., the nodes that can reach n_i traversing just one edge).

Computing the distance between two chains of space-time prisms. First, we introduce an adaptation to our problem of the Dijstra's algorithm [20]. We want to find the shortest path to all nodes in the network, from a given node p, (a point in a trajectory, matched to a vertex in the network), provided that the traveling time is less than a given threshold. For each p, we apply Algorithm 1.

Algorithm 1. *DijkstraSource($p, maxTime, nbrVertices$)*

Output: *output$_P$ = { a list of nodes $N_m = N \cup N_1$, where $N = (n_1, \dots n_k)$ is the set of nodes reachable from p in at most $maxTime$, and $N_1 = \{n | n \notin N$ and $\exists\ n_i \in N, (n_i, n) \in G,$ and $d(p, n) > maxTime\}$; distlist = $\{d_1, d_2, 0, d_n\}$, a list containing the distances from p to the nodes in N_m; a list O_e of the form $n, \langle e_1, \dots e_f \rangle$ representing the edges outgoing from the nodes in N_m}*

```
 1: ToProcess = {(p, 0)};
 2: distlist = {∞, ∞, 0, ∞, }, a list of length —V—, with a zero in the position
    of the input node p.
 3: Processed = {};
 4: N_m = {};
 5: while ToProcess.notEmpty() && ToProcess[1].[2] < maxTime do
 6:     current := ToProcess[1];
 7:     append current.[1].[1] to N_m;
 8:     append (current[1].[1], ⟨rowNonZeroes[current]⟩) to O_e;
 9:     delete ToProcess[1];
10:     for each node n_j in rowNonZeroes[current] do
11:         build a pair (n_j, distlist[current.[1]] + d_{current.[1],j});
12:         distlist[j] := min(distlist[j], distlist[current.[1]] + d_{current.[1],j});
13:         if n_j ∉ Processed then
14:             append (n_j, d_j) to ToProcess;
```

```
15:        end if
16:     end for
17:     sort ToProcess by time;
18:     append current.[1] to Processed;
19: end while
```

An analogous algorithm, denoted DijkstraDestination, computes the shortest path from all nodes in the network, to a given node q, such that q is a point in a trajectory, matched to a vertex in the network, and provided that the traveling time is less than a given threshold. The main difference with Algorithm 1 is that $rowNonZeroes[current]$ is replaced by $columnNonZeroes[current]$, and that N_m contains the nodes that can reach q (i.e., instead of outgoing edges we work with incoming edges).

Algorithm 2. *DijkstraDestination($q, maxTime, nbrVertices$)*

Output: *$output_Q = \{$ a list of nodes $N_m = N \cup N_1$, where $N = (n_1, ...n_k)$ is the set of nodes that can reach q in at most $maxTime$, and $N_1 = \{n | n \notin N$ and $\exists\ n_i \in N, (n, n_i) \in G$, and $d(n, q) > maxTime\}$; distlist $= \{d_1, d_2, 0, d_n\}$, a list containing the distances to q from the nodes in N_m; a list I_e of the form $n, \langle e_1, ...e_f \rangle$ representing the edges incoming to the nodes in $N_m\}$*

```
 1: ToProcess = {(q, 0)};
 2: distlist = {∞, ∞, 0, ∞, }, a list of length —V—, with a zero in the position
       of the input node q;
 3: Processed = {};
 4: Nm = {};
 5: while ToProcess.notEmpty() && ToProcess[1].[2] < maxTime do
 6:     current := ToProcess[1];
 7:     append current.[1].[1] to Nm;
 8:     append (current[1].[1], ⟨columnNonZeroes[current]⟩ to Oe;
 9:     delete ToProcess[1];
10:     for each node nj in columnNonZeroes[current] do
11:        build a pair (nj, distlist[current.[1]] + d_current.[1],j);
12:        distlist[j] := min(distlist[j], distlist[current.[1]] + d_current.[1],j);
13:        if nj ∉ Processed then
14:           append (nj, dj) to ToProcess;
15:        end if
16:     end for
17:     sort ToProcess by time;
18:     append current.[1] to Processed;
19: end while
```

Now, given two points p and q in a trajectory, we compute their space-time prisms $prism_1$ and $prism_2$, using the output of Algorithms 2 and 3. From them, we

compute their union and intersection, from which the distance follows straightforwardly.

Algorithm 3. $\underline{\mathrm{prism}(p_1, t_{p_1}, q_1, t_{q_1}, p_2, t_{p_2}, q_2, t_{q_2})}$

Output: *The polygons and their surface, in the two prisms and their intersection (for the prisms defined by the two pairs of points in the input).*

1: $DijkstraSource(p_1, maxTime, nbrVertices)$;
2: $DijkstraDestination(q_1, maxTime, nbrVertices)$;
3: $DijkstraSource(p_2, maxTime, nbrVertices)$;
4: $DijkstraDestination(q_2, maxTime, nbrVertices)$;
5: **if** $p_1 \in output_{Q_1}[1]$ **and** $q_1 \in output_{P_1}[1]$ **then**
6:    $prism_1 = \{N_{m_{p_1}} \cap N_{m_{q_1}}, distlist_{p_1}, distlist_{q_1}, O_{e_{p_1}} \cap I_{e_{q_1}}\}$;
7: **end if**
8: **if** $p_1 \in output_{Q_2}[1]$ **and** $q_1 \in output_{P_2}[1]$ **then**
9:    $prism_2 = \{\{N_{m_{p_2}} \cap N_{m_{q_2}}, distlist_{p_2}, distlist_{q_2}, O_{e_{p_2}} \cap I_{e_{q_2}}\}$;
10: **end if**
11: **if** $prism[1] \cap prism_2[1] == \emptyset$ **then**
12:    *Compute only the surface of the polygons;*
13: **else**
14:    compute $polys = \{distlist_{p_1}, distlist_{q_1}, distlist_{p_2}, distlist_{q_2}, prism_1[1] \cup prism_2[1], prism_1[4] \cup prism_2[4]\}$;
15:    **for** every edge in polys **do**
16:       *Compute the polygons of $prism_1$, $prism_2$, and $prism_1 \cap prism_2$, using the algorithm in [11];*
17:    **end for**
18: **end if**

The final step of the algorithm computes the intersection and union of two
chains of prisms. We explain this through an example, for the sake of clarity.
Let us consider two trajectories:

$$T_1 = \{(p_1, t_1)(p_2, t_2)(p_3, t_3)...(p_n, t_n)\}$$
$$T_2 = \{(p'_1, t'_1)(p'_2, t'_2)(p'_3, t'_3)...(p'_m, t'_m)\}$$

We now compute the interval where the trajectories overlap, and compute the
surface with Algorithm 3. For example, if the time instants are such that
$t_3 < t'_1 < t_4 < t'_2 < t'_3 < t_5....$, we first use the points $(p_3, t_3), (p_4, t_4)$ and
$(p'_1, t'_1), (p'_2, t'_2)$, and compute the intersection and union of the two prisms.

The algorithm is based on a sliding window approach. Thus, we next
compute the intersection of the prisms corresponding to $(p_4, t_4), (p_5, t_5)$ and
$(p'_1, t'_1), (p'_2, t'_2)$ (note that there is a non-empty intersection here too); and, analogously, (p_4, t_4), (p_5, t_5) and (p'_2, t'_2), (p'_3, t'_3).

A first optimization in the computation of the distance function is that if
the intersection is empty, we do not compute the surface of the polygons in the

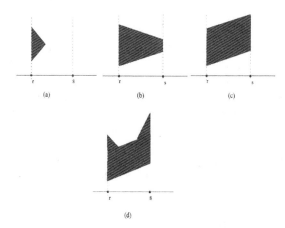

Fig. 6. Space-time prism components

extremes (for computing the union). A second optimization we implemented in our algorithm (besides the pre-computation step already mentioned), is that we keep the results of the previous iteration when computing the intersection of the chains of prisms (for example, we keep the prism for $(p'_1, t'_1), (p'_2, t'_2)$ above).

Intersection of space-time prisms. The distance defined in Section 4 requires the computation of the intersection between two space-time prisms. For this computation, in short, the algorithm described in [11] iterates over every edge that has at least one polygon in both space-time prisms, and compute the intersection as an intersection between polygons. However, since we use this intersection to compute the distance between two chains of space-time prisms, a subtle problem appears here. A prism on a road network is basically composed by the three geometries depicted in Figure 6 (a) through (c). This yields a geometry like the one in Figure 6 (d). A naive computation of the intersection would imply intersecting each component of one prism, with all the components of the other one. It is clear that this may result in an intersection larger than the union. Thus, we only intersect each component in one prism, with its analogous in the other prism. In this way, if we have two prisms B_1 and B_2, such that $B_1 = B_2$, then $B_1 \cap B_2 = B_1 \cup B_2$. Note that, in addition of allowing to use our distance function, this way of computing the intersection provides an appropriate and natural semantics to the space-time prisms. In two-way roads, and given the points r and s of Figure 6, the triangle corresponding to r represents the trajectories going from r to r. Analogously, the triangle corresponding to s represents the trajectories going from s to s. The parallelogram (Figure 6 (b)) represents the trajectories going from r to s. In one-way roads (like in the case we are studying), only the latter is considered.

6   Preliminary Experimental Results

We performed (very) preliminary experiments using a set of 52 trajectories obtained from cars moving in the city center of Milan, in one day, from 1PM to 2PM, that is, all in the same short period, increasing the possibility of moving together. We used k-mediods with k=2..6 as clustering algorithm. We only consider trajectories with at least 5 different points. Also, we split a trajectory if there is a gap of more then 6 minutes between two consecutive points.

Results for k=6 are shown in Figure1 7. The blue cluster represents all trajectories that, according to principle of alibi query [11], could not have met. Thus, the distance between any two trajectories in the blue cluster is 1. The cluster composition is shown in the next table.

Cluster	Color	# of trajectories
0	blue	26
1	red	13
2	green	4
3	black	2
4	cyan	1
5	orange	6

Figure 8 shows the clusters for k=4, overlayed with the road network of Milan, to provide a more realistic way of displaying the results. We see that the blue clusters are similar to the ones obtained using k=6. More precisely, for all clustering runs (with k=2..6), the blue cluster contained 26 trajectories, i.e., the trajectories that could have never met each other are always together in one cluster, no matter the parameter k used.

Prism-based clustering allows to draw some additional conclusions, useful for traffic analysis. Note that the closer to the speed limits the objects move (i.e., fluent traffic), the less likely their trajectories are to get clustered together. In our case, our results suggest that cars in the sample move at speeds higher than

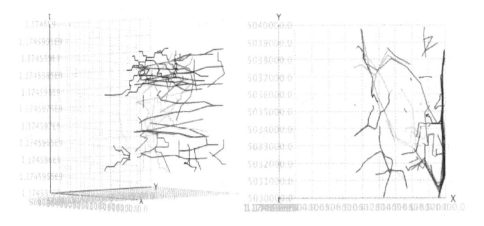

Fig. 7. Result of clustering with k=6 (left); Projection over x-y coordinates (right)

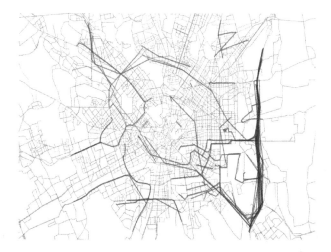

Fig. 8. Result of clustering Milan data with k=4 overlayed with the road network

the speed limit. This follows from observing that, initially, using the city's maximum speed limit in downtown, many empty prisms are obtained. Adjusting these limits, increasing the speed by 20%, the number of empty prisms dropped considerably, but prisms were narrow, mainly yielding empty intersections. However, we remark that this assertion always needs further insight into the dataset, since also the results are influenced by the map matching process. In our case, we use a *geometric* approach, which maps a point to the closest street segment, which sometimes could yield imprecise results.

7 Conclusion

We have presented a new distance function for clustering trajectory samples, that accounts for the inherent uncertainty contained in GPS observations. Instead of computing the distance between the trajectories themselves, we measure the relation between the union and the intersection of the space-time prisms associated to each trajectory (using the speed limit of the road segment), in a way such that if the intersection is empty, the two trajectories could not have met, and the distance equals '1'. To the best of our knowledge, this is the first proposal in this sense. We proved that this distance is actually a metric, sketched the algorithm to compute it, and presented preliminary experimental results, using real data obtained from cars moving in the city of Milan.

Acknowledgments. This research has been partially funded by the Research Foundation Flanders (FWO- Vlaanderen), Research Project G.0344.05, the European Union under the FP6-IST-FET programme, Project n. FP6-14915, GeoP-KDD: Geographic Privacy-Aware Knowledge Discovery and Delivery, and the Argentina Scientific Agency, project PICT 2004 11-21.350.

References

1. Agrawal, R., Lin, K., Sawhney, H.S., Shim, K.: Fast similarity search in the presence of noise, scaling, and translation in time-series databases. In: VLDB, pp. 490–501 (1995)
2. Ankerst, M., Breunig, M., Kriegel, H.P., Sander, J.: Optics: Ordering points to identify the clustering structure. In: SIGMOD Conference, pp. 49–60 (1999)
3. Egenhofer, M.: Approximation of geopatial lifelines. In: SpadaGIS, Workshop on Spatial Data and Geographic Information Systems (2003)
4. Ester, M., Kriegel, H.P., Sander, J., Xu, X.: A density-based algorithm for discovering clusters in large spatial databases with noise. In: KDD, pp. 226–231 (1996)
5. Faloutsos, C., Lin, K.L.: Fastmap: A fast algorithm for indexing, data-mining and visualization of traditional and multimedia datasets. In: SIGMOD Conference, pp. 163–174 (1995)
6. Gaffney, S., Smyth, P.: Trajectory clustering with mixtures of regression models. In: KDD, pp. 63–72 (1999)
7. Güting, R.H., Schneider, M.: Moving Objects Databases. Morgan Kaufmann, San Francisco (2005)
8. Hägerstrand, T.: What about people in regional science? Papers of the Regional Science Association 24, 7–21 (1970)
9. Han, J., Kamber, M.: Data Mining: Concepts and Techniques. Morgan Kaufmann, San Francisco (2000)
10. Ketterlin, A.: Clustering sequences of complex objects. In: KDD, pp. 215–218 (1997)
11. Kuijpers, B., Othman, W.: Modelling uncertainty of moving objects on road networks via space-time prisms. International Journal of Geographical Information Science (to appear, 2009)
12. Miller, H.: Modeling accessibility using space-time prism concepts within geographical information systems. International Journal of Geographical Information Systems 5, 287–301 (1991)
13. Miller, H., Wu, Y.: GIS software for measuring space-time accessibility in transportation planning and analysis. GeoInformatica 4, 141–159 (2000)
14. Nanni, M.: Clustering Methods for Spatio-Temporal Data. PhD thesis, Computer Science Departrment, University of Pisa (2002)
15. Nanni, M., Pedreschi, D.: Time-focused clustering of trajectories of moving objects. Journal of Intelligent Information Systems 27(3), 267–289 (2006)
16. Nanni, M., Kuijpers, B., Körner, C., May, M., Pedreschi, D.: Spatiotemporal data mining. In: Mobility, Data Mining and Privacy, pp. 267–296 (2008)
17. Pfoser, D., Jensen, C.S.: Capturing the uncertainty of moving-object representations. In: Güting, R.H., Papadias, D., Lochovsky, F.H. (eds.) SSD 1999. LNCS, vol. 1651, pp. 111–132. Springer, Heidelberg (1999)
18. Rinzivillo, S., Pedreschi, D., Nanni, M., Giannotti, F., Andrienko, N., Andrienko, G.: Visually driven analysis of movement data by progressive clustering. Information Visualization 7, 225–239 (2008)
19. Vlachos, M.G.D., George Kollios, G.: Discovering similar multidimensional trajectories. In: ICDE, pp. 673–684 (2002)
20. Dijkstra, E.W.: A note on two problems in connexion with graphs. Numerische Mathematik 1, 269–271 (1959)
21. Wolfson, O.: Moving objects information management: The database challenge. In: Halevy, A.Y., Gal, A. (eds.) NGITS 2002. LNCS, vol. 2382, pp. 75–89. Springer, Heidelberg (2002)

Route Search over Probabilistic Geospatial Data

Yaron Kanza[1,*], Eliyahu Safra[2], and Yehoshua Sagiv[3,**]

[1] Technion—Israel Institute of Technology
kanza@cs.technion.ac.il
[2] Environmental Systems Research Institute
esafra@esri.com
[3] The Hebrew University of Jerusalem
sagiv@cs.huji.ac.il

Abstract. In a *route search* over geospatial data, a user provides terms for specifying types of geographical entities that she wishes to visit. The goal is to find a route that *(1)* starts at a given location, *(2)* ends at a given location, and *(3)* travels via geospatial entities that are relevant to the provided search terms. Earlier work studied the problem of finding a route that is *effective* in the sense that its length does not exceed a given limit, the relevancy of the objects is as high as possible, and the route visits a single object from each specified type. This paper investigates route search over *probabilistic geospatial data*. It is shown that the notion of an effective route requires a new definition and, specifically, two alternative semantics are proposed. Computing an effective route is more complicated, compared to the non-probabilistic case, and hence necessitates new algorithms. Heuristic methods for computing an effective route, under either one of the two semantics, are developed. (Note that the problem is NP-hard.) These methods are compared analytically and experimentally. In particular, experiments on both synthetic and real-world data illustrate the efficiency and effectiveness of these methods in computing a route under the two semantics.

1 Introduction

The recent growth of the world-wide web has made geographical applications prevalent and accessible to many different users. This has raised the need for geographical applications that are adapted to the environment of the web in the following three aspects. First, applications on the web should be simple, in order to be suitable for novice users. Secondly, in many cases, the applications should be able to deal with heterogeneous data, and in particular with data that is inaccurate and incomplete. Thirdly, applications should be highly efficient in order to be provided as web services. Efficiency is essential also when considering

---

* The work of this author was supported by the German-Israeli Foundation for Scientific Research & Development (Grant 2165-1738.9/07) and by a grant from the Goldstein UAV and Satellite Center at the Technion.
** The work of this author was supported by the German-Israeli Foundation for Scientific Research & Development (Grant 973-150.6/2007).

N. Mamoulis et al. (Eds.): SSTD 2009, LNCS 5644, pp. 153–170, 2009.

geographical applications that should run on devices with a limited computation power, such as cellular phones or car navigation systems.

One of the geographic applications that is recently being adapted to the web is a *route search*. In a route search, the goal is to find a route that starts at a given location, ends at a given location and travels via geographical entities that satisfy the search terms. The search terms define which geographical entities the route should visit in order to satisfy the needs of the user—*i.e.*, they define the types of these entities and additional constraints about them. For example, suppose that Alice has to go from her office to a meeting at a certain place, and on her way she wants to stop by at an ATM, a restaurant that serves a specific kind of food and a pharmacy. A route-search application should receive these requirements as a query and then return a route that starts at the office of Alice, ends at the location of the meeting, and travels via an ATM, a restaurant and a pharmacy, not necessarily in this specific order.

Data on the web might be inaccurate and incomplete. This can be caused by changes in the real world that have not yet been updated in the database. For instance, a restaurant has been closed, but still appears in the database. It can also be the result of errors caused by incorrect integration of data that arrive from different sources on the web. An integration allows utilizing different sources of information; however, an incorrect join leads to an inaccurate result. For instance, if one restaurant is erroneously joined with the menu of another restaurant, query results over such data may not be correct.

Uncertainty is affected not just by the data, but also by the accuracy of the query. Commonly, queries on the web are simple and imprecise, which leads to uncertainty regarding whether a given geographical entity satisfies the user requirements. For example, when Alice arrives at the restaurant, she might find that it does not serve food that complies with her diet. Similarly, the pharmacy may be closed or may not include the medicine she needs.

For coping with inaccuracy and incompleteness, we use a probabilistic model. In a probabilistic model, each geographic object is assigned a *probability-of-success*. The probability-of-success indicates how likely it is that the geographical entity represented by the object will satisfy the user specifications. For example, when a user asks for a route that travels via a place that sells pizza, a place called "Pizza Hut" will receive a very high probability of success. Similarly, a place called "Italian Restaurant" will also receive a high probability of success. However, for a place called "Japanese Restaurant," the assigned probability should be much lower. This assigned probability can be based on collected statistics, on information-retrieval techniques and on user profiling, *e.g.*, for a user whose profile indicates he has a low income, the probability-of-success of expensive restaurants should be reduced in comparison to the probability assigned to economy restaurants. How exactly to assign these statistics is, however, beyond the scope of this paper.

When computing a route over probabilistic data, two approaches are possible. In the first approach, the route should travel via exactly one object of each required type, while attempting to go via the objects whose probability-of-success is as high as possible; *e.g.*, in the above example, the generated route will go via exactly one restaurant, one ATM and one pharmacy. This approach is problematic in the following sense. In this approach, the probability-of-success for a route is the probability that all the objects on this route will satisfy the user. Yet, if for one particular type of entities, all the objects of that type have low probabilities, then the probability of the entire route will also be low. For instance, if in the route search of Alice, all the restaurant objects in the database have a low probability, the system will not be able to create a route that has a high probability of success.

In the second approach, the route can go via several objects of each type and the probability-of-success is the probability that for each type, the route has at least one object of that type that meets the user requirements. For instance, a route for Alice may travel via several ATMs to increase the probability that Alice will visit a functioning ATM. In this paper, we investigate only this approach. Note that in order for the route to be effective, it should not be too long. Thus, the set of objects that the route travels through should be chosen carefully.

In this paper, we propose two semantics for route-search queries over probabilistic data. Under the *bounded-length semantics*, a distance limit ℓ is provided and the goal is to find the route with the highest probability of success among the routes that satisfy the query and whose length does not exceed ℓ. Under the *bounded-probability semantics*, a probability value p is given, and the goal is to find the shortest route among the routes that satisfy the query and provide probability of success not smaller than p.

The bounded-length semantics is appropriate when the length of the route must not exceed a given value. For example, a customer of a rental-car agency would like not to exceed the traveling distance that is included in the base price. The bounded-probability semantics is suitable when the user can be flexible with the length, yet would like to guarantee a certain probability of success.

As a web application, a route-search algorithm should be highly efficient. However, computing a route under either the bounded-length semantics or the bounded-probability semantics is an NP-hard problem—simple reductions from the traveling-salesman problem (TSP) show that. Hence, in this paper, we present heuristics for the problems.

The paper is organized as follows. In Section 2, we present the formal framework and define the two semantics of route search over probabilistic data. In Section 3, we survey recent papers on route search, and we compare the computation of a route under the two semantics to similar problems that exist in the literature. Heuristic algorithms for computing a route under the two semantics are presented in Section 4. In Section 5, we present the results of experiments that illustrate the effectiveness and efficiency of our algorithms. Finally, we conclude in Section 6.

2 Probabilistic Route Search

In this section, we present our formal framework and define the concept of a *probabilistic route search*.

2.1   Geographical Datasets

A *geo-spatial dataset* is a collection of geo-spatial objects. Each object represents a real-world geographical entity and has a location—the location of an object is the location of the entity it represents. An object may have additional spatial and non-spatial attributes. Height and shape are examples of spatial attributes. Address and name are examples of non-spatial attributes. We assume that locations are points and are unique, *i.e.*, different objects have different locations. For objects that are represented by a polygonal shape and do not have a specified point location, we consider the center of mass of the polygonal shape to be the point location.

For simplicity, we measure the distance between two objects in terms of the Euclidean distance between their point locations. However, our algorithms do not assume that distances are Euclidean and, hence, they are applicable also when movement is constrained to a road network. We denote the distance between two objects o_1 and o_2 by $distance(o_1, o_2)$. Similarly, if o is an object and l is a location, then $distance(o, l)$ is the distance from o to l.

2.2   Search Queries

Users specify what entities they would like to visit by *search queries*. Formally, a query is a pair $Q = (W, C)$, where *(1)* W is a set of keywords, *and (2)* C is a set of constraints having the form $A \diamond v$, such that A is an attribute name, v is a value and $\diamond$ is a comparison symbol among $=, <, >, \neq, \leq$ and $\geq$. For instance, Hotel, Wireless Internet Access, rank ≥ 3, price ≤ 100 specify that the user would like to go via a hotel that provides an Internet wireless connection, has a ranking of at least three stars and a rate that does not exceed \$100.

In a non-probabilistic setting, there is a clear-cut notion of when an object o satisfies a given query. In our framework, each object has some degree of *relevancy* to a given search query Q. That degree is stated as a *probability-of-success* (or *probability*, for short), which is a value between 0 and 1. This probability indicates how high are the chances that o satisfies the user requirements. We denote the probability of an object o by $Pr(o)$.

Methods for determining the probability of an object are beyond the scope of this paper. In a nutshell, initial probabilities are determined when objects are created or updated in the database. Those probabilities may be derived by estimating the reliability of the information sources, as well as other factors. In addition, rules could be applied in order to adjust the probability of an object according to the query at hand. For example, if an attribute of a restaurant object contains the string "Italian food," then a rule may determine that the restaurant serves pizza with probability 0.9. Finally, the underlying model determines how to compute the probability of joint events. For example, the probability that a restaurant object serves pizza and charges moderate prices could be computed by assuming that these two events are independent. In summary, we assume that the probability of each object is given and incorporates all the relevant factors.

2.3   Route-Search Queries

In a *route-search query*, the user specifies a *source location*, a *target location* and
the entities that the route should visit. We represent a route-search query as a
triplet $R = (s, t, Q)$, where s is a source location, t is a target location and Q is
a set of search queries.

Example 1. Consider again the route-search task of Alice presented in Section 1.
A suitable route-search query for this task should include *(1)* the location s of
the office of Alice, *(2)* the location t where the meeting should be held, and
(3) the following three search queries: $Q_1 = \{\texttt{restaurant esoteric food}\}$,
where "esoteric food" refers to the type of diet Alice needs; $Q_2 = \{\texttt{ATM}\}$; and
$Q_3 = \{\texttt{pharmacy}\}$.

Consider a route-search query $R = (s, t, Q)$, and let $Q_1, \dots, Q_m$ be the search
queries of Q. The *result* of Q_i, denoted by A_i, comprises the objects of the
database that are relevant to Q_i. We assume that the A_j are pairwise disjoint.
In other words, distinct search queries of Q refer to different types of objects.
For example, one search query is about hotels, another is concerning restaurants,
etc. A *pre-answer* to R is a route that starts at s, ends at t and goes via objects
of the results $A_1, \dots, A_m$. That is, a route is a sequence $s, o_1, \dots, o_k, t$, such that
each o_i belongs to some A_j. The *length* of the route is the sum of the distances
between consecutive objects, that is,

$$distance(s, o_1) + \Sigma_{i=1}^{k-1} distance(o_i, o_{i+1}) + distance(o_k, t).$$

As mentioned earlier, each object o_i on the route has a probability $Pr(o_i)$ that
indicates its relevancy to the corresponding search query. Let ρ denote the above
route, namely, $s, o_1, \dots, o_k, t$. The *probability-of-success* of ρ (or *probability*, for
short), denoted by $Pr(\rho)$, is given by the following equation.

$$Pr(\rho) = \Pi_{j=1}^{m}(1 - \Pi_{o_i \in A_j}(1 - Pr(o_i))).$$

This equation is derived as follows. First, $1 - Pr(o_i)$ is the probability that object
$o_i \in A_j$ does not satisfy the requirements implied by the search query Q_j. So,
for a given A_j, the product $\Pi_{o_i \in A_j}(1 - Pr(o_i))$ is the probability that A_j has no
relevant object on the route ρ. Hence, $1 - \Pi_{o_i \in A_j}(1 - Pr(o_i))$ is the probability
of the complement event, namely, A_j has at least one relevant object o_i on the
route ρ. The final product over all the A_j is the probability of the route ρ.

The *answer* to a route-search query is a pre-answer chosen according to a specific semantics. In this paper, we present two semantics for route-search queries
over probabilistic data.

Bounded-Length Semantics. Let ℓ be a given distance limit. Under the
bounded-length semantics, the answer is the pre-answer that has the highest
probability of success among the pre-answers whose length does not exceed ℓ.

Bounded-Probability Semantics. Let p be a given probability threshold.
Under the *bounded-probability semantics*, the answer is the pre-answer that has
the shortest length among the pre-answers whose probability of success is not
smaller than p.

3   Related Work

Some variations of route search have been investigated in earlier works. Safra et al. [13] studied the case of a route search over uncertain data where all the objects belong to the same set. This is similar, but not identical, to a special case of the route search presented in this paper, where there is a single search query in R and the bounded length semantics is used. This special case also has some similarity to the *orienteering problem*. In the orienteering problem, the input consists of a distance limit, a start location and a set of objects where each object has a score. The problem is to compute a route that (1) starts at the given starting location, (2) have a length that does not exceed the given distance limit and (3) goes via objects whose total score is maximal. The orienteering problem has been studied extensively [2,5] and several heuristics [1,4,8,9,16] and approximation algorithms [12] were proposed for it.

There are two main differences between orienteering and the problem of computing a route under the bounded-length semantics. First, in a route search, the objects are divided into sets (the sets are the answers to the search queries) and an object from each set must be visited in order to have a probability-of-success greater than zero. In the orienteering problem, objects differ only in their location and score.

Secondly, in the orienteering problem, when an object is added to a route, its effect on the total score is always equal to its score, while in a probabilistic route search, the effect of an object on the probability depends on both the probability of the object and the probabilities of objects in the route. For instance, an object with score 0.5 will always add 0.5 to the score of an orienteering sequence. However, in a probabilistic route search, an object of set A with probability 0.5 will have a different effect in each of the following three cases: a route that does not contain any object of A, a route that contains an object of A with probability 0.4, and a route that contains an object of A with probability 1.

Because of these differences, there is no simple way of using heuristic or approximation algorithms of the orienteering problem to solve a route search, even when the route search comprises a single search query.

A route search over objects that are partitioned into sets has been studied by Kanza et al. [7]. However, they considered a semantic of route search that is different from the semantics suggested in this paper. First, under their semantics, the route must visit exactly one object of each set while in the probabilistic case, a route can visit several objects of each set. Secondly, as explained above, in the non-probabilistic case, the effect of an object on a route depends on the object, whereas in the probabilistic case, it depends on both the object and the route. Therefore, algorithms and heuristics for the non-probabilistic case do not solve a route search over probabilistic data. For instance, consider a search for a phone booth and the following two routes, having the same length. A route that goes via a single place which has a phone booth with probability 0.6, and a route that goes via 10 places where a phone booth can exist with probability 0.5. A non-probabilistic search will prefer the first route while the probability-of-success is much greater in the second route.

Several papers [3,10,11,14] study route-search queries over datasets in which objects have neither scores nor probabilities. The work of [15] investigates two variants of the shortest-route problem. In one, there should be exactly k intermediate points. In the second, the distance between any two consecutive points should not exceed a given bound. In the framework of [15], there are no scores, probabilities, or different types of points. In [6], they consider a user who travels a predetermined route in a road network. Their goal is to modify the route so that a new point is visited in a manner that optimizes some spatial preferences.

4   Algorithms

Under the bounded-length semantics, even a restricted version of the problem is NP-hard. To prove it, consider an instance of the following decision problem: does the traveling-salesman route has a length of at most ℓ. We construct a database D and a route-search query $R = (s, t, \mathcal{Q})$ as follows. The objects of D are those of the traveling-salesman problem. An arbitrary object o of D is chosen as both s and t. $\mathcal{Q}$ has a single search query Q, such that every object of $D \setminus \{o\}$ is relevant to Q with probability 0.5. The distance bound is ℓ. Clearly, the length of the traveling-salesman route is at most ℓ if and only if the answer to the route-search query (under the bounded-length semantics) has probability $1 - 0.5^{n-1}$, where n is the number of objects in D. Under the bounded-probability semantics, we can use the same reduction in order to compute the length of the traveling-salesman route. For that, we have to choose $1 - 0.5^{n-1}$ as the probability threshold.

4.1   Heuristics for the Bounded-Length Semantics

Since route search is a hard problem, we present heuristics instead of algorithms that provide exact answers. In this section, we give four heuristics for route search under the bounded-length semantics. These heuristics were devised to be efficient, which is a necessary requirement when developing web applications.

Throughout this section, $R = (s, t, \mathcal{Q})$ is the given route-search query, where s is the start location, t is the target location, and $\mathcal{Q} = \{Q_1, \ldots, Q_m\}$ are the search queries. D is the dataset over which R is posed and $A_1, \ldots, A_m$ are the results of the search queries, i.e., $A_i = Q_i(D)$ for all $1 \leq i \leq m$. The probability-of-success of each object $o \in D$ is denoted by $Pr(o)$. Finally, ℓ is the length limit.

Ratio-Greedy Heuristic. The *Ratio-Greedy Heuristic* (RGreedy, for short) is our baseline for comparison with the other, more advanced heuristics. It is presented in Fig. 1. Initially, it generates a route ρ that *(1)* goes from s to t, *(2)* visits exactly one object of each set A_i, and *(3)* is as short as possible. The task of computing this initial route is by itself a hard problem, and thus, it is solved heuristically. In our implementation, we used the Infrequent-First Heuristic of [7] for this task.

After the initial step of Line 3, the RGreedy algorithm iteratively increases the probability-of-success of the route, using a greedy approach. While the length of ρ does not exceed the length limit ℓ, the algorithm repeatedly extends ρ by adding

---

Ratio Greedy $((s, t, Q_1, \ldots, Q_m), \ell, D, Pr())$

Input: Source location s, target location t, search queries $Q_1, \ldots, Q_m$, limit distance ℓ, a dataset D, a probability function $Pr()$
Output: A route starting at s, ending at t, not exceeding a length of ℓ and attempting to provide the highest probability-of-success

1: **for** $i = 1$ to m **do**
2:     $A_i \leftarrow Q_i(D)$
3: $\rho \leftarrow$ ShortestRouteHeuristic$((s, t, Q_1, \ldots, Q_m), D)$ // the initial short route

4: *Candidates* $\leftarrow \{o \mid$ there exists an i such that $length(Add_i(\rho, o)) \leq \ell\}$
5: **while** *Candidates* $\neq \emptyset$ **do**
6:     let i be an index and o_m be an object of *Candidates* such that for all $o \in$ *Candidates* and index j, it holds that $Ratio_i(\rho, o_m) \geq Ratio_j(\rho, o)$
7:     $\rho \leftarrow Add_i(\rho, o_m)$
8:     *Candidates* $\leftarrow \{o \mid$ there exists an i such that $length(Add_i(\rho, o)) \leq \ell\}$
9: **return** ρ

Fig. 1. The Ratio-Greedy Heuristic for answering route-search queries under the bounded-length semantics

a new object in each iteration of Line 5. We use $Add_i(\rho, o)$ to denote the route that is obtained by adding the object o between the i-th and the $(i+1)$-st objects of ρ. That is, if $\rho = s, o_1, \ldots, o_k, t$ then $Add_i(\rho, o) = s, \ldots, o_i, o, o_{i+1}, \ldots, t$. When $i = 0$, we add o between s and o_1, and similarly, when $i = k$, we add o between o_k and t. An object o is a *candidate* for extension if after adding o to ρ, the length of the new route does not exceed ℓ, that is, there is an i such that $length(Add_i(\rho, o)) \leq \ell$.

The candidate that is actually added to the current route is the one that gives the largest ratio of the increase in the probability-of-success to the increase in the length of the route. That is, let

$$Ratio_i(\rho, o) = \frac{Pr(Add_i(\rho, o)) - Pr(\rho)}{length(Add_i(\rho, o)) - length(\rho)}$$

be this ratio when inserting o at position i of the current route ρ. Each iteration of Line 5 chooses the maximal $Ratio_i(\rho, o)$ and adds o at position i of ρ. Note that o is added at the position where it causes the smallest increase in the length of the route, because the probability does not depend on the length. Hence, the length limit is not exceeded.

We also experimented with two other criteria for adding objects. One is choosing the object that maximizes the increase in the probability-of-success. The second is picking the object that minimizes the increase in the length. However, in the experiments, these two criteria were found to be inferior to choosing according to the largest ratio.

Increase-Decrease $((s, t, Q_1, \ldots, Q_m), \ell, D, Pr())$

Input: Source location s, target location t, search queries $Q_1, \ldots, Q_m$, limit distance ℓ, a dataset D, a probability function $Pr()$
Output: A route starting at s, ending at t, not exceeding a length of ℓ and attempting to provide the highest probability-of-success

1: **for** $i = 1$ to m **do**
2:    $A_i \leftarrow Q_i(D)$
3: $\rho \leftarrow$ RGreedy$((s, t, Q_1, \ldots, Q_m), \ell, D, Pr())$ // ρ is the initial route
4: let k be the number of objects in the current route, i.e., $\rho = s, o_1, \ldots, o_k, t$
5: $r \leftarrow \lceil \frac{k}{2} \rceil$
6: **while** $r > 0$ **do**
7:    **for** $i = 0$ to $k - r$ **do**
8:        let ρ_i^- be the result of removing from ρ the objects $o_{i+1}, \ldots, o_{i+r}$, i.e.,

$$\rho_i^- = \begin{cases} s, o_{r+1}, \ldots, o_k, t & : \quad i = 0 \\ s, o_1, \ldots, o_i, o_{i+r+1}, \ldots, t & : \quad 0 < i < k - r \\ s, o_1, \ldots, o_{k-r}, t & : \quad i = k - r \end{cases}$$

9:    let ρ_m^- be a route in $\{\rho_0^-, \ldots, \rho_{k-r}^-\}$ whose length is minimal, *i.e.*, for every
      $0 \leq i \leq k - r$ it holds that $length(\rho_m^-) \leq length(\rho_i^-)$
10:   let $(\rho_m^-)^+$ be the result of adding objects to ρ_m^- by applying the increase
      stage of the Ratio-Greedy Heuristic (Lines 4-9 of Fig. 1)
11:   **if** $Pr((\rho^-)^+) > Pr(\rho)$ **then**
12:     $\rho \leftarrow (\rho^-)^+$
13:   **else**
14:     $r \leftarrow r - 1$
15: **return** ρ

Fig. 2. The Increase-Decrease Heuristic for answering route-search queries under the bounded-length semantics

Increase-Decrease Heuristic. The Ratio-Greedy Heuristics only adds objects. An object o that is inserted into the route, either in the initial step or during the iterative stage, remains in the route even if its utility is eventually diminished (this may happen when subsequent iterations add many objects that are far away from o). The *Increase-Decrease Heuristic* (Inc-Dec, for short) changes that by considering the option of removing objects from the current route and replacing them with some other objects.

Increase-Decrease is presented in Fig. 2. Initially, Line 3 generates a route using the Ratio-Greedy Heuristic. Then, each iteration of Line 6 picks r adjacent objects as candidates for removal. In particular, Line 9 chooses the r contiguous objects that yield the largest decrease in the length of the current route. These r objects are actually removed if the Ratio-Greedy Heuristic can add some other objects, such that the new route has a higher probability-of-success and does not exceed the length limit (Lines 10–12). If the objects are indeed removed, r is not changed; otherwise, it is decreased by one. In either case, the next iteration of Line 6 tries again to replace r objects. The algorithm terminates when $r = 0$.

Extra-Length Heuristic. The former heuristics are unlikely to add objects that are not close to the initial route, because of the length limit. Yet, in some cases, it may be beneficial to add objects that are far away from the initial route if those objects have high probabilities and are clustered together in a small area. The approach of the *Extra-Length Heuristic* (Ext-Len, for short), presented in Fig. 3, is to start by applying the Ratio-Greedy Heuristic with a *pseudo* length limit that is larger than the actual one (*e.g.*, multiply the given ℓ by five). This makes it possible to generate intermediate routes that visit far-away objects. Clearly, the route obtained in Line 4 is too long. Hence, objects are repeatedly removed in the loop of Line 5. In each iteration, the removed object is the one that gives the largest ratio of the decrease in the length to the decrease in the probability-of-success. For an object o, this ratio is

$$RemRatio(\rho, o) = \frac{length(\rho) - length(Remove(\rho, o))}{Pr(\rho) - Pr(Remove(\rho, o))}$$

where $Remove(\rho, o)$ is the route obtained by removing o from ρ. Since this may lead to removing "too much," a final stage of adding objects using the Ratio-Greedy approach is applied (Line 8).

When using a large length limit, the probability-of-success of the current route can become close to one. Consequently, all the candidates for addition to

---

Extra Length $((s, t, Q_1, \ldots, Q_m), \ell, D, Pr())$

Input: Source location s, target location t, search queries $Q_1, \ldots, Q_m$, limit distance ℓ, a dataset D, a probability function $Pr()$
Output: A route starting at s, ending at t, not exceeding a length of ℓ and attempting to provide the highest probability-of-success

1: **for** $i = 1$ to m **do**
2:     $A_i \leftarrow Q_i(D)$
3:     $\ell_{extra} \leftarrow 5\ell$
4:     $\rho \leftarrow RGreedy^\varepsilon((s, t, Q_1, \ldots, Q_m), \ell_{extra}, D, Pr())$ /* ρ is the initial route generated by using the Ratio-Greedy Heuristic while guaranteeing that the probability-of-success does not exceed $1 - \varepsilon$ */
5: **while** $length(\rho) > \ell$ **do**
6:     let o_r be an object in ρ such that $RemRatio(\rho, o_r) \geq RemRatio(\rho, o)$ for every object o in ρ
7:     $\rho \leftarrow Remove(\rho, o_r)$
8: add objects to ρ by applying the increase stage of the Ratio Greedy Heuristic (Lines 4-9 of Fig. 1)
9: **return** ρ

Fig. 3. The Extra-Length Heuristic for answering route-search queries under the bounded-length semantics

---

Universal Start $((s, t, Q_1, \ldots, Q_m), \ell, D, Pr())$

Input: Source location s, target location t, search queries $Q_1, \ldots, Q_m$, limit distance ℓ, a dataset D, a probability function $Pr()$

Output: A route starting at s, ending at t, not exceeding a length of ℓ and attempting to provide the highest probability-of-success

1: **for** $i = 1$ to m **do**
2:    $A_i \leftarrow Q_i(D)$
3: $\rho_m \leftarrow s, t$
4: **for each** o in D **do**
5:    generate an initial route $\rho = s, d_1, \ldots, d_m, o, t$ where o is an object of D and $d_1, \ldots, d_m$ are dummy objects all located at s, satisfying $d_i \in A_i$, for $1 \leq i \leq m$ and with probability-of-success that is a small ε (*i.e.*, the dummy objects do not affect the length of the route and just guarantee that the probability-of-success of ρ is not zero, even before adding an object of each group among $A_1, \ldots, A_m$)
6:    add objects to ρ by applying the increase stage of the Ratio Greedy Heuristic (Lines 4-9 of Fig. 1)
7:    **if** $Pr(\rho) > Pr(\rho_m)$ **then**
8:        $\rho_m \leftarrow \rho$
9: **return** ρ_m

Fig. 4. The Universal-Start Heuristic for answering route-search queries under the bounded-length semantics

the current route yield similar ratios, and the Ratio-Greedy Heuristic may add objects that have very low probabilities. In order to avoid that, we use a variation RGreedy$^\varepsilon$ of the Ratio-Greedy Heuristic that prevents the current route from reaching a probability-of-success that is too close to one. This variant stops when either the given length limit is reached or the probability-of-success exceeds $1 - \varepsilon$.

Universal-Start Heuristic. All of the former heuristics may fail to consider some of the objects that are needed for generating a route that is (close to) the best. The *Universal-Start Heuristic* (UStart, for short), shown in Fig. 4, surmounts this problem by applying the Ratio-Greedy Heuristic multiple times. In particular, in the loop of Line 4, UStart considers all the objects o that are within the length limit. For a given o, Line 5 creates an initial route consisting of the start location, the target location and o itself. Line 6 applies the Ratio-Greedy Heuristic to this initial route. The output is the route with the highest probability-of-success. In order for the initial route to have a nonzero probability-of-success, Line 5 adds dummy objects that do not change the length and have an ε effect on the probability, where ε is a very small constant. These placeholders are removed after adding the relevant objects, *i.e.*, the dummy object of A_i is deleted from the current route when an object of A_i is added to the route.

4.2   Heuristics for the Bounded-Probability Semantics

With some minor modifications, the above heuristics for the bounded-length semantics can also be used for the bounded-probability semantics. We now describe these modifications.

In the case of the the Ratio-Greedy Heuristic, we only need to change the termination condition. In other words, this heuristic repeatedly extends the current route, as described earlier, until the probability-of-success of the current route exceeds the threshold p. Clearly, this modification applies to the original Ratio-Greedy Heuristic (Fig. 1) as well as to the application of this heuristic by the other three algorithms.

The Extra-Length Heuristic is changed into an Extra-Probability Heuristic as follows. We first generate an initial route that has a much higher probability than the given threshold p. We do it by multiplying p by some constant and then applying the modified Ratio-Greedy Heuristic. In addition, we change the termination condition in the phase of removing objects. This phase continues while the probability-of-success is greater than the probability bound p.

No changes are needed in the Increase-Decrease and the Universal-Start Heuristics, except for the obvious fact that they apply the Ratio-Greedy Heuristics with the modification described above.

4.3   Complexity Analysis

We now discuss the complexity of the four heuristics. For simplicity, we consider the heuristics only under the bounded-length semantics.

First, we note that some of the objects of the underlying dataset may not be relevant to the computation of a given route-search query. Specifically, given a query $R = (s, t, \mathcal{Q})$ and a length limit ℓ, only the objects in the elliptic area $\{u \mid distance(s, u) + distance(u, t) \leq \ell\}$ are relevant to the computations. Objects outside this area cannot be in the result. When a spatial data structure exists for the dataset over which R is computed, it can be used for retrieving the relevant objects before the computation. Otherwise, the algorithms can start by reading the entire dataset and filtering out irrelevant objects. Hence, in the rest of this section, we will consider D to be a dataset of merely n relevant objects.

In the Ratio-Greedy Heuristic, the complexity of constructing the initial route depends on the algorithm being used. The iterative stage has $O(n^3)$ time complexity. To see that, first note that there are at most n iterations of adding an object to the route. In each iteration, at most n objects are considered as candidates for addition. The computation where to add an object and the computation of the probability are proportional to the length of the sequence, which is at most n. Note that if we consider k to be the maximal number of objects in a route, then the time complexity is actually $O(nk^2)$.

For the Increase-Decrease Heuristic, again our analysis considers only the phase after the initial route has been computed. In this algorithm, the main iteration is performed at most 2^n times. To see why, observe that in each iteration, we replace a subset of the route by a new subset. Since the probability of the route constantly increases, no two routes can have exactly the same set of objects. Hence, there are at most 2^n possible replacements. Now, in each iteration, the iterative stage of the Ratio-Greedy Heuristic is executed (and as shown above, it has $O(nk^2)$ time complexity). Thus, the time complexity of this heuristic is $O(n2^n k^2)$.

The time complexity of the Extra-Length Heuristic is similar to that of the Ratio-Greedy Heuristic, $i.e.$, $O(nk^2)$. However, the former does more iterations than the latter, because it uses a larger length limit and, moreover, it has an additional phase of removing objects. Finally, the Universal-Start Heuristic applies the Ratio-Greedy Heuristic n times and, thus, has $O(n^2 k^2)$ time complexity.

5   Experiments

We have tested our algorithms over both synthetically-generated datasets and real-world datasets. The goal of our experiments was to compare our methods in terms of efficiency and the quality of the results. The experiments were conducted on a PC equipped with a Core 2 Duo processor 2.13 GHz (E6400), 2 GB of main memory and Windows XP Professional operating system.

5.1   Tests on Real-World Data

We extracted real-world data from a digital map of the City of Tel-Aviv. A fragment of that data is shown in Fig. 5. Specifically, we used objects of the "Point Of Interest" (POI) layer of the map. This layer represents many different types of geographical entities. The extracted dataset comprises 103 objects of three different types (20 cinemas, 29 hotels, 54 post offices). As a query, we used a particular choice of source and target locations, and assumed that all the objects are in the answer to the query. The objects received probabilities that are normally distributed, with mean of 0.5. We ran additional experiments on more queries as well as another dataset with five types of objects, and the results were similar.

Evaluation under the Bounded-Length Semantics. Fig. 7 and Fig. 8 present the test results for the four heuristics over the real-world dataset, under the bounded-length semantics. Fig. 7 shows the quality of the answer of each algorithm as a function of the length limit ℓ. In this graph, the x-axis shows the factor by which the length limit ℓ is larger than the length of the initial route created by RGeedy. Recall that this initial route has exactly one object from each A_i and is constructed by the Infrequent-First Heuristic of [7]. For instance, when x is 1.8, it means that $\ell = 1.8 \times a$, where a is the length of the initial route. The y-axis is the probability-of-success of the answer. It can be seen that UStart provides the best answers. Ext-Len provides answers that are almost as good. The quality of the answers of RGreedy are the worst among the four heuristics.

Fig. 8 presents the running times of the methods as a function of the length limit ℓ. The x-axis is similar to the x-axis in Fig. 7. The y-axis shows the running time of the computation, in milliseconds. RGreedy is the most efficient. Ext-Len

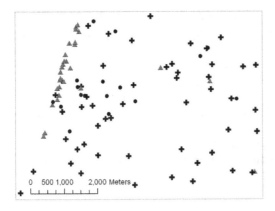

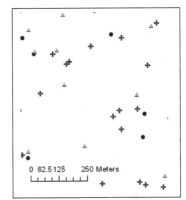

Fig. 5. A fragment of the real-world dataset

Fig. 6. A fragment of the synthetic dataset

is almost as efficient as RGreedy. Note that its running time does not change as ℓ grows, because even when ℓ is small, the computation is with respect to a larger length (i.e., 5ℓ). The Inc-Dec heuristic is efficient for small values of ℓ, but its running time climbs sharply as ℓ grows, and then it drops again. The reason for that is that the size of the replaced subsets, as well as their number, grows when the route becomes longer. However, the probability of the generated route also grows when the length limit increases. In the implementation, Inc-Dec stops when it finds a route with a probability that is very close to one. This happens early when the length limit is large and, therefore, the running time drops.

The running times of UStart are not shown in Fig. 8, because they are much greater than those of the other methods. In order to compare them to the running times of the other three methods, they are presented in Table 1. For large values of the length limit, the running time of UStart drops, because UStart quickly finds a route with a probability that is very close to one (and in the implementation, UStart stops when that happens).

Table 1. The running times (milliseconds) under the bounded-length semantics

Length	RGreedy	Inc-Dec	Ext-Len	UStart
1.01	2	7	81	34
1.4	7	43	90	551
1.8	10	61	82	2193
2.2	24	331	100	209
2.6	45	179	80	178

Evaluation under the Bounded-Probability Semantics. Fig. 9 and Fig. 10 present the test results of the algorithms over the real-world dataset, under the bounded-probability semantics. Fig. 9 shows the length of the answer as a function of the probability threshold. Not surprisingly, these results are analogous

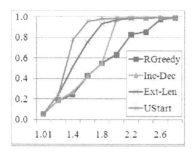

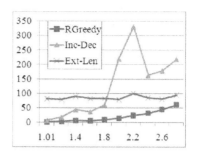

Fig. 7. Probability versus length limit

Fig. 8. Running times (millisecond) versus length limit

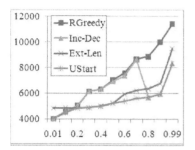

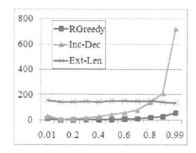

Fig. 9. Length versus probability bound

Fig. 10. Running times (millisecond) versus probability bound

to the results under the bounded-length semantics. UStart provides the best answers, Ext-Len is almost as good as UStart, while RGreedy provides the longest route among the four methods. The test results of Inc-Dec are similar to those of RGreedy for low probabilities, and similar to those of UStart for high probabilities. The reason for that is that when the probability threshold increases, the generated route becomes longer. Hence, the size and number of the replaced subsets is greater, thereby creating more opportunities for improving the route.

Fig. 10 shows the running time of each methods as a function of the probability threshold. Here, as well, the results are similar to the results under the bounded-length semantics. RGreedy is the most efficient. Ext-Len is almost as efficient as RGreedy and its running time does not change much as a function of the probability threshold. Inc-Dec is very efficient for small and medium probability thresholds, but is very inefficient in the case of high thresholds, because it makes many iterations of replacement in this case.

5.2   Tests on Synthetic Data

In order to have control over our experiments, and for testing our methods over datasets with specific properties, we used synthetically generated datasets. In a synthetic dataset, we have control over the distribution of the locations of objects

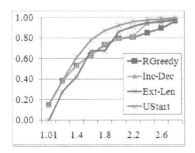

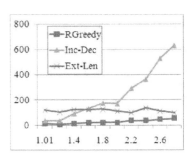

Fig. 11. Probability versus length bound

Fig. 12. Running times (milliseconds) versus length bound

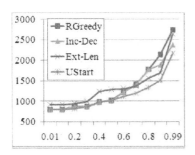

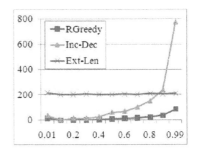

Fig. 13. Length versus probability bound

Fig. 14. Running times (milliseconds) versus probability bound

in the area of the map, the way that the objects are partitioned into sets, etc. For generating synthetic datasets, we implemented a random-dataset generator. Our generator is a two-step process. First, the objects are generated. The locations of the objects are randomly chosen according to a given distribution, in a square area. In the second step, we partition the objects into sets and a probability value is attached to each object. The partitioning of objects into sets can be uniform or according to a distribution specified by the user.

The user provides the following parameters to the dataset generator. The number of objects, the size of the square area in which the objects are located and the minimal distance between objects; for simulating search results, the user provides the distribution of object probabilities, and the distribution of the size of the sets in the partition. These parameters allow a user to generate tests with different sizes of datasets and different partitions of the datasets into sets.

We conducted experiments over several synthetically generated datasets and multiple choices of queries. We present a few, typical results. Figures 11, 12, 13 and 14 are for a particular query and a dataset of 150 objects distributed over an area of 1300 × 1300 square meters (a fragment of this dataset is shown in Fig. 6). The objects in this dataset are partitioned into three types having 25, 50 and 75 objects. All the objects are in the answer to the query.

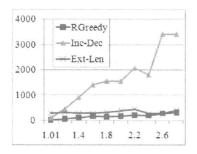

Length	RGreedy	Inc-Dec	Ext-Len	UStart
1.01	14	75	191	216
1.4	118	898	227	11333
1.8	159	1546	407	47521
2.2	216	2045	602	8965
2.6	282	3392	579	1858

Fig. 15. Running times (millisecond) versus length bound over large dataset

Fig. 16. The time for computing a route over dataset with 389 features

Fig. 11 shows the probability-of-success of the answer as a function of the length limit, under the bounded-length semantics. Fig. 12 depicts the running time as a function of the length limit, under the bounded-length semantics. The differences between the methods are not as significant as in the case of real-world data. The reason for that is the uniform distribution of the objects in the area that is queried. Also note that the Ext-Len method usually provides answers that are almost the best, and it does so in time that is almost the shortest. However, it may fail, as can be seen in Fig. 11 when the length limit is very small. In this case, Ext-Len returns zero as the probability of success, because during the removal stage, it removes all the objects of one of the sets.

The results over the synthetic data under the bounded-probability semantics are depicted in Fig. 13 and in Fig. 14. In general, the results over the synthetic data show that Inc-Dec efficiently computes good answers when the bound (of either the length or the probability) is small. For medium or large bounds, Ext-Len is efficient and provides good answers, in comparison to the other methods.

We also tested the scalability of the four heuristics. For that, we used a larger dataset with 389 objects and groups of sizes 31, 169 and 189. The results, which are summarized in Fig. 15 and Fig. 16, show that RGreedy and Ext-Len are scalable. Inc-Dec remains efficient only when the length limit is not high.

6   Conclusion

In this paper, we investigated the problem of route search over probabilistic datasets. We explained why algorithms and heuristics for non-probabilistic route search are inapplicable to probabilistic data, and then, we gave four heuristics for the problem. We showed how these heuristics can be applied under the two proposed semantics. In these heuristics, there is a tradeoff between the efficiency and the quality of the answers. Our experiments over both real-world and synthetically-generated data illustrated this tradeoff and showed that the UStart heuristic provides the best answers at the cost of a relatively long computation time. RGreedy is the most efficient method; however, usually its answers are inferior to those of the other methods. Inc-Dec provides a good tradeoff between the computation time and the quality of the answers when the bounds (i.e., the length limit and the probability threshold) are small. Finally, Ext-Len provides a good tradeoff between the computation time and the quality of the answers in all cases, except when the bounds are small. For future work, we plan to investigate how to use these algorithms in an adaptive way in cases where the user can send or receive information while traveling.

References

1. Chao, I., Golden, B., Wasil, E.: A fast and effective heuristic for the orienteering problem. European Journal of Operational Research 88(3), 475–489 (1996)
2. Chao, I., Golden, B., Wasil, E.: The team orienteering problem. European Journal of Operational Research 88, 464–474 (1996)
3. Chen, H., Ku, W.S., Sun, M.T., Zimmermann, R.: The multi-rule partial sequenced route query. In: GIS, pp. 1–10 (2008)
4. Golden, B., Wang, Q., Liu, L.: A multifaceted heuristic for the orienteering problem. Naval Research Logistics 35, 359–366 (1988)
5. Golden, B.L., Levy, L., Vohra, R.: The orienteering problem. Naval Research Logistics 34, 307–318 (1987)
6. Huang, X., Jensen, C.S.: In-route skyline querying for location-based services. In: Kwon, Y.-J., Bouju, A., Claramunt, C. (eds.) W2GIS 2004. LNCS, vol. 3428, pp. 120–135. Springer, Heidelberg (2005)
7. Kanza, Y., Safra, E., Sagiv, Y., Doytsher, Y.: Heuristic algorithms for route-search queries over geographical data. In: GIS, pp. 1–10 (2008)
8. Keller, P.C.: Algorithms to solve the orienteering problem: A comparison. European Journal of Operational Research 41, 224–231 (1989)
9. Leifer, A.C., Rosenwein, M.S.: Strong linear programming relaxations for the orienteering problem. European J. of Operational Research 73, 517–523 (1994)
10. Li, F., Cheng, D., Hadjieleftheriou, M., Kollios, G., Teng, S.H.: On trip planning queries in spatial databases. In: Bauzer Medeiros, C., Egenhofer, M.J., Bertino, E. (eds.) SSTD 2005. LNCS, vol. 3633, pp. 273–290. Springer, Heidelberg (2005)
11. Ma, X., Shekhar, S., Xiong, H., Zhang, P.: Exploiting a page-level upper bound for multi-type nearest neighbor queries. In: GIS, pp. 179–186 (2006)
12. Ramesh, R., Yoon, Y., Karwan, M.: An optimal algorithm for the orienteering tour problem. ORSA Journal on Computing 4(2), 155–165 (1992)
13. Safra, E., Kanza, Y., Dolev, N., Sagiv, Y., Doytsher, Y.: Computing a k-route over uncertain geographical data. In: Papadias, D., Zhang, D., Kollios, G. (eds.) SSTD 2007. LNCS, vol. 4605, pp. 276–293. Springer, Heidelberg (2007)
14. Sharifzadeh, M., Kolahdouzan, M.R., Shahabi, C.: Optimal sequenced route query. The VLDB Journal 17(8), 765–787 (2008)
15. Terrovitis, M., Bakiras, S., Papadias, D., Mouratidis, K.: Constrained shortest path computation. In: Bauzer Medeiros, C., Egenhofer, M.J., Bertino, E. (eds.) SSTD 2005. LNCS, vol. 3633, pp. 181–199. Springer, Heidelberg (2005)
16. Tsiligirides, T.: Heuristic methods applied to orienteering. Journal of the Operational Research Society 35(9), 797–809 (1984)

Utilizing Wireless Positioning as a Tracking Data Source

Spiros Athanasiou[1], Panos Georgantas[2], George Gerakakis[2], and Dieter Pfoser[1]

[1] Institute for the Management of Information Systems
"Athena" Research Center
G. Mpakou 17, 11524Athens, Greece
{spathan,pfoser}@imis.athena-innovation.gr
[2] School of Electrical and Computer Engineering
National Technical University of Athens
Greece, 15780
{pgeor,ggera}@dblab.ece.ntua.gr

Abstract. Tracking data has become a valuable resource for establishing speed profiles for road networks, i.e., travel-time maps. While methods to derive travel time maps from GPS tracking data sources, such as floating car data (FCD), are available, the critical aspect in this process is to obtain amounts of data that fully cover all geographic areas of interest. In this work, we introduce Wireless Positioning Systems (WPS) based on 802.11 networks (WiFi), as an additional technology to extend the number of available tracking data sources. Featuring increased ubiquity but lower accuracy than GPS, this technology has the potential to produce travel time maps comparable to GPS data sources. Specifically, we adapt and apply readily available algorithms for (a) WPS (centroid and fingerprinting) to derive position estimates, and (b) map matching to derive travel times. Further, we introduce map matching as a means to improve WPS accuracy. We present an extensive experimental evaluation on real data comparing our approach to GPS-based techniques. We demonstrate that the exploitation of WPS tracking data sources is feasible with existing tools and techniques.

Keywords: wireless positioning, map matching, tracking, FCD.

1 Introduction

Incorporating travel times into road network information, i.e., travel time maps, is an important prerequisite for a large number of spatiotemporal tasks. Examples include shortest path computation, traffic avoidance, emergency response, etc. Solutions typically rely on collected floating car data (FCD) that sample the overall traffic conditions [16, 5] in a given region. FCD capture temporal variations in achievable vehicle speeds throughout the road network. For example, speeds during the rush-hour are considerably lower than during night traffic. Then, in a post-processing step termed map-matching [4, 19], tracking data is accurately related to the road network and travel times are extracted. It is critical that *large amounts* of FCD are available for long periods of time and geography, so that the extracted speed profiles are accurate. Currently, all methods use GPS for tracking the position of vehicles.

N. Mamoulis et al. (Eds.): SSTD 2009, LNCS 5644, pp. 171–188, 2009.

1.1 The Case for GPS vs. WPS

While GPS is the most popular positioning technique, it has several drawbacks. First, it requires the use of specific hardware limiting the number of vehicles or users that can collect and provide tracking data. Second, there are occasions where GPS is inadequate (e.g., limited coverage, interference of high frequency electronic equipment). This is especially true for "urban canyons", i.e., areas in urban environments where line-of-sight with the GPS satellites is obscured, leading to inaccurate readings or no coverage at all. As demonstrated by LaMarca et al. [11], the average availability of GPS in an urban environment is only 4.5% during a user's daily schedule. In contrast, wireless networks, such as WiFi and GSM, are available on average 94.5% and 99.6% respectively. Third, the addition of extra integrated or autonomous GPS modules lead to increased power consumption, and thus limit the user's mobility or application of GPS.

These drawbacks of GPS have led to the rise of Wireless Positioning Systems (WPS), where the user location is estimated with the help of other, readily available wireless networks. As a technology, WPS delivers less accurate results (e.g., ~40m for WiFi/outdoors), but provides greater coverage characteristics (e.g., above 90% of a user's time). Further, WPS can be integrated in practically any computing device that incorporates a wireless network interface, and with a negligible burden on the interface's power consumption. So while WPS is less accurate than GPS, for typical everyday applications it can efficiently augment or even replace GPS.

Lately, WPS capable devices and applications are becoming a common place for end users, with examples like the iPhone, Android, Google Gears, Mozilla Firefox 3.1, etc. In addition, the integration of WPS in GPS and WiFi chipsets (e.g., SiRF, Broadcom, Texas Instruments) will result in a state where practically all mobile devices will have WPS capabilities. This argument is a fact, rather than a prediction, with great implications on spatiotemporal data management in general. In combination with the emerging usage of geolocation Web APIs (e.g., W3C Geolocation) we anticipate that in the near future there will be an abundance of readily available WPS positioning data.

Consequently, the technical advance of WPS is leading to new challenges and potential gains for numerous applications, where the scale and amount of positioning data will require corresponding advances in algorithmic solutions. Further, repurposing this sort of data by accommodating their particularities (e.g., varying levels of accuracy, ubiquitous coverage, etc.) in order to extract hidden knowledge, will be another area of great interest.

Our work is therefore extremely relevant in this newly established context, and applied to the specific issue of creating travel time maps. Currently, the creation of travel time maps from actual travel data is based solely on FCD. While this guarantees the use of position readings of high accuracy, it also limits the availability of such data for extended periods of time and geography. However, by successfully exploiting WPS, we would have access to data (a) whose size is several orders of magnitude greater, (b) temporally span bigger periods, and (c) extend to larger geographic areas. One could argue that WPS is only feasible in urban areas. While this observation is true, it actually strengthens our argument; urban areas are *exactly* where travel time maps are valuable resources for routing solutions.

1.2 Contributions

In this work, we advocate the use of WPS to complement and/or replace GPS tracking data sources to produce travel time maps. This increases the potential number of data providers and ultimately the quality of the resulting travel times. To the best of our knowledge, this is the first attempt of repurposing WPS tracking data to produce travel time maps. Our contributions are:

- We adapt and extend the two most important classes of WPS algorithms (centroid and fingerprinting) for our setting (WiFi network, outdoors operation).
- We experimentally evaluate the optimal parameters of the various classes of WPS algorithms and identify an optimal solution in terms of accuracy and coverage under realistic settings.
- We adapt an online map-matching algorithm to WPS tracking data as a post-processing step to improve WPS accuracy.
- We adapt a global map-matching algorithm to extract travel time maps from historic WPS tracking data and compare the results to GPS derived travel time maps.
- We demonstrate that for high sampling frequencies, WPS derived travel times are comparable to GPS in absolute terms. Further, even for low sampling frequencies, the results in terms of speed profiles (categories) are useful as well.

The remainder of this paper is structured as follows. Section 2 introduces techniques for wireless positioning. Section 3 briefly introduces the map-matching algorithm used for deriving travel times from tracking data. Section 4 gives an experimental evaluation of WPS techniques and travel times derived from WPS data. Finally, Section 5 presents our conclusions and directions for future research.

2 Wireless Positioning

Wireless Positioning Systems (WPS) provide a position estimate based on the radio signals received at a given location (*measurement*), and a known *radio map* of the environment. In the case for 802.11 (WiFi) wireless networks, the measurement consists of a set of the visible access point ids (BSSID), and their corresponding *received signal strength* (RSS[1]). The measurement is then compared to the radio map through a *distance* metric, and a position estimate is calculated.

Different wireless positioning algorithms exist, which imply different forms and means to create the radio maps, as well as distance metrics to provide an estimate. In all cases, the radio maps for a given region are produced by training data, typically collected through *wardriving*. Wardriving is the process of massively collecting geocoded RSS measurements when driving through a certain geographic area. For a given measurement period (e.g., 5sec), we perform a scan of the available WiFi networks in the environment (BSSID, RSS) and obtain the position of this scan through GPS.

In this section, we present the outline of two classes of WPS algorithms we adapted and implemented for our experiments, i.e., centroid and fingerprinting. For both classes, numerous approaches and variations exist, depending on the wireless

---

[1] Note that we always refer to the absolute value of RSS.

network (e.g., [13, 12, 17, 7, 18]) and environment (e.g., indoors/outdoors [10, 2, 3, 8]). We have either adopted these variations as is, or properly adapted and extended them to suit our case.

2.1 Centroid

Centroid is the simplest and the fastest method for wireless positioning. In centroid, the radio map consists of a set of the available APs and their positions, i.e., <*BSSID, X, Y*>. Consequently, centroid depends on having the true locations of the AP positions. Since this information is practically not available, nor feasible to produce, we must create the radio map from the training data, essentially *estimating* the position of the APs. Therefore, for each AP in the training data, we find all the positions it was visible, and estimate the AP's position as the arithmetic mean of these coordinates. Having established the radio map, a position estimate is provided in a similar manner. Given a measurement from the environment where certain APs are visible, we calculate the arithmetic mean of their coordinates, as provided by the radio map.

In order to improve accuracy when creating the radio map and/or calculating an estimate, we adopted *weighted* centroid from [6] and proposed two new heuristics: *k-max* and *thresholds*. Specifically:

- *Weighted*. The simple arithmetic mean is substituted by a weighted arithmetic mean, where the weight is based on the RSS.
- *K-max*. We apply the arithmetic mean on only the k APs with the lowest RSS (low RSS values correspond to strong received signal).
- *Thresholds*. We define three thresholds $t_1 \leq t_2 \leq t_3$ which split the RSS space in four regions. If there are APs which fall in the first threshold (RSS$\leq t_1$), then we use only them in the arithmetic mean and ignore the rest. If there no APs in the first threshold, we use the ones in the second ($t_1 \leq$RSS$\leq t_2$), and so forth. In case there are APs only in the last threshold ($t_3 \leq$RSS), then the algorithm does not provide an estimate since we consider the measurement to provide highly inaccurate readings.

Consequently, for centroid, there are a total of 16 different combinations of techniques to create the radio map and to provide an estimate: 4 to create the radio map, and 4 to provide an estimate. A specific centroid technique will be denoted as *centroid <radio map, estimation>*, where *radio map* and *estimation* can be one of the following: arithmetic mean (*am*), weighted (*w*), k-max (*k=n*), and thresholds (t_1-t_2-t_3). For example, centroid <k=2, 60-70-80>, means that the radio map was built with the *k-max* technique with k=2, and the estimation is provided with the *thresholds* technique with t_1=60, t_2=70, and t_3=80.

2.2 Fingerprinting

Fingerprinting assumes that the APs and associated RSS observed at a particular location are stable over time. Consequently, a measurement at a given location, i.e., the list of visible APs and RSS, can be considered as the *unique* fingerprint of that location. Thus, in fingerprinting, the training data themselves comprise the radio map.

To estimate the position, the algorithm calculates the Euclidean distance in the signal strength space between the current fingerprint and all available fingerprints in the radio map that contain the same APs. It then selects the *k-nearest* fingerprints in terms of distance, and returns as an estimate the arithmetic mean of their coordinates. This comparison is possible only if the current fingerprint and the fingerprints in the radio map contain exactly the same APs. Otherwise, calculating their distance in the Euclidean space is not possible.

However, in realistic conditions the current fingerprint may not contain exactly the same APs with the ones in the radio map. For example, some of the APs may have been turned off or removed, new APs may have been deployed, or the network interface may not provide APs with an RSS below a given threshold (typical behavior of Windows 802.11 hardware drivers).

To account for this situation, we calculate the distance between the current fingerprint and the ones in the radio map based on a *subset* of common APs. In particular, we extended the algorithm in [6] so that the subset is defined by two parameters:

- l: We compare the current fingerprint with fingerprints that contain *at most l less* APs. For example, suppose the WiFi scan $<(AP_1, RSS_1), (AP_2, RSS_2), (AP_3, RSS_3)>$. For $l=1$, a fingerprint $<x_a, y_a, (AP_1, RSS_1), (AP_2, RSS_2)>$ would be included in the position estimation, in contrast with $<x_b, y_b, (AP_2, RSS_2)>$ which would be ignored since there are two missing APs.
- m: We compare the current fingerprint with fingerprints that contain *at most m more* APs. For example, suppose the WiFi scan $<AP_1, RSS_1>$. For $m=1$, the fingerprint $<x_a, y_a, (AP_1, RSS_1), (AP_2, RSS_2), (AP_3, RSS_3)>$ would be excluded from the estimation due to the two extra APs.

Consequently, fingerprinting is modified as follows. The algorithm calculates the Euclidean distance in the signal strength space between the current fingerprint and all fingerprints in the radio map that contain at most l less and m more APs. It then selects the *k-nearest* fingerprints in terms of distance in the signal space, and returns as an estimate the arithmetic mean of their coordinates. As a result, there are many instances of the fingerprinting algorithm based on different parameters of k, l, and m. During the rest of the paper, we will use the notation *fingerprinting $<k, l, m>$* to denote a specific instance of the fingerprinting algorithm.

3 Map Matching

Deriving travel times from tracking data implies the alignment of the tracking data with a respective trajectory in the road network, i.e., finding the actual roads the vehicle has traversed. Now, provided that the tracking data is precise, this task would be simple. However, tracking data is obtained by sampling a vehicle's movement, typically with GPS and in our case with WPS. Unfortunately, both *GPS and WPS are not precise* due to the *measurement error* caused by the limited positioning accuracy, and the *sampling error* caused by the sampling rate, i.e., not knowing where the moving object was in between position samples [14]. Therefore, a processing step is needed that matches tracking data to the road network. This technique is commonly referred to as *map matching*.

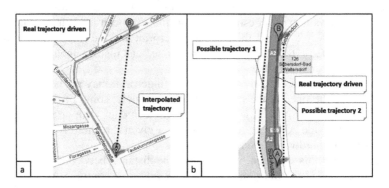

Fig. 1. Map-Matching example

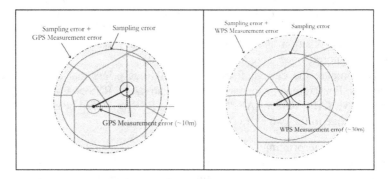

Fig. 2. Sampling error and measurement error

To illustrate these errors and the map-matching problem in general, Fig.1 gives two examples of measured positions and the possible trajectory the vehicle could have taken. Fig. 1a shows the interpolated path in between position samples A and B and the actual path with respect to the road segment. Further, as evident in Fig.1b, the positioning error becomes significant when facing several parallel roads close by. Specifically, in the case of WPS (Fig.2), the measurement error might grow quite large. This significantly increases the challenge for proper map-matching, since with a large measurement error, one is presented many alternative paths in the road network to map the sampled movement to. Thus, we expect that at least minimizing the sampling error by using high sampling rates will prove to be important.

3.1 Theoretical Considerations

Most map-matching algorithms are tailored towards mapping *current* positions onto a vector representation of a road network. Onboard systems for vehicle navigation utilize, besides continuous positioning, dead reckoning to minimize the positioning error and to produce accurate positions that can be easily matched to a road map. However, for the purpose of processing tracking data collected over a period of time, the entire trajectory, given as a sequence of historic position samples, needs to be mapped.

The algorithm we utilize in this work is the *global map-matching algorithm* of [4, 19], which employs the Fréchet distance measure for curves [1]. A popular illustration of the Fréchet distance is the following. Suppose a person is walking his dog, the person is walking on the one curve and the dog on the other. Both are allowed to control their speed but they are not allowed to go backwards. The *Fréchet distance* of the curves is the minimal length of a leash that is necessary for both to walk the curves from beginning to end. Using this distance measure, our global map-matching algorithm tries to match the tracking data geometry to a respective path in the road network by comparing it to the shapes of all possible paths in the road network. Although conceptually quite an elaborate task, this can be accomplished in O($mn\log mn$) time, with m being total number of nodes and edges of the road network and n the size of the tracking data to be matched [4].

The global map-matching algorithm is therefore a *shape-matching algorithm* that matches one curve, the tracking data trajectory, to another curve, the road network path that most closely resembles the tracking trajectory. As such, the algorithm is predestined for matching historic data.

Consider now the *online map matching* case, in which tracking data is matched as it is collected, i.e., in real time. Here, we apply the same global map matching algorithm, but instead of exploiting the complete trajectory (which is not known), we take advantage of the available historic data, i.e., the tracking data available so far. Experimentation showed that typically a trajectory consisting of 10 position samples collected with a sampling rate of 30s can be matched with the same accuracy as longer trajectories, i.e., 10 position samples represent a reasonably large enough curve for the global map-matching algorithm to produce a good quality match when applied to the online case. Hence, to perform online map matching, we apply the global map matching algorithm on the trajectory formed by the current position estimate and the 9 last position estimates.

3.2 Deriving Travel Times

Having mapped the tracking data to the road network, travel times are derived by mapping the travel times contained in the tracking data to the respective portions of the road network. The map-matching algorithm performs essentially shape matching and tries to find a path in the road network that most closely resembles the trajectory, i.e., the tracking data (cf. dotted line in Fig. 4). In the process, it maps all position samples (circles in Fig. 4) to the road network and all nodes along the corresponding path to the tracking data trajectory. Since the original tracking data contained the timestamp they were received, this information is transferred to the map-matched tracking data along the road network. The former can be seen as an effort to rediscover where on the road network the position samples would have been originally recorded. As such, these mappings are the ideal means for assigning travel times to the respective road network edges. Overall, the approach we employ is to uniformly map the time recorded between two consecutive position samples (e.g., $t_{i+1} - t_i$) in Figure 4, to the respective portions of the road network.

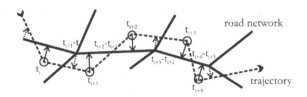

Fig. 3. Distance and travel time assignment

4   Experimental Evaluation

The primary scope of our experimental evaluation is to establish the suitability of WPS data as a source to provide travel times. First, to provide a complete examination of the relevant technologies and potential uses, we provide an evaluation of WPS accuracy and coverage and also introduce map matching as a means to improve WPS accuracy.

4.1   Experimental Setup

The experimentation was carried out in the Zografou neighborhood of Athens, Greece. The area was selected (i) due its to geographical characteristics (mix of flat areas and hills), (ii) varying levels of WiFi AP density (0-15 APs/m^2), (iii) typical urban structure with a mix of shops and residential areas, and (iv) fluctuating traffic.

4.1.1   Data Collection

Data was collected through wardriving over a period of two months in an area covering approximately 100,000m^2. For data collection typical road speeds and driving habits were maintained. Driving speeds varied from 0kph (stationary for more than 5mins) to 70kph. Fig. 5 shows a respective map of the Zografou area and the sampled locations on the road network where at least one WiFi AP was visible.

Our data set consists of records of the form $<tid, x, y, t, AP>$, where *tid* is the unique id of a trajectory, *x and y* are the GPS coordinates, *t* is the timestamp of the measurement, and *AP* is a list of the APs (BSSID) and their respective received signal strength (RSS). The sampling rate during data collection (i.e., every when a measurement is taken from the environment) was 5sec. In total, we collected roughly 200MBs of data, and we divided them (70%-30%) into two separate sets: (a) the *training* data, which were used to create the maps for the WPS techniques, and (b) the *testing* data, which were used to assess the WPS accuracy and to calculate travel times.

Concerning the chosen wardriving approach, instead of multiple passes from each road segment (which may reveal more APs, produce more samples for an AP, etc.), we performed at most one pass. This implies that the collected data set may be less complete than it could be, but resembles a realistic *large scale* mapping effort to create the radio map of any given region.

The equipment that was used comprised an Intel Core Duo laptop with a single 802.11a/b/g NIC and two Bluetooth GPS devices, all situated in the passenger compartment. We used Kismet [9] with a set of custom add-ons to extract geocoded WiFi measurements. All wardriving logs were later offloaded to a PostGIS database. Our WPS algorithms (centroid, fingerprinting) were developed in C/C++ and the map-matching algorithm was implemented in Java. Certain auxiliary processing/visualization tools were developed in PHP, Python, and Java. Accurate map data for the road network of Zografou were provided by Eratosthenis S.A. The experiments were executed by three Windows 2000 servers over a period of two weeks. Visualization of the results was performed with QGIS [15].

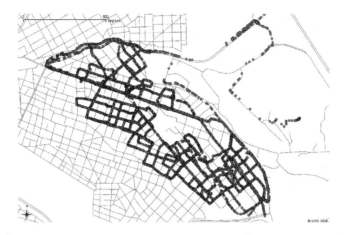

Fig. 4. Zografou map and WiFi AP locations

4.1.2 WPS Feasibility
The following interesting observations can be made with respect to the data. First, the total number of unique APs discovered was 2,184, and on average we observed 5 APs for each sampled location. Considering the covered geographic area, this yields 2.1 APs per 100m^2. Second, in most cases when WiFi was not available, then GPS was not available as well (e.g., under a bridge, near a large building). Third, almost all APs were available 24/7. Overall, these facts confirm the increased penetration of WiFi networks in urban environments and constitute a foundation for the proliferation of WiFi-based WPS as a ubiquitous and dependable alternative to GPS.

4.2 WPS Positioning Accuracy

4.2.1 WPS Accuracy and Coverage
The following experimentation evaluates WPS techniques in terms of accuracy and coverage. In particular, we used our training data to create the radio maps and the testing data to calculate the position estimates based on these maps. We experimented with all permutations of means described in Section 2. For each point in the testing data, the position estimate provided by each WPS algorithm for specific parameter settings is compared to the respective GPS measurement taken (ground truth).

Table 1 shows the results concerning *accuracy* using a ranking based on the average error of the WPS estimates. In addition, for each result its respective coverage (i.e., the percent of times the technique can provide an answer) is stated. For each class of WPS algorithms (centroid, fingerprinting) the best three accuracy achieving parameter settings are presented. What can be observed is that given the right parameters, fingerprinting achieves the best positioning accuracy (25.24m). However, the results overall only differ slightly. What is of interest is the respective coverage that can be achieved with each method. For example, the best performing fingerprinting method has a coverage of 56%, i.e., the technique cannot provide a position estimate 44% of the time. This behavior is caused by the WPS algorithms themselves and by our wardriving approach to collect training data. For example, centroid<k=1, 60-80-90> provides an estimate based only on APs with RSS below 60. The estimate will be more accurate because the required RSS threshold is low, but since this is also highly selective, there are many instances where RSS below 60 is not available.

Table 1. WPS accuracy compared to GPS

	Average Error (m)	Coverage (%)
Centroid <k=1, 60-70-80>	26.61	74
Centroid <k=1, 65-80-80>	26.65	82
Centroid <k=1, 75-85-90>	26.82	64
Fingerprinting <6-1-5>	**25.24**	**56**
Fingerprinting <6-1-4>	26.40	54
Fingerprinting <6-1-6>	26.57	56

Table 2 ranks WPS techniques based on their *coverage* values. As expected, the techniques producing the best coverage underperform in terms of average error. To design an actual wireless positioning system one needs to consider this trade-off between accuracy and coverage, i.e., is providing a more accurate estimate better than always providing a crude estimate?

Table 2. WPS coverage

	Average Error (m)	Coverage (%)
Centroid <k=1, weighted>	35.52	94
Centroid <k=1, 70-80-85>	47.11	93
Centroid <k=1, 65-75-80>	47.15	92
Fingerprinting <6-2-6>	36.45	82
Fingerprinting <2-4-6>	51.53	81
Fingerprinting <6-6-1>	48.93	78

One conclusion to the above question is to provide a *hybrid* WPS technique for centroid and fingerprinting. In particular, we obtain an estimate from the best performing technique in terms of accuracy, but should the said technique not be available (coverage), we obtain an estimate from the technique with the best coverage. These hybrid WPS techniques have high coverage (> 96%) with an acceptable increase in average error (cf. Table 3).

Table 3. Average error and coverage of the hybrid WPS techniques

	Average Error (m)	Coverage (%)
Hybrid Centroid	32.77	99
Hybrid Fingerprinting	28.40	96

Unless stated otherwise, hybrid WPS techniques will be used through the rest of our experiments, denoted as *WPS-C* and *WPS-F* for hybrid centroid and hybrid fingerprinting respectively.

4.2.2 Map Matching to Improve WPS Accuracy

Map-matching is known as a technique to relate tracking data to a map dataset. One can also see it as a method for imposing geometric constraints (shapes of paths in the road network) to tracking data. As such, this technique might be a viable means to "correct" WPS data and improve its accuracy. In this experiment, we utilize two map-matching algorithms, a simple one (called *naive*) that maps position samples to the closest point on the road network and the online algorithm presented in Section 4.2, which exploits shape information. To compare the various approaches in terms of accuracy, we calculated the average error and standard deviation for the complete WPS dataset with respect to the GPS measurements.

The results are given in Table 4 and confirm the findings in the relevant literature, with fingerprinting providing more accurate results than centroid. However, note that in both cases the average error is roughly 30m. Further, while the naïve map-matching algorithm only marginally reduces the average error (~1m), the shape-based map-matching algorithm reduces the average error by 37% (WPS-C) and 25% (WPS-F). This happens, because in contrast to a naïve map-matching approach, the shape-based algorithm exploits past WPS estimates to produce a trajectory that best fits the road network. Hence, an extremely important side-effect of proper map-matching, stemming from its inherent robustness towards *inaccurate* data, is the improvement of the accuracy provided by WPS. Combining WPS with map-matching reduces the average error of WPS (~20m) very close to the average error of GPS in urban environments (5-15m). This observation clearly opens the room for more research and experimentation, since in the WPS literature GPS is always considered as the *ground truth* for calculating the average error. Obviously, this is something needed to be questioned given our findings. Our future work and current experimentation is focused on exploiting GNSS available in Greece of greater accuracy (<1m), such as Galileo CS [20] and HEPOS [21].

Table 4. WPS average error and standard deviation

	Avg. Error (m)	Stdev. (m)	Avg. Error with naïve mm (m)	Stdev. with naïve mm (m)	Avg. Error with mm (m)	Stdev. with mm (m)
WPS-C	32.77	49.80	31.74	48.34	**20.47**	19.74
WPS-F	28.40	42.48	28.36	41.68	**21.15**	22.16

Moreover, we performed a set of experiments to assess the impact of the data collection speed, and AP density, towards WPS accuracy. In particular, to assess the impact of the data collection speed (i.e. frequency of collecting measurements from the environment), we removed records from the collected data to simulate frequencies ranging from 2Hz to 0,2Hz (Fig.5a). Further, we sampled our entire data set to randomly remove APs in order to simulate densities up to only 25% of the original one (Fig.5b). Our results illustrate that centroid is the most robust technique, maintaining an acceptable average error at all times.

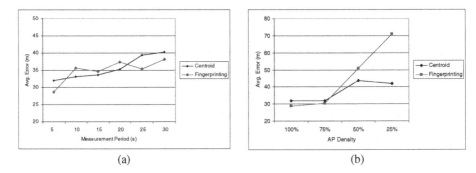

(a)   (b)

Fig. 5. Average error dependence from (a) measurement period and (b) AP density

4.3   Extracting Travel-Time Maps

To establish the feasibility of using WPS data to derive travel times, we compared the travel times produced from GPS data to the ones produced from WPS data for the same trajectories. The format of the collected testing data was *<tid, x, y, t, AP>*, where *tid* the trajectory id, *x* and *y* the GPS coordinates, *t* the timestamp of the measurement, and *AP* the WiFi-related measurements, i.e., AP BSSIDs and RSS. For the testing data, WPS-C and WPS-F were used to produce WPS estimates, resulting in trajectory data of the form of *<tid, x, y, t, xc, yc, xf, yf>*, where *xc, yc, xf,* and *yf* are the coordinates produced by the centroid and fingerprinting algorithms respectively. For the three types of trajectory data, GPS, WPS-C, and WPS-F, global map-matching was applied, and using the approach detailed in Section 3, the respective *travel times* were derived for each case. Consequently, for each road segment in our network, we established three different travel time estimates, (i) GPS, (ii) WPS-C and (iii) WPS-F. Versions of the travel time dataset were produced for sampling rates of 5, 10, 20, and 30secs.

4.3.1   Qualitative Evaluation
In order to compare the trajectories produced by GPS and WPS position data, we will define the measures of *recall* and *precision*. Let $G_i = \{g\}$, be the set of vertices produced by the map-matching algorithm on GPS data, for trajectory i. Also, let $W_i = \{w\}$ be the set of vertices produced by the map matching algorithm on WPS data for the same trajectory. The intersection $G_i \cap W_i$ contains the vertices the two sets have in common. Recall R and precision P can be defined as follows:

$$R_i = \frac{|G_i \cap W_i|}{|G_i|} \qquad P = \frac{|G_i \cap W_i|}{|W_i|} \tag{1}$$

R indicates the fraction of road segments covered by GPS trajectories that is also covered by WPS. Ideally, R should be equal to 1, i.e., WPS returns all the road segments GPS does (but possibly more). Further, P indicates the fraction of road segments covered by WPS trajectories that is also covered by GPS. Again, we want P to be equal to 1, i.e., WPS does not produce road segments not produced by GPS.

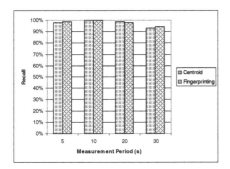 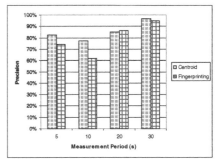

Fig. 6. Recall and precision for the WPS derived trajectories in our entire data set

In Fig.6, the values of recall and precision for our entire data set using varying sampling rates are shown. Common to all cases, recall is high, close to 100%. Notice that recall is optimal for a sampling rate of 10s while precision is best for a sampling rate of 30s. This was expected, as for low sampling rates, the sampling error dominates the measurement error in the map matching process. Thus, *both WPS and GPS produce practically the same trajectories.*

Fig.7 illustrates the above by giving a sample trajectory that accurately represents our findings for the entire data set. Fig.7(a) shows raw GPS tracking data while Fig.7(b) shows the WPS estimates derived by the WPS-C technique. Notice that although the 'noise' in WPS estimates is apparent (with several outliers as well), the trajectory can easily be distinguished. Fig.7(c),(d) show the produced trajectories after applying our map matching algorithm using a sampling rate of 30s. Fig. 7(e),(f) show details of the trajectory, highlighting specific map-matching cases.

4.3.2 Quantitative Evaluation

Having established how trajectories produced by WPS fare in comparison to GPS, in the following, we compare the respective travel times derived from these approaches. Given the set of links for which WPS and GPS derived travel times are available, we calculated the average error of WPS compared to GPS derived travel times, as shown in Fig.8. What can be readily observed is that the optimal sampling period is 10s, with no real difference between the two WPS techniques. For a period of 30s, the errors are 80.3% (WPS-C) and 125.4% (WPS-F). This could be interpreted as a serious problem for map matching based on WPS data for lower sampling frequencies, since most travel time databases are calculated from fleet management logs with sampling periods of 20-30s.

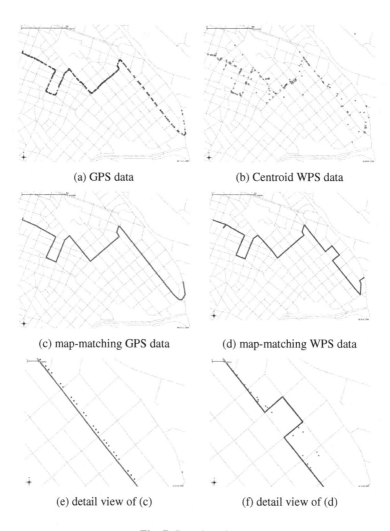

(a) GPS data   (b) Centroid WPS data

(c) map-matching GPS data   (d) map-matching WPS data

(e) detail view of (c)   (f) detail view of (d)

Fig. 7. Sample trajectory

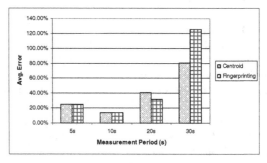

Fig. 8. Average error of WPS derived travel times compared to GPS derived travel times

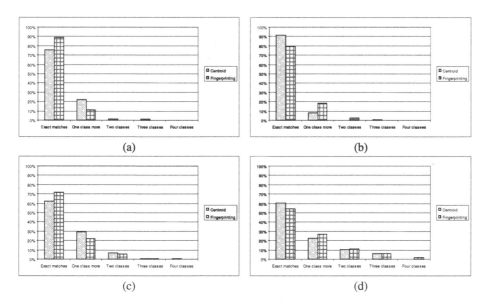

Fig. 9. Speed profile matches for GPS and WPS derived travel times, for various sampling periods: (a) 5sec, (b) 10sec, (c) 20sec, and (d) 30sec

However, for the creation of *dynamic road network profiles*, travel times are used to classify road network links. For example, suppose that a road category is defined as including speeds ranging from 10-20kph. Here two road links with respective travel times of 10.5 and 19.5kph will be subsumed under the same category. This *quantization* is beneficial, because it results to lower storage requirements, faster route calculation, and routes of similar quality.

We experimented with such quantization in travel time speeds and introduced for our experiments five road categories characterized by the following speeds (in kph): [0-10), [0-20), [20-30), [40-50), [50,∞). We classified all road links based on GPS and WPS data, and for various sampling frequencies. Further, for each road link in our network, we compared the classification produced from GPS, WPS-C, and WPS-F. Our results are shown in Fig. 9. For example, in Fig. 9(a), 75% of the road links are classified under the same category for WPS-C, compared to GPS. For WPS-F, this number is close to 90%. From Fig.9, we can also observe that for sampling rates of 5s and 10s, at least one of the two WPS techniques derives the same road categories for 90% of the road links. As the sampling rate decreases, this percentage is reduced to roughly 60%, with additional 25% of the road links classified to one category higher or lower. Therefore, we can conclude that for higher sampling rates, WPS produces very accurate travel times which are indeed comparable to GPS. For lower sampling rates (30s) the results are encouraging, since at least 80% of the derived travel times fall within the same or a directly neighboring category.

What follows in Fig.10 is the *actual link classification* based on GPS and WPS. Fig.10 shows the percentage of road links that fall in one of our five categories for GPS, WPS-C and WPS-F. It is evident that for small and high sampling rates alike, a WPS derived classification is very similar to a GPS classification.

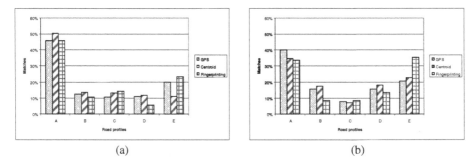

(a)   (b)

Fig. 10. Road segment classification for (a) 5sec and (b) 30sec

5   Conclusions and Future Work

We have evaluated the use of WPS data as an alternative data source for extracting travel times for road networks. We adapted and evaluated various classes of the centroid and fingerprinting WPS algorithms. Further, we applied map matching as a post processing filter to improve WPS accuracy and demonstrating significant gains. In addition, we extracted travel times from GPS and WPS data with a map-matching algorithm. Our evaluation demonstrated that for measurement periods up to 10sec, the produced travel times are practically identical to the ones derived from GPS data. Further, when applying a typical speed profile classification on travel times, even for sampling rates of up to 30sec, the produced travel times are still of respectable quality. Finally, we showed that through our analysis of WPS data, the distribution of road segments to speed profiles can be accurately discovered.

Our ongoing work evolves around further exploring and manifesting the benefit and potential uses of huge amounts of crowd-sourced WPS data. In this respect, our efforts are focused on three fronts. First, improve the accuracy of WPS techniques by integrating map matching into the WPS algorithms. Second, explore different uses for WPS data, such as routing (by fully replacing GPS), and automatic road network construction. Third, we aim to model and accommodate the inherent inaccuracy of wireless positioning data sources into spatiotemporal tasks and algorithms.

Acknowledgements

This work was partially supported by the project "*TALOS: Task Aware Location Based Services for Mobile Environments*", funded by the European Community, Framework Programme 7, Research for the benefit of SMEs. We would like to thank Dimitris Sacharidis, Kostas Patroumpas, Theodore Dalamagas, and Panagiotis Bouros for their valuable comments.

References

1. Alt, H., Godau, M.: Computing the Fréchet distance between two polygonal curves. Int. J. Comput. Geom. Appl. 5, 75–91 (1995)
2. Bahl, P., Padmanabhan, V.N.: RADAR: An In-Building RF-Based User Location and Tracking System. In: 9th IEEE Conference on Computer Communications, pp. 775–784. IEEE Press, Los Alamitos (2000)
3. Bahl, P., Padmanabhan, V.N., Balachandran, A.: Enhancements to the Radar User Location and Tracking System. Technical Report, Microsoft Research MSR-TR-00-12 (2000)
4. Brakatsoulas, S., Pfoser, D., Salas, R., Wenk, C.: On map-matching vehicle tracking data. In: 31st Very Large Data Bases Conference, pp. 853–864. ACM, New York (2005)
5. Brockfeld, E., Wagner, P., Passfeld, B.: Validating travel times calculated on the basis of Taxi Floating Car Data with test drives. In: 14th World Congress on Intelligent Transport Systems (2007)
6. Cheng, Y., Chawathe, Y., LaMarca, A., Krumm, J.: Accuracy Characterization for Metropolitan-scale Wi-Fi Localization. In: 3rd International Conference on Mobile Systems, Applications, and Services, pp. 233–245. ACM, New York (2005)
7. Chen, M.Y., Sohn, T., Chmelev, D., Hightower, D.H.J., Hughes, J., LaMarca, A., Potter, F., Smith, I., Varshavsky, A.: Practical metropolitan-scale positioning for GSM phones. In: Dourish, P., Friday, A. (eds.) UbiComp 2006. LNCS, vol. 4206, pp. 225–242. Springer, Heidelberg (2006)
8. Hightower, J., Consolvo, S., LaMarca, A., Smith, I., Hughes, J.: Learning and recognizing the places we go. In: Beigl, M., Intille, S.S., Rekimoto, J., Tokuda, H. (eds.) UbiComp 2005. LNCS, vol. 3660, pp. 159–176. Springer, Heidelberg (2005)
9. KISMET, http://www.kismetwireless.net/
10. Krishnan, P., Krishnakumar, A.S., Ju, W., Mallows, C., Ganu, S.: A system for LEASE: Location estimation assisted by stationary emitters for indoor RF wireless network. In: 23rd IEEE Conference on Computer Communications, pp. 1001–1011 (2004)
11. LaMarca, A., Chawathe, Y., Consolvo, S., Hightower, J., Smith, I., Scott, J., Sohn, T., Howard, J., Hughes, J., Potter, F., Tabert, J., Powledge, P., Borriello, G., Schilit, B.: Place lab: Device positioning using radio beacons in the wild. In: Gellersen, H.-W., Want, R., Schmidt, A. (eds.) PERVASIVE 2005. LNCS, vol. 3468, pp. 116–133. Springer, Heidelberg (2005)
12. Laitinen, H., Lahteenmaki, J., Nordstrom, T.: Database correlation method for GSM location. In: Proceedings of the 53rd IEEE Vehicular Technology Conference, pp. 2504–2508. IEEE Press, Los Alamitos (2001)
13. Otsason, V., Varshavsky, A., LaMarca, A., Lara, E.D.: Accurate GSM Indoor Localization. In: Beigl, M., Intille, S.S., Rekimoto, J., Tokuda, H. (eds.) UbiComp 2005. LNCS, vol. 3660, pp. 141–158. Springer, Heidelberg (2005)
14. Pfoser, D., Jensen, C.S.: Capturing the Uncertainty of Moving-Object Representations. In: Güting, R.H., Papadias, D., Lochovsky, F.H. (eds.) SSD 1999. LNCS, vol. 1651, pp. 111–132. Springer, Heidelberg (1999)
15. Quantum GIS Project, http://www.qgis.org/
16. Schaefer, R.P., Thiessenhusen, K.U., Wagner, P.: A Traffic Information System by Means of Real-time Floating-car Data. In: 9th World Congress on Intelligent Transport Systems (2002)
17. Sohn, T., Varshavsky, A., LaMarca, A., Chen, M.Y., Choudhury, T., Smith, I., Consolvo, S., Hightower, J., Griswold, W.G., Lara, E.D.: Mobility Detection Using Everyday GSM Traces. In: Dourish, P., Friday, A. (eds.) UbiComp 2006. LNCS, vol. 4206, pp. 212–224. Springer, Heidelberg (2006)

18. Varshavsky, A., Chen, M., Lara, E.D., Froehlich, J., Haehnel, D., Hightower, J., LaMarca, A., Potter, F., Sohn, T., Tang, K., Smith, I.: Are GSM phones THE solution for localization? In: 7th IEEE Workshop on Mobile Computing Systems and Applications, pp. 20–28. IEEE Press, Los Alamitos (2006)
19. Wenk, C., Salas, R., Pfoser, D.: Addressing the Need for Map-Matching Speed: Localizing Global Curve-Matching Algorithms. In: 19th Scientific and Statistical Database Management Conference, pp. 379–388 (2006)
20. European Space Agency - Galileo, http://www.esa.int/esaNA/galileo.html
21. Hellenic Positioning System, http://www.hepos.gr/

Indexing Moving Objects Using Short-Lived Throwaway Indexes

Jens Dittrich[1], Lukas Blunschi[2], and Marcos Antonio Vaz Salles[3]

[1] Saarland University
jens.dittrich@cs.uni-saarland.de
[2] ETH Zurich
lukas.blunschi@inf.ethz.ch
[3] Cornell University
vmarcos@cs.cornell.edu

Abstract. With the exponential growth of moving objects data to the Gigabyte range, it has become critical to develop effective techniques for indexing, updating, and querying these massive data sets. To meet the high update rate as well as low query response time requirements of moving object applications, this paper takes a novel approach in moving object indexing. In our approach we do *not* require a sophisticated index structure that needs to be adjusted for each incoming update. Rather we construct conceptually simple *short-lived throwaway indexes* which we only keep for a very short period of time (sub-seconds) in main memory. As a consequence, the resulting technique MOVIES supports at the same time high query rates *and* high update rates and trades this for query result staleness. Moreover, MOVIES is the first main memory method supporting time-parameterized predictive queries. To support this feature we present two algorithms: non-predictive MOVIES and predictive MOVIES. We obtain the surprising result that a predictive indexing approach — considered state-of-the-art in an external-memory scenario — does not scale well in a main memory environment. In fact our results show that MOVIES outperforms state-of-the-art moving object indexes like a main-memory adapted B^x-tree by orders of magnitude w.r.t. update rates and query rates. Finally, our experimental evaluation uses a workload unmatched by any previous work. We index the complete road network of Germany consisting of 40,000,000 road segments and 38,000,000 nodes. We scale our workload up to 100,000,000 moving objects, 58,000,000 updates per second and 10,000 queries per second which is unmatched by any previous work.

1 Introduction

Indexing support for moving objects is a crucial requirement in domains such as car tracking [18], airplane surveillance [45], mobile phone tracking [1], emergency services [10], social networking [24], and gaming engines [49]. In these applications an *update* may be a car/airplane/phone sending a message on its new position. A *query* may be a range query, a nearest-neighbor query, or a time-parametrized range query asking for predicted positions of moving objects at a future time t_q. Queries are issued either by car/air traffic control or by users themselves. All of the above applications face a principal problem: how to support efficient query processing under high update rates.

N. Mamoulis et al. (Eds.): SSTD 2009, LNCS 5644, pp. 189–207, 2009.

Traditionally, index creation has been considered an extremely costly process. For that reason, research on moving object indexes has been centered around creating sophisticated index structures. These indexes are created once, kept, and then modified according to incoming updates. This has led to a plethora of complex index structure proposals in the past. However, with the rise of large main memories and fast multi-core CPUs this "natural law" of keeping a moving object index can be questioned.

We present a novel main-memory method termed MOVIES (MOVing objects Indexing using frEquent Snapshots) that supports time-parameterized (predictive) queries and is at the same time space-, query-, update-, and multi-CPU-efficient. At its core our method *MOVIES* resembles the approach taken by a cinematographer: as it is impossible to capture continuously moving data with any camera in one image, a cinematographer has to take a series of *still* images at a given *frame rate*. As long as the frame rate exceeds the inertia of the human eye (i.e., at least 24 frames per second), an illusion of continuous movement is created. We follow exactly the same approach: we try to provide as many still *index images* of the data as possible. For a very short period of time we use that index to answer incoming queries. After that, we throw that index away. As long as the *index build rate* is high, an illusion of a continuously up-to-date index will be created. We will show that — surprisingly — index creation can be a matter of subseconds even for datasets comprising hundreds of thousands of moving objects. For instance, we will demonstrate that index creation for 1 million moving objects (a common data set size used in recent moving objects studies, see Section 6.4) takes as little as 0.16 seconds on a single computing core allowing for six index rebuilds per second. The price we have to pay for these features is slightly out-of-date (stale) query results, i.e., even though queries are executed immediately in our approach, query results may not consider the most recent updates. However, we will show that even for massive data sets this query result staleness may be reduced to (sub-)seconds. This meets by far the demands of real applications. For instance, state-of-the-art flight control in Europe currently works with a staleness of 5 seconds [39].

1.1 Contributions

In summary, this paper makes the following contributions:

1. We provide a novel approach termed MOVIES (MOVing objects Indexing using frEquent Snapshots) to effectively index moving objects. As described above MOVIES resembles the approach taken by a cinematographer by creating a series of different indexes each second. Like that we provide at all times a read-optimized index not suffering from update handling.
2. MOVIES is the first main-memory moving object index to support time-parameterized queries. This allows users to pose predictive queries asking for predictive results at a future time t_q. Previous main memory approaches such as [51] did not support this type of query. We will present two different MOVIES variants to support these type of queries: *Non-Predictive Indexing MOVIES (NPI)* and *Predictive Indexing MOVIES (PI)*. We will show that MOVIES NPI performs better than MOVIES PI for high update rates. This is a surprising result as predictive indexing approaches are considered state-of-the-art for external memory methods.

3. We present techniques to make update handling efficient. As we collect incoming updates in a buffer, the cost for collecting the updates is very small. We will show that different buffer organizations have different impact on the overall performance of MOVIES. Therefore, we will propose two more variants of our algorithm: *Logged* MOVIES and *Aggregated* MOVIES.

4. We provide a thorough experimental study of MOVIES using standard hardware and realistic data sets unmatched by any previous work. Our experiments show that MOVIES scales well up to 25 million moving objects on a single machine. We show that MOVIES provides excellent query response times while at the same time being able to process huge amounts of updates. In addition, we show that MOVIES outperforms state-of-the-art indexing methods like the B^x-tree by orders of magnitude w.r.t. the number of queries and updates being handled — even though the latter methods have been adapted to work effectively in main memory. Finally, we evaluate a distributed implementation of MOVIES indexing 100 million moving objects on a small cluster of shared-nothing machines. Note that the data sets used in our experiments are 10 times larger than in the biggest study available [23] and at least 100 times larger than in all other studies.

This paper is structured as follows. The following section presents preliminaries. Sections 3&4 present MOVIES. Section 5 presents our experimental evaluation. Section 6 discusses related work and its relationship to MOVIES.

2 Preliminaries

2.1 Problem Statement

We consider a data set of N moving objects in a two-dimensional space of data of a domain $|X| \times |Y|$ where $|X|$ (resp. $|Y|$) represents the number of different positions in the horizontal (resp. vertical) dimension. Extending our technique to more dimensions is straightforward. Similarly to [17], we assume a discrete space of $2^{16} \times 2^{16} = 2^{32}$ different positions. Each moving object is identified by a unique key termed an *OID*. Each moving object emits updates on its current location (x, y) and its speed vector $\overrightarrow{sv}$ by sending an $(x, y, \overrightarrow{sv}, OID)$-tuple to central indexing server(s). Like in [17], we assume that objects travel at a maximum speed S_{max} and are guaranteed to send updates at least every $t_{\Delta max}$ seconds. We assume that indexes are queried with two-dimensional predictive range queries $Q(r, t_q)$ specifying a range $r = [x_{low}; x_{high}] \times [y_{low}; y_{high}]$ and a query time t_q. Note that other query types such as predictive k-nearest-neighbor may easily be derived from predictive range queries (see e.g. [17]).

2.2 Formal Argument

In this section we provide a formal argument to illustrate the core benefit of our approach. We do not strive to provide a full-blown cost model but rather focus on the key aspects. For realistic moving objects scenarios the amount of updates will be in the tens of millions per second. We will develop a method that does not trade query performance for update performance as done in several existing methods. Consider a

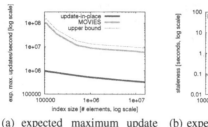 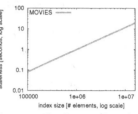

(a) expected maximum update   (b) expected staleness (the price)
rate (the gain)

Fig. 1. Expected performance of MOVIES versus update-in-place

simple index structure organizing a sorted mapping *spatialposition* $\mapsto$ *OID* (binary
range search on a B+tree or any cache-optimized tree). We assume that the spatial position is linearized using a linearization function (see Section 3.3). The cost for both
querying and updating in-place are of the order $O(log\,N)$ where N is the number of
entries. An update in a positional index consists of deleting the old entry and creating a new entry. Thus in the worst case we need two logarithmic traversals. We derive
a cost formula $C_{\text{update-in-place}} = 2 \cdot c_1 \cdot \log_2 N$ where c_1 is a hardware-dependent constant. Similarly, the initial cost for bulkloading for an index is of the order $O(N\,log\,N)$,
which translates to a cost formula $C_{\text{bulkloading}} = c_2 \cdot N \log_2 N$ where c_2 is a hardware-
dependent constant. Now let's assume that instead of performing updates in-place we
collect W updates in a separate structure with $O(1)$, i.e., $C_{\text{array update}} = c_3$. We will peri-
dodically rebuild a new index from that structure. The cost for this is $C_{\text{collect and rebuild}} =
W \cdot C_{\text{array update}} + C_{\text{bulkloading}}$. When will this be cheaper than update-in-place? We ob-
tain $W \cdot c_3 + c_2 \cdot N \log_2 N \leq W \cdot 2c_1 \cdot \log_2 N \Rightarrow c_2 \cdot N \log_2 N \leq W \cdot (2c_1 \log_2 N - c_3) \Rightarrow
W \geq (c_2 \cdot N \log_2 N)/(2c_1 \log_2 N - c_3)$. Now, we may estimate upper bounds for the con-
stants assuming a single core and the index to be limited to 16 million elements as
$c_1 = 73.6ns$, $c_2 = 8.9ns$, and $c_3 = 112.5ns$. Thus we receive $W \geq (8.9 \cdot N \log_2 N)/(2 \cdot
73.6 \cdot \log_2 N - 112.5)$. For an index of $N = 1,000,000$, the collect and rebuild approach
will already be cheaper when the number of updates reaches $W = 62,872$. Also note
that the query processing costs in both approaches are *exactly the same*. We just ar-
gued on how to improve update cost without touching query cost. On the contrary,
the collect and rebuild approach could even be improved to build read-optimized indexes. That would additionally improve the query response time over an update-in-
place approach. Now let's examine the maximum number of updates supported by the
different methods. How many updates can we expect to support in a collect and re-
build approach? We may rebuild the index every $T_{\text{frame time}} > C_{\text{collect and rebuild}}$ seconds.
This can be rewritten to $C_{\text{collect and rebuild}}/T_{\text{frame time}} \leq 1 \Rightarrow W \cdot C_{\text{array update}} + C_{\text{bulkloading}} \leq
T_{\text{frame time}} \Rightarrow W \leq (T_{\text{frame time}} - C_{\text{bulkloading}})/(C_{\text{array update}})$. The maximum number of up-
dates processed per second can then be computed as $U^{max} = W/T_{\text{frame time}}$ which is lim-
ited by the *upper bound* $1/C_{\text{array update}}$. Assume we allow for a $T_{\text{frame time}}$ of $3C_{\text{bulkloading}}$,
then we obtain the function displayed in Figure 1(a). The figure shows that we may
expect to *gain* an order of magnitude over update-in-place. The *price* we pay for that is

query result staleness which is limited by $2T_{\text{frame time}}$. Figure 1(b) shows that even for an index of 1,000,000 elements staleness will remain below a second even when using only a single computing core.

3 MOVIES

This section presents the MOVIES indexing algorithm (MOVing objects Indexing using frEquent Snapshots). As stated in the Introduction, our method resembles the approach taken by a cinematographer: we try to create as many still index images as possible. This generates the illusion of a continuously up-to-date index.

3.1 Algorithmic Walkthrough

The MOVIES algorithm is based on *index frames*. Each index frame is active during a short time interval called the *frame time* $T_i = [t_i; t_{i+1})$ where i denotes the ID of the frame and t_i denotes the moment in time when frame i started. During each index frame, e.g., time interval T_{45} for Frame 45 in Figure 2, we use a read-only index, I_{44}, to answer all incoming queries. We also keep an update buffer U_{45} collecting all updates arriving during T_{45}. In addition, we build a new read-only index I_{45} based on the up-

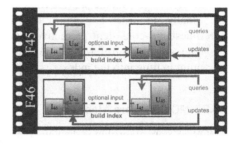

Fig. 2. Two index frames of MOVIES core algorithm

dates collected in update buffer U_{44} during T_{44} (see arrow ⟵). Depending on whether the update buffers contain updates for all OIDs, this index build has to consider information available in index I_{44} (see arrow ⟵ - -). As soon as the new index I_{45} is built, we start a new frame, e.g., Frame 46. In this frame we use the newly built read-only index I_{45} from Frame 45 to answer all incoming queries. We keep an update buffer U_{46} to collect incoming updates. In addition, we build a new read-only index I_{46} based on the updates collected in U_{45} during T_{45}. Again, depending on whether the update buffers contain updates for all OIDs this index build has to consider information available in index I_{45}. As soon as the index is built, we start a new frame, e.g., Frame 47 (not shown) which is similar to Frame 45.

3.2 Comparison to Differential Files

The idea of collecting updates in a separate space and applying them in a batch was first used in the context of relational databases more than 30 years ago [38]. The idea of that paper was to collect changes in a separate *differential file* and merge that file regularly with the existing external memory index. Since then differential files were extended in multiple ways [29,16,28] and became state-of-the-art for *read-mostly* environments like data warehouses (DWH) [46] as well as desktop [25], enterprise [46], and web search [11] engines.

In contrast to all of these approaches MOVIES differs as follows:

1. For a moving objects application the query result staleness of a file-based method as followed in other applications [46,25,11] would be unacceptable. For moving object indexing we require query result staleness to be below a few seconds (e.g., for an aircraft surveillance scenario it should be below 5 seconds [39]). Therefore we have to optimize our algorithm for keeping staleness low. This can only be achieved by keeping the data entirely in main-memory.
2. In our scenario moving objects are guaranteed to send an update at least every $t_{\Delta max}$ seconds. This was used in similar studies, e.g. [17]. Therefore for certain situations, e.g. $t_{\Delta max} < T_i$ we may completely ignore the information available in the old index: we simply need to build an index image from the update buffer. Therefore, in contrast to differential file-based approaches [29,16,28], in MOVIES there is no need to perform a costly *merge* with the previous index. An index merge will only be used as a fallback.
3. Finally, in order to support time-parameterized queries we need to introduce timestamp-consistent predictive indexes (MOVIES PI). Thus, instead of indexing data as-is as in differential file-based approaches, we will predict data to a future point in time into the index.

3.3   MOVIES Core Algorithm

MOVIES core algorithm is displayed in Figure 3.

It takes as input a stream of updates $StreamU$, a stream of queries $StreamQ$, and a *bootstrapTime* interval. The algorithm starts by creating a new update buffer U_0 (Line 1). Then the stream of updates $StreamU$ is routed to that buffer (Line 2). An empty index I_0 is created in Line 3.

Then the algorithm waits for a certain amount of time specified by a *bootstrapTime*. During that time, however, incoming updates are collected in update buffer U_0. Lines 5–15 show the indexing loop used to create the sequence of index frames. This loop will be repeated until a global flag should_terminate is set to true (Line 5). The loop keeps a counter *currentFrameID* for the current frame ID. Inside the loop an index frame starts by creating a new update buffer $U_{currentFrameID}$ (Line 6). The stream of updates is routed to the new update buffer (Line 7). The stream of queries is routed to the index created in the previous iteration (Line 8). For the first loop iteration this index will be the empty index I_0. In Line 9 we check whether the *currentFrameID* is two or higher. If that is the case, we destroy the index built in index

```
Input:  Stream of updates StreamU
        Stream of queries StreamQ
        TimeInterval bootstrapTime
1  U₀.create()
2  StreamU.setDestination(U₀)
3  I₀.create()
4  wait(bootstrapTime);
5  for (Integer currentFrameID = 1; ¬should_terminate) do
6      U_currentFrameID.create()
7      StreamU.setDestination(U_currentFrameID)
8      StreamQ.setDestination(I_currentFrameID−1)
9      if currentFrameID ≥ 2 then
10         I_currentFrameID−2.destroy()
11     end
12     if ("may ignore old index") then
13         I_currentFrameID ← buildIndex(U_currentFrameID−1)
14     else
15         I_currentFrameID ← buildIndex(U_currentFrameID−1, I_currentFrameID−1)
16     end
17     U_currentFrameID−1.destroy()
18     currentFrameID = currentFrameID + 1
19 end
```

Fig. 3. MOVIES Core Algorithm

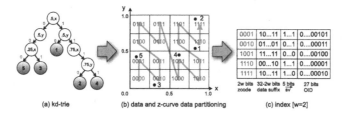

(a) kd-trie   (b) data and z-curve data partitioning   (c) index [w=2]

Fig. 4. Kd-trie mapped to a pointer-free linearization

frame $currentFrameID - 2$ (Line 10). After that we check whether we may ignore the data available in the old index (Line 12), e.g., this may happen if all elements in the old index became outdated by elements in the update buffer. If that check succeeds, we simply call the buildIndex operation on the update buffer (Line 13) otherwise we create a new index $I_{currentFrameID}$ also using information from the old index (Line 15). After that we destroy the update buffer filled in the previous frame (Line 17). Finally, we increase the $currentFrameID$ counter by one and continue looping (Line 18).

Organization of Update Buffers. This section describes the organization of update buffers. As we want to handle high update rates, we have to make sure that the update buffers do not exceed the available main memory. We solve this as follows. For high update rates the update buffers may contain several updates for the same OID, i.e., the update buffer may be considerably larger than the original index. As the aggregate of updates to an OID is sufficient for query processing (e.g., the most current position of a moving object), it makes sense to implement update buffers U_i by *aggregation buffers* $\hat{U}_i$, organized using $OIDs$ as keys. For each OID, $\hat{U}_i$ only keeps a MIN aggregate, i.e., the last update received for this moving object. In our approach updates are written to aggregation buffers $\hat{U}_i$ immediately when they arrive. We implemented aggregation buffers using arrays of size N where the slot at position i stores the aggregate of object $i = OID$. Note that other implementations are possible, e.g., any hash table. This ensures constant insert time for updates. We refer to this variant as *Aggregated MOVIES*. The algorithm based on FIFO update buffers is termed *Logged MOVIES*.

Organization of Read-Only Indexes. As read-only index we use a state-of-the-art spatial indexing method. We focus on a technique that is at the same time simple and efficient. Therefore we have chosen linearized kd-tries [47,32][1]. This index was used in many papers in different variants (e.g., [36,31,9,17]) and was shown to outperform competing approaches. The core idea of a linearized kd-trie is to *simulate* a pointer-based index structure. This is achieved by assigning each node of a virtual kd-trie a unique identifier termed a *locational code*. Locational codes are based on a space-filling curve like the z-curve [47,32] (see Figure 4(b)) or the Hilbert curve [14]. These recursive space-filling curves enumerate a multi-dimensional space (i.e., the nodes of the kd-trie, see Figure 4(a)) with a one-dimensional curve (see Figure 4(b)). For each object that needs to be managed by an index, it then suffices to compute its locational code, i.e., the

---

[1] Following the terminology of Donald Knuth [20] a *trie* partitions the data space whereas a *tree* partitions the data.

virtual node it belongs to in the kd-trie. This calculation is independent of the locational codes of other objects. Therefore, at no point in time it is required to actually create a pointer-based kd-trie. As the locational codes are one-dimensional, they are ordered to provide efficient query processing. The resulting codes plus the data are then stored in a sorted index (see Figure 4(c) using $w = 2$ bits per dimension). Note that locational codes may be inlined with the data to avoid extra storage cost. Both point and range queries are efficiently supported. The latter are crucial for our scenario as several other types of queries such as nearest neighbor queries may be based on range queries (see e.g. [17]). Moreover, kd-tries are not limited to two dimensions but also work well for high dimensional spaces. Also recall that in contrast to grid-based indexes (e.g., as used in [51]) which would need to store an exponential number $(1/grid\_length)^d$ of pointers to inclusion lists, an approach based on locational codes will for all d only require N indexing slots. Thus, our method can easily be extended to higher dimensional data.

4   MOVIES Query Processing

4.1   Time-Parameterized Query Processing

A time-parameterized query $Q(t_q)$ asks for object positions in the range $([x_{low};x_{high}] \times [y_{low};y_{high}])$ at a time t_q. In order to answer these queries, we must transport objects to their predicted positions at time t_q. This can be done at indexing time, which we term *Predictive MOVIES (PI MOVIES)* or at querying time, which we term *Non-Predictive MOVIES (NPI MOVIES)*. Both strategies work for both Logged and Aggregated MOVIES resulting in a total of four different combinations.

Predictive MOVIES (PI) Indexing Strategy. For each index build we index all data w.r.t. a *single* point in time $t_{index} > t_u$. We term t_{index} the *index time*. Thus, for every incoming update u we index the moving object at a predicted position $(x,y) + \vec{sv}(t_{index} - t_u)$. Here, we may avoid any extra storage space by translating objects immediately when an update arrives. However, for each incoming update we have to compute the predicted position — which may be costly. After that, the timestamp for the update may be dropped. If during fallback an object is encountered that has not received an update (Line 22 of the buildIndex algorithm), that object is simply translated to a new position using the new index time.

Query Strategy. As t_q may be either larger or smaller than t_{index} we have to consider three cases:

1. $t_q < t_{index}$: the objects have to be translated to an earlier time,
2. $t_q > t_{index}$: objects have to be translated to a later time (see also [17]),
3. $t_q = t_{index}$: objects do not have to be translated.

For cases (1)&(2) we rewrite $Q(t_q)$ to consider the maximum distance $\varepsilon := S_{max} \cdot |t_q - t_{index}|$ an object may have travelled relative to the index time. Every $Q(t_q)$ is rewritten to $Q(t_q)' := [x_{low} - \varepsilon; x_{high} + \varepsilon] \times [y_{low} - \varepsilon; y_{high} + \varepsilon]$. $Q(t_q)'$ is then postfiltered w.r.t. t_q and their respective speed vectors $\vec{sv}$ as obtained from the index.

Choice of Index Time. Let $\bar{Q} = \{Q_1(t_q^1), .., Q_k(t_q^k)\}$ be the set of queries arriving during a single indexing frame of MOVIES. To minimize query enlargements and therefore the performance penalty for large range queries, we wish to minimize the sum of query window areas added due to enlargement, or alternatively the term $\sum_{i=1}^{k} |t_q^i - t_{index}|^2$. Assume, for simplicity, that all queries ask for a fixed time into the future, i.e., $t_q = t_{now} + \Delta t$. After being rebuilt, an index receives queries during one frame time $T_{frametime}$. To produce balanced query enlargements, the index time should be $\Delta t + T_{frame\ time}/2$ ahead of the time the index is ready for querying. In order to achieve that, we must set the time of the update buffer that will be used to build a new index appropriately. An update buffer is used to collect updates two frames before being used to build a new index (see Figure 3). Therefore, the index time to be used for an update buffer collecting updates in the next frame should be set to $t_{index} = t_{now} + 2T_{frame\ time} + (\Delta t + T_{frame\ time}/2) = t_{now} + 2.5T_{frame\ time} + \Delta t$.

Non-Predictive MOVIES (NPI) Indexing Strategy. For each index build we index each index entry w.r.t its timestamp t_u. In order to do this, we need to keep for each update its corresponding timestamp. Thus, we require slightly more storage space, but do not have to compute predicted positions at indexing time.

Query Strategy. We use query enlargement like with Predictive MOVIES. However, as objects are indexed with different last update times, we must perform query enlargement with respect to the time of the oldest update considered in the index, i.e., $\varepsilon := S_{max} \times |t_q - t_{min}|$. As calculating t_{min} by scanning the timestamp array has prohibitive cost, we provide a bound on t_{min}. When an update arrives, it takes at most two times the frame time $T_{frame\ time}$ for it to appear in the index used for querying (see Figure 3). As an update for each object must arrive within $t_{\Delta max}$, then $t_{min} \geq t_{now} - (t_{\Delta max} + 2T_{frame\ time})$.

5   Experiments

This section presents a thorough experimental analysis of MOVIES. We explore various aspects of our approach and compare it with state-of-the-art approaches. The goals of our experiments are:

1. Determine the maximum supported update rate of MOVIES when scaling the index size (Section 5.3).
2. Determine query throughput of MOVIES when scaling the update rate (Section 5.4).
3. Determine the performance of MOVIES when implemented on a cluster of shared-nothing machines (Section 5.5).

5.1   Setup

All experiments were performed on servers having each two 2.4 GHz Dual Core AMD Opteron 280 processors, i.e., four cores in total, and 6 GB of main memory. We used two separate servers M1 and M2 to generate updates and queries. These updates and queries were sent over the network and received by servers termed *processing nodes* (PN1–PN4) that did the actual indexing. M1 and M2 were each connected to the switch by one network cable and in addition with one network cable each directly to PN3 (PN4 respectively). All network links supported 1 Gbit/s.

For the single instance experiment we used one processing node PN1. For the parallelization experiment we used up to four processing nodes PN1–PN4. All code used for the experiments was implemented in Java 5. In our implementation, we avoided complex object types whenever possible and used primitive Java types. To make maximal use of the four cores provided by each machine, we implemented a multi-threaded variant of MOVIES.

parameter	setting
index size N	100,000 … **6,400,000** … 100,000,000
update rate V	0 /s … 58,000,000 /s
query rate	0 /s … **1,000** /s … 10,000 /s
query window size q_w	**1 km × 1 km** .. 10 km × 10 km
# road network segments	39,509,805
# road network nodes	37,967,339
data region	640 km × 863 km
index granularity	26.3 m × 26.3 m
S_{max}	60 m/s
$t_{\Delta max}$	$\leq N/V$

Fig. 5. Settings

For the experiments in Sections 5.3 to 5.5 evaluating MOVIES, we waited until at least 8 re-indexing phases were completed. Then we measured at least 10 re-indexing phases and report the average.

5.2 Data and Queries

Our experiments were inspired by the scenario *'index all cars in Germany'* which comprises 58M cars [22]. We obtained a commercial data set containing the complete road network of Germany [44]. This data set consists of 38 million nodes and 40 million road segments. The geography of Germany covers 640 km x 863 km. We assumed cars to travel at a maximum speed of $S_{max} = 60$m/s= 216km/h. As in similar studies [26,27,34] we initially used the moving object generator of [5]. However, it turned out that that generator does not scale for the massive workloads considered in this paper. In particular, large number of nodes and road segments are not possible. Therefore we developed our own generator based on the ideas of [5]. This means that we used the same moving object placement strategy *network-based approach (NB)*. It places cars using the same

Fig. 6. Road network of Germany

skewed distribution as the roads and nodes themselves. We then moved cars on the network by assigning each car a constant random direction which it would try to follow on the network. This avoids the overheads of doing Dijkstra-computations for each car. This generates similar traces as in [5], but at much lower cost. Our trace generator is open source and available from sourceforge at *http://moto.sourceforge.net/*. If not mentioned otherwise, a data set of 6.4 million moving objects was used on MOVIES experiments with a single processing node. For scalability experiments on a single processing node, we used up to 25 million moving objects (raw size = 200MB).

For the parallelization experiment we used up to 100 million moving objects (raw size = 800MB). As outlined in the Introduction, we assumed all data and indexes for all methods to fit into main memory. We evaluate time-parameterized predictive range queries. Note again that other query types such as time-parameterized predictive k-nearest-neighbors may be inferred from predictive range queries (see [17]). We used a query window size corresponding to the size of a small town center like Oldenburg which has an extension of roughly 1,000 m $\times$ 1,000 m. Bigger query windows did not substantially change our results and therefore we do not show those results. Query centers were chosen using the NB strategy [5] thus creating a skewed distribution on queries. If not mentioned otherwise, we set a query rate of 1,000 queries per second and a query time $t_q = t_{now} + 1.5t_{\Delta max}$. Figure 5 summarizes the settings.

5.3 Scalability in Index Size

The goal of this experiment is to understand the maximum update rates supported by MOVIES when scaling the size of the data set. We compare MOVIES to baseline index structures, including binary search trees and B^+-trees. Furthermore, we compare MOVIES against a state-of-the-art moving object index: the B^x-tree [17]. As all tree-based methods are hard to parallelize without considerably sacrificing performance (locking), we parallelized all tree-based methods to obtain lock-free methods as follows: we partitioned the data by *OID* into four disjoint partitions, and used a separate tree and thread to index each partition as suggested in [40]. Thus, the tree-based methods could make maximal use of the four cores available on a server. All methods evaluated in this and the following experiments could make use of the same amount of main memory which was set to 5.5 GB. In particular, all tree-based methods resided completely in main memory. Therefore, at no point any disk-I/O was performed. We tuned the node size of the trees to obtain the best possible performance in a separate experiment. Only the best tree-based methods are displayed.

Figure 7(a) shows the results of a scalability experiment where the index size is varied up to 25.6 million moving objects. We kept a fixed query rate of 1,000 time-parameterized queries/s and display the maximum update rate supported by each indexing method. The results show that both variants of MOVIES outperform all other methods. Figure 7(a) shows that all tree-based methods degrade sharply with growing index sizes. The binary search tree was not able to scale beyond 3.2 M objects as then it could not meet the query rate anymore. For the B^+-tree, we experimented with several values of k and k^*. However, the best B^+-tree we could devise ($k = k^* = 16$) was not able to scale beyond 12.8M moving objects. For 12.8M the B^+-tree could only handle 0.6M updates/s. Similarly, for the B^x-tree we performed a separate experiment varying

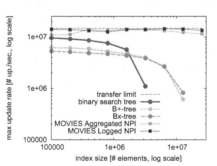
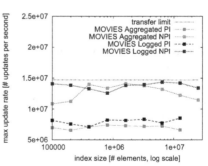

(a) max update rate: comparison of MOVIES, binary search tree, B^+-tree, and B^x-tree

(b) max update rate: comparison of four different variants of MOVIES

Fig. 7. Scalability in index size for high update rates [query rate=1,000/s], single processing node

the number of phases n and display only the best version of the B^x-tree we could devise ($k = 16$ and $n = 2$). Interestingly, the B^x-tree was also not able to scale beyond 12.8M moving objects. The B^x-tree even performed slightly worse than the best B^+-tree except for N=12.8M. This is due to the fact that the B^x-tree incurs overhead compared to the B^+-tree as it has to compute predictions for each incoming update at indexing time.

In the experiment, in contrast to all other methods, MOVIES Logged with Non-Predictive Indexing (NPI) shows update rates of around 14 million updates/s for index sizes up to 25.6M objects. This value is close to the network limit of 14.7 million updates/s. MOVIES Aggregated NPI shows in average a slighlty smaller update rate ranging from 11M to 14M updates/s. However, MOVIES Aggregated NPI performs still better than all tree-based methods. For an index of size of 12.8 M objects the improvement of the best MOVIES variant over the best B^+-tree and B^x-

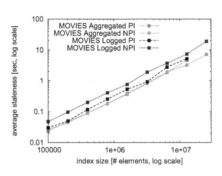

Fig. 8. Average staleness when scaling index size: comparison of four different variants of MOVIES [query rate=1,000/s]

tree is factor 15. For an index size of 25.6 M only MOVIES was able to index the data meeting the query rate. Interestingly, in this experiment the binary search tree performs even better for small indexes than the B^x-tree. This is due to the high cost for computing predictions for each incoming update. This is also evidenced when we compare the four different variants of MOVIES using different indexing methods. Figure 7(b) displays the update throughput for MOVIES when using different indexing schemes, i.e., either Aggregated or Logged MOVIES (Section 3.3), and either non-predictive (NPI) or predictive indexing (PI) (Section 4.1). The figure shows that Logged MOVIES NPI has the best performance. In contrast, Logged MOVIES PI achieves only half of the throughput. This is due to the fact that for predictive indexing each update has to be translated to a new position. This is CPU-intensive and also explains why the B^x-tree performs worse than a standard B^+-tree: at extreme update rates, the computational cost of predictions offsets the gain obtained by smaller query window enlargements.

Figure 8 shows the average staleness observed in the scaling experiment. The results show that the staleness grows for larger index sizes. That is expected as an important component of frame time is the time to sort the data in a new index. For MOVIES Logged NPI the staleness grows up to 19 sec, for MOVIES Aggregated NPI it increases up to 7 sec. If the staleness has to be reduced, this can be achieved by scaling out on multiple processing nodes. This is explored in Section 5.5.

5.4 Scalability in Update Rate

The goal of this experiment is to understand the maximum query rate supported by MOVIES when scaling the update rate. We keep the index size constant at 6.4M objects and vary the update rate. Figure 9 shows the result. The figure shows that the binary search tree is only able to sustain a very low query rate. For update rates above 2.1M updates/s this method is not able to execute any more queries and thus fails to scale beyond this point. Similarly, we observe that for update rates between 0.1M to 1M the best B^+-tree is able to execute between 3,000 and 2,000 queries/s, respectively. For higher update rates, however, the B^+-tree degrades sharply: for an update rate of 4M updates/s, the best B^+-tree is only able to execute a small amount of queries and thus fails to scale beyond this point. The B^x-tree has better query performance than the B^+-tree for update rates up to about 3M updates/s, but also fails to scale beyond 4M updates/s. The MOVIES variants show an interesting behavior: The predictive variants outperform the non-predictive variants in terms of query performance up to an update rate of 4M updates/s, exhibiting query rates around 3,000 queries/s. Above 4M updates/s, however, the query performance of the non-predictive variants sharply increases up to 9,200 queries/s. This behavior can be explained by analyzing the trade-off between indexing predictions and performing query window enlargements. For modest update rates, non-predictive methods must perform bigger query enlargements to compensate for the relatively large value of $t_{\Delta max}$. These enlarged queries impose significant computational overhead. Predictive methods, on the other hand, aggressively reduce query window enlargements by computing predictions for each update applied to the index. At high update rates, however, we observe the opposite effect: predictive methods pay a high computational cost for predicting every update applied to the index. The gain in query

window enlargements is not enough to offset these costs, because as $t_{\Delta max}$ is relatively low, the enlargements performed by non-predictive methods are also relatively small. In addition, non-predictive methods have lower computational cost for collecting updates. Another effect may be observed for the non-predictive variants: at an update rate of about 10M updates/s, the query rate drops to around 6,000 queries/s. This slight drop in the query rate may be explained by the fact that at very high update rates, the cost to collect updates starts to become significant, draining CPU resources from both query processing and index rebuilding.

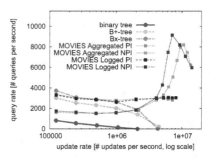

Fig. 9. Scalability in update rate: Comparison of MOVIES with binary search tree, B^+-tree and B^x-tree for high update rates [index size=6.4E6]

The staleness for Logged MOVIES NPI stayed constant around 3 sec up to 7M updates/s. For higher updates rates it increased linearly up to 7.2 sec. For Aggregated MOVIES NPI the staleness was constant around 2.5 sec. The predictive variants could not be scaled beyond 8M updates/s and their average staleness stayed between 2 to 3 sec. The relatively high staleness for low update rates can be explained as follows: If during one index rebuild MOVIES receives only few updates, then MOVIES has to retrieve the old data for many objects from the old index. This leads to many random accesses to the old index and therefore hurts rebuild performance. The time needed to lookup old data goes down as the update rate increases and reaches zero around 5M updates/s. Even though this effect would lead to decreasing staleness, the staleness stays about constant, because processing the updates becomes more expensive. In summary, this experiment shows that the MOVIES variants scale well for high update rates. Of all methods, only MOVIES was able to scale up to 14M updates/s. Note again that all methods completely resided in main memory.

5.5 Shared-Nothing Scale-Out

The goal of this experiment is to examine how MOVIES scales when increasing the number of processing nodes. In order to adapt the different methods to a shared-nothing landscape, we horizontally hash-partitioned the data by OID. We keep the index size constant at 25.8M and vary the number of processing nodes PN from one to four. As our experiments with a single processing node have shown, the transfer limit imposed by the network is a serious bottleneck. Therefore, we required a special network setup as described in Section 5.1. With that setup we could transfer up to 58M updates/s to four processing nodes while still being able to distribute queries. Figure 10(a) shows the results. The NPI MOVIES variant Aggregated (resp. Logged) scales up to 47M (resp. 54M) updates/s. Figure 10(b) displays an experiment where we keep the index size constant at 25.8M and keep the maximum update rate at 5M updates/s, which is supported by the worst MOVIES method. We display the average staleness. The figure shows that staleness goes down almost linearly if we increase the number of processing nodes. For four processing nodes staleness goes below 3 seconds for all four variants of MOVIES. In summary, this experiment shows that MOVIES scales linearly w.r.t. the maximum number of updates and linearly w.r.t. to the average query result staleness.

In another experiment we used all four processing nodes for indexing. Figure 10(c) shows the results. Similarly to the single instance experiment MOVIES outperforms all other methods. All tree-based methods, including the B^x-tree degrade sharply for growing index sizes. The tree-based methods fail to scale beyond an index of size 51M, i.e., 12.8M moving objects per processing node. In contrast, MOVIES scales up to 102M moving objects. Furthermore, for index sizes up to 51M, Logged MOVIES sustains an update rate close to the network limit of 58M updates/s. For 51M moving objects the improvement of MOVIES over the best B^+-tree is factor 15; the improvement over the best B^x-tree is factor 11. The average staleness of all the MOVIES variants is the same

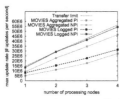 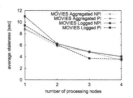 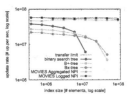

(a) max update rate: effects of scaling number of processing nodes [index size = 25.8E6]

(b) average staleness: effects of scaling number of processing nodes [index size = 25.8E6]

(c) Scalability in index size on four shared-nothing servers: Comparison of MOVIES with binary search tree, B^+-tree, and B^x-tree

Fig. 10. Shared-nothing performance [query rate = 1,000/s]

as shown for the single instance experiment, but the index size is four times larger. See Figure 8. For example, the staleness of MOVIES Logged NPI is 21 seconds for an index with 102M elements.

6   Related Work

Considerable work has been done in the area of moving objects. The existing methods can be classified into two groups: methods with or without time-parameterized (TP) queries. General design issues for moving object indexes can be found in [30].

6.1   Methods with TP Queries

External Memory. Many approaches are centered around extending external memory structures like the B^+-tree, R-tree[13], or R^*-tree[3]. All of these methods assume that data would not fit into main memory. Examples include the TR-tree and TB-tree [35], the TPR-tree [48], the TPR*-tree [42], the STP-tree [41] and the R^{PPF}-tree [34]. The most relevant work to our work is the B^x-tree [17] as, conceptually, it has some similarities to the MOVIES indexing strategy. The core idea of the B^x-tree is to map three-dimensional data (two spatial and one temporal dimension) to a one-dimensional space. This is done by using a recursive space-filling curve and mapping data to a B^+-tree very similarly to [32]. However, in contrast to the latter approach, the B^x-tree also partitions data into phases corresponding to future time intervals. For each phase it uses a separate subtree to index moving objects and predicted positions. As a consequence, prediction queries are supported. As the B^x-tree is based on a B^+-tree, it is very easy to integrate it into existing DBMSs. The B^x-tree was shown to outperform competing methods such as the TPR-tree [48]. However, in contrast to MOVIES the B^x-tree does not rebuild the index based on updates buffers but rather follows an update strategy similar to update-in-place. Also the partitioning into phases used by the B^x-tree leads to relatively high query cost (as observed in our experiments) which is avoided by MOVIES. Other methods index moving objects by transforming them to a higher dimensional space. This includes STRIPES [33] and [21] which transform d-dimensional space to $2d$-dimensional Hough-X space [15]. The recently proposed B^{dual}-tree [50] uses the same idea; however it maps the Hough-X space back to a one-dimensional space using a Hilbert curve. [43] presents a study on dual methods concluding that if query efficiency is required (as required in this paper), dual methods are not competitive. Interestingly, in the concluding remarks of [43] it is suggested that it could be beneficial to rather reconstruct a non-dual method periodically. Exactly this approach is followed by MOVIES.

Main Memory. The approach of [7] partitions data into sets of active objects that stay in a main memory buffer and inactive objects that reside on external memory. Therefore that work is more of a buffering scheme for moving object indexing. It is orthogonal to the techniques presented here and can be applied on top of any moving objects index.

6.2   Methods without TP Queries

Main Memory. Relevant to our work are methods that use main memory for monitoring queries. The method of [19] uses a fix-sized grid where the grid-size is chosen w.r.t. the average query window sizes. Each grid cell maintains pointers to two lists with query results. Query results are periodically reevaluated and query results are delivered with a time delay Δt. [51] extends [19] to k-NN queries. [26] improves [51] to only update grid-cells that are affected by an incoming update. However, none of the former methods provides any support for time-parameterized queries. Also [19,51,26] do neither provide any means how to scale for cases when the main memory is exhausted nor provide any parallelization scheme. In contrast MOVIES provides solutions for all of these issues. [27] focusses on k-NN in road networks where the distance among objects is not the euclidean distance but rather the length of the shortest path on the network. Therefore the latter method will not work for objects not following roads, e.g., planes, ships, people's phones. In contrast, MOVIES supports all of these scenarios.

6.3   Extensions for Efficient Updates

External Memory. Frequent update handling in R-trees was treated in [23,4]. A general survey on how to optimize B-trees for high update rates was recently presented by Graefe [12]. Several of these optimizations may be traced back to Lars Arge's buffer tree [2]. Graefe also mentions differential files [38] as an effective means to trade query performance for update performance. However, [12] does not mention that one could trade query result staleness and keep *both* queries and updates efficient as in MOVIES.

Main Memory. Batching updates in a similar way to Lars Arge's buffer tree [2] was also considered for main memory optimized trees such as [52,8] however trading query for update performance. In contrast, MOVIES does not trade query performance for update performance. Other cache-efficient trees are the CSS-tree [37] and the FPB+-tree [6]. An interesting challenge would be to extend both the B^x-tree and MOVIES to include these optimizations. However, as pointed out in Section 2.2, the query processing performance is not affected by MOVIES. Therefore, the general trade-off of update-in-place versus collect and rebuild as used by MOVIES will remain unchanged. Rather, as MOVIES may build read-only indexes at each index frame, MOVIES could even improve overall query performance by building read-only cache-aware indexes.

6.4   Experimental Studies

Moving object scenarios comprise a large number of objects and a large number of updates. As mentioned above, the number of cars in Germany is about 58,000,000 [22]. Assume every car sends an update on its position every 2 seconds, then this boils down to 29,000,000 updates per second. If we were to index not only cars but also planes, people's cellular phones, etc., we would face even higher data and update volumes. In this work we are interested in supporting these large scale scenarios. Therefore we are considering data sets of up to 100,000,000 moving objects. This is 10 times larger than in the biggest study available [23] and by at least two orders of magnitude larger than in all other studies, e.g., [26,27,17,19,51,7]. We think it is important to scale to such large data sets in order to understand the limits of the different methods.

7 Conclusions

This paper has proposed a novel approach to time-parameterized moving object indexing of massive data sets under very high update rates. Our approach is based on frequently building short-lived throwaway indexes. This keeps at the same time query throughput high, query response time low, and update performance high. The price we have to pay is slightly out-of-date (stale) query results, which is acceptable in several applications including aircraft control [39]. We have shown that this price can be reduced to be as small as a few seconds even for very large data sets of up to 100,000,000 moving objects. Our experiments have demonstrated the feasibility of our approach even for massive realistic data sets. We have presented results of an experimental study using the entire road network of Germany: a network size unmatched by any previous work. In our study we scale up to 100,000,000 moving objects and 58,000,000 updates per second. MOVIES shows order of magnitude improvements over state-of-the-art approaches like the B^x-tree, as well as several baseline methods w.r.t. supported update rates and query rates. One general conclusion is that the popular pattern of *keeping and modifying an index* should be dropped for moving object scenarios. Another surprising conclusion of our study is that the idea of indexing predictions for time-parameterized queries as done by some external memory indexes does only work well in main memory for low update rates. In terms of future work we plan to examine the trade-off of scalability and stalenesss in more detail. Another research direction would be to extend our approach to consider cache-aware B^+-trees, e.g. [37]. However, as shown by our formal analysis, the general trade-off of update-in-place versus collect-and-rebuild would even be improved in favor of MOVIES.

References

1. Anderson, I., et al.: Shakra: Tracking and Sharing Daily Activity Levels with Unaugmented Mobile Phones. Mobile Networks and Applications 12(2-3) (2007)
2. Arge, L.: The Buffer Tree: A New Technique for Optimal I/O-Algorithms (Extended Abstract). In: Sack, J.-R., Akl, S.G., Dehne, F., Santoro, N. (eds.) WADS 1995. LNCS, vol. 955. Springer, Heidelberg (1995)
3. Beckmann, N., Kriegel, H.-P., Schneider, R., Seeger, B.: The R*-Tree: An Efficient and Robust Access Method for Points and Rectangles. In: SIGMOD (1990)
4. Biveinis, L., Šaltenis, S., Jensen, C.S.: Main-Memory Operation Buffering for Efficient R-Tree Update. In: VLDB (2007)
5. Brinkhoff, T.: A Framework for Generating Networkbased Moving Objects. GeoInformatica 6(2), 153–180 (2002)
6. Chen, S., Gibbons, P.B., Mowry, T.C., Valentin, G.: Fractal Prefetching B+trees: Optimizing Both Cache and Disk Performance. In: SIGMOD (2002)
7. Cui, B., Lin, D., Tan, K.-L.: Towards Optimal Utilization of Main Memory for Moving Object Indexing. In: Zhou, L.-z., Ooi, B.-C., Meng, X. (eds.) DASFAA 2005. LNCS, vol. 3453, pp. 600–611. Springer, Heidelberg (2005)
8. Dittrich, J.-P., Fischer, P.M., Kossmann, D.: AGILE: Adaptive Indexing for Context-Aware Information Flters. In: SIGMOD (2005)
9. Dittrich, J.-P., Seeger, B.: GESS: a Scalable Similarity-Join Algorithm for Mining Large Data Sets in High Dimensional Spaces. In: SIGKDD (2001)

10. Enhanced 911, `http://www.fcc.gov/pshs/911`
11. Google Web Search, `http://www.google.com`
12. Graefe, G.: B-tree indexes for high update rates. SIGMOD Rec. 35(1) (2006)
13. Guttman, A.: R-Trees: A Dynamic Index Structure for Spatial Searching. In: SIGMOD (1984)
14. Hilbert, D.: Über die stetige Abbildung einer Linie auf ein Flächenstück. Mathematische Annalen 38, 459–460 (1891)
15. Hough, P.: Method and means for recognizing complex patterns. United States Patent No. 3069654 (1962)
16. Jagadish, H.V., et al.: Incremental Organization for Data Recording and Warehousing. In: VLDB (1997)
17. Jensen, C.S., Lin, D., Ooi, B.C.: Query and Update Efficient B+-Tree Based Indexing of Moving Objects. In: VLDB (2004)
18. Jensen, C.S., Pakalnis, S.: TRAX - Real-World Tracking of Moving Objects. In: VLDB (2007)
19. Kalashnikov, D.V., Prabhakar, S., Hambrusch, S.E.: Main Memory Evaluation of Monitoring Queries Over Moving Objects. Distributed and Parallel Databases 15(2), 117–135 (2004)
20. Knuth, D.E.: The Art of Computer Programming. Sorting and Searching, vol. III. Addison-Wesley, Reading (1973)
21. Kollios, G., Papadopoulos, D., Gunopulos, D., Tsotras, J.: Indexing mobile objects using dual transformations. VLDB Journal 14(2), 238–256 (2005)
22. Kraftfahrt-Bundesamt. Number of Vehicles in Germany over time, `www.kba.de/Abt3_neu/FZ/Bestand/Themen_jaehrlich_pdf/bki1_2008.pdf`
23. Lee, M.-L., Hsu, W., Jensen, C.S., et al.: Supporting Frequent Updates in R-Trees: A Bottom-Up Approach. In: VLDB (2003)
24. Loopt, `http://www.loopt.com`
25. Apache Lucene, `http://lucene.apache.org/java/docs`
26. Mouratidis, K., Papadias, D., Hadjieleftheriou, M.: Conceptual Partitioning: An Efficient Method for Continuous Nearest Neighbor Monitoring. In: SIGMOD (2005)
27. Mouratidis, K., Yiu, M.L., Papadias, D., Mamoulis, N.: Continuous Nearest Neighbor Monitoring in Road Networks. In: VLDB (2006)
28. Muth, P., O'Neil, P.E., Pick, A., Weikum, G.: The LHAM Log-Structured History Data Access Method. VLDB J. 8(3-4), 199–221 (2000)
29. O'Neil, P.E., Cheng, E., Gawlick, D., O'Neil, E.J.: The Log-Structured Merge-Tree (LSM-Tree). Acta Inf. 33(4) (1996)
30. Ooi, B.C., Tan, K.L., Yu, C.: Frequent Update and Efficient Retrieval: an Oxymoron on Moving Object Indexes? In: WISE Workshops 2002 (2002)
31. Orenstein, J.A.: An Algorithm for Computing the Overlay of k-Dimensional Spaces. In: Günther, O., Schek, H.-J. (eds.) SSD 1991. LNCS, vol. 525. Springer, Heidelberg (1991)
32. Orenstein, J.A., Merrett, T.H.: A Class of Data Structures for Associative Searching. In: PODS (1984)
33. Patel, J.M., Chen, Y., Chakka, V.P.: STRIPES: An Efficient Index for Predicted Trajectories. In: SIGMOD (2004)
34. Pelanis, M., Šaltenis, S., Jensen, C.S.: Indexing the Past, Present, and Anticipated Future Positions of Moving Objects. ACM TODS 31(1), 255–298 (2006)
35. Pfoser, D., Jensen, C.S., Theodoridis, Y.: Novel Approaches to the Indexing of Moving Object Trajectories. In: VLDB (2000)
36. Ramsak, F., Markl, V., et al.: Integrating the UB-Tree into a Database System Kernel. In: VLDB (2000)
37. Rao, J., Ross, K.A.: Making B+-Trees Cache Conscious in Main Memory. SIGMOD 29(2) (2000)

38. Severance, D.G., Lohman, G.M.: Differential Files: Their Application to the Maintenance of Large Databases. ACM TODS 1(3), 256–267 (1976)
39. Personal communication with Skyguide Flight Control
40. Stonebraker, M., Madden, S., et al.: The End of an Architectural Era (It's Time for a Complete Rewrite). In: VLDB (2007)
41. Tao, Y., Faloutsos, C., et al.: Prediction and Indexing of Moving Objects with Unknown Motion Patterns. In: SIGMOD (2004)
42. Tao, Y., Papadias, D., Sun, J.: The TPR*-Tree: An Optimized Spatio-Temporal Access Method for Predictive Queries. In: VLDB (2003)
43. Tao, Y., Xiao, X.: Primal or dual: which promises faster spatiotemporal search? VLDB J. 17(5) (2008)
44. Tele Atlas MultiNet Europe Q4/2006. Germany
45. Thirde, D., et al.: Evaluation of Object Tracking for Aircraft Activity Surveillance. In: 2nd Joint IEEE International Workshop on VS-PETS (2005)
46. Thomas Legler, A.R., Lehner, W.: Data Mining with the SAP Netweaver BI Accelerator. In: VLDB, pp. 1059–1068 (2006)
47. Tropf, H., Herzog, H.: Multimensional Range Search in Dynamically Balanced Trees. Ang. Informatik 23(2), 71–77 (1981)
48. Šaltenis, S., Jensen, C.S., et al.: Indexing the Positions of Continuously Moving Objects. In: SIGMOD (2000)
49. White, W.M., Demers, A.J., Koch, C., Gehrke, J., Rajagopalan, R.: Scaling Games to Epic Proportion. In: SIGMOD (2007)
50. Yiu, M.L., Tao, Y., Mamoulis, N.: The Bdual-Tree: indexing moving objects by space filling curves in the dual space. VLDB J. 17(3) (2008)
51. Yu, X., Pu, K.Q., Koudas, N.: Monitoring k-Nearest Neighbor Queries over Moving Objects. In: ICDE (2005)
52. Zhou, J., Ross, K.A.: Buffering Accesses to Memory-Resident Index Structures. In: VLDB (2003)

Indexing the Trajectories of Moving Objects in Symbolic Indoor Space

Christian S. Jensen[1], Hua Lu[1], and Bin Yang[1,2]

[1] Department of Computer Science, Aalborg University, Denmark
[2] School of Computer Science, Fudan University, China
{csj,luhua,yang}@cs.aau.dk

Abstract. Indoor spaces accommodate large populations of individuals. With appropriate indoor positioning, e.g., Bluetooth and RFID, in place, large amounts of trajectory data result that may serve as a foundation for a wide variety of applications, e.g., space planning, way finding, and security. This scenario calls for the indexing of indoor trajectories. Based on an appropriate notion of indoor trajectory and definitions of pertinent types of queries, the paper proposes two R-tree based structures for indexing object trajectories in symbolic indoor space. The RTR-tree represents a trajectory as a set of line segments in a space spanned by positioning readers and time. The TP^2R-tree applies a data transformation that yields a representation of trajectories as points with extension along the time dimension. The paper details the structure, node organization strategies, and query processing algorithms for each index. An empirical performance study suggests that the two indexes are effective, efficient, and robust. The study also elicits the circumstances under which our proposals perform the best.

1 Introduction

People spend large parts of their lives in indoor spaces such as office buildings, shopping centers, conference facilities, airports, and other transport infrastructures. At the same time, such spaces are becoming increasingly large and complex. For example, the New York City Subway has 468 stations and a network of 842 miles. Each day, the subway serves more than 6 million users, totalling 2+ billion annually.

With the deployment of indoor positioning based on technologies such as RFID [23], Bluetooth [8], and Wi-Fi [3], large volumes of tracking data are becoming available that enable a range services akin to those enabled by GPS-based positioning in outdoor settings. Example services include indoor navigation, personal security, and those providing insight into how and how much the indoor space is being used, which is important in planning applications and for the pricing of advertisement space and store rentals. Motivated by these observations, this paper provides two techniques for the indexing of the trajectories of objects moving in symbolic indoor space.

Over the past decade, much research has been devoted to outdoor applications involving moving objects [9,24], and substantial research concerns the indexing and querying of the positions of moving objects [2,6,12,13,14,18,20,21,25] and their trajectories [5,17]. However, the outcomes of this body of research is not easily applicable in indoor scenarios. First, indoor space is typically modeled differently from outdoor space, where either

N. Mamoulis et al. (Eds.): SSTD 2009, LNCS 5644, pp. 208–227, 2009.

Euclidean space or a spatial network is typically assumed. Indoor space is characterized by entities such as doors, rooms, and hallways that enable and constrain movement. This renders movement more constrained than outdoor Euclidean movement. Consequently, geometric representations, e.g., the linear model that is widely adopted for describing outdoor movement, are not suitable for describing indoor movement. Further, indoor movement is less constrained than outdoor spatial-network movement, where the position of an object is constrained to a position on a polyline. As a result, symbolic models, rather than geometric models, of indoor space are often used [4], which renders indexes for outdoor moving objects inapplicable.

Second, indoor positioning technologies differ fundamentally from those typically assumed in outdoor settings. Unlike GPS and cellular technologies, which are capable of continuously reporting the position and velocity of an object with varying accuracies, we assume indoor positioning technologies that rely on proximity analysis [11] and are able to report neither velocities nor exact locations. In particular, an indoor moving object is detected only when it enters the sensing or activation range of a positioning device, e.g., an RFID reader, or a Bluetooth base station.

We propose two R-tree based structures for indexing the trajectories of objects moving in symbolic indoor space, together with algorithms for the processing of pertinent queries, including spatiotemporal range and topological queries. We assume a symbolic representation of indoor space and use RFID for positioning. The trajectory of an object is then represented as a series of records, each of which indicates that the object is within the activation range of a specific RFID reader during a period of time.

The RTR-tree, similar to the R-tree, uses a specific node organization. It organizes a trajectory as a set of line segments in the plane spanned by positioning readers and time. For each type of query, a corresponding geometrical representation is derived and used to search the RTR-tree similarly to how an R-tree is searched.

The TP^2R-tree uses the same underlying space, but applies a data transformation that turns a trajectory into a set of points in the plane, augmented with temporal extents. The objective is to obtain a better node organization. When a node overflows during insertion, splitting is handled by taking the time extents into consideration such that fewer node accesses are expected by subsequent queries. Efficient query processing algorithms for the TP^2R-tree are also detailed.

A comprehensive performance study on synthetic and real data sets is conducted to evaluate the two indexes, together with relevant query processing algorithms. A wide range of parameter settings are applied. The results show that tree construction is scalable with respect to the trajectory data size and that query processing is quite efficient.

The paper's contributions may be summarized as follows. The paper formalizes a moving-object trajectory model and trajectory-related queries in symbolic indoor space. It presents two index structures with different node organization strategies that apply specifically to indoor trajectories. It presents accompanying algorithms for the processing of pertinent queries. Finally, it presents the results of a comprehensive and favorable performance study of the paper's proposals.

The rest of this paper is organized as follows. Section 2 describes existing indexes for moving objects. Section 3 presents the assumed data model, the assumed representation of indoor trajectories, and the queries considered. Sections 4 and 5 detail the RTR-tree and and TP^{2}R-tree, respectively. Section 6 reports on the relevant experimental study. Section 7 concludes and offers research directions.

2   Existing Moving-Object Indexes

Most research on outdoor moving-object indexing assumes, explicitly or implicitly, the availability of GPS-type positioning. A GPS receiver can continuously (typically each second) report location (longitude and latitude) and velocity. Hence, the trajectory of a GPS-equipped outdoor moving object is usually modeled as a polyline in three-dimensional space-time space.

The TB-tree (Trajectory Bundle) [17] extends the R-tree to allow the indexing of such trajectories by allowing a leaf node to contain line segments only from the same trajectory, which facilitates retrieval of the trajectory of an individual object, but adversely affects spatiotemporal range queries. SETI (Scalable and Efficient Trajectory Index) [5] is based on a static partitioning of the spatial dimensions. The aim is that trajectory line segments in the same index partition belong to the same trajectory. Then, within each partition, all line segments are indexed by an R-tree. The performance of SETI depends heavily on the partitioning function used. Because indoor positioning technologies only determine the presence of moving objects at predefined locations, the polyline representation used for outdoor trajectories is inapplicable for indoor trajectories. As a result, the TB-tree and SETI cannot be applied to indoor trajectories.

In addition to indexes directly targeting trajectories, proposals exist for the indexing of historical positions of outdoor moving objects in order to support spatiotemporal range queries. The 3D R-tree [22] treats the time dimension as yet another spatial dimension. The HR-tree (historical R-tree) [16] adds a temporal dimension to the R-tree by means of replication when updates occur, so that multiple R-trees result, each corresponding to a time interval. A tree node belongs to multiple consecutive trees if the data in it remains constant during the time interval associated with the trees. The MV3R-tree (Multi-version 3D R-tree) [20] integrates the HR-tree and the 3D R-tree. The PPR-tree (Partially-Persistent R-tree) [14] indexes both historical and current positions of animated objects by applying multi-versioning.

The idea behind the 3D R-tree, of treating time simply as an additional dimension, can be used to index indoor trajectories. Our RTR-tree with its basic node organization strategies (see Section 4.2) adopts this approach. However, our studies show that this straightforward approach yields poor performance. The HR-tree is unsuitable for indexing indoor trajectories for three reasons. First, a mapping from each indoor positioning device to its sensing range is needed for indoor positioning data to be indexed by an HR-tree. Second, the limited device sensing ranges cause high data volatility, as few objects remain for long within a range. This considerably reduces the overlapping among tree nodes that the HR-tree exploits. Third, the HR-tree prefers time point queries over time interval queries. Our proposals aim to support both query types efficiently.

Other indexes for the current and future positions of outdoor moving objects [2,6,12,13,18,25] are unsuitable for indoor trajectory indexing. Moreover, almost all assume a linear movement model, which is unavailable for indoor objects. In particular, although the BB^x-tree [15] and the R^{PPF}-tree [19] index the historical, current, and future positions of moving objects, they both assume a linear movement model.

3 Data Model, Trajectory Representation, and Queries

3.1 Data Model and Queries

As suggested earlier, the geometric polyline representation used for outdoor trajectories is not ideal for indoor trajectories. For example, assume that an object moves from one room to another so that two consecutive location reports are in different rooms. (Due to the limitations of indoor positioning, locations between these two reports are not obtained.) Using the polyline representation, the line segment between the two reported locations is unlikely to intersect with the door, but will go through the wall between the rooms. This means that the trajectory has the moving object going through the wall, which contradicts reality and renders the trajectory of limited use.

Typical queries on indoor trajectories contain symbolic references such as room numbers rather than simply geometric locations. For example, we may be interested in determining which room a moving object was in at a specific time, or the sequence of rooms a moving object visited during a given time interval. Such considerations lead to an indoor trajectory model that is composed of trajectory records in the format $(objectID, symbolicID, \mathbb{T})$. Here, $objectID$ is the identifier of a moving object; $symbolicID$ is the identifier for a specific indoor space region (e.g., a room); and $\mathbb{T}$ indicates the time, either a time instant or a time interval.

We consider two types of queries. Given an indoor spatial extent E_s and a temporal extent E_t, an *indoor range query* $Q(E_s, E_t)$ returns all trajectory records that intersect the spatiotemporal region defined by E_s and E_t: $Q(E_s, E_t) \rightarrow \{trajectory\ records\}$.

Specifically, E_s can be represented by either a subset of symbolic references (e.g., room numbers) or a subset of Euclidean space (e.g., a polygon or a circle in a floor plan). Next, E_t indicates a temporal extent, which is either a time instant or a time interval. For example, query $Q(room_1, [1:00\ \text{p.m.}, 1:15\ \text{p.m.}])$ returns all trajectory records indicating that the corresponding objects were in $room_1$ at some time between 1:00 p.m. and 1:15 p.m. A Euclidean spatial constraint in a query can be transformed to one or more symbolic references. This is covered in detail in Section 4.3.

Next, given an indoor space partition (e.g., a room), a temporal extent, and a topological predicate, an *indoor topological query* returns all objects whose trajectories satisfy the given predicate with respect to the space partition within the given temporal extent: $Q(E_s, E_t, P) \rightarrow \{objectID\}$.

Here, E_t is as before, and E_s is an indoor space partition with the property that positioning devices are deployed such that the satisfaction of predicate P can be detected or inferred from the positioning data. (Issues related to deployments are addressed in Section 4.3.) Moreover, P denotes a topological predicate such as (a) *enter*, (b) *leave*, or (c) *cross* [7,17]. We use the cross predicate to mean enter and then leave, or leave and then enter. For example, query $Q(room_1, [1:00\ \text{p.m.}, 1:15\ \text{p.m.}], enter)$ returns all objects that entered $room_1$ between 1:00 p.m. and 1:15 p.m.

The range and topological queries can work as building blocks for constructing more complex queries. Such queries can be processed by combining the algorithms presented in this paper.

3.2 RFID Based Indoor Positioning

Our focus is on positioning by means of RFID technology [23], and we assume a setting where RFID readers are deployed at fixed locations, e.g., at building entrances, doors, hallways, while RFID tags are attached to moving objects. An RFID reader employs proximity analysis [11] to detect an RFID tag when the tag (with the object to which it is attached) enters the reader's activation range. Different readers support different sensing ranges. The position of each reader is recorded in the database after deployment.

Each RFID reader continuously detects and reports tags with a frequency determined by its sampling rate, a hardware specific parameter that usually varies from 1 to 3 times per second [23]. We use T_S to denote the RFID reader sampling period. The raw RFID readings are of the format $(readerID, tagID, t)$, meaning that a reader $readerID$ detects the moving object with tag $tagID$ in its activation range at timestamp t. Our proposals accommodate the possibility of a tag being detected simultaneously by multiple readers.

Unlike GPS that reports accurate geographic locations, RFID-based positioning only reports the presences of objects in readers' activation ranges. An object's location during the time in-between consecutive RFID readings cannot be inferred from the reading sequence without additional information such as the floor plan. Also, different deployments of RFID readers in the same indoor space generally result in different positioning accuracies. Positioning reader deployment is beyond the scope of this paper. Rather, our focus is to enable indexing of whatever RFID positioning data is available.

Given a raw RFID reading sequence, a trajectory table can be constructed that contains records of the following format: $(recordID, tagID, readerID, t_s, t_e)$. Here, $recordID$ is the identifier of each trajectory record, and t_s (t_e) indicates the first (last) time point when reader $readerID$ detects tag $tagID$ in its activation range. This representation removes the "duplicate" readings caused by the sampling. A table containing trajectories is shown in Table 1.

We proceed to propose two indexes for indoor trajectories. In Section 4 we detail the RTR-tree which, treating trajectories as sets of line segments, extends the R-tree with specific node organization strategies. In Section 5, we detail the TP^{2}R-tree that is intended to improve the tree organization by representing trajectories as points with time extension parameters.

Table 1. Indoor Trajectories

recordID	tagID	readerID	t_s	t_e
rd_1	tag_1	$reader_1$	t_1	t_3
rd_2	tag_3	$reader_2$	t_2	t_4
rd_3	tag_2	$reader_3$	t_3	t_5
rd_4	tag_3	$reader_2$	t_6	t_8
rd_5	tag_2	$reader_2$	t_7	t_9
rd_6	tag_1	$reader_4$	t_{10}	t_{11}
rd_7	tag_3	$reader_3$	t_{11}	t_{12}

4 RTR-Tree: Reader-Time R-Tree

The *Reader-Time R-tree* (*RTR-tree*) that is essentially a two dimensional R-tree [10] on the Reader-Time space. The vertical axis, which we call the *reader axis*, represents reader identifiers. The horizontal axis, which we call the *time axis*, represents time. Using this space, an indoor trajectory record becomes as a horizontal line segment.

4.1 RTR-Tree Index Structure

Leaf nodes in the RTR-tree contain index entries of the form $(MBR, recordID)$, where MBR is the minimum bounding rectangle of the record, identified by $recordID$. An

MBR in a leaf node is a line segment of the form $MBR(readerID, t_s, t_e)$, where $readerID$ indicates a reader on the Reader axis and t_s and t_e (coming from the corresponding trajectory record) indicate the (closed) time interval $[t_s, t_e]$. A leaf entry implies that a specific tag is detected by $readerID$ between timestamps t_s and t_e.

Non-leaf nodes of the RTR-tree have entries of the form (MBR, cp), where cp is a pointer to a child node in the RTR-tree and MBR is the minimum bounding rectangle that contains all MBRs in the child node's entries. Each MBR in a non-leaf node is in the form $MBR(readerID_{min}, readerID_{max}, t^\vdash, t^\dashv)$, where $readerID_{min}$ and $readerID_{max}$ indicate the interval $[readerID_{min}, readerID_{max}]$ on the Reader axis and $t^\vdash, t^\dashv$ indicates the time interval $[t^\vdash, t^\dashv]$. Therefore, a non-leaf entry implies that between times $t^\vdash$ and $t^\dashv$, some tags are detected by some readers whose identifiers fall inclusively into the range $[readerID_{min}, readerID_{max}]$.

Since the spatial information is represented by readers, it is beneficial to order the reader identifiers on the Reader axis according to their spatial proximity. Space-filling curves, e.g., the Hilbert curve, are often used to map objects in a higher dimensional space to one-dimensional space in order to preserve their proximity as best as possible [6,12]. However, space-filling curves are less attractive for use in indoor space because the obstacles prevalent in indoor settings and the constrained topology of indoor space tend to result in locations being close in a Euclidean sense not actually being close. For example, two close locations on either side of a wall may be relatively far apart when taking into account the topology.

Instead, we propose to take into account the indoor topology when assigning identifiers to readers. Our objective is to assign consecutive identifiers to readers in the same partition, to readers in adjacent partitions, and to readers on the same floor. Given a specific indoor space and reader deployment, a variety of different orderings can be envisioned that target this objective.

4.2 Node Organization Strategies

Insertion of new index entries into an RTR-tree is carried out as in the R-tree: new index entries are added into leaf nodes, nodes that overflow are split, and splits may propagate up the tree. The R-tree's strategies for identifying an appropriate leaf node and for node splitting are applied here, using the area of each MBR to guide node organization.

We calculate the area of an MBR in the RTR-tree in two different ways: (1) $Area = (readerID_{max} - readerID_{min}) * (t^\dashv - t^\vdash)$; (2) $Area^+ = (readerID_{max} - readerID_{min} + 1) * ((t^\dashv - t^\vdash)/T_S + 1)$. The $Area$ formula computes the geometric area and is identical to the method used by the original R-tree. In this formula, the area of any MBR of records with the same reader is 0 as the time dimension is neglected. Consequently, records with the same reader are put into the same node (resulting from leaf node choosing and node splitting), even if they are considerably apart in the time dimension. We call this the basic strategy.

The $Area^+$ formula takes into account the number of possible raw readings. Each point in the Reader-Time coordinate system indicates a possible raw reading. By adding 1 (and not any other arbitrary positive number), the $Area^+$ formula actually calculates the number of possible raw readings in an MBR. This calculation renders each node to

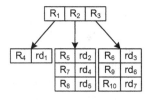

Fig. 1. RTR-Tree, Basic

have as few raw readings as possible. This way, we avoid too many zero-area MBRs that disable the use of area differences for leaf node selection and node splitting. We call this strategy *RTR-tree Area$^+$*.

For example, let the capacity of each node be 3. After all the trajectory records shown in Table 1 are inserted into the RTR-tree using the basic strategy, the the RTR-tree and MBRs are shown in Figures 1 and 2, respectively. Using the Area$^+$ strategy results in the MBRs shown in Figure 3. Note that each r_i on the Reader axis corresponds to a reader identifier $reader_i$.

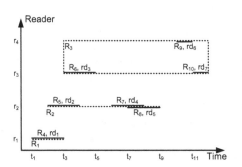

Fig. 2. MBRs of the RTR-Tree, Basic

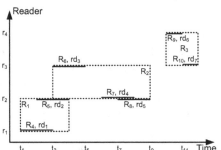

Fig. 3. MBRs of the RTR-Tree, Area$^+$

4.3   Query Processing

Given a query as described in Section 3.1, it is first transformed according to its type such that a corresponding geometric representation g is obtained that describes the query in the Reader-Time coordinate system and is used to process the query. Particularly, a recursive depth-first tree search is used on the RTR-tree to process the query. The search involves all nodes whose MBR overlaps with g. Due to space limitations, we omit the details of the search algorithm. We proceed to detail how to obtain the geometric representations of all query types.

Range Queries Transformation.   Given a range query $Q(E_s, E_t)$, two possibilities exist regarding the representation of the spatial extent E_s.

Symbolic Representation.   If the spatial extent E_s is represented in symbolic space, i.e., by *readerID*s, no matter which form the temporal extent E_t takes, the query can be transformed into basic geometric shapes, which is described in Table 2. *Single ReaderID* indicates that E_s is represented as a single *readerID*; *Continuous ReaderIDs* indicates that E_s is represented as a set of *readerID*s that are continuous on the vertical Reader axis. A query with a single reader identifier (i.e., query types QT_1 and QT_2) and continuous reader identifiers (i.e., query types QT_3 and QT_4) can be executed directly on the RTR-tree. Example representations for queries with different combination of E_s and E_t are shown in Table 2 and Figure 4.

Table 2. Symbolic Range Query Transformation for RTR-Tree

Time		Single ReaderID	Continuous ReaderIDs
Instant	Query Format	Query Type 1: $QT_1(readerID, t)$	Query Type 3: $QT_3([readerID_m, readerID_n], t)$
	Geometry Representation	Point	Vertical line segment
	Example	$Q_1(reader_2, t_3)$	$Q_3([reader_1, reader_3], t_6)$
Interval	Query Format	Query Type 2: $QT_2(readerID, [t_i, t_j])$	Query Type 4: $QT_4([readerID_m, readerID_n], [t_i, t_j])$
	Geometry Representation	Horizontal line segment	Rectangle
	Example	$Q_2(reader_3, [t_2, t_4])$	$Q_4([reader_1, reader_3], [t_7, t_9])$

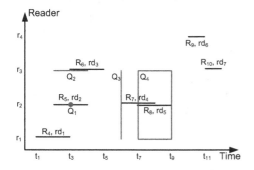

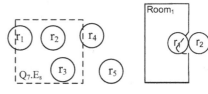

Fig. 4. Symbolic Range Queries in the RTR-Tree

Fig. 5. Euclidean Range Queries in the RTR-Tree

Fig. 6. Reader Deployment

Euclidean Representation. If the spatial extent E_s is represented in Euclidean space, it must be mapped to symbolic space. For example, the spatial extent of query Q_5 is represented as a dashed rectangle in Figure 5; the circle surrounding each reader identifier indicates the reader's activation range. If a reader's activation range is covered by E_s then the reader-observed tags are definitely in E_s, e.g., $reader_2$ and $reader_3$. Next, if a reader's activation range overlaps with E_s then the reader-observed tags are possibly in E_s, e.g., $reader_1$ and $reader_4$. Finally, if a reader's activation range is disjoint from E_s then the reader-observed tags are definitely not in E_s, e.g., $reader_5$.

Thus, we transform E_s to two separate sets of reader identifiers: S_p contains all readers partially overlapping E_s, and S_d contains all readers fully covered by E_s. In Figure 5, $S_p = \{reader_1, reader_4\}$; $S_d = \{reader_2, reader_3\}$. If the temporal extent of Q_5 is a time interval, Q_5 will be transformed into a type 4 query $Q_5'([reader_1, reader_4], E_t)$. After Q_5' is processed, a refinement can be applied to distinguish results obtained from different reader sets.

The example implicitly assumes that the reader identifiers are ordered according to their spatial proximity as discussed in Section 4.1, which results in a single type 4 query. Generally, an Euclidean range query like Q_5 can be transformed into several queries of type 2 and/or type 4 depending on the reader identifiers involved. More transformed queries are expected to incur higher processing costs. We investigate the effect of reader identifer ordering on Euclidean range queries in Section 6.3.

Indoor Topological Queries Transformation. Similarly to how Euclidean range queries are handled, indoor topological queries can be transformed to symbolic range queries. However, the transformations are usually more complex than for Euclidean range queries because knowledge of the specific RFID reader deployment is needed in order to do such transformations for indoor topological queries.

To be specific, a pair of readers are required to be deployed on a door to detect the movement direction of an object. Based on this, entering or leaving a room can be determined from the RFID readings. A possible deployment is shown in Figure 6. If an object enters $room_1$, it will be observed by $reader_2$ and then $reader_1$; if the object leaves $room_1$, it will be observed first by $reader_1$.

With this type of reader deployment, we are able to process indoor topological queries. An indoor topological query $Q(E_s, E_t, P)$ can be transformed and processed by the procedure described in Algorithm 1. Here, the given topological query is transformed into two type 2 range queries (lines 2–3) on two readers. If an object appears in the results of both queries and the corresponding time intervals overlap, it is added to the result of the given topological query (lines 4–8).

For different types of topological queries, as indicated by the predicate P, we need to input the two readers in the correct order when calling EnterLeaveQuery. We let the reader inside the room (door) be rin and let the one outside be $rout$. If P is $enter$, $rout$ should be used as $reader_{first}$ followed by rin as $reader_{second}$. If P is $leave$, the two readers should be switched.

Algorithm   1.   EnterLeaveQuery(ReaderID $reader_{first}$, ReaderID $reader_{second}$, Timestamp t_i, Timestamp t_j)

1: TagIDSet $result$;   RecordSet $R_1 \leftarrow \emptyset$;   RecordSet $R_2 \leftarrow \emptyset$;
2: Execute query $Q_2(reader_{first}, [t_i, t_j])$, get result into R_1;
3: Execute query $Q_2(reader_{second}, [t_i+T_S, t_j])$, get result into R_2;
4: **for** each record $r_i \in R_1$ **do**
5:     **for** each record $r_j \in R_2$ **do**
6:         **if** $r_i.tagID = r_j.tagID$ **and** $r_j.t_s \leq r_i.t_e + T_S \leq r_j.t_e$ **then**
7:             Add $tagID$ to $result$
8: **return** $result$;

Refer to the example shown in Figure 6. An indoor topological query Q($room_1$, [t_i, t_j], $enter$), which is intended to find those objects that enter $room_1$ during time period [t_i, t_j], can be executed as EnterLeaveQuery($reader_2$, $reader_1$, t_i, t_j). Similarly, a leave query Q($room_1$, [t_i, t_j], $leave$) can be executed as EnterLeaveQuery($reader_1$, $reader_2$, t_i, t_j). A cross query Q($room_1$, [t_i, t_j], $cross$) can be executed as an intersection of two queries: EnterLeaveQuery($reader_1$, $reader_2$, t_i, t_j) $\cap$ EnterLeaveQuery ($reader_2$, $reader_1$, t_i, t_j), which returns those objects that entered and then left the room within the given time period.

5   TP^{2}R-Tree: Time Parameter Point R-Tree

An indoor trajectory record is a horizontal line segment in Reader-Time space. The efficiency of processing a query in an R-tree based index depends on the areas of the

MBRs in the index. It is thus generally beneficial to minimize the MBRs. To achieve this, we represent the horizontal line segments as more compact points with a time parameter indicating their lengths along the time axis. Then we index the resulting points using an R-tree that is modified to take the time parameters into account.

5.1   TP²R-Tree Index Structure

A leaf node in the TP²R-tree contains index entries of the form $(MBR, \Delta t, recordID)$, where MBR indicates the minimum bounding rectangle of the corresponding record identified by $recordID$, and Δt is a time parameter that indicates the duration of the continuous reading by the same reader. For each trajectory record, Δt thus equals $t_e - t_s$. Since each record is represented as a point, a leaf-node MBR in is a point of the form $MBR(readerID, t_s)$, where $readerID$ is a reader on the Reader axis and t_s is a time point on time axis.

The non-leaf nodes contain index entries of the form $(MBR, \Delta t, cp)$, where MBR covers all rectangles in the child node's entries, formatted as in the RTR-tree; cp is a child pointer; and Δt is a time parameter. If cp points to a leaf node N_l, Δt is represented as follows:

$$\max_{\forall e_i \in N_l} (e_i.MBR.t_s + e_i.\Delta t) - \max_{\forall e_j \in N_l} (e_j.MBR.t_s)$$

If cp points to a non-leaf node N_n, Δt is represented as follows:

$$\max_{\forall e_i \in N_n} (e_i.MBR.t^{\dashv} + e_i.\Delta t) - \max_{\forall e_j \in N_n} (e_j.MBR.t^{\dashv})$$

This way, Δt indicates the tightest bound that covers all subnodes on the time axis.

5.2   Node Organization Strategies

Insertion is done using the same framework as for the RTR-tree. We consider three node organization strategies. The first two are based on the least area enlargement. The area calculation of the first strategy is based on the $Area$ formula; we call this the basic strategy.

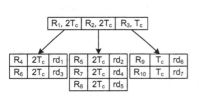

Fig. 7. TP²R-Tree, Basic

Given a node capacity of 3, the tree and MBRs for the basic strategy are exemplified in Figures 7 and 8. For simplicity, T_c denotes $t_{i+1} - t_i$ for any two consecutive timestamps.

The second strategy, TP²R-tree $Area^+$, calculates areas using the $Area^+$ formula. Figure 9 illustrates the MBRs resulting from the TP²R-tree $Area^+$ strategy.

The third strategy, TP²R-tree Split2, uses specific **ChooseLeaf** and **SplitNode** algorithms. We first need to define the virtual minimum bounding rectangle (VMBR) of an entry e in non-leaf node of the TP²R-tree: $VMBR(readerID_{min}, readerID_{max}, t^{\vdash}, t^{\dashv})$. Here, $readerID_{min}$ and $readerID_{max}$ are the same as the MBR of the entry, $VMBR.t^{\vdash}$ equals $MBR.t^{\vdash}$, and $VMBR.t^{\dashv}$ equals $MBR.t^{\dashv} + e.\Delta t$. The VMBR of MBR R_1 in Figure 8 is shown in Figure 10.

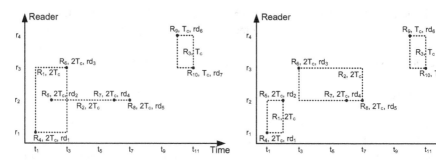

Fig. 8. MBRs of the TP²R-Tree, Basic   **Fig. 9.** MBRs of the TP²R-Tree, Area⁺

Referring to the analysis of the query types in Section 4.3, the basic queries are query type 1 and query type 2. These queries involve merely one specific reader. Thus, we try to place entries with same or near-same readerID in the same node. The TP²R-tree Split2 strategy chooses the leaf node whose MBR needs the least Reader dimension enlargement to include the new index entry, and it resolves ties by choosing the entry whose VMBR areas need the least area enlargement (using $Area^+$ formula) to include the new index entry. Due to space limitations we omit the ChooseLeaf Algorithm.

The basic criterion for node splitting is the same as for the R-tree, minimizing the probability that both new nodes will be examined in subsequent queries. The pseudo code of **SplitNode** is shown in Algorithm 2. It discriminates on the Reader dimension first, selects two entries containing the biggest (smallest) readerID, and assigns them to two groups as seeds (lines 2–3). The assignment of the remaining entries gives priority to satisfying the requirement of having a minimum number of objects in each group (lines 5–6). Next, if there is an entry with a readerID range in that of one of the groups, the entry is assigned to that group (lines 9–12). If such an entry is chosen, a new iteration of the while loop will be invoked, ensuring no groups with underflow.

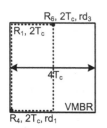

Fig. 10.   VMBR,
TP²R-Tree

If no such entries exist, both Reader and time dimensions are considered using the $Area^+$ definition. For each remaining entry, the VMBR $Area^+$ increment of either group needed to include the entry is calculated (lines 15–16). From all remaining entries, we choose the one with the maximum difference of two VMBR $Area^+$ increments, and we assign it to the group that needs least VMBR Area⁺ enlargement (lines 17–18). The resulting MBRs of the TP²R-tree example are shown in Figure 11.

Similarly to the query transformation used in the RTR-tree, queries here can also be transformed into the Reader-Time coordinate representation. However, the TP²R-tree differs from the RTR-tree in that its entries have both an MBR and a VMBR. A query not overlapping with an entry's MBR may overlap with the entry's VMBR, qualifying some records in that entry for the query. We thus need to modify the RTR-tree's search algorithm to accommodate the time parameter and the VMBRs in the TP²R-tree.

Algorithm 2. SplitNode (EntrySet $SetE$)

1: EntrySet $GroupB \leftarrow \emptyset$, $GroupS \leftarrow \emptyset$;
2: Move the entry e_b with the biggest $readerID$ from $SetE$ into $GroupB$;
3: Move the entry e_s with the smallest $readerID$ from $SetE$ into $GroupS$;
4: **while** $SetE \neq \emptyset$ **do**
5:    **if** One group has so few entries that all the rest must be assigned to it in order for it to have the minimum number **then**
6:       Move all entries from $SetE$ to that group; **break**;
7:    Boolean $flag \leftarrow$ FALSE;
8:    **for** each entry e in $SetE$ **do**
9:       **if** The reader range of $e.MBR$ is in the reader range of $GroupB.MBR$ **then**
10:          Move e from $SetE$ into $GroupB$; $flag \leftarrow$ TRUE; **break**;
11:       **else if** The reader range of $e.MBR$ is in the reader range of $GroupS.MBR$ **then**
12:          Move e from $SetE$ into $GroupS$; $flag \leftarrow$ TRUE; **break**;
13:    **if** $\neg flag$ **then**
14:       **for** each entry e in $SetE$ **do**
15:          Calculate d_1, the VMBR Area$^+$ increment of $GroupB$ to include the entry $e.MBR$
16:          Calculate d_2, the VMBR Area$^+$ increment of $GroupS$ to include the entry $e.MBR$
17:       Choose the entry e_m with the maximum $|d_1 - d_2|$.
18:       Move e_m from $SetE$ to the group whose VMBR Area$^+$ needs the least enlargement.

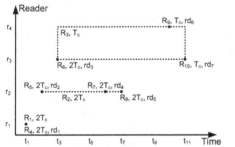

Fig. 11. MBRs of the TP^{2}R-Tree, Split2

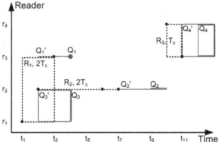

Fig. 12. Query Expansion for TP^{2}R-Tree

5.3 Expansion-Based Query Processing

The idea is to expand the query geometry in the temporal dimension while still using depth-first search. In particular, when checking whether a TP^{2}R-tree entry e overlaps with the given query geometry g, we use an expanded version of g. Table 3 shows in detail how a query geometry g is expanded with respect to an encountered entry e. For each query type, the query geometry representation in the Reader-Time coordinate system is expanded to the left horizontally by the Δt value in a given entry. The examples in the table are also illustrated in Figure 12, with the top-level MBRs from the TP^{2}R-tree with the basic strategy (shown in Figure 8).

Table 3. Query Geometry Expansion for TP^{2}R-Tree

Original Query	Expanded Query
Type 1: $QT_1(readerID, t)$	Type 2: $QT_2(readerID, [t - e.\Delta t, t])$
E.g., $Q_1(reader_3, t_4)$	E.g., $Q_1'(reader_3, [t_2, t_4])$
Type 2: $QT_2(readerID, [t_i, t_j])$	Type 2: $QT_2(readerID, [t_i - e.\Delta t, t_j])$
E.g., $Q_2(reader_2, [t_9, t_{10}])$	E.g., $Q_2'(reader_2, [t_7, t_{10}])$
Type 3: $QT_3([readerID_m, readerID_n], t)$	Type 4: $QT_4([readerID_m, readerID_n], [t - e.\Delta t, t])$
E.g., $Q_3([reader_1, reader_2], t_4)$	E.g., $Q_3'([reader_1, reader_2], [t_2, t_4])$
Type 4: $QT_4([readerID_m, readerID_n], [t_i, t_j])$	Type 4: $QT_4([readerID_m, readerID_n], [t_i - e.\Delta t, t_j])$
E.g., $Q_4([reader_3, reader_4], [t_{12}, t_{13}])$	E.g., $Q_4'([reader_3, reader_4], [t_{11}, t_{13}])$

6   Experimental Study

6.1   Experimental Settings

Both indexes, together with all the node organization strategies presented, and the query processing algorithms are implemented in Java. The index implementations are based on the Spatial Index Library [1]. A computer with Windows XP professional, a 2.66GHz Core2 Duo CPU, and 3.25GB main memory is used to run all experiments.

We set the the page size (i.e., the tree node size) to 4096 bytes. This yields 204 (170) entries per non-leaf node and 256 (256) entries per leaf node in the RTR-tree (the TP^{2}R-tree). We investigate both tree construction costs and query processing costs. For the former, the total running time is measured; for the latter, the total number of tree node accesses are measured, as this is proportional to the dominant cost in query processing.

We generate moving objects using a 3-floor building plan with 30 rooms and 3 staircases on each floor. All rooms and staircases are connected by doors to a hallway in a star-like way. An RFID reader is deployed by the door of each room. In addition, readers are also deployed along the hallway and in the staircases. The reader identifiers are assigned as follows. First, multiple readers within a partition (e.g., a room or a hallway) are assigned consecutive identifiers whose ordering represent their physical proximity. Second, on each side of a hallway, adjacent partitions (e.g., rooms) are assigned consecutive identifiers and/or identifier ranges. Third, adjacent floors are assigned consecutive identifier ranges. All objects move according to two rules: 1) an object in a room or a staircase can go to the hallway through a door, or move inside the same room or staircase; 2) an object in the hallway can move in the hallway, move to a staircase, or move into a room through a door.

Three data-related parameters are varied. Table 4 lists the settings of these parameters, with default values given in bold. With the default minimum object lifespan and reader activation range, the number of moving objects varies from 1,000 to 50,000, resulting in indoor trajectory tables consisting of between

Table 4. Data Parameter Settings

Parameters	Settings
Object number	1K, 5K, **10K**, 20K, . . . , 50K
Minimum object lifespan	50, **100**, 150, 200, 250 (sec)
Reader activation range	**100**, 150, 200, 250 (cm)

25K and 1,973K records. When the minimum object lifespan varies between 50 and 250 seconds and with the other parameters set default, the trajectory tables consist of between 253K and 973K records. Accordingly, the simulation period for all experiments varies from 10,650 to 12,586 seconds. In addition, the variation of reader

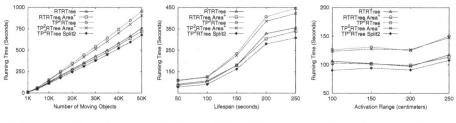

(a) Effect of Object Number   (b) Effect of Object Lifespan   (c) Effect of Activation Range

Fig. 13. Tree Construction Costs

activation range between 100 and 250 centimeters results in trajectory tables consisting of between 292K and 322K records.

6.2 Tree Construction

Tree construction costs are reported in Figure 13. The first two strategies of the TP^2R-tree yield the longest running time for tree construction, as they both involve large numbers of area calculations and ΔT calculations. However, the hybrid strategy, TP^2R-tree Split2, incurs the least cost for almost all settings. This indicates that the Split2 strategy is effective at simplifying the node splitting during TP^2R-tree construction.

As the number of moving objects increases, the trajectory data size increases accordingly. Therefore, the tree construction cost also increases, as reported in Figure 13(a). Referring to Figure 13(b), a longer object lifespan causes a higher tree construction cost, because a longer lifespan also causes the trajectory data size to increase.

As the reader activation range increases, two different effects occur. On one hand, the time that an object is within a reader's range increases. This may decrease the trajectory data size in the trajectory table, although the numbers of raw RFID readings increases. On the other hand, readers with larger activation ranges tend to detect more objects, thus producing more raw readings and probably more trajectory records. Consequently, the tree construction cost remains steady or decreases slightly as the activation range varies from 100 cm to 200 cm, while the cost increases as the activation range reaches 250 cm, as reported in Figure 13(c).

6.3 Query Processing

Next, we study the query processing costs, and we consider the effects of the ordering of reader identifiers along the Reader dimension.

Results on Range Queries. In each batch of experiments in this section, we generate 100 random queries. Unless explicitly stated otherwise, each query is issued with random query parameters. In particular, the time in a type 1 or type 3 query is a random value between 0 and the largest timestamp in the involved set of trajectories. The readerID in a type 1 or type 2 query is chosen randomly among all readerID values. The time interval in a type 2 or type 4 query is a random sub-interval between 0 and the largest timestamp in the data set. The readerID range in a type 3 or type 4 query is a random subrange within the range of readerIDs.

Effect of Object Number. We first fix the minimum object lifespan to 100 s and the RFID reader activation range to 100 cm; we then vary the object number from 1,000 to 50,000 to see the effect on the query performance.

For query type 1 and type 2, the TP^{2}R-tree using the Split2 strategy outperforms all the other trees, as reported in Figure 14(a) and Figure 15(a). The Split2 strategy separates entries with different readerID (ranges) into different nodes after splitting; therefore, the node overlap along the Reader axis is reduced. As a result, the point selections on the Reader axis in query type 1 and type 2 involve fewer tree nodes.

For query type 3 and type 4, the selection on the Reader axis is a range selection, and hence the Split2 strategy loses its clear advantage. As reported in Figure 16(a) and Figure 17(a), both Area$^+$ strategies are always the best (or among the best) as they minimize the number of raw RFID readings in each MBR, which benefits range selections.

Effect of Minimum Object Lifespan. We also vary the minimum object lifespan from 50 to 250 s; the results are reported in Figures 14(b) to 17(b). As a whole, increasing the object lifespan increases the query processing cost for each query type. This is attributed to the increased trajectory data size and the corresponding index sizes that result from the longer object lifespans. Note that the TP^{2}R-tree constructed using the Split2 strategy still outperforms the others for query types 1 and 2, as reported in Figures 14(b) and 15(b). While trees constructed using Area$^+$ strategy prefer query types 3 and 4, as reported in Figures 16(b) and 17(b).

Effect of Reader Activation Range. We then vary the RFID reader activation range from 100 to 250 cm. Referring to Figure 15(c) and Figure 17(c), varied reader activation ranges cause the query processing cost to fluctuate for query type 2 and type 4. This is attributed to two reasons. First, larger reader activation ranges make objects stay longer within a reader's range, thus increasing the time length of each trajectory record and the chance of a record to be in the query result. Second, the range selection on the time axis in query types 2 and 4 then tends to find the answer in fewer nodes as each node becomes "wider" along the time axis.

For query types 1 and 3, as reported in Figures 14(c) and 16(c), the results are relatively more steady as the queries use a point selection on the time axis. For all query types, the Split2 and Area$^+$ strategies outperform the basic ones in almost all settings.

Effect of Query Parameters. As the final experiment with range queries, we use data collected in Copenhagen Airport. Although we have assumed RFID technology, this data is obtained using Bluetooth technology (for which the paper's proposals are also applicable). The data set contains more than 500,000 tracking records from 25 Bluetooth hotspots per day. We extract the tracking data on the most active day from April 2008 to October 2008. We vary the query parameters for query types 2, 3 and 4.

For both query types 2 and 4, we vary the time interval from 1% to 20% of the total time span of the set of trajectories. Each query time interval starts at a random time point and is fully within the total time span. According to the results reported in

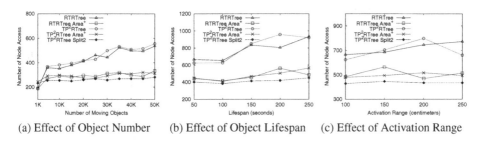

(a) Effect of Object Number   (b) Effect of Object Lifespan   (c) Effect of Activation Range

Fig. 14. Performance of Query Type 1

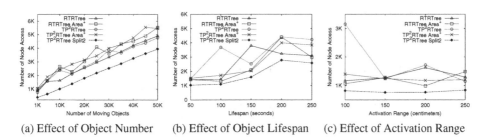

(a) Effect of Object Number   (b) Effect of Object Lifespan   (c) Effect of Activation Range

Fig. 15. Performance of Query Type 2

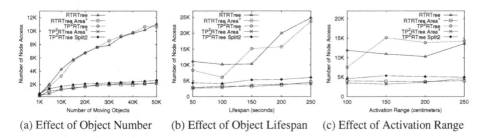

(a) Effect of Object Number   (b) Effect of Object Lifespan   (c) Effect of Activation Range

Fig. 16. Performance of Query Type 3

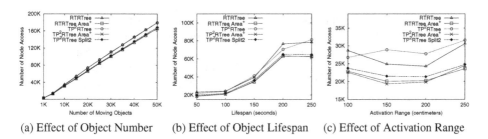

(a) Effect of Object Number   (b) Effect of Object Lifespan   (c) Effect of Activation Range

Fig. 17. Performance of Query Type 4

Figure 18(a) and (b), the larger time interval range causes more node access. This is because the large time interval results in a larger query geometry: longer horizontal line segments for type 2 and larger rectangles for type 4. We see that the RTR-Tree Area$^+$ and and TP^{2}R-tree Split2 strategies perform better and degrade more slowly than do the other strategies.

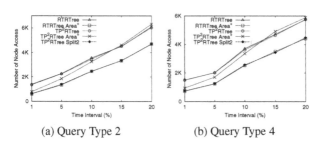

(a) Query Type 2   (b) Query Type 4

Fig. 18. Query Performance vs. Time Interval

We also vary the readerID range for query types 3 and 4, with a random length of 1% to 5% of the difference between the maximum and minimum readerIDs. The results are reported in Figures 19(a) and (b). Larger readerID ranges involve more readers and therefore cause more node accesses. However, the cost increase is moderate for the trees using the Area$^+$ and Split2 strategies.

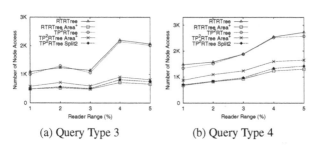

(a) Query Type 3   (b) Query Type 4

Fig. 19. Query Performance vs. Reader Range

As a whole, the RTR-trees constructed with the Area$^+$ strategy and the TP^{2}R-trees with the Area$^+$ and Split2 strategies perform steadily compared to their alternatives. When using them, the query processing does not degrade markedly as query parameters scale up. This indicates that these node organization strategies result in index trees that are robust to changing query work loads.

Summary. Three important observations follow from the experiments with symbolic range queries. First, compared to the basic node organization strategies, our Area$^+$ and Split2 strategies result in more efficient and robust indexes. Second, the Area$^+$ strategy performs the best for query types 3 and 4, while the Split2 strategy performs the best for query types 1 and 2. Third, the TP^{2}R-tree with Split2 strategy is a quite balanced choice for efficient query processing for all query types.

Topological Queries. We also investigate different predicates in topological queries. We pick 35 doors on the first floor as those that are deployed with paired readers. We use a data set with 10,000 objects, 100 cm RFID reader activation ranges, and 100 s minimum object lifespans. For each predicate *enter*, *leave*, and *cross*,

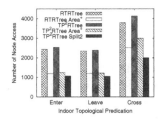

Fig. 20. Effects of Predicates

we generate 3 queries against each room. In each query, the time interval starts at a random time point, and the interval is set to the lifespan of each object. The resulting node accesses in query processing are reported in Figure 20. The TP^{2}R-tree Split2 strategy outperforms all others because topological queries are transformed to type 2 range queries favored by the Split2 strategy. The savings are especially apparent for *cross*, as it involves more range queries after the transformation.

Effects of Reader Identifier Ordering. We finally consider the effects of varying the reader identifier ordering used in the tree proposals. We generate two data sets using the default settings: 10,000 moving objects, a 100 cm RFID reader activation range, and a 100 s minimum object lifespan. In one set, all RFID reader identifiers are ordered as described in Section 4.1. In the other set, the RFID readers are ordered randomly.

We then generate 10 random Euclidean range queries. The spatial extent E_s in each query is a random rectangle within the floor plan, with an area between 5% and 20% of the floor area. The temporal extent E_t in each query starts at a random time point and lasts 100 s. The results are reported in Figure 21.

It is seen that a careful ordering of the RFID reader identifiers improves the query performance considerably. Remember that each Euclidean range query is transformed into a symbolic range query as described in Section 4.3. After the transformation, each set of

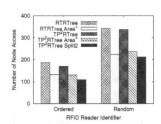

Fig. 21. Effects of ReaderID Ordering on Interval Queries

consecutive reader identifiers will involve only one type 4 query, while any other individual identifier will involve a separate query of type 2. If all reader identifiers are ordered according to spatial proximity, more identifiers are likely to be included in type 4 queries. This results in fewer symbolic queries being executed. For random reader identifiers, the number of symbolic queries after the transformation tends to be large, which yields higher query costs.

7   Conclusion and Future Work

Because of the uniqueness of indoor space and indoor positioning technologies, indoor moving object trajectories are represented differently than traditional outdoor trajectories. Efficient queries against such indoor trajectories require novel indexes. The paper proposes two R-tree based indexes for trajectories of moving objects in symbolic indoor spaces, supporting both spatiotemporal range queries and topological queries. Range queries are processed directly using the indexes; while topological queries are efficiently transformed into range queries. A comprehensive experimental study on both synthetic and real data sets discloses the following findings. First, the two trees are effective indexes for moving object trajectories in symbolic indoor space. Their specific node strategies make them efficient and robust for a wide range of settings. Second, our transformation from topological queries to range queries is effective and efficient. Third, the spatial proximity based ordering of reader identifiers in our trees improves query performance.

Several relevant research directions exist. First, it is possible to extend the TP^2R-tree to accommodate on-line trajectories and the current locations of indoor moving objects, as the time extensions in the TP^2R-tree can be used to maintain on-line information. Second, it is of interest to apply data mining techniques to indoor trajectories in order to find useful movement patterns for different purposes. The indexes proposed in this paper can be used to facilitate the mining. Third, it is also of interest to adapt the paper's proposals to other indoor positioning technologies like Wi-Fi, and it is of interest to employ multiple positioning technologies in the same indoor space, such that queries can return more accurate answers.

Acknowledgments. This research was partially supported by the Indoor Spatial Awareness project of the Korean Land Spatialization Group and BK21 program. C. S. Jensen is currently a Visiting Scientist at Google Inc.

References

1. Spatial Index Library, http://research.att.com/~marioh/spatialindex/
2. Agarwal, P.K., Arge, L., Erickson, J.: Indexing Moving Points. In: Proc. PODS, pp. 175–186 (2000)
3. Bahl, P., Padmanabhan, V.N.: RADAR: An In-Building RF-Based User Location and Tracking System. In: Proc. INFOCOM, pp. 775–784 (2000)
4. Becker, C., Dürr, F.: On Location Models for Ubiquitous Computing. Personal Ubiquitous Computing 9(1), 20–31 (2005)
5. Chakka, V.P., Everspaugh, A., Patel, J.M.: Indexing Large Trajectory Data Sets With SETI. In: Proc. CIDR (2003)
6. Chen, S., Ooi, B.C., Tan, K.-L., Nascimento, M.A.: ST^2B-Tree: A Self-Tunable Spatio-Temporal B^+-Tree Index for Moving Objects. In: Proc. SIGMOD, pp. 29–42 (2008)
7. Erwig, M., Schneider, M.: Developments in Spatio-Temporal Query Languages. In: Proc. DEXA Workshop STDML, pp. 441–449 (1999)
8. Feldmann, S., Kyamakya, K., Zapater, A., Lue, Z.: An Indoor Bluetooth-Based Positioning System: Concept, Implementation and Experimental Evaluation. In: Proc. ICWN, pp. 109–113 (2003)
9. Güting, R.H., Böhlen, M.H., Erwig, M., Jensen, C.S., Lorentzos, N.A., Schneider, M., Vazirgiannis, M.: A Foundation for Representing and Quering Moving Objects. ACM Trans. Database Syst. 25(1), 1–42 (2000)
10. Guttman, A.: R-Trees: A Dynamic Index Structure for Spatial Searching. In: Proc. SIGMOD, pp. 47–57 (1984)
11. Hightower, J., Borriello, G.: Location Systems for Ubiquitous Computing. IEEE Computer 34(8), 57–66 (2001)
12. Jensen, C.S., Lin, D., Ooi, B.C.: Query and Update Efficient B^+-Tree Based Indexing of Moving Objects. In: Proc. VLDB, pp. 768–779 (2004)
13. Kollios, G., Gunopulos, D., Tsotras, V.J.: On Indexing Mobile Objects. In: Proc. PODS, pp. 261–272 (1999)

14. Kollios, G., Tsotras, V.J., Gunopulos, D., Delis, A., Hadjieleftheriou, M.: Indexing Animated Objects Using Spatiotemporal Access Methods. IEEE Trans. Knowl. Data Eng. 13(5), 758–777 (2001)
15. Lin, D., Jensen, C.S., Ooi, B.C., Šaltenis, S.: Efficient Indexing of the Historical, Present, and Future Positions of Moving Objects. In: Proc. MDM, pp. 59–66 (2005)
16. Nascimento, M.A., Silva, J.R.O.: Towards Historical R-Trees. In: Proc. SAC, pp. 235–240 (1998)
17. Pfoser, D., Jensen, C.S., Theodoridis, Y.: Novel Approaches in Query Processing for Moving Object Trajectories. In: Proc. VLDB, pp. 395–406 (2000)
18. Šaltenis, S., Jensen, C.S., Leutenegger, S.T., Lopez, M.A.: Indexing the Positions of Continuously Moving Objects. In: Proc. SIGMOD, pp. 331–342 (2000)
19. Pelanis, M., Šaltenis, S., Jensen, C.S.: Indexing the Past, Present, and Anticipated Future Positions of Moving Objects. ACM TODS 31(1), 255–298 (2006)
20. Tao, Y., Papadias, D.: MV3R-Tree: A Spatio-Temporal Access Method for Timestamp and Interval Queries. In: Proc. VLDB, pp. 431–440 (2001)
21. Tao, Y., Papadias, D., Sun, J.: The TPR*-Tree: An Optimized Spatio-Temporal Access Method for Predictive Queries. In: Proc. VLDB, pp. 790–801 (2003)
22. Theodoridis, Y., Vazirgiannis, M., Sellis, T.K.: Spatio-Temporal Indexing for Large Multimedia Applications. In: Proc. ICMCS, pp. 441–448 (1996)
23. Want, R.: RFID Explained: A Primer on Radio Frequency Identification Technologies. Synthesis Lectures on Mobile and Pervasive Computing 1(1), 1–94 (2006)
24. Wolfson, O., Xu, B., Chamberlain, S., Jiang, L.: Moving Objects Databases: Issues and Solutions. In: Proc. SSDBM, pp. 111–122 (1998)
25. Yiu, M.L., Tao, Y., Mamoulis, N.: The B^{dual}-Tree: Indexing Moving Objects by Space Filling Curves in the Dual Space. VLDBJ 17(3), 379–400 (2008)

Monitoring Orientation of Moving Objects around Focal Points

Kostas Patroumpas and Timos Sellis

School of Electrical and Computer Engineering
National Technical University of Athens, Hellas
{kpatro,timos}@dbnet.ece.ntua.gr

Abstract. We consider a setting with numerous location-aware moving objects that communicate with a central server. Assuming a set of focal points of interest, we aim at continuously monitoring object orientations and hence detect situations where many objects get closer to or move away from any such site. Towards this goal, we propose a streaming approach that delegates part of the processing to objects, which relay positional updates upon significant deviations at their course. The central processor maintains the changing distribution of current object headings around each focal point and may issue alerts once it observes many objects moving along a direction (e.g., increased northbound traffic near the stadium). To efficiently answer such navigational queries, we introduce a novel access method that indexes object headings influencing a specific site. Furthermore, we extent this scheme to examine trajectory movements around sites over the recent past. Experimental results verify that this framework is able to cope with scalable numbers of objects at reduced communication cost, while offering instant notification of important trends along diverse directions for multiple focal points.

1 Introduction

Proliferation of location-based applications has led into efficient algorithms for processing typical continuous queries, such as range or k-nearest neighbor search [2,5,11], dealing with current coordinates of monitored objects (e.g., humans, vehicles, devices etc.). Still, less attention is given to observing evolving trajectories or mutable motion patterns, such as abrupt velocity variations or unexpectedly increasing concentration of objects in particular regions across time.

In this work, we turn our focus on studying movement from a navigational perspective, by examining significant changes in object headings. In navigation, the *heading* (a.k.a. *bearing*) of a moving object is its orientation, expressed as an angle from a known direction, usually north. By collecting heading information from streaming positional updates of numerous objects, it could be feasible to observe their mode of progression. But objects usually move at diverse directions amenable to sudden changes (e.g., turns), so perhaps no safe conclusion on movement patterns can be drawn from such a volatile variety of orientations. Even a fact like "40% of objects currently move eastbound" scarcely offers a valuable knowledge, as relevant objects may be located anywhere in the monitored area.

N. Mamoulis et al. (Eds.): SSTD 2009, LNCS 5644, pp. 228–246, 2009.
© Springer-Verlag Berlin Heidelberg 2009

In our view, it seems much more useful to maintain the distribution of object orientations with respect to selected *focal points* or *sites* of interest, like terminal stations, sporting venues, traffic junctions etc. For each such site (say, a stadium), we would like to detect orientation trends *online* and also distinguish influencing objects that converge to or diverge from that site; e.g., whether a large number of vehicles are currently moving westbound and may be approaching the stadium soon. Typically for vehicles, ships, aircrafts etc., we implicitly assume that each object follows a consistent movement, hence is not arbitrarily displaced, but is moving towards –more or less– the same direction over a time interval.

In effect, we suggest a framework that acts like a constellation of radars (one per site), offering better insight along frequently-followed directions at progressively finer resolution. This mechanism can answer continuous *orientation-based queries* like "identify trucks bound for the port from the west at a distance less than 2km" or "issue an alert once a squadron of aircrafts are heading towards Athens from southeast over the past 10 minutes". To assist efficient evaluation of such requests, we propose a novel index structure that organizes the detection range of a specific site as a hierarchical tree. Influencing objects are assigned to tree nodes that represent sectors at gradually refined angles and extents. This index supports multiple orientation-based queries associated with a common focal point, each inspecting a diverse range and direction around it.

Since the entire mechanism must work in a streaming fashion to keep in pace with the bulk of incoming geospatial data, we adopt a collaborative scheme, where objects are capable of communicating with a central server and also have minimal processing capabilities to retain their recent positions and update their heading. A set of fixed focal points are allocated in the monitoring area; for each observation site, the server maintains the current distribution of headings based on the most recent status of objects detected within its area of interest. Reducing communication overhead is a major concern, so frequent positional updates referring to slight changes in objects' movement should better be avoided.

To the best of our knowledge, this is the first work on monitoring streaming orientations of moving objects. Our contribution can be summarized as follows:

- We introduce a novel spatiotemporal access method, namely PolarTree, which can effectively maintain object headings of interest to a given focal point.
- We propose a stream-based processing scheme that can provide real-time response to an important –yet largely neglected– class of navigational queries.
- We further extent this mechanism by employing sliding windows, practically examining the general heading for evolving portions of objects' trajectories.
- We evaluate empirically the robustness and efficiency of the framework with scalable numbers of moving objects and various settings for focal sites.

The remainder of this paper is organized as follows: Section 2 discusses fundamental concepts concerning focal points and object headings. Section 3 introduces the structure of PolarTree and presents its properties and operations. The processing framework for monitoring object headings is described in Section 4. Experimental results are reported in Section 5. Section 6 briefly reviews related work. Finally, Section 7 offers conclusions and future research directions.

2   Preliminaries

2.1   Scope of Focal Points

We assume a finite set $F = \{f_1, f_2, \ldots, f_n\}$ of stationary focal points (sites), which can monitor a large number of location-aware objects continuously moving on the 2-d Euclidean space $\mathbf{E}$. Each site $f_i \in F$ has a *focal scope* that represents its maximum range for detecting objects moving in its vicinity. In this setting, we assume that the scope of each focal point f_i practically translates into a circle $O(f_i, R_i)$ of a given radius R_i centered at the fixed location of f_i. In fact, every focal point specifies an *advanced range search*, aiming not just to observe objects inside its circular scope, but also to distinguish their orientations.

 We do not assume any particular allocation of sites on plane $\mathbf{E}$, so they can be distributed randomly, evenly, but typically depending on the application. For instance, a traffic monitoring system may configure focal sites at major junctions along arterial roads and highways, while an environmental application may opt for observation points near wildlife habitats. Hence, the scopes of any two focal points $f_i, f_j \in F$ may intersect, signifying a common interest on area $O(f_i, R_i) \cap O(f_j, R_j) \neq \emptyset$. Each focal point may also designate a different radius, as depicted in Fig. 1. It may occur that $\mathbf{E}$ is not covered in its entirety, i.e., $\mathbf{E} \neq \bigcup_{i=1}^{n} O(f_i, R_i)$, meaning that some areas of $\mathbf{E}$ may not be monitored at all. Finally, we make no assumption on the total count n of sites, although we expect that a few hundred focal points of adequate scopes are more than sufficient to monitor a large geographical area (e.g., a city or a national park).

2.2   Object Headings and Focal Distances

Each moving object o is aware of its current timestamped location $\langle x, y, t \rangle$, where x, y are the coordinates (on plane $\mathbf{E}$) of a point position measured at time instant t. An object also knows its *heading* with reference to a previously recorded position $\langle x_0, y_0, t_0 \rangle$. That previous position of object o can be either its last recorded location or an *anchor point* representing its origin, a designated position or even a shifting location somewhere along its route (Section 4.3). Anyway, the heading signifies the direction of movement and can be represented as an angle θ with respect to a fixed direction; if this angle is measured from north it is commonly known as *azimuth*. For facilitating geometric calculations, in our model we measure headings counterclockwise with respect to the positive x-axis. This reflects the *slope* of the line segment that connects these two locations, expressed as an angle $\theta \in [0, 2\pi)$ on the trigonometric circle (as indicated for object i in Fig. 1). Formally:

$$\theta = \begin{cases} atan2(y - y_0, x - x_0), & \text{if } y \geq y_0 \\ atan2(y - y_0, x - x_0) + 2\pi, & \text{if } y < y_0 \end{cases}$$

 In fact, we use the variant function $atan2$ instead of $arctan(\frac{y - y_0}{x - x_0})$, such that the calculated slope θ is also mapped to the correct quadrant of the trigonometric circle, thus signifying the direction of the vector from previous position to the current one. Since function $atan2$ takes values strictly in $(-\pi, \pi]$, we add the term

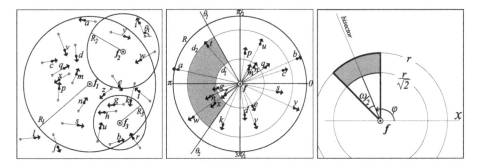

Fig. 1. Focal sites & scopes   **Fig. 2.** Polar mapping   **Fig. 3.** Polar sector

2π for negative slopes, hence always $\theta \in [0, 2\pi)$. Figure 1 illustrates the current locations and headings of several objects moving in the vicinity of three focal points. Note that the heading of an object depends solely on its own movement from a previously recorded position and has nothing to do either with the focal point or the movement pattern of its neighboring objects.

Objects are moving freely, but we assume that their heading does not change abruptly at each positional update. Otherwise, had they been allowed to move towards random directions at each timestamp, there would be no reason in monitoring their incoherent orientations. Hence, objects are expected to follow a consistent course for a while, before heading towards another direction (e.g., making a turn). This motion pattern is frequently observed in several occasions of interest to location-aware applications, including vehicles, aircrafts, ships, migratory birds, etc.; so by no means is it limited to objects moving in fixed networks.

For an object o within the scope of a site f, its *focal distance* d is its Euclidean distance from f. Obviously, an object that influences multiple focal points, has different focal distances with respect to each one of them. For instance, object y in Fig. 1 is within the scope of both f_1 and f_2, but is currently closer to f_2.

2.3 Polar Mapping of Objects

To get better insight on the distribution of object headings around a given site f, we perform a mapping based on focal distance d and heading θ for every object o within the scope of f. As illustrated in Fig. 2 for the case of site $f \equiv f_1$, every qualifying object is mapped into a polar circle with center f (*pole*) and radius R (*scope*). Each object o is abstracted into a point at distance d from f and at an angle equal to its heading θ with respect to the positive x-axis. Locations beyond scope of f are ignored, as it happens for objects i, l, j, r in Fig. 1. Thus, each site f gets a clearer view of influencing objects and their distribution around f in terms of their focal distance and heading. We stress that this is not the usual mapping from Cartesian into *polar coordinates*, as our main concern is not just object positioning in relation to a specific point f, but their orientations around f instead. For example, the mapping of object t in Fig. 2 does not convey that t is at the northwest of nearby site f, but that t moves towards northwest.

Consequently, *orientation-based* queries related to focal sites can be expressed more easily. Formally, a query q related to a focal site f will search for objects:

$$\{o \mid o.heading \in [\theta_1, \theta_2) \wedge distance(o, f) \in (d_1, d_2]\}$$

i.e., all objects within the specified ranges for headings ($0 \leq \theta_1 \leq \theta_2 < 2\pi$) and focal distances ($0 < d_1 < d_2 \leq R$) from f. Intuitively, such a navigational query is transformed into a slice of a ring around site f (the shaded area in Fig. 2).

3   The PolarTree Index

In this section, we introduce a *main-memory* access method that can be used to index frequently updated object orientations observed around a single site f. A PolarTree partitions the focal scope of f into non-overlapping *polar sectors*, each denoting a specific range of object headings, since these are measured as angles. Then, movement directions (towards east, north, northwest, etc.) can be identified using suitable angular ranges. Recursive subdivision of the circular scope into smaller convex sectors, serves as the guiding principle for assigning objects at the nodes of this hierarchical tree, as exemplified in Fig. 4.

3.1   Index Structure

More specifically, the PolarTree is a binary tree with the following properties:

- The root node represents the entire scope R of focal site f and has no entries.
- Every node corresponds to a *polar sector* of a circle centered at f. Each polar sector is uniquely characterized by a *radius* r and its *bisector*, expressed as an angle ϕ on the trigonometric circle (Fig. 3). For instance, G is the only sector in Fig. 4a with radius $r = \frac{R}{2}$ and bisector at $\phi = \frac{\pi}{8}$.
- An internal node (i.e., not a leaf) with radius r has exactly two children, denoted as *leftChild* and *rightChild*, each with radius $\frac{r}{\sqrt{2}}$. The root (at level $l = 0$) has always two children, each with a radius equal to focal scope R.
- At an internal node, the central angle ω of its sector is bisected into two equal parts that characterize its children (Fig. 3). The size of angle ω is determined by the level of that node in the tree. For instance, nodes at level $l = 1$ are the children of the root and have $\omega = \pi$, nodes at level $l = 2$ have $\omega = \frac{\pi}{2}$ and so on. Therefore, the *angular range* $[\theta_{min}, \theta_{max})$ of each sector is unique, where $\theta_{min} = \phi - \frac{\omega}{2}$ and $\theta_{max} = \phi + \frac{\omega}{2}$. Consequently, the angular range[1] of its left child is $[\phi - \frac{\omega}{2}, \phi)$ and that of its right child is $[\phi, \phi + \frac{\omega}{2})$.
- Every node (excluding the root) maintains a catalogue of entries. Each entry denotes an object o with heading θ falling within the angular range of the respective sector and whose distance d from f is less than sector radius r. Object o is uniquely assigned to a single sector s, such that $\theta \in [s.\theta_{min}, s.\theta_{max})$ $\wedge d \leq s.radius$ and $\nexists s' \neq s$, $\theta \in [s'.\theta_{min}, s'.\theta_{max})$ $\wedge d \leq s'.radius < s.radius$.

---

[1] The angular range is not actually an attribute of a node, as it can be easily calculated from its bisector ϕ and its level l in the tree (i.e., the size of angle ω). However, this notion is used in the sequel for better exposition of the algorithms.

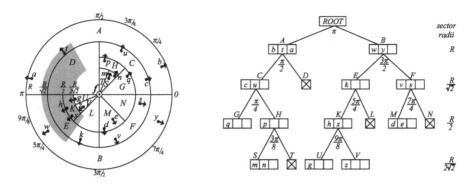

Fig. 4. (a) Headings assigned to polar sectors   (b) Polar tree with $M = 3$

- Let M be the maximum number of entries allowed in a leaf node. A bisection occurs when a leaf has more than M entries (*overflow*), so it becomes an internal node itself with two new children. Entries of the original node may be assigned to its children, if their headings and distances from f qualify to the specifications of the new leaf nodes, as explained later on.
- Collapsing is applicable to leaf nodes only. If both descendants of an internal node are leaves and the total count of entries in these three nodes is less than a limit m (*underflow*), then the leaves are eliminated and their entries get merged with the entries of their parent, which now becomes a leaf itself.

Figure 4a illustrates the partition of the scope of focal point f for the example given in Fig. 2, assuming that at most $M = 3$ objects may be assigned at a leaf node. The shape of the corresponding PolarTree is shown in Fig. 4b, with the angles below each node signifying their distinctive bisectors ϕ.

Data is associated with both internal nodes and leaves (terminal nodes) of the PolarTree. Leaves stand for sectors originating from the focal point, while internal nodes practically represent *truncated sectors*. Indeed, an internal node with radius r represents a portion bounded by two "rings" at radii r and $\frac{r}{\sqrt{2}}$ around the focal point (shaded area in Fig. 3); in short, the area covered by its subtrees is cut off. Therefore, the focal scope is subdivided into non-overlapping portions, which are either truncated sectors assigned to internal nodes (e.g., sector F in Fig. 4) or circular sectors assigned to leaf nodes (e.g., sectors M, N).

By design, an internal node is responsible to monitor orientations occurring in the outer (truncated) part, while its descendants inspect the inner subsectors in more detail. Indeed, under a uniform distribution of headings, an object has equal probability to be monitored from a parent node or one of its descendants. So, a node and the unified set of its subdivisions purposely have equal shares of the focal scope, as illustrated in Fig. 3. Trivially, we can verify that:

Lemma 1. *For an internal node e in* PolarTree, *the circular area covered by both subtrees of e has size equal to the truncated sector assigned to node e.*

Leaf nodes may appear at any level $l \geq 1$, so the tree never collapses into a single leaf. In case that no object headings are currently found within the scope of f, its respective tree degenerates into a root node with two empty leaves corresponding to the two semicircles. A circular sector having no entries, remains as an empty leaf in the tree if its sibling is an internal node. For example, in Fig. 4, sector D exists as an empty leaf, since its sibling C is not a leaf.

Each entry is a tuple $\langle oid, addrH, fDistance, sign \rangle$, essentially pointing to the memory address $addrH$ where both the heading and location of object oid are actually maintained. Attribute $fDistance$ keeps track of the current focal distance of that object from site f, while $sign$ denotes whether the object converges to $(+)$ or diverges $(-)$ from f with respect to its previously known location. We stress that object headings and locations are maintained as items in a separate structure (e.g., an array $\mathbf{H}$ in our implementation), whereas entries of tree nodes just point to them. An object may possibly be found within the scope of multiple focal points, but its current location and heading are the same for each one of them. Hence, apart from simplifying the tree structure and reducing its space requirements, such a design decision makes a clear distinction between data concerning each individual object (all stored in array $\mathbf{H}$) and information referring to its influence on several focal points (maintained in separate PolarTrees).

3.2   Index Operations

The PolarTree index is inherently dynamic. Insertions and deletions may occur arbitrarily, while the tree always remains adjusted to assist searching for ranges of headings. Specifically, a PolarTree supports the following operations:

SearchSector. This search operation descends the PolarTree $\mathbf{T}$ of a focal point f in order to identify a sector s that corresponds to the given heading θ and focal distance d. Obviously, this sector s is unique among the contents of $\mathbf{T}$ and it is the strictest one that satisfies both $\theta \in [s.\theta_{min}, s.\theta_{max})$ and $d \leq s.radius$. Searching starts from the root and follows a single path, checking with the bisector and radius of each node to determine whether it should continue at one of its subtrees (Algorithm 1). Note that search may end up at an internal node, in case the specified heading and focal distance fall inside a truncated sector.

Insert. Insertions index object headings at suitable internal nodes or leaves of the PolarTree $\mathbf{T}$ built for a focal point f. First, *SearchSector* is invoked to identify sector s corresponding to the specified heading θ and focal distance d. Provided that such a sector exists (i.e. object o is not beyond the focal scope of f), a new entry for o is inserted into the catalogue of s. In case that s refers to a leaf and the new entry causes an overflow, procedure *Bisect* is invoked to split that leaf.

Delete. A deletion removes a given object o from the tree of site f. Again, *SearchSector* is called with the known heading θ and focal distance d of o to find its corresponding sector s. If s is a leaf, after removing o from its entries, a check for underflows is made by invoking operation *Merge* for the parent of s.

Algorithm 1. PolarTree Operations

1: **Function** *SearchSector* (focal site f, angle θ, distance d)
2: **Input:** PolarTree **T** maintained for focal site f;
3: **Output:** the strictest sector s of **T**, s.t. $\theta \in [s.\theta_{min}, s.\theta_{max})$ and $d \leq s.radius$;
4:    $s \leftarrow$ **T**.root ;   //*Initialize sector and start descending* **T** *following a single path*
5: **if** $s.radius < d$ **then**
6:      **return** nil;   //*Beyond the scope of site f*
7: **end if**
8: **while** $s \,!= $ nil **do**
9:    **if** s is leaf **then**
10:      **return** s ;
11:    **else if** $\theta < s.bisector$ **and** $d \leq s.radius/\sqrt{2}$ **then**
12:      $s \leftarrow s.$leftChild ;   //*Search left subtree*
13:    **else if** $\theta \geq s.bisector$ **and** $d \leq s.radius/\sqrt{2}$ **then**
14:      $s \leftarrow s.$rightChild ;   //*Search right subtree*
15:    **else**
16:      **return** s;   //*Search may end up at an internal node*
17:    **end if**
18: **end while**
19: **End Function**

Bisect. Overflows are checked *for leaves only*, so a bisection does not affect upper tree levels. As already mentioned, it effectively partitions an existing sector s into three parts: a truncated sector that becomes an internal node and two new circular sectors as its children (leaves). Original entries of s are also partitioned, checking their focal distance and heading against the bounds of the three nodes.

Merge. This operation collapses two sibling leaves and appends their entries to their parent node, which becomes a leaf itself. As a precondition, the parent should not be the root, while the total count of entries at the three original nodes must be less than threshold m. In our setting, we have chosen $m = \frac{3}{4}M$ so as to avoid a possible bisection soon after a few subsequent insertions, but other values $m < M$ are also acceptable. Collapsing of leaves may propagate further up in the tree, as long as an underflow is discovered with respect to the new leaf node, its sibling (if also a leaf) and their parent node.

Update. To update the PolarTree **T** of site f with the current heading θ and focal distance d of an object o, we must first identify the sector s where o has been assigned before. As shown in Algorithm 2, an invocation to *SearchSector* is made with the previous heading θ' and distance d' of o (retrieved from array **H**). In case that current values of θ and d fall beyond the bounds of sector s, then o must be removed from the catalogue of that node and should be inserted into a suitable node of **T** by invoking an *Insert* operation (Lines 10-12). Note that if this insertion fails, object o is surely beyond the scope of f. But if o remains in the same sector, then only its focal distance should be updated in the catalogue of s. During an update, it is also determined whether o gets closer or farther from f (attribute *sign*), by comparing its focal distances d' and d (Lines 5-9).

Algorithm 2. PolarTree Operations (*continued*)

1: **Function** *Update* (focal site f, object o, angle θ, distance d, angle θ', distance d')
2: **Input:** PolarTree **T** maintained for focal site f;
3: **Output:** sector s of **T** where object o is assigned to;
4:   $s \leftarrow SearchSector(f, \theta', d')$; //*Sector where o resides due to previous assignment*
5: **if** $d' \le d$ **then**
6:     $o.\text{sign} \leftarrow -$;   //*o is moving away from f*
7: **else**
8:     $o.\text{sign} \leftarrow +$;   //*o is approaching f*
9: **end if**
10: **if** $\theta \notin [s.\theta_{min}, s.\theta_{max})$ **and** $d \notin (\frac{s.radius}{\sqrt{2}}, s.radius]$ **then**
11:     Remove o from the catalogue maintained at s; //*o has moved into another sector*

12:     $s \leftarrow Insert(f, o, \theta, d)$;   //*Insert o into a suitable sector*
13: **else**
14:     $o.fDistance \leftarrow d$;   //*o remains in the same sector, but update its focal distance*
15: **end if**
16: **return** s;
17: **End Function**

18: **Procedure** *RangeSearch* (sector s, angle θ_1, angle θ_2, distance d_1, distance d_2)
19: **if** $s! = \text{nil}$ **then**
20:     **if** $\theta_1 < s.bisector$ **and** $d_1 \le s.radius/\sqrt{2}$ **then**
21:         RangeSearch ($s.\text{leftChild}, \theta_1, \theta_2, d_1, d_2$);   //*Search left subtree of s*
22:     **end if**
23:     **for** each object entry o in the catalogue of s **do**
24:         **if** $\theta_1 \le o.heading < \theta_2$ **and** $d_1 < o.fDistance \le d_2$ **then**
25:             Report o;   //*o is a qualifying object at sector s*
26:         **end if**
27:     **end for**
28:     **if** $\theta_1 \ge s.bisector$ **and** $d_1 \le s.radius/\sqrt{2}$ **then**
29:         RangeSearch ($s.\text{rightChild}, \theta_1, \theta_2, d_1, d_2$);   //*Search right subtree of s*
30:     **end if**
31: **end if**
32: **End Procedure**

RangeSearch. This method offers response to orientation-based queries associated to site f that specify a range $[\theta_1, \theta_2)$ for headings and another $(d_1, d_2]$ on focal distances, as mentioned in Section 2.3. Since many paths of the tree may be probably visited, this procedure (pseudo-code given in Algorithm 2) is called for the root node and recursively performs a depth-first search. When visiting an internal node that represents a sector s, the algorithm must decide whether to further descend to a subtree by comparing the bisector of s with the given angle range and also checking if d_1 is less than the radius of its children (Lines 20-22 and 28-30). When backtracking, any qualifying entries of a visited node s with headings and focal distances falling within the given ranges are reported as results (Lines 23-27). For the query specified with the shaded area in Fig. 4, nodes A, D, B, E, K, L will be visited (in that order). With a small variation, this method can also distinguish between objects approaching site f and those moving away from it, by simply checking their respective *sign* values.

3.3   Discussion

A PolarTree arranges all headings of interest to a focal point f into compact sectors, which can get recursively refined for better monitoring of movements closer to f. The initial subdivision of focal scope may not necessarily be carried out with the x-axis as *prime bisector* (at the root), but across any arbitrary direction. For instance, if headings were measured as azimuths or a focal point was mainly concerned with east- or west-bound orientations, then the y-axis should be used as prime bisector. We opted for a scheme with its prime bisector at angle $\phi = \pi$, because all derived subsectors are mapped to well-known portions of the trigonometric circle with obvious advantages on geometric calculations.

Bisection adheres to a repetitive pattern applied to all tree levels. This strategy decomposes the initial scope into finer partitions for progressively obtaining higher resolution of movements that occur closer to the focal point. The less the radius of a sector, the more segmented the scope across that direction, thus offering more detailed tracking of orientation trends. Besides, overflow threshold M represents the maximum capacity of leaves and reflects the level of detail prescribed for orientations close to the focal point. In effect, M specifies the *resolution* at which a focal point wishes to observe movements in its vicinity.

The tree is usually unbalanced, since a uniform distribution of headings and focal distances could be observed only rarely. For a skewed distribution, where most objects head towards certain directions, the respective sectors would be gradually subdivided at very tiny angles. Nodes may be unevenly filled, and even internal ones may occasionally be left empty. Even under a uniform distribution, larger-area sectors expectedly contain many more entries (far from site f) than a tiny sector monitoring a small range of headings in the close vicinity of f.

The height of the tree for a focal point f depends on threshold M, the number N of objects currently within scope, but also on their distribution around f. But:

Lemma 2. *A PolarTree has height at least* $\frac{1}{2}(1 + \log_2 \lceil \frac{N}{M} \rceil)$.

Proof. Apparently, the more uniform the distribution of headings around f, the lower the tree height. According to Lemma 1 and assuming a uniform distribution, the count of entries assigned to a subtree at level l is $1/4$ of those assigned to a subtree at $l-1$, i.e., proportional to their respective area. So, it turns out that a leaf at level $l > 0$ has $\frac{N}{2^{2l-1}}$ entries. Due to uniformity, all leaves are at the same level, since branching factor is 2 and applies to all nodes. In order for a leaf not to be split, it suffices that $\frac{N}{2^{2l-1}} \leq M$, which yields the lower bound.   □

4   Processing Streaming Orientations of Moving Objects

4.1   System Model

System infrastructure for processing orientations consists of a central server that communicates with numerous moving objects via a cellular network. A number of *base stations* are merely used for relaying messages between the server and objects located in their cell, so we shall ignore their role on data processing.

Each object o is identified by its *oid* and has enough resources to retain its current position $\langle x, y, t \rangle$ and to calculate its velocity, i.e., its speed v and heading θ. Normally, each object notifies the server about its *status* by sending a tuple $\langle oid, x, y, t, v, \theta \rangle$ at a specified frequency, i.e., every τ_0 time units. Duplicate, delayed or out-of-order status updates are not considered, so all messages stream synchronously into the server at a sequential pattern for each object.

But a status update should be sent instantly, once the heading or speed deviate significantly (e.g., a sudden slow down or a turn) from the values conveyed to the server with the latest message. We employ a simple detection method that utilizes two system-wide parameters λ and $d\theta$ specifying *thresholds* for acceptable deviations in speed and heading, respectively. In particular, if an object o changes its speed from v to v', an update should be sent if $|v - v'| > \lambda v$, denoting that the object accelerated or decelerated more than $\lambda\%$ compared to its previous speed. Similarly, the server must be notified if the heading of o changes from θ to θ' and it occurs that $|\theta - \theta'| > d\theta$. Such lightweight calculations can be performed by every object with negligible overhead, although a more sophisticated dead-reckoning [12] or threshold-guided strategy [7] could also be applied.

The server registers a set F of focal points and at each time instant t inspects movements in their scope according to streaming object statuses. Status information is retained in an array **H** for all objects, but not all statuses get updated concurrently, since some objects may report more frequently than others, depending on their motion pattern. At any rate, no object status can be more than τ_0 time units older compared to timestamp t of a newly received message.

A focal point f_i with scope at fixed radius R_i may be dynamically registered with the server and remains active for a duration of δ_i time units, until it gets eventually suspended and possibly resumed after some time. For instance, observation points at highways may be turned on at rush hours or switched off at night. A server-resident PolarTree is dedicated to retain the distribution of object orientations related strictly with a single f_i, so the server keeps track of $|F|$ separate trees. In effect, each pole f_i maintains at its own PolarTree a *"polar chart"* (Fig. 4a) that always reflects objects' movement as observed from the perspective of that particular f_i. At any given timestamp, the shape of this tree is independent of the order that headings were inserted or deleted.

Several continuous orientation-based queries may be specified at each focal point (Section 2.3). Without loss of generality, we assume that queries get activated when their corresponding pole f is registered. At any instant t, query evaluation (with *RangeSearch*) is based on current entries at the PolarTree of f.

4.2   Continuous Monitoring of Object Headings

In order to successfully maintain current object orientations around focal sites, the server operates at each execution cycle (i.e., distinct timestamp t) in two phases: (i) processing all status updates currently received from objects, and (ii) refreshing status for the remaining objects that did not send updates.

i) Update phase. Incoming statuses get processed one by one, in strict arrival order. Once the server receives such a message, it attempts to identify affected focal points and update their PolarTree (Algorithm 3). But a single status update may influence multiple focal points, if that object is currently located within their intersecting scopes (e.g., object g in Fig. 1). To quickly identify affected sites, focal scopes are indexed with a regular *grid partitioning* of the entire area $\mathbf{E}$ into $c \times c$ square cells (Fig. 5). Each grid cell c_i maintains a list of pointers to every site with a circular scope intersecting cell c_i. As soon as an update arrives from object o, its location is hashed against the grid to identify the corresponding cell c_i, thus determining that only the subset $S \subset F$ of *candidate* focal points indexed at c_i need be probed. In Fig. 5, when object k sends update, sites $S = \{f_1, f_3\}$ should be examined, since their scopes overlap its (dark-shaded) cell.

However, a status update may also signify that object o has just fallen beyond the scope of a site f, so any reference to o in its respective PolarTree must be eliminated. Thus, any focal sites influenced by the previous status of o should be probed as well. By identifying the grid cell c' corresponding to the last known location of o, we get an additional set S' of candidate sites indexed at c' that also need examination. Figure 5 reveals that object k has just become of no interest to f_2, by only checking focal points $S' = \{f_1, f_2\}$ indexed at the light-shaded cell of the old location of k. Note that a site may appear in both S and S', in case that its scope overlaps with cells c and c' (which may be a single cell).

So it suffices that the status of object o is only checked against each candidate $f \in S \cup S'$ to detect changes in affected sites (Lines 8-25). There are four possible situations: (i) if o has just entered the scope of f, it must be inserted into its respective PolarTree, (ii) in case that o remains within scope of f, the PolarTree might need updating when o changes sector or its focal distance is modified, (iii) if o has just passed out of scope for f, then it must be removed from its tree, and (iv) if o stays beyond the scope of f, no further action is needed.

Besides, upon message receipt from an object at time t, the server also estimates the next time instant $t + \tau_e$ this object should relay its status again, so as to maintain a consistent distribution in relevant PolarTrees. We distinguish two cases that an object status should be renewed (Fig. 6), depending either on focal site(s) currently influenced or those that might soon be affected:

Departure Forecast. Let an object q be within the focal scope R of site f, and the server has just been informed for its current speed v and heading θ. Assuming that q continues the same course until further notice, the server is able to forecast when this object will fall out of scope R. Figure 6 depicts the expected course α of object q until it crosses the scope of f at location E_1. With simple trigonometric manipulations, it can be easily verified that $R^2 = \alpha^2 + \beta^2 + 2\alpha\beta \cos(\theta - \phi)$,

---

Algorithm 3. Server Operations

1: **Function** *UpdateStatus* (object o)
2: **Input:** Server-resident array **H** of most recent status for all monitored objects;
3: **Output:** the next time instant τ_e that o should send a status update;
4: $\tau_e \leftarrow \tau_0$;   // *Initialize refresh time to default value*
5: $q \leftarrow$ **H**[o].location; $p \leftarrow$ o.location;   // *Previous and current location of o*
6: $c' \leftarrow gridHash(q)$; $c \leftarrow gridHash(p)$; // *Grid cells of last and current location*
7: *candSites* $\leftarrow$ focal points with scopes overlapping cells c and c';
8: **for** each focal point $f \in$ *candSites* **do**
9:     $R \leftarrow f$.radius;   // *Focal scope of site f*
10:    $h \leftarrow$ *distance*(f.location, q); $d \leftarrow$ *distance*(f.location, p); // *Focal distances for o*

11:    **if** $d \le R$ **then**
12:        **if** $h \le R$ **then**
13:            *Update*(f, o, o.heading, d, **H**[o].heading, h); // *o already monitored by f*
14:        **else**
15:            *Insert*(f, o, o.heading, d); // *Object o has just become of interest to site f*
16:        **end if**
17:        $\tau \leftarrow$ Estimate time that o will fall beyond the scope of f; $\tau_e \leftarrow \min(\tau, \tau_e)$;
18:    **else**
19:        **if** $h \le R$ **then**
20:            *Delete*(f, o, o.heading, d);   // *Object o just gone outside the scope of f*
21:        **else if** $d < h$ **then**
22:            $\tau \leftarrow$ Estimate time that o could reach the scope of f; $\tau_e \leftarrow \min(\tau, \tau_e)$;
23:        **end if**
24:    **end if**
25: **end for**
26: Update **H**[o] with current heading and location of o;
27: **if** $\tau_e < \tau_0$ **then**
28:    **return** τ_e; // *Earliest time for renewal, as estimated from all sites affected by o*
29: **else**
30:    **return** nil; // *No need to change default object settings*
31: **end if**
32: **End Function**

---

where β is the focal distance of q and ϕ is the slope of segment β. This equation is always valid for all possible configurations of the heading and location of q at any quadrant inside the scope of f. It can be proven that the positive root α^+ corresponds to the distance from q to E_1, while the negative root α^- to E_1' (i.e., towards the opposite direction). Hence, after at most $\tau = \lceil \frac{\alpha^+}{v} \rceil$ time units, object q is expected to be found beyond the scope of f (Line 17).

Arrival Forecast. As shown in Fig. 6 for l, an object may not currently affect a site f, but it seems approaching; if l continues moving along ξ, it might soon fall within scope at E_2. Yet, to safely predict the earliest time τ_e that l could cross the scope of any focal site, is not an easy task. Indeed, it could also involve inspection of sites $f \notin S \cup S'$, with scopes perhaps closer to object l, but indexed in cells neighboring to that of its current location. Sooner or later, l should send

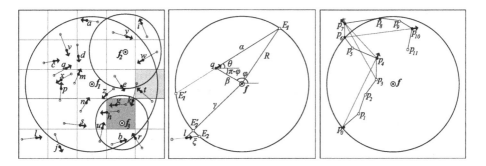

Fig. 5. Grid partitioning   **Fig. 6.** Forecasting   **Fig. 7.** Trajectory headings

a status update (at most after τ_0 time units), so we opt for a simplified strategy that only examines candidate sites $f \in S \cup S'$ with negligible overhead. Given current focal distance γ and speed v of an object l, this approximation makes an eager forecast $\tau = \lfloor \frac{\gamma - R}{v} \rfloor$ of the time it takes to reach the scope of site f at E_2', by ignoring actual heading as if l were directed straight towards f (Line 22).

Similar forecasts τ are made for all sites influenced by the currently examined object o. Among them, the smallest τ_e anticipates the earliest time that o may cause a change to current orientations (at time t) of any focal site. In case the interval τ_e is less than the prescribed renewal period τ_0, a message is sent to that object, specifying the time $t + \tau_e$ of its expected next update (Lines 27-31).

ii) Refresh phase. To maintain reliable object distributions around each site, the server *approximately* adjusts entries in PolarTrees for objects that have not currently relayed their status. Although velocity v of each such object o is deemed unchanged until further notice, its location p is not; hence, its focal distance from influenced sites changes. Assuming the known status of o is Δt units old, its expected position is $p + \Delta t \cdot v$; accordingly, the server rearranges entries in respective tree(s), in a fashion similar to the update phase. But, at this stage, no forecasting of status renewal times is made, so no messages are sent to objects.

4.3   Examining Trajectory Headings

Up to this point, processing only examines current object headings. However, it would be more insightful to monitor orientations referring to the most recent portion of every trajectory. With a *sliding window* of w time units that considers locations recorded during this evolving period, we can repetitively calculate for each object o its *trajectory heading* from the two extreme positions of o within that window, as indicated in Fig. 7 for $w = 5$ units and location updates every time unit. Such a setting does not actually change server-side processing, but it only modifies computation of headings separately at each object, provided they have enough memory to retain a finite portion of their own recent positions.

But for a realistic application, it seems more suitable to consider that sliding windows are specified along with focal points and their orientation-based queries.

Hence, we assume that a sliding window w_i refers to a site f_i and is applied to all queries associated to f_i, thus affecting objects found within the scope of f_i with their headings computed from readings received during the past w_i time units (e.g., 10 min). Queries may specify various ranges for headings and focal distances, whereas an object may become of interest to multiple sites with diverse window extents. Thus, computation of trajectory headings has to be performed at the server, which should now retain a series of recent statuses for each object; still, objects send their status regularly, but also when a deviation is detected or in due time upon server request (displayed as black spots in Fig. 7).

A single status update from object o may cause changes to multiple trajectory headings maintained for o at diverse sites, because each f_i can specify a different window w_i, thus potentially returning a different anchor point. Each trajectory heading shall be derived from available object statuses received over the past w_i time units. Although not completely accurate, such an orientation still conveys the movement trend of every object, provided that $w_i > \tau_0$ to guarantee frequent renewal of all indexed headings. In Fig. 7, the trajectory heading at p_{10} will be correct due to availability of p_6, while the heading at p_7 will be slightly tilted by using p_4 as anchor point instead of non-relayed p_3. Obsolete statuses are purged from the server, when they cease to fall inside the window extent of any site.

5   Experimental Evaluation

Next, we report indicative results from an empirical validation of our framework for monitoring orientations. Due to space constraints, we refrain from discussing index performance and threshold calibration, and focus on server-side operations.

Experimental Setup. We generated synthetic datasets for varying numbers P of objects moving at diverse speeds along the road network of Athens (area $\sim$250 km^2). After calculating shortest paths between randomly chosen network nodes (i.e., origin and destination of objects), we took point samples at 200 concurrent timestamps along each such route. We also randomly selected diverse sets of n focal points at various radii R, which remain active all the time ($\delta = 200$ units).

All processing takes place in main memory. Algorithms were implemented in C++ and experiments with diverse parameter settings were simulated on an Intel Core 2 Duo 3GHz CPU running GNU/Linux with 2GB of main memory. Results are averages of actual measurements over 200 timestamps. Table 1 summarizes experimentation parameters and their ranges; default values are in bold.

Table 1. Experiment parameters

Parameter	Values	Parameter	Values
Number P of objects	10k, 20k, 50k, **100k**	Grid granularity (c)	50, **100**, 500, 1000
Number n of sites	100, **200**, 500, 1000	Heading deviation ($d\theta$)	10°, 20°, **30°**
Focal radius (R)	0.5, 1, **2**, 3, 4 km	Speed deviation (λ)	0.05, 0.1, **0.2**
Leaf capacity (M)	100, **200**, 500	Window extent (w)	40, **50**, 100 units

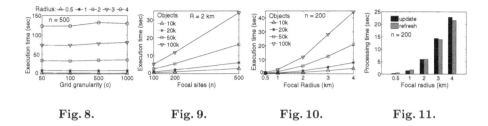

Fig. 8.   Fig. 9.   Fig. 10.   Fig. 11.

System Configuration. For specifying granularity c of grid index for focal scopes, we measured the per cycle execution cost (sum of update and refresh times) for the most demanding case with 100k objects and diverse scope sizes for $n = 500$ sites (Fig. 8). As expected, grid partitioning proves more useful for larger scopes with higher degree of overlaps. We fix $c = 100$ in the sequel, as such a reasonably fine grid seems to offer better performance at all scope sizes.

Object-side parameters only control frequency of status updates. Next, we stipulate that objects should relay new status at most every $\tau_0 = 30$ timestamps, while we set thresholds $\lambda = 0.2$ and $d\theta = 30°$, typically for moving vehicles.

Experimental Results. The main part of experiments refer to the efficiency and scalability of our approach. As shown in Fig. 9, the per cycle cost at the server depends on object count and is linear in the number of sites (their scopes fixed at 2km), since each PolarTree is maintained separately. But execution time escalates for larger scopes, as depicted in Fig. 10 for $n = 200$ sites, because the probability that an object influences multiple sites with intersecting scopes increases as well. This incurs additional overhead on forecasting and causes transmission of extra status updates from relevant objects, due to frequent crossing of scope boundaries. This is also verified from Fig. 11 that plots a breakdown of execution times per phase: handling incoming updates and forecasting arrivals and departures from focal scopes is often more costly than refreshing existing object statuses, especially for wider areas of interest. As it turns out, performance is sensitive to the size of scopes, but chiefly depends on their mutual overlaps. Nonetheless, for realistic radii (less than 3km) this scheme can always provide quick notification about observed orientations in less than 30 seconds.

Regarding communication cost, Fig. 12 illustrates the percentage of message savings for several scope sizes, i.e., the fraction of positional readings that did not cause any status change and hence were not relayed to the server. For small radii, the reduction in message transmission is considerable and may exceed 70%. But for larger scopes, the advantage of threshold-guided detection of motion changes is gradually annihilated, as an object becomes of interest to many sites and must report its status over and over due to their alternating demands.

In practice, focal scopes should be leveraged with appropriate choice of leaf capacity M. After all, it is improbable that a long-range site wishes to monitor movement trends at the finest resolution. As suggested in Lemma 2 and verified in Fig. 13, by increasing M the tree becomes shorter with wider sectors; a tree with 6 levels corresponds to central angles of $5.625°$ at its bottommost leaves.

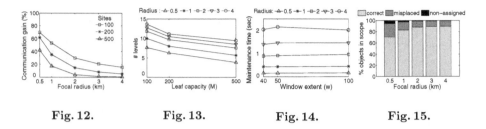

Fig. 12.   Fig. 13.   Fig. 14.   Fig. 15.

Concerning trajectory headings, they are handled exactly like current object orientations, but in addition require maintenance of sliding windows (Section 4.3). Figure 14 reveals that this maintenance overhead is proportional to scope size, but almost independent of the window extent. Anyway, such cost is negligible and can be compensated with valuable knowledge of recent orientations.

Finally, our processing scheme was designed to provide an approximate, yet consistent view of movements close to focal points. To assess the quality of such monitoring, we issued orientation queries based on polar sector boundaries (i.e., one query per polar sector). We then compared their approximate answers with those returned from an exhaustive evaluation where all objects relay their status at each timestamp, and thus always get mapped to correct sectors. In Fig. 15 the accuracy of answers is displayed for a single PolarTree (similar results obtained at multiple sites). At any time instant, less than 5% of qualifying objects are not reported within scope, but the majority of them (more than 70% at the worst case) are correctly assigned. Although monitored, another 15% of objects are misallocated to a neighboring sector due to small variations (that rarely exceed 8°) in their assumed heading. Indeed, smaller circles are subdivided in very tiny sectors, so it is more likely that an object be misplaced; yet, the wider the focal scope, the greater the accuracy of answers. Overall, polar charts prove able to offer a reliable insight into the actual distribution of object orientations.

6   Related Work

A taxonomy of spatiotemporal queries has been proposed in [8], distinguishing between *coordinate-based* queries, such as range or k-nearest neighbor search, and *trajectory-based* queries. This latter class includes *navigational* queries involving derived information of trajectories, like speed, heading, traveled distance etc. Index structures introduced in [8] aim at trajectory preservation, but no technique is suggested for maintaining object headings. Another type of spatial requests inspects *object-based directional* relationships [4,10], e.g., identifying objects to the north of a given landmark. But such directional queries deal with relative positions of static features, and not with their movement and orientation.

In spatiotemporal databases, *dead-reckoning* policy suggests that an object should send a positional update when it deviates from its known motion vector, thus reducing communication cost. Two such schemes were introduced in [12] and adjust the uncertainty threshold at each update according to the current motion pattern. From a streaming perspective, in [7] we employed threshold-guided policies for online detection of movement changes in order to maintain concise trajectory synopses. All these approaches are orthogonal and can be easily integrated into our framework, as they only control object update frequency. Velocity vectors were also used in [3] to construct motion-sensitive bounding boxes for indexing moving objects. Although such structures can make predictions about future object positions, they are tailored for coordinate-based queries only.

Centralized or distributed techniques for managing streaming locations offer scalable techniques mostly for range [2,6] or k-NN search [5,11], by examining only current object positions. We are not aware of other research work on processing object orientations in a streaming fashion. The proposed PolarTree is a hierarchical structure reminiscent of *space-driven* access methods for indexing multidimensional features [1]. Similarly to a quadtree [9], which is based on successive subdivision of areas into four equal-sized quadrants, a PolarTree utilizes *angle bisection* as its underlying design principle. Of course, our objective is not indexing locations of spatial features, but their changing orientations instead.

7   Concluding Remarks

In this paper, we have introduced a novel, simple, yet versatile, access method that can greatly assist continuous monitoring of movement orientations in suitably divided sectors around selected focal points of interest. We have also empirically evaluated the robustness and scalability of a processing scheme that offers real-time response to multiple requests with reduced communication cost.

In the future, we plan to study a variant tree structure with dynamic division in dissimilar sectors according to the observed density of object headings. Besides, distributed processing of orientations at designated base stations may further exhibit the powerfulness of the proposed spatiotemporal index.

References

1. Gaede, V., Günther, O.: Multidimensional Access Methods. ACM Computing Surveys 30(2), 170–231 (1998)
2. Gedik, B., Liu, L.: Mobieyes: A Distributed Location Monitoring Service using Moving Location Queries. Transactions on Mobile Computing 5(10), 1384–1402 (2006)
3. Gedik, B., Wu, K.-L., Yu, P., Liu, L.: Processing Moving Queries over Moving Objects Using Motion-Adaptive Indexes. IEEE TKDE 18(5), 651–668 (2006)
4. Liu, X., Shekhar, S., Chawla, S.: Object-Based Directional Query Processing in Spatial Databases. IEEE TKDE 15(2), 295–304 (2003)
5. Mouratidis, K., Hadjieleftheriou, M., Papadias, D.: Conceptual Partitioning: An Efficient Method for Continuous Nearest Neighbor Monitoring. ACM SIGMOD, 634–645 (June 2005)
6. Prabhakar, S., Xia, Y., Kalashnikov, D., Aref, W., Hambrusch, S.: Query Indexing and Velocity Constrained Indexing: Scalable Techniques for Continuous Queries on Moving Objects. IEEE Transactions on Computers 51(10), 1124–1140 (2002)

7. Potamias, M., Patroumpas, K., Sellis, T.: Sampling Trajectory Streams with Spatiotemporal Criteria. In: SSDBM, July 2006, pp. 275–284 (2006)

8. Pfoser, D., Jensen, C., Theodoridis, Y.: Novel Approaches in Query Processing for Moving Objects. In: VLDB, September 2000, pp. 395–406 (2000)

9. Samet, H.: The Quadtree and Related Hierarchical Data Structures. ACM Computing Surveys 16(2), 187–260 (1984)

10. Skiadopoulos, S., Sarkas, N., Sellis, T., Koubarakis, M.: A Family of Directional Relation Models for Extended Objects. IEEE TKDE 19(8), 1116–1130 (2007)

11. Wu, W., Guo, W., Tan, K.-L.: Distributed Processing of Moving k-Nearest-Neighbor Query on Moving Objects. In: ICDE, April 2007, pp. 1116–1125 (2007)

12. Wolfson, O., Sistla, P., Chamberlain, S., Yesha, Y.: Updating and Querying Databases that Track Mobile Units. Distributed and Parallel Databases 7(3), 257–287 (1999)

Spatial Skyline Queries:
An Efficient Geometric Algorithm

Wanbin Son, Mu-Woong Lee, Hee-Kap Ahn, and Seung-won Hwang

Pohang University of Science and Technology, Korea
{mnbiny,sigliel,heekap,swhwang}@postech.ac.kr

Abstract. As more data-intensive applications emerge, advanced retrieval semantics, such as ranking and skylines, have attracted attention. Geographic information systems are such an application with massive spatial data. Our goal is to
efficiently support skyline queries over massive spatial data. To achieve this goal,
we first observe that the best known algorithm VS^2, despite its claim, may fail
to deliver correct results. In contrast, we present a simple and efficient algorithm
that computes the correct results. To validate the effectiveness and efficiency of
our algorithm, we provide an extensive empirical comparison of our algorithm
and VS^2 in several aspects.

1   Introduction

With the advent of data-intensive applications, advanced query semantics, which enable efficient and intelligent access to large scale data, have been actively studied. Geographic information systems (GISs) are such an application, which aim to support
efficient access to massive spatial data, as Example 1 illustrates.

Example 1. Consider a hotel search scenario for a business trip to Aalborg, where the
user marks two locations of interest, e.g., the conference venue and an airport, as Fig. 1a
illustrates. Given these two query locations, one option is to identify hotels that are
close to both locations. To better illustrate this problem, Fig. 1b rearranges the hotels
with respect to the distance to each query point. From this figure, we can claim that
hotel H3 is more desirable than H10, because H3 is closer to both query points than
H10 is. Such advanced retrieval, by *ranking* the hotels using the aggregate distance to
the given query points, or by finding *skyline* hotels, will enable intelligent access to the
underlying hotel datasets.

In particular, this paper focuses on supporting *skyline queries* [1,2,3,4,5] to identify the
objects that are "not dominated" by any other objects, i.e., no other object is closer to all
the given query points simultaneously. For instance, in Fig. 1b, H3 is a skyline object,
while H10 is dominated by H3 and does not qualify as a skyline object.

Skyline queries have gained attention lately, as formulating such queries is highly intuitive, compared to ranking where users are required to identify ideal distance functions
to minimize. However, most of the existing skyline algorithms have not been devised
for spatial data and thus do not consider spatial relationships between objects.

N. Mamoulis et al. (Eds.): SSTD 2009, LNCS 5644, pp. 247–264, 2009.

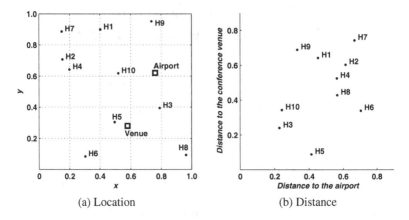

Fig. 1. Hotel search scenario

Our goal is to efficiently support skyline queries over spatial data. This problem has already been studied by Sharifzadeh and Shahabi [6] and they presented two algorithms for the problem, one of which, VS^2, is known to be the most efficient solution thus far. We claim, however, that VS^2 may fail to identify the correct results. In a clear contrast, we propose an algorithm for the problem that can identify the exact results in $O(|P|(|S|\log|\mathcal{CH}(Q)|+\log|P|))$ time, for the given set P of data points, set Q of query points, set S of spatial skyline points, and the *convex hull* of Q, denoted by $\mathcal{CH}(Q)$.

Our contributions can be summarized as follows:

- We study the spatial skyline query processing problem, which enables intelligent and efficient access to massive spatial data.
- We show that the best known algorithm is incomplete in the sense that it may not return all the skyline points.
- We propose a novel and correct spatial skyline query processing algorithm and analyze its complexity.
- We extensively evaluate our framework using synthetic data and validate its effectiveness.

The remainder of this paper is organized as follows. In Section 2, we provide a brief survey of related work. In Section 3, we observe the drawbacks in the best known algorithm, and propose a new algorithm in Section 4. Section 5 discusses the details of our implementation of the proposed algorithm. In Section 6, we report our evaluation results, and Section 7 concludes this paper.

2 Related Work

This section provides a brief survey on work related to (1) skyline query processing, and (2) spatial query processing.

2.1 Skyline Computation

Skyline queries were first studied as maximal vectors [1]. Later, Börzsönyi et al.[2] introduced skyline queries in database applications. A number of different algorithms for skyline computation have been proposed, for example, progressive skyline computation using auxiliary structures [3], nearest neighbor algorithm for skyline query processing [7], branch and bound skyline (BBS) algorithm [4], sort-filter-skyline (SFS) algorithm leveraging pre-sorted lists [5], and linear elimination-sort for skyline (LESS) algorithm with attractive average-case asymptotic complexity [8].

Recently, there have been active research efforts to address the "curse of dimensionality" problem of skyline queries [9,10,11] using inherent properties of skyline points such as *skyline frequency*, *k-dominant skylines*, and *k-representative skylines*. All these efforts, however, do not consider spatial relationships between data points.

2.2 Spatial Query Processing

The most extensively studied spatial query mechanism is ranking neighboring objects by the distance to a single query point [12,13,14]. For multiple query points, Papadias et al. [15] studied ranking by the "aggregate" distance, for a class of monotone functions aggregating the distances to multiple query points. As these nearest neighbor queries require a distance function, which is often cumbersome to define, another line of research studied skyline query semantics which do not require such functions.

For a spatial skyline query with a single query point, Huang and Jensen [16] studied the problem of finding spatial locations that are not dominated with respect to the *network distance* to the query point. For such a query with multiple query points, Sharifzadeh and Shahabi [6] proposed two algorithms that identify the skyline locations to the given query points such that no other location is closer to all query points. While the proposed problem enables intelligent access to spatial data, we later show that a solution proposed in [6] is incorrect. In contrast, we present a correct and exact algorithm.

3 Preliminaries

In this section, we introduce some geometric concepts, and define our problem. Then we discuss how the best known algorithm fails to identify exact answers.

3.1 Convex Hull

A subset S of the plane is *convex* if and only if for every two points $p, q \in S$ the whole line segment $\overline{pq}$ is contained in S. The *convex hull* $\mathcal{CH}(S)$ of a set S is the intersection of all convex sets that contain S [17]. The *upper chain* of $\mathcal{CH}(S)$ is the part of the boundary of $\mathcal{CH}(S)$ from the leftmost point to the rightmost point in clockwise order. The *lower chain* is the part of the boundary of $\mathcal{CH}(S)$ from the rightmost point to the leftmost point in counterclockwise order.

3.2   Voronoi Diagram and Delaunay Graph

For a set P of n distinct points in the plane, the Voronoi diagram of P, denoted by $\text{Vor}(P)$, is the subdivision of the plane into n cells [17]. Each cell contains only one point of P, which is called the *site* of the cell. Any point q in a cell is closer to the site of the cell than any other site. The Delaunay graph of a point set P is the dual graph of the Voronoi diagram of P [17]. Two points of P have an edge in the Delaunay graph if and only if the Voronoi cells of these points share an edge in $\text{Vor}(P)$.

3.3   Problem Definition

In the spatial skyline query problem, we are given two point sets: one is a set P of data points, and the other is a set Q of query points. The points in P and Q have d-dimensional coordinate attributes in $\mathbb{R}^d$ space. Distance function $d(p, q)$ returns the Euclidean distance between a pair of points p and q, which obeys the triangle inequality. Before we set the goal of the problem, we need the following definitions.

Definition 1. *We say that p_1 spatially dominates p_2 if and only if $d(p_1, q) \leq d(p_2, q)$ for every $q \in Q$, and $d(p_1, q') < d(p_2, q')$ for some $q' \in Q$.*

Definition 2. *A point $p \in P$ is a* spatial skyline point *with respect to Q if and only if p is not spatially dominated by any other point of P.*

The goal of the problem is to retrieve all the spatial skyline points from P with respect to Q. We denote S as the set of spatial skyline points of P.

3.4   Existing Approaches

Though there is a lot of work on skyline queries in the literature, little is known about skyline queries for spatial data. Recently, Sharifzadeh and Shahabi [6] studied the spatial skyline query problem and proposed two algorithms that compute S: Branch-and-Bound Spatial Skyline Algorithm (B^{2}S^2) and Voronoi-based Spatial Skyline Algorithm (VS^2).

In VS^2, they employed two well-known geometric structures, the *Voronoi diagram* of P and the *convex hull* of Q, and claimed that these structures reflect the spatial dominance to some extent, and therefore the algorithm efficiently computes S. In fact, their experiments show that VS^2 runs $2 \sim 3$ times faster than B^{2}S^2, and VS^2 is known to be the most efficient solution thus far.

VS^2, however, may fail to find all the spatial skyline points: In Lemma 4 of [6], to verify VS^2 they claimed that, for some $p \in P$, if all its Voronoi neighbors and all their Voronoi neighbors are spatially dominated by other points, p is not a spatial skyline. Therefore VS^2 simply marks p as *dominated* and does not consider it afterwards. But this is not necessarily true.

Fig. 2 shows a counter example to their claim. There are 3 query points (q_0, q_1, q_2) and 9 data points. Note that all the data points, except three (p_0, p_1 and p_2), are spatially dominated by p_0 or p_1. That is, all the Voronoi neighbors of p_2 are spatially dominated, and VS^2 thus simply marks p_2 as "dominated" and does not consider it again. However,

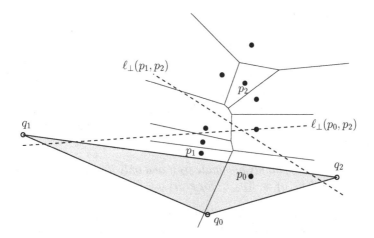

Fig. 2. VS^2 fails to find p_2 even though p_2 is a spatial skyline point

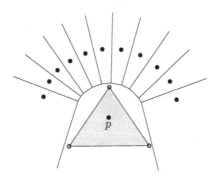

Fig. 3. A point can have many Voronoi neighbors

in fact, p_2 is a spatial skyline point, as the *bisector* $\ell_\perp(p_1, p_2)$ of p_1 and p_2, i.e., a perpendicular line to the line segment $\overline{pq}$, intersects $\mathcal{CH}(Q)$. This implies that there is a query point (q_2) closer to p_2 and therefore p_2 is not spatially dominated by p_1, as we will discuss more formally later in Lemma 4. Similarly, p_2 is not spatially dominated by p_0, because $\ell_\perp(p_0, p_2)$ intersects $\mathcal{CH}(Q)$. Since every bisecting line of p_2 and other points intersects $\mathcal{CH}(Q)$, we conclude that p_2 is a spatial skyline point.

Moreover, the asymptotic time complexity analysis of VS^2 in [6] is incorrect. The authors assumed implicitly that VS^2 tests only $O(|S|)$ points and claimed that it finds S in time $O(|S|^2|\mathcal{CH}(Q)| + \sqrt{|P|})$. However, a skyline point p can have at most $O(|P|)$ Voronoi neighbors that are all spatially dominated by p, as Fig. 3 illustrates. Since it also calls $|P|$ heap operations during the iteration, each of which takes $\log |P|$, the correct worst-case time complexity of VS^2 must be $O(|P|(|S||\mathcal{CH}(Q)| + \log |P|))$.

4   Computing Spatial Skylines

We first propose a progressive algorithm for the spatial skyline problem, which retrieves all the spatial skyline points of P with respect to Q, then we improve this algorithm by using the Voronoi diagram of the dataset. We assume the dimensionality d of data and query points is $d = 2$ for now, which can be extended for arbitrary dimension (as we will discuss in Section 7). Before we explain our algorithms, we show some properties of spatial skyline that will be used later on. The following lemma is the contraposition of Definition 1.

Lemma 1. p_1 *does not spatially dominate* p_2 *if and only if either* $d(p_1, q) > d(p_2, q)$ *for some* $q \in Q$, *or* $d(p_1, q) = d(p_2, q)$ *for every* $q \in Q$.

Lemma 2. *Let* p_1, p_2 *and* p_3 *be three data points such that* p_2 *spatially dominates* p_3. *If* p_1 *does not spatially dominate* p_3, *it does not spatially dominate* $p_2 \in P$.

Proof. Since p_1 does not spatially dominate p_3, either (1) $d(p_3, q') < d(p_1, q')$ for some $q' \in Q$, or (2) $d(p_3, q) \leq d(p_1, q)$ for every $q \in Q$ by Lemma 1.

Case (1). By Definition 1, $d(p_2, q) \leq d(p_3, q)$ for every $q \in Q$. This implies that $d(p_2, q') \leq d(p_3, q') < d(p_1, q')$. Therefore, p_1 does not spatially dominate p_2 by Lemma 1.

Case (2). Since p_2 spatially dominates p_3, there exists a point $q \in Q$ satisfying $d(p_2, q) < d(p_3, q)$, which implies that $d(p_3, q) \leq d(p_1, q)$. Therefore, p_1 does not spatially dominate p_2 by Lemma 1.   ⌑

Lemma 3. *If some data point* p_1 *is not a spatial skyline point, there always exists a spatial skyline point* p_2 *that spatially dominates* p_1.

Proof. Since p_1 is not a spatial skyline point, there exists some data point that spatially dominates p_1. Let P' be the set of data points that spatially dominate p_1, and let p_2 be the point which has the minimum sum of distances to all $q \in Q$ among points in P'. Then it is not difficult to see that for every point $p' \in P'$, there always exists some query point q such that $d(p_2, q) < d(p', q)$. Therefore, p_2 is not spatially dominated by any point in P'. By Lemma 2, p_2 is not spatially dominated by any data point which does not spatially dominate p_1. This means that p_2 is not spatially dominated by any other data points, so p_2 is a spatial skyline point.   ⌑

We now move on to discuss how to use these properties to reduce (1) time required for each dominance test, and (2) number of dominance tests.

4.1   Efficient Spatial Dominance Test

Sharifzadeh and Shahabi [6] showed that we can determine spatial dominance by using just the convex hull of Q instead of all query points in Q: If $p \in P$ is not dominated by any other point in P with respect to the vertices of $\mathcal{CH}(Q)$, then p is a spatial skyline point. In fact, we can interpret this property in a geometric setting as follows.

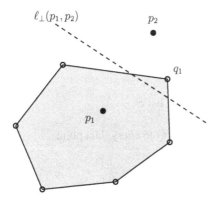

Fig. 4. $\mathcal{CH}(Q)$ intersect the bisector of two data points

Lemma 4. *The bisector of two data points intersects the interior of $\mathcal{CH}(Q)$ if and only if they do not spatially dominate each other.*

Proof. If the bisector of two data points intersects the interior of $\mathcal{CH}(Q)$, then for each of the data points, there exists a vertex of $\mathcal{CH}(Q)$ closer to it than the other. For example, in Fig. 4, the bisector of p_1 and p_2 intersects $\mathcal{CH}(Q)$, so at least one query point is closer to one of each data point than the other. Therefore they do not dominate each other. If the bisector does not intersect the interior of $\mathcal{CH}(Q)$, all the vertices of $\mathcal{CH}(Q)$ (therefore all the query points) are closer to one data point than the other. It means one data point spatially dominates the other point.   ⊟

As we can determine whether a line intersects the convex hull or not in $O(\log |\mathcal{CH}(Q)|)$ time by using a binary search technique, the dominance test can be done in the same time.

Lemma 5. *When $\mathcal{CH}(Q)$ is given, the dominance test for a pair of data points can be done in $O(\log |\mathcal{CH}(Q)|)$ time.*

4.2 Bounding the Number of Dominance Test

To make the algorithm faster, we reduce the number of dominance tests. For some vertex q of $\mathcal{CH}(Q)$, we keep the sorted list $\mathcal{A}$ of all the data points in the ascending order of distance from q. With this list, we can determine that, if a data point p_1 is located before p_2 in $\mathcal{A}$, then p_2 does not spatially dominate p_1 using Lemma 1. Therefore, together with Lemma 3, it is sufficient to perform the dominance test on p only with the spatial skyline points that are located before p in $\mathcal{A}$, as we formally state below.

Lemma 6. *For a data point p, if we have the set of all the spatial skyline points located before p in $\mathcal{A}$, we can determine whether p is a spatial skyline or not by $O(|S|)$ dominance tests.*

If there are two data points with the same distance from q, we can break the tie by computing the distances from another vertex of $\mathcal{CH}(Q)$. Since no two points have the same distance from three vertices of $\mathcal{CH}(Q)$, we only need to do this at most three times. Our algorithm for retrieving all the spatial skyline points is given below:

Algorithm. *SpatialSkyline*
Input: P, Q
Output: S
1.   initialize the array $\mathcal{A}$ and the list S
2.   compute the $\mathcal{CH}(Q)$
3. $\mathcal{A}\leftarrow$the distances from $q_1 \in Q$ to every data point
4.   sort $\mathcal{A}$ in ascending order
5.   **for** $i \leftarrow 0$ **to** $|P| - 1$
6.   **do if** $\mathcal{A}[i]$ is not spatially dominated by S
7.   **then** insert $\mathcal{A}[i]$ to S
8.   **return** S

In Line 2, the convex hull can be constructed in $O(|Q| \log |Q|)$ time [17]. Line 4 takes $O(|P|)$ time and sorting in Line 5 can be done in $O(|P| \log |P|)$ time. In Line 8, we perform the dominance test $O(|S|)$ times, each of which takes $O(\log |\mathcal{CH}(Q)|)$ time. As the **for** loop in Lines 6-9 repeats $|P|$ times, the entire loop takes $O(|P||S|| \log |\mathcal{CH}(Q)|)$ time. Since $|Q| < |P|$ in most realistic skyline models, the total time complexity is $O(|P|(|S| \log |\mathcal{CH}(Q)| + \log |P|))$.

4.3   Bypassing Dominance Tests Using the Voronoi Diagram

In this section, we discuss how we can further reduce dominance tests by identifying a subset of skyline results, which we call *seed skyline points*, that can be identified as skyline points with no dominant test. That is, before we perform the algorithm *SpatialSkyline*, we can quickly retrieve the seed skyline points to improve the performance of the algorithm dramatically, by bypassing dominance tests on these skyline points.

To achieve this goal, we first discuss a relationship of the Voronoi diagram $\mathrm{Vor}(P)$ of a dataset P and $\mathcal{CH}(Q)$. Theorem 1 describes this relationship between $\mathrm{Vor}(P)$ and $\mathcal{CH}(Q)$.

Theorem 1 (Seed Skyline). *For given a set P of data points and a set Q of query points, if the Voronoi cell $\mathcal{V}(p)$ of $p \in P$ intersects with the boundary of $\mathcal{CH}(Q)$ or $\mathcal{CH}(Q)$ contains $\mathcal{V}(p)$, then p is a skyline point [6].*

Proof. See the proofs of Theorem 1 and 3 in [6].   ▫

We now present an efficient algorithm to identify the seed skyline points, as the starting point to perform the algorithm *SpatialSkyline* to identify the rest of the skyline points. To retrieve seed skyline points efficiently, we first find a Voronoi cell that contains a vertex of $\mathcal{CH}(Q)$ by using typical point location query [17] on $\mathrm{Vor}(P)$. From this Voronoi cell, we follow the edges of $\mathcal{CH}(Q)$ and find the Voronoi cells that intersect the edges. Then we find Voronoi cells that lie inside $\mathcal{CH}(Q)$ by traversing the Delaunay graph [17]. Our enhanced algorithm works as follow. Let $e_i = (q_i, q_{i+1})$ denote the i-th edge along the boundary of $\mathcal{CH}(Q)$.

Algorithm. *SeedSkyline*
Input: P, Q
Output: S_{seed}
1.   initialize S_{seed}
2.   compute $\mathcal{CH}(Q)$ and $\text{Vor}(P)$
3.   find a Voronoi cell $\mathcal{V}(p)$ containing q_0
4.   **for** $i \leftarrow 0$ **to** $|\mathcal{CH}(Q)| - 1$
5.   find all the Voronoi cells $\mathcal{V}(p)$ intersecting e_i and insert p to S_{seed}
6.   find all the Voronoi cells $\mathcal{V}(p)$ lying in $\mathcal{CH}(Q)$ by traversing Delaunay graph and insert p to S_{seed}
7.   **return** S_{seed}

Note that, we can compute $\mathcal{CH}(Q)$ and $\text{Vor}(P)$ in $O(|Q| \log |Q|)$ time and in $O(|P| \log |P|)$ time (Line 2), respectively, and locate the Voronoi cell $\mathcal{V}(p)$ containing the query point q_0 in $O(\log |P|)$ time by point location query on $\text{Vor}(P)$ (Line 3).

To find all the Voronoi cells intersecting an edge $e_0 = (q_0, q_1)$ in Line 5, we follow the procedure below (also illustrated in Fig. 5). We first compute the intersection r of e_0 with the boundary of $\mathcal{V}(p)$, which can be done in time $O(\log |P|)$ using binary search because $\mathcal{V}(p)$ is a convex polygon and since we store its edges sorted along the boundary, as we will discuss later in Section 5.1. Because r lies on a boundary edge shared by two neighboring Voronoi cells, we can get the pointer to the neighboring Voronoi cell $\mathcal{V}(p')$ in constant time from the Delaunay graph. We repeat this until we reach the other endpoint q_1. Then we proceed to the next convex hull edge $e_1 = (q_1, q_2)$ and repeat the above process until we find all the Voronoi cells intersecting the boundary of $\mathcal{CH}(Q)$.

Note that a Voronoi cell may contain an edge of $\mathcal{CH}(Q)$ in its interior or intersect several edges of $\mathcal{CH}(Q)$ – the number of the intersection tests is thus bounded by the larger of $O(|S|)$ and $O(|\mathcal{CH}(Q)|)$, i.e., at most $O(|S| + |\mathcal{CH}(Q)|)$. Combining the number and cost of intersection tests, the overall worst-case time complexity becomes $O((|S| + |\mathcal{CH}(Q)|) \log |P|)$. Traversing Delaunay graph can be done in $O(|S|)$ time (Line 6). Therefore the total time complexity of *SeedSkyline* is $O((|S| + |\mathcal{CH}(Q)|) \log |P|)$ if $\mathcal{CH}(Q)$ and $\text{Vor}(P)$ are given.

By combining the algorithms *SpatialSkyline* and *SeedSkyline*, we can retrieve all spatial skyline points more efficiently than by *SpatialSkyline* alone. Instead of testing

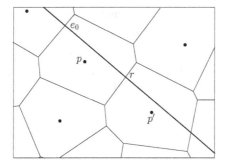

Fig. 5. Two Voronoi cells share the intersection

dominance for all data points we can find seed skyline points using *SeedSkyline*, and then find the other skyline points using *SpatialSkyline*. We present the combined algorithm *EnhancedSpatialSkyline* from this idea as follows:

Algorithm. *EnhancedSpatialSkyline*
Input: P, Q
Output: S
1. initialize the array $\mathcal{A}$ and the list S
2. compute the $\mathcal{CH}(Q)$
3. $S \leftarrow SeedSkyline(P, Q)$
4. $\mathcal{A} \leftarrow$ the distances from $q_1 \in Q$ to every data point
5. sort $\mathcal{A}$ in ascending order
6. **for** $i \leftarrow 0$ **to** $|P| - 1$
7.   **do if** $\mathcal{A}[i]$ is not in S
8.   **then if** $\mathcal{A}[i]$ is not spatially dominated by S
9.   **then** insert $\mathcal{A}[i]$ to S
10. **return** S

The asymptotic time complexity of *EnhancedSpatialSkyline* is the same as that of *SpatialSkyline*. In practice, however, by bypassing the dominance tests for seed skyline points, it shows better performance than *SpatialSkyline*.

5   Implementation

In this section, we discuss the details of the implementation of the algorithms, including how to compute and store the Voronoi diagram (Section 5.1) and the query convex hull (Section 5.2) to optimize the implementation of our proposed algorithm.

5.1   Voronoi Diagrams

First, we discuss how we construct the Voronoi diagram and the Delaunay graph of the data points. As both are extensively studied structures, many algorithms and codes are available, including 'Qhull [18]' which we adopt for our implementation.

However, it is challenging to store the resulting diagram and graph in such a way that the spatial skyline query computation can be optimized. Toward the goal, we store the Voronoi cells and Delaunay graph edges as follows:

–   **cells:** As each Voronoi cell is a convex region, we take advantage of this convexity and store the vertices of each cell in increasing angular order from one point, which preserves the adjacency of vertex pairs in the cell.
–   **edges:** Every edge of a Voronoi cell is shared by a neighboring Voronoi cell. To represent the Delaunay graph, for each edge $\overline{v_i v_{i+1}}$, from a vertex v_i of a Voronoi cell, we need to store the pointer to the neighboring cell sharing the edge.

Using this structure, we can exploit the convexity of a Voronoi region and the Delaunay graph discussed above, by reading only one Voronoi cell block from the file. To find a specific Voronoi cell block, we maintain a file pointer for each Voronoi cell block.

5.2 Convex Hull

To compute the convex hull $\mathcal{CH}(Q)$, we use the *Graham's scan algorithm* [17]. By using a binary search technique, the dominance test can be done in $O(\log |\mathcal{CH}(Q)|)$ time, as discussed in Lemma 5. We implement the test as follows.

Remember that we denote the bisector of two data points, p_1 and p_2, by $\ell_\perp(p_1, p_2)$. As discussed in Section 4.1, we can determine the dominance of two data points by testing whether $\ell_\perp(p_1, p_2)$ intersects $\mathcal{CH}(Q)$ or not. If $\ell_\perp(p_1, p_2)$ intersects $\mathcal{CH}(Q)$, at least one vertex of the upper chain of $\mathcal{CH}(Q)$ lies above $\ell_\perp(p_1, p_2)$, and at least one vertex of the lower chain of $\mathcal{CH}(Q)$ lies below $\ell_\perp(p_1, p_2)$ (Fig. 4). Let e_i and e_{i+1} be two edges of the upper chain sharing a vertex q_i such that $\ell_\perp(p_1, p_2)$ has a slope in between the maximum and the minimum of the slopes of e_i and e_{i+1}. If $\ell_\perp(p_1, p_2)$ intersects $\mathcal{CH}(Q)$, then q_i lies strictly above $\ell_\perp(p_1, p_2)$ by convexity of $\mathcal{CH}(Q)$. We can use a similar argument for the lower chain of $\mathcal{CH}(Q)$. Because the upper and the lower chain of $\mathcal{CH}(Q)$ is sorted in the increasing order of the slopes of edges, we can find these two vertices by using binary search on the slopes of edges. After finding these two vertices in $O(\log |\mathcal{CH}(Q)|)$, we can determine the dominance in constant time. When $\mathcal{CH}(Q)$ is small, a linear search may outperform binary search, and we use a linear search in this case.

5.3 VS^2

As a baseline to compare with our proposed algorithm, we use VS^2 proposed in [6]. As the authors could not provide the code, we implement the algorithm using the same implementation of R*-tree [19] and the Voronoi diagram we used to implement our proposed algorithm, to ensure the fairness in empirical comparison.

For constructing the convex hull, we share the same implementation used for our proposed algorithm, except that, to accommodate the dominance test of complexity $O(|\mathcal{CH}(Q)|)$ discussed in [6], we use linear scan.

In our implementation, R*-tree is used to find the closest point to one query point. The leaves of a R*-tree index contain Voronoi cells which are packed by MBRs for each, such that we can easily obtain candidate Voronoi cells containing a query point.

However, as shown in Section 3.4, VS^2 may fail to find all the spatial skyline points in some cases. Our implementation of VS^2 is revamped to eliminate these cases. Specifically, we remove one condition. For some $p \in P$, if all its Voronoi neighbors and all their Voronoi neighbors are spatially dominated by other points, then the original VS^2 does not test p, but we implement VS^2 to test this point for finding all skyline points.

5.4 Enhanced Spatial Skyline (*ES*)

Our enhanced algorithm *ES* works as follows. We compute the Voronoi diagram and the Delaunay graph of the data points, and store them in the form of the file mentioned in Section 5.1. To find the point closest to one query point, R*-tree is used. Then *ES* computes the Voronoi cells intersecting the boundary of the query convex hull and finds all the Voronoi cells lying in the convex hull by traversing the Delaunay graph. As we only need to see each Voronoi cell at most once during traversing the Delaunay graph of the data points, we read it from the file when it is required and deallocate it from memory after passing it by.

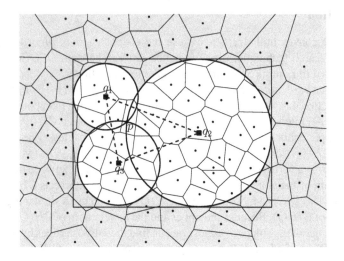

Fig. 6. p dominates all the points in the shaded region, with respect to the query $Q = \{q_1, q_2, q_3\}$

In this process, we restrict the region to search for the rest of the skyline points to the bounding box containing $|Q|$ circles for $|Q|$ query points (Fig. 6). More precisely, we set the bounding box as the intersection of all bounding boxes defined by the skyline subset found so far. After that, we get a list of the candidates in this bounding box by using R*-tree. We sort the list in ascending order of the candidates' distances to a query point and process them one by one in this order. When we find a new skyline point, we reduce the size of the bounding box by taking the intersection of the current bounding box with the bounding box of this new skyline point. During the process, if some candidate point is not contained in the bounding box, then we can simply skip the dominance test.

6   Experimental Evaluation

In this section, we report our experimental settings (Section 6.1) and evaluation results to validate the efficiency of our framework (Section 6.2). We compared our algorithm for spatial skylining with VS^2 in several aspects. As datasets, we used both synthetic datasets and a real dataset of points of interest (POI) in California.[1]

6.1   Experimental Settings

Synthetic dataset. A synthetic dataset contains up to one million uniformly distributed random locations in a 2D space. The space of datasets is limited to a unit space, i.e., the upper and lower bound of all points are 0 and 1 for each dimension respectively. More precisely, we used five synthetic datasets with 50K, 100K, 200K, 500K, and 1M uniformly distributed points.

---

[1] Available at http://www.cs.fsu.edu/~lifeifei/SpatialDataset.htm

Using synthetic datasets, we investigated the effect of the number of points in a query $|Q|$, distribution of the points in a query σ, and cardinalities of the datasets $|P|$. Parameters used in the experiments are summarized in Table 1.

Table 1. Parameters used for synthetic datasets

Parameter	Setting	Default
Dimensionality	2	
Distribution of data points	Independent	
Dataset cardinality	50K, 100K, 200K, 500K, 1M	500K
The number of points in a query	5, 10, 15, 20, 40	15
Standard deviation of points in a query	0.01, 0.02, 0.04, 0.06, 0.08	0.06

Queries were generated through the following steps: (1) we randomly generate a *center point*, then (2) generate the query points, normally distributed around the center. In particular, for each dimension, we generate points that are normally distributed, with mean as the center point and deviation as user-specified parameter σ, which varies between 0.01, 0.02, 0.04, 0.06, and 0.08 as listed in Table 1. We generated one hundred queries (each consisting of up to 40 query points) for each setting and measured average response times of all algorithms.

POI dataset. We also validate our proposed framework using a real-life dataset. In particular, we use a POI dataset, which consists of 104,770 locations of 63 different categories in California. Fig. 7 shows the characteristics of this POI dataset.

For this POI dataset, we investigated the effect of $|Q|$ and σ. We similarly generated the queries, by randomly picking one data point as a center point and generating query points to be normally distributed around the center point, in the same way we generated synthetic points. The reason why we pick the center point among data points, instead of generating a random point, is to avoid generating queries to regions with no data points (such as blank regions in Fig. 7). We generate one hundred queries for each setting,

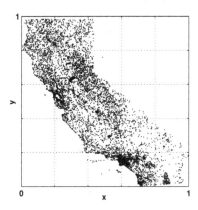

Fig. 7. 10,000 sampled points from the California's POI dataset

varying the number of query points in the range from 5 to 40 and the standard deviation from 0.01 to 0.08, just as in our synthetic data generation.

We carry out our experiments on a Pentium IV PC running on Linux with Pentium IV 3.2GHz CPU and 1GB memory, and all the algorithms were coded in C++.

6.2 Efficiency

We validate the efficiency of our framework, over varying $|P|$, $|Q|$, and σ.

Fig. 8 shows the effect of the dataset cardinality to response time (Fig. 8a), I/O cost measured as the number of accessing (reading) Voronoi cells and R*-tree nodes (Fig. 8b), and the number of dominance tests (Fig. 8c). From Fig. 8a, our proposed algorithm ES outperforms VS^2 by an order of magnitude. Similarly in Fig. 8c, ES performs a remarkably smaller number of dominance tests than VS^2, by bypassing the dominance tests for the skyline points whose Voronoi cells intersect the boundary of $\mathcal{CH}(Q)$. Such saving is more significant between skyline points, as the number of dominance tests for skyline points is significantly higher.

Fig. 8b shows the I/O costs of the two algorithms, the two algorithms perform the same number of I/Os on the index of Voronoi cells, because each algorithm only uses the index to find a Voronoi cell containing a query point. To find non-seed skyline points, ES uses the index of data points, which incurs less I/Os (random accesses) than VS^2. ES, though the size of each I/O (R*-tree node) is larger than that of VS^2 (a Voronoi cell), outperforms VS^2 by reducing the "number" of I/Os, each of which incurs a random access, the cost of which dominates the overall access cost, in our scenario of performing many random accesses of smaller size.

Fig. 9 shows the effect of $|Q|$ to response time, I/O cost, and the number of dominance tests. We observe similar trends as in Fig. 8, except that the response time and I/Os scale more gracefully over increasing $|Q|$. This can be explained by the fact that all two algorithms use $\mathcal{CH}(Q)$, instead of using Q itself, the size of which grows much slower than that of Q. For instance, even when $|Q|$ is doubled, the size of convex hull may not change much, if the deviation σ stays the same.

Fig. 10 shows the effect of σ. Similarly to prior results, ES significantly outperforms VS^2 in terms of response time, dominance tests, and I/Os while VS^2 outperforms our algorithm when query points are crowded in a very small area. This phenomenon can be explained as ES performs more I/Os than VS^2 when the size of $\mathcal{CH}(Q)$ is very small (Fig. 10b). However, ES starts to outperform VS^2 as the size of $\mathcal{CH}(Q)$ grows.

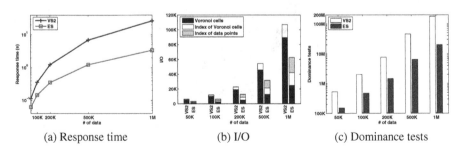

(a) Response time　　　　　　(b) I/O　　　　　　(c) Dominance tests

Fig. 8. Effect of the dataset cardinality for synthetic datasets

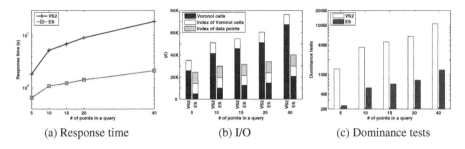

(a) Response time (b) I/O (c) Dominance tests

Fig. 9. Effect of the number of query points for synthetic datasets

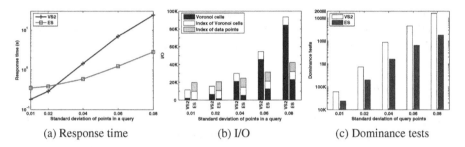

(a) Response time (b) I/O (c) Dominance tests

Fig. 10. Effect of σ of a query for synthetic datasets

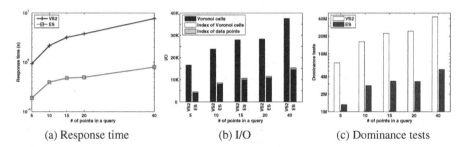

(a) Response time (b) I/O (c) Dominance tests

Fig. 11. Effect of the number of query points for the POI dataset

The other slight difference to note is that the response time of the algorithms increase relatively faster as σ increases, as the size of $\mathcal{CH}(Q)$ may increase quadratically as σ increases. For example, when σ changes from 0.04 to 0.08 (two-fold), the circle area containing the points within the 95% confidence interval increases four-fold (i.e., quadratic), and also the area of $\mathcal{CH}(Q)$ may increase quadratically. As such points are guaranteed to be skyline points, this observation suggests why the number of skyline points increases quadratically as σ increases.

We perform the same sets of experiments on the POI dataset, varying the size of query and σ, reported in Fig. 11 and 12, respectively. Our observations of these evaluations are roughly consistent with the corresponding evaluation for synthetic datasets. However, in these experiments, I/Os on Voronoi cells are dominant parts of the I/O

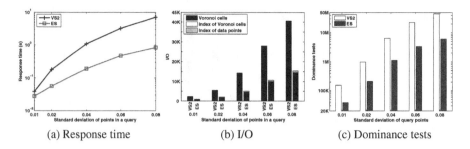

(a) Response time   (b) I/O   (c) Dominance tests

Fig. 12. Effect of σ of a query for the POI dataset

cost. The reason is that, as the cardinality of the dataset is relatively smaller, the depth of the R*-tree is also small, thus incurring less index I/Os. A similar phenomenon can be observed in Fig. 8b, when the dataset cardinality is small (50K).

7 Conclusion

We have studied spatial skyline query processing and presented an efficient and correct exact algorithm. We showed that our algorithm can identify the correct result in $O(|P|(|S|\log|\mathcal{CH}(Q)| + \log|P|))$ time, while the best known algorithm may fail to compute the correct result. Lastly, we empirically validated our proposed algorithm.

So far we have assumed that the points lie in 2-dimensional space, and shown how to efficiently retrieve spatial skyline points using some geometric structures such as the convex hull and the Voronoi diagram of points in the plane. We now turn our attention to higher dimensional skyline queries. All the definitions, lemmas, and algorithms described in this paper generalize to higher dimensions: For the set of n points in d-dimensional space, the Voronoi diagram of them has $\Theta(n^{\lceil d/2 \rceil})$ combinatorial complexity [20] and can be computed in $O(n \log n + n^{\lceil d/2 \rceil})$ time [21,22,23]. The convex hull of those points has $\Theta(n^{\lfloor d/2 \rfloor})$ combinatorial complexity (by the so-called *Upper Bound Theorem*) and can be computed in $\Theta(n^{\lfloor d/2 \rfloor})$ expected time [17]. Dominance test, the intersection query of a line with a convex polygon used in Section 4.1, can be generalized for higher dimensions, as an intersection query of a hyperplane with a convex polyhedron in higher dimensions. Similarly, the intersection of an edge with the Voronoi diagram can also be generalized as the intersection of a $(d-1)$-face with the Voronoi diagram in d-dimensional space.

For future work, we will study how our algorithms can be extended to support queries over urban road networks with additional constraints.

Acknowledgement

The second and last authors were supported by Engineering Research Center of Excellence Program of Korea Ministry of Education, Science and Technology (MEST) / Korea Science and Engineering Foundation (KOSEF), grant number R11-2008-007-03003-0.

References

1. Kung, H.T., Luccio, F., Preparata, F.: On finding the maxima of a set of vectors. Journal of the Association for Computing Machinery 22(4), 469–476 (1975)
2. Börzsönyi, S., Kossmann, D., Stocker, K.: The skyline operator. In: ICDE 2001: Proc. of the 17th International Conference on Data Engineering, p. 421 (2001)
3. Tan, K., Eng, P., Ooi, B.C.: Efficient progressive skyline computation. In: VLDB 2001: Proc. of the 27th International Conference on Very Large Data Bases, pp. 301–310 (2001)
4. Papadias, D., Tao, Y., Fu, G., Seeger, B.: An optimal and progressive algorithm for skyline queries. In: SIGMOD 2003: Proc. of the 2003 ACM SIGMOD International Conference on Management of Data, pp. 467–478 (2003)
5. Chomicki, J., Godfery, P., Gryz, J., Liang, D.: Skyline with presorting. In: ICDE 2007: Proc. of the 23rd International Conference on Data Engineering (2007)
6. Sharifzadeh, M., Shahabi, C.: The spatial skyline queries. In: VLDB 2006: Proc. of the 32nd International Conference on Very Large Data Bases, pp. 751–762 (2006)
7. Kossmann, D., Ramsak, F., Rost, S.: Shooting stars in the sky: An online algorithm for skyline queries. In: VLDB 2002: Proc. of the 28th International Conference on Very Large Data Bases, pp. 275–286 (2002)
8. Godfrey, P., Shipley, R., Gryz, J.: Maximal vector computation in large data sets. In: VLDB 2005: Proc. of the 31st International Conference on Very Large Data Bases, pp. 229–240 (2005)
9. Chan, C.Y., Jagadish, H., Tan, K., Tung, A.K., Zhang, Z.: On high dimensional skylines. In: Ioannidis, Y., Scholl, M.H., Schmidt, J.W., Matthes, F., Hatzopoulos, M., Böhm, K., Kemper, A., Grust, T., Böhm, C. (eds.) EDBT 2006. LNCS, vol. 3896, pp. 478–495. Springer, Heidelberg (2006)
10. Chan, C.Y., Jagadish, H., Tan, K.L., Tung, A.K., Zhang, Z.: Finding k-dominant skylines in high dimensional space. In: SIGMOD 2006: Proc. of the 2006 ACM SIGMOD International Conference on Management of Data (2006)
11. Lin, X., Yuan, Y., Zhang, Q., Zhang, Y.: Selecting stars: The k most representative skyline operator. In: ICDE 2007: Proc. of the 23rd International Conference on Data Engineering, pp. 86–95 (2007)
12. Roussopoulos, N., Kelley, S., Vincent, F.: Nearest neighbor queries. In: SIGMOD 1995: Proc. of the 1995 ACM SIGMOD international conference on Management of data, pp. 71–79 (1995)
13. Berchtold, S., Böhm, C., Keim, D.A., Kriegel, H.P.: A cost model for nearest neighbor search in high-dimensional data space. In: PODS 1997: Proc. of the 16th ACM SIGACT-SIGMOD-SIGART symposium on Principles of database systems, pp. 78–86 (1997)
14. Beyer, K.S., Goldstein, J., Ramakrishnan, R., Shaft, U.: When is "nearest neighbor" meaningful? In: Beeri, C., Bruneman, P. (eds.) ICDT 1999. LNCS, vol. 1540, pp. 217–235. Springer, Heidelberg (1998)
15. Papadias, D., Tao, Y., Mouratidis, K., Hui, C.K.: Aggregate nearest neighbor queries in spatial databases. ACM Transactions on Database Systems 30(2), 529–576 (2005)
16. Huang, X., Jensen, C.S.: In-route skyline querying for location-based services. In: Kwon, Y.-J., Bouju, A., Claramunt, C. (eds.) W2GIS 2004. LNCS, vol. 3428, pp. 120–135. Springer, Heidelberg (2005)
17. de Berg, M., Cheong, O., van Kreveld, M., Overmars, M.: Computational Geometry: Algorithms and Applications, 3rd edn. Springer, Heidelberg (2008)
18. Qhull code for convex hull, delaunay triangulation, voronoi diagram, and halfspace intersection about a point. World Wide Web electronic publication (May 1995),
http://www.qhull.org/

19. Beckmann, N., Kriegel, H.P., Schneider, R., Seeger, B.: The R*-tree: An efficient and robust access method for points and rectangles. In: SIGMOD 1990: Proc. of the 1990 ACM SIGMOD international conference on Management of data, pp. 322–331 (1990)
20. Klee, V.: On the complexity of d-dimensional Voronoi diagrams. Archiv der Mathematik 34, 75–80 (1980)
21. Chazelle, B.: An optimal convex hull algorithm and new results on cuttings. In: Proc. 32nd Annu. IEEE Sympos. Found. Comput. Sci., pp. 29–38 (1991)
22. Clarkson, K.L., Shor, P.W.: Applications of random sampling in computational geometry. II. Discrete Comput. Geom. 4, 387–421 (1989)
23. Seidel, R.: Small-dimensional linear programming and convex hulls made easy. Discrete Comput. Geom. 6, 423–434 (1991)

Incremental Reverse Nearest Neighbor Ranking in Vector Spaces

Tobias Emrich, Hans-Peter Kriegel, Peer Kröger, Matthias Renz, and Andreas Züfle

Institute for Informatics, Ludwig-Maximilians-Universität München
Oettingenstr. 67, 80538 München, Germany
{emrich,kriegel,kroegerp,renz,zuefle}@dbs.ifi.lmu.de
http://www.dbs.ifi.lmu.de

Abstract. In this paper, we formalize the novel concept of incremental reverse nearest neighbor ranking and suggest an original solution for this problem. We propose an efficient approach for reporting the results incrementally without the need to restart the search from scratch. Our approach can be applied to a multidimensional feature database which is hierarchically organized by any R-tree like index structure. Our solution does not assume any preprocessing steps which makes it applicable for dynamic environments where updates of the database frequently occur. Experiments show that our approach reports the ranking results with much less page accesses than existing approaches designed for traditional reverse nearest neighbor search applied to the ranking problem.

1 Introduction

While the reverse nearest neighbor (RNN) search problem, i.e. finding all objects in a database that have a given query q among their corresponding k-nearest neighbors, has been studied extensively in the past years, considerably less work has been done so far to support an RNN ranking of objects of a database. An RNN ranking sorts the objects o of the database according to the number of other objects in the database that are more similar to o than q. Thus, if an object o has a ranking score of i w.r.t. a query q, object o would also be a reverse k-nearest neighbor of q for all $k \geq i$ but not a reverse k-nearest neighbor of q for all $k < i$.

Initially, the RNN ranking query reports those objects having the smallest ranking scores in a non-deterministic way since several objects may have the same minimal ranking score. Thereby, the results are reported on demand whenever a function called getNext() is invoked. In other words, each consecutive call of getNext() reports one object with minimal ranking score until all objects have been reported.

The major challenge for algorithms that support rankings in general and RNN rankings in particular is that the result of each getNext()-call should be computed incrementally rather than from scratch, i.e. the current state after each getNext()-call needs to be stored and serves as a starting point to compute the results of the next call. The advantage of an incremental ranking method in general is that no parameter k has to be specified for the query in advance and the first (most relevant) results are reported immediately without the overhead of simultaneously computing less relevant results. In

N. Mamoulis et al. (Eds.): SSTD 2009, LNCS 5644, pp. 265–282, 2009.
© Springer-Verlag Berlin Heidelberg 2009

addition, if the initial results are not sufficient due to any application specific reasons, further results can be requested on demand by calling the getNext() function.

The reminder of this paper is organized as follows. In Section 2 we formally define the RNN ranking problem we want to solve here and discuss related work. Section 3 explores our novel solution to this problem. In Section 4 we present an experimental evaluation. Last but not least, Section 5 concludes the paper.

2 Survey

2.1 Problem Formalization

In the following, we assume that $\mathcal{D}$ is a database of n feature vectors and $dist$ is the Euclidean distance[1] on the points in $\mathcal{D}$. In addition, we assume that the points are indexed by any traditional aggregate point access method like the aR-Tree family [1,2]. The set of *k-nearest neighbors* of a point q is the smallest set $NN_k(q) \subseteq \mathcal{D}$ that contains at least k points such that $\forall o \in NN_k(q), \forall \hat{o} \in \mathcal{D} - NN_k(q) : dist(q,o) < dist(q,\hat{o})$. The point $p \in NN_k(q)$ with the highest distance to q is called the *k-nearest neighbor* (*k*NN) of q. The distance $dist(q,p)$ is called kNN distance of q.

The set of *reverse k-nearest neighbors* (R*k*NN) of a point q is then defined as

$$RNN_k(q) = \{p \in \mathcal{D} \mid q \in NN_k(p)\}.$$

Here, we will be interested in computing a *ranking* of reverse nearest neighbors (RNNs) w.r.t. a query object q rather than in computing the R*k*NN of q for a fixed value of k. Let the function $R : \mathcal{D} \to \mathbb{N}$ return for an object $o \in \mathcal{D}$ the number of objects which are closer to o than the query q, i.e. formally,

$$R(o) = |\{p \in \mathcal{D} : dist(p,o) < dist(q,o)\}|.$$

Obviously, it holds that $o \in RNN_k(q)$ iff $R(o) \leq k$.

The problem of a *reverse nearest neighbor ranking* is to return incrementally all objects $o \in \mathcal{D}$ in increasing order of the values of $R(o)$ by calling the method getNext(). In case of ties, getNext() may report any qualifying object, i.e. we will allow for nondeterminism. Let us note that the i-th call of getNext() not necessarily returns an object that is an R*i*NN of q, because for a fixed value of k the set $RNN_k(q) - RNN_{k-1}(q)$ generally may contain an (even empty) set of points. In other words, the i-th call of getNext() may report an object o with $R(o) \neq i$. As a consequence, as additional information, the result of each of the ranking steps should include not only the actual object o but also its *ranking count (ranking score)* $R(o)$.

2.2 Related Work

The problem of supporting reverse k-nearest neighbor (R*k*NN) queries efficiently, i.e. computing for a given query q and a number k the R*k*NNs of q, has been studied extensively in the past years. Existing approaches for Euclidean R*k*NN search can be

---

[1] The concepts described here can also be extended to any L_p-norm.

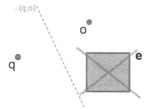

Fig. 1. TPL pruning ($k = 1$)

classified as self-pruning approaches or mutual-pruning approaches. *Self-pruning approaches* like the RNN-Tree [3] and the RdNN-Tree [4] are usually designed on top of a hierarchically organized tree-like index structure. They try to estimate the kNN distance of each index entry e. If the kNN distance of e is smaller than the distance of e to the query q, then e can be pruned. Thereby, self-pruning approaches do not usually consider other entries (database points or index nodes) in order to estimate the kNN distance of an entry e, but simply precompute kNN distances of database points and propagate these distances to higher level index nodes. Since the kNN distances need to be materialized, both approaches are limited to a fixed value of k and cannot be generalized to answer RkNN-queries with arbitrary values of k. In addition, approaches based on precomputed distances can generally not be used when the database is updated frequently. *Mutual-pruning approaches* such as [5,6,7] use other points to prune a given index entry e. The most general and efficient approach called TPL is presented in [7]. It uses any hierarchical tree-based index structure such as an R-Tree to compute a nearest neighbor ranking of the query point q. The key idea is to iteratively construct Voronoi hyper-planes around q w.r.t. to the points from the ranking. Points and index entries that are beyond k Voronoi hyper-planes w.r.t. q can be pruned and need not to be considered for Voronoi construction anymore. The idea of this pruning is illustrated in Figure 1 for $k = 1$. Entry e can be pruned, because it is beyond the Voronoi hyper-plane between q and candidate o, denoted by $\perp(q, o)$. For the general case, e can be pruned if e is beyond k hyper-planes w.r.t. all current candidates. If e cannot be pruned, it is refined, or, if e is already a database object, e is a new candidate and the hyperplane $\perp(q, e)$ will be considered for pruning in the following. If the ranking queue is empty, the remaining candidate points must be refined, i.e. for each of these candidates, a kNN query must be launched.

Recently, a method for ranked RkNN search has been proposed in [8]. In fact, the authors provide a method for computing the results of an RkNN query with fixed k that are ranked according to k, i.e. the RiNNs are ranked higher than the RjNNs if $i < j \leq k$. This problem is obviously different to the problem of computing an incremental RNN ranking which will be adressed here.

Beside solutions for Euclidean data, solutions for general metric spaces (e.g. [9,10,11]) usually implement a self-pruning approach. Typically, metric approaches are less efficient than the approaches tailored for Euclidean data because they cannot make use of the Euclidean geometry.

3   Incremental RNN Ranking

Our approach is based on an index structure $\mathcal{I}$ for point data which is based on the concept of minimal-bounding-rectangles, e.g. the R-tree family like [12,13,14]. In particular, we use multi-resolution aggregate versions of these indexes as described in [1,2] that e.g. aggregate for each index entry e the number of objects that are stored in the subtree with root e. The set of objects managed in the subtree of an index entry $e \in \mathcal{I}$ is denoted by *subtree*(e). Note that the entry e can be an intermediate node in $\mathcal{I}$ or a point, i.e. an object in $\mathcal{D}$. In the latter case, *subtree*$(e) = \{e\}$.

The general idea of our solution is based on the TPL-like [7] pruning of entries that are beyond a given number of Voronoi hyperplanes. However, instead of pruning an index entry e, we need to estimate the ranking count value $R(o)$ for all points $o \in subtree(e)$. The key observation is that if an index entry e is beyond a Voronoi hyperplane w.r.t. q, then we know that for all $o \in subtree(e)$, the value of $R(o)$ can be increased by one. For example, in Figure 1, entry e is beyond the Voronoi hyperplane between q and x, denoted by $\perp(q, x)$. Thus, x will have a smaller distance to all objects $o \in subtree(e)$ than q, i.e. all objects $o \in subtree(e)$ will have a ranking count $R(o)$ of at least 1. Simply speaking, the ranking count $R(o)$ of any object $o \in \mathcal{D}$ equals the number of Voronoi hyperplanes (including $\perp(q, o)$) that divide the data space such that o and q are in different half spaces.

In the following, we will extend this idea in several important aspects:

- First, we will extend the concept of Voronoi hyperplanes presented in [7] to higher levels of the index. Originally, the TPL approach considers only Voronoi hyperplanes between the query q and another database object, i.e. at least one leaf entry of the index needs to be fully refined before any Voronoi hyperplane is constructed for pruning. Analogously, this would mean that we can only estimate the ranking count values of objects by means of other objects. This will obviously result in a large overhead of unnecessary page accesses. Rather, we will extend the idea of Voronoi-based pruning/ranking to intermediate entries of the index, i.e. we will also consider Voronoi hyperplanes between the query and intermediate index entries.
- Second, we will also integrate the idea of self-pruning in order to estimate the ranking count of objects within a given subtree.
- Third, we further improve the ranking count estimation by taking also partial hyperplane - entry coverings into account. Hyperplanes where an intermediate index entry e is partially beyond them w.r.t. q can also be used to estimate the ranking count of e.

The above estimation strategies give us better estimations of the ranking counts which will be important for the ranking algorithm. Last but not least, we will present a ranking algorithm based on the two previously mentioned ideas to estimate the ranking count that incrementally computes the next object of an RNN ranking on demand without recomputing the entire ranking from scratch.

3.1   Ranking Count Estimation

Now we explore strategies for estimating the ranking count based on the hyperplane concept. The basic idea of our approach is to apply the ranking count strategy mentioned

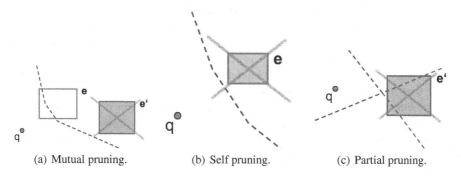

(a) Mutual pruning.   (b) Self pruning.   (c) Partial pruning.

Fig. 2. Ranking count estimation based on different pruning strategies

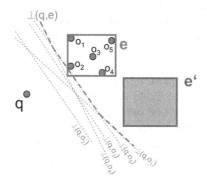

Fig. 3. Conservative approximation $\bot(q, e)$ of the hyperplanes associated with all objects of an index entry e

above during the traversal of the index, i.e. to identify candidates with high ranking counts as early as possible in order to reduce the I/O costs by saving unnecessary page accesses for the computation of the first results. The ability to push candidates to higher ranking positions already at the directory level of the index implies that a directory entry is used to push itself or other entries.

Estimation based on mutual pruning. First, we want to consider the case that a directory entry is used to push other entries back to higher ranking positions by increasing its ranking count. This is similar to the mutual-pruning idea used for RkNN query processing. Generally, the ranking count of an index entry $e \in \mathcal{I}$ can be increased by k according to another entry $e' \in \mathcal{I}$ if there are at least k objects in *subtree*(e') such that e is behind the Voronoi hyperplane between q and e', denoted by $\bot(q, e')$. In the following a hyperplane *associated with* an entry/object e is denoted by $\bot(q, e)$.

The key idea of the directory-level-wise ranking count estimation is to identify a hyperplane $\bot(q, e)$ which can be associated with an index entry e and which conservatively approximates the hyperplanes associated with all objects o_i in the subtree of e, i.e. $o_i \in$ *subtree*(e). Figure 3 illustrates the idea of this concept. We say that the hyperplane associated with an index entry e is *related to* the set of objects in the subtree of e. Since we assume that the number of objects stored in the subtree of an index entry e is known, if we exploit the indexing concept as proposed in [1], we also know for the hyperplane associated with that index entry e, $\perp(q, e)$, how many objects this hyperplane relates to. This means that if an entry/object e' is behind a hyperplane $\perp(q, e)$ associated with an index entry e, the entry e' is also behind all hyperplanes $\perp(q, o)$ associated with the objects $o \in subtree(e)$. We can use this information in order to increase the ranking count of entries according to e without accessing the child entries of e. Consequently, the ranking count of an entry/object e' which is behind a hyperplane $\perp(q, e)$ can be increased by $|subtree(e)|$. In Figure 3, the ranking count of entry e' can be increased by 5 because $subtree(e)$ contains five points, i.e. $|subtree(e)| = 5$.

Estimation based on self pruning. In addition, we can use these considerations also for increasing the ranking count of an intermediate index entry e by itself. This is similar to the self-pruning idea used for RkNN query processing. If an entry $e \in \mathcal{I}$ is behind its own hyperplane, then the ranking count of e can be increased by $|subtree(e)| - 1$, because each object $o \in subtree(e)$ would be behind the hyperplanes associated with all other objects in $subtree(e)$.

Estimation based on partial pruning. The ranking count estimation of an intermediate index entry e can also be based on hyperplanes that do not fully cover e. For example, if one part of e is beyond one hyperplane and the other part of e is beyond another hyperplane. In this case, each point in e is at least behind one hyperplane such that the complete entry can be safely moved to a higher ranking position. In general terms, assume that an entry $e \in \mathcal{I}$ is intersected, but not fully covered by n hyperplanes $\perp(q, e_0), \ldots, \perp(q, e_{n-1})$ associated with index entries $e_0 \ldots, e_{n-1}$. Now, the points in e are covered by different numbers of hyperplanes. The ranking count of e can be increased by the minimal number of hyperplanes a point of e is covered by.

Note that the partial pruning based estimation can become very expensive in higher-dimensional spaces. The reason is that the determination of the minimal coverage of an index entry w.r.t. all hyperplanes requires complex spatial segmentation operations in order to find the subregions having the same amount of hyperplanes they are behind. This can be very costly in higher-dimensional spaces. In this paper, we propose an efficient approach for the partial pruning based ranking count estimation for the two-dimensional space.

In the following, we present solutions for the ranking count estimation according to the above three strategies. First, we show in Section 3.2 how the ranking count estimations can be efficiently computed based on the mutual and self pruning strategies. Next, in Section 3.3, we propose an efficient 2D solution for the partial pruning based estimation.

3.2   Ranking Count Updates w.r.t. Intermediate Index Entry Hyperplanes

We first need to determine an entry $e' \in \mathcal{I}$ is completely or partially behind a hyperplane $\perp(q, e)$ associated with an entry $e \in \mathcal{I}$. An important observation is that a hyperplane associated with an object o represents all points p which have the same distance to the query point q and to o, formally:

$$\perp(q, o) = \{p \in \mathbb{R}^d : dist(p, q) = dist(p, o)\}.$$

In addition, we know that all objects stored in the subtree of an index entry e are located inside the minimum bounding hyper-rectangle $e.mbr$ that defines the page region of e. Thus, we can determine a conservative hyperplane representation of all points stored in the subtree of entry e if we replace the distances between the hyperplane points $p \in \perp(q, e)$ and $o \in subtree(e)$ by the maximum distance between p and the mbr-region of e. Consequently, the hyperplanes of all objects $o \in subtree(e)$ are conservatively approximated by a hyperplane representation consisting of all points in the vector space that fulfill the following condition:

$$\perp(q, e) = \{p \in \mathbb{R}^d : dist(p, q) = MaxDist(p, e.mbr)\}.$$

In general, a hyperplane representation H is called *conservative approximation* of a set of hyperplanes H', if all objects behind H are definitely behind each hyperplane $h' \in H'$, formally:

$$(o \text{ behind } H) \Rightarrow (\forall h' \in H' : o \text{ behind } h')$$

We can assign such a hyperplane representation to each intermediate entry of our index.

In consideration of the above equations, an index entry $e' \in \mathcal{I}$ is defined to be behind a hyperplane $\perp(q, e)$ if the following condition holds:

$$\forall p \in e.mbr : dist(p, q) > MaxDist(p, e.mbr).$$

Figure 3 illustrates the conservative approximation $\perp(q, e)$ of all hyperplanes $\perp(q, o)$ for all objects $o \in subtree(e)$.

In the following we briefly discuss how this conservative approximation $\perp(q, e)$ can be associated with an index entry e. An important observation is that a hyperplane associated with an object o represents all points p which have the same distance to the query point q and to o. In addition, we know that all objects stored in the subtree of an index entry e are located inside the minimum bounding hyper-rectangle (mbr) that defines the page region of e. Thus, we can determine a conservative hyperplane representation of all points stored in $subtree(e)$ if we replace the distances between the hyperplane points $p \in \perp(q, e)$ and $o \in subtree(e)$ by the maximum distance between p and the mbr-region

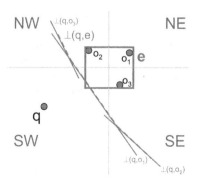

Fig. 4. Computation of conservative hyperplane approximations

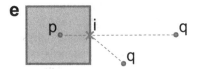

Fig. 5. Example to Lemma 2

of e. Figure 4 illustrates the computation of such a conservative approximation for a given index entry e in a 2D feature space. First, we have to specify the maximum distance between the mbr-region of the index entry e and any point in the vector space. It suffices to find for each point p in the vector space the point $o \in subtree(e)$ which is within the mbr-region of e having the maximum distance to p. This can be done by considering partitions of the vector space which are generated as follows: in each dimension the space is split paraxially at the center of the mbr-region. As illustrated for the 2D example in Figure 4, we obtain partitions denoted by NW, NE, SE and SW. In each of these partitions P, the vertex point of the mbr-region which lies within the diagonal-opposite partition is the mbr-region point which has the maximum distance to all points in P. In our example, for any point p in SW the maximum distance of p to e is the distance between p and point o_1 in partition NE. Consequently, the hyperplane $\perp(q, o_1)$ is a conservative approximation of all hyperplanes between points within the mbr-region of e and the points within the partition SW. In our example, the hyperplane associated with e is composed by the three hyperplanes $\perp(q, o_2)$, $\perp(q, o_1)$ and $\perp(q, o_3)$.

3.3   Efficient Spatial Partial Pruning

The basic idea of the partial pruning as introduced in 3.1 is to find for an intermediate index entry e the point in e with the lowest ranking count w.r.t. the hyperplanes that intersect e. An example of this situation is given in Figure 6 where e is intersected by three hyperplanes $\perp(q, e_1)$, $\perp(q, e_{n2})$ and $\perp(q, e_3)$ and e_1 corresponds to an intermediate index entry containing five points, whereas e_2 and e_3 correspond to data points. $\perp(q, e_1)$, $\perp(q, e_{n2})$ and $\perp(q, e_3)$ partition e into segments. For any segment s, the ranking count of all points $p \in s$ is equal. Thus, the minimum number of hyperplanes any point in e is covered by, is the minimal ranking count of all segments of e. Therefore, the minimal ranking count of e is 1 in our example. Thus, the ranking key of e is increased by one.

However, this computation requires to examine $O(2^m)$ segments, where m is the number of hyperplanes intersecting e. In [7], a similar approach is used, that requires to examine an exponential number of pruning intersection areas.

Next, we propose an efficient approach for partial pruning that is based on the following observations:

Lemma 1. *If an intermediate index entry e contains the query point q, then the ranking count of e is always 0.*

The above lemma is obvious, because if e contains q, then e may contain points that can be arbitrarily close to q, and thus, can have q as their nearest neighbour regardless of the distance of q to other points in the database.

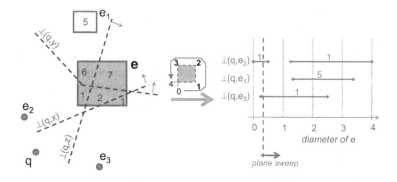

Fig. 6. Partial pruning based ranking count estimation strategy

Lemma 2. *If an intermediate index entry e does not contain the query point q, then the minimal ranking key of all points of e is equal to the minimal ranking key of all points on the edges (i.e. the boundary) of e.*

Proof. Assume an intermediate index entry e and a query point q outside of e. Let p be an arbitrary point inside of e and let i denote the intersection of segment $[p, q]$ with the boundary of e. Assume that the ranking count of i is k. Thus, by definition, at least k points $p_0, \ldots, p_{k-1}$ are closer to i than to q, i.e. $dist(i, q) > dist(i, p_i) \forall p_i \in \{p_0, \ldots, p_{k-1}\}$. For each $p_i \in \{p_0, \ldots, p_{k-1}\}$, the following inequation holds:

$$dist(p, q) = dist(p, i) + dist(i, q) > dist(p, i) + dist(i, p_i)$$

and the triangle inequality yields:

$$dist(p, i) + dist(i, p_i) > dist(p, p_i)$$

Therefore, any point that is closer to i than to q, is also closer to p than to q. Thus the ranking count of p is at least the ranking count of i. And since i is located on the boundary of e, its ranking count is at least the ranking count of the minimum of the ranking counts of all points on the boundary of e.

For two dimensional data, we can use the above lemma to efficiently compute the partial pruning count of an index entry e using a plane sweep algorithm if q is outside of the page region of e.

Since only the boundary of an intermediate index entry e is required to determine its partial ranking count, we linearize e to the interval $[0, 4]$. Points on the lower edge of e are represented by their relative position on that edge, points on the right edge of e are presented by the relative position on that edge plus one, and so on (c.f. Figure 6). For each hyperplane approximation $\perp (q, e_i)$ of an index entry e_i that intersects e, we determine all intersections of e_i with e^2 and the respective regions of

---

[2] Note that a hyperplane approximation can have at most four intersection with an mbr, due to its convex shape.

the boundary of e that is pruned by e_i. In Figure 6 the boundary of e is pruned by e_2 in the interval [2.3,3.4] and by e_1 in the interval [1.2,0.5], which is split into two intervals [1.2,4.0] and [0.0,0.5]. Now, the task is to find the minimum ranking count of all points on the boundary of e. The ranking count of a point p on the boundary can be obtained by summing up all the ranking counts of all hyperplanes for which the corresponding interval covers p. Now the minimum ranking count of all points on the boundary can be computed using a plane sweep technique as depicted in Figure 6. During the sweep the global minimum of the ranking count is computed.

3.4   Best-First Search Based Incremental RNN Ranking Algorithm

In this section, we show how we explore the index such that the first results can be reported early without causing unnecessary page accesses. We start with an informal description of our solution before we present implementation details and pseudo code.

Similar to the TPL approach for RkNN queries our approach is based on a best-first search method exploiting a priority queue organizing the index entries to be explored. In contrast to the TPL approach, we propose to give the priorities to the index entries according to the estimated ranking count, i.e. entries with low ranking counts are ranked higher than entries with high ranking count. This means that entries containing objects with a low expected ranking position are explored before entries containing objects with a high expected ranking position. The rational for this strategy is that in this way we try to explore those entries first which contain potential candidates to be reported next from the ranking query.

For the organization of the index entries during the traversal of the index we maintain a priority queue Q storing entries with the corresponding estimated ranking count which are sorted in ascending order according to their estimated ranking count. Thereby we assume that the ranking count of each entry in this queue was generated by taking all current entries in the queue into account using the aforementioned strategies for increasing the ranking count.

The top element of the queue is the entry which has to be explored next. Whenever an entry e is explored, i.e. e is loaded from disk and is refined, we have to perform the following two steps: first, we have to update the ranking counts of all elements in the queue according to the children of e and, second, the ranking counts of e's child elements have to be computed before we insert them into Q.

For the first step, we have to determine those entries in Q which could be affected by the refinement of e, i.e. for which the ranking count might be increased after refining e. Obviously, those entries which are completely behind the hyperplane representation of e, $\perp(q,e)$, must also be behind the hyperplane representations of each child of e and, thus, their ranking count is not affected by the refinement of e. In the example shown in Figure 7, entry e_3 is not affected by the refinement of entry e due to the above considerations. Furthermore, we can ignore those entries e' which cannot be behind a hyperplane of any object within $subtree(e)$, e.g. entry e_1 in the example in Figure 7, i.e. those entries e' for which the following statement holds:

$$\exists p \in e'.mbr : dist(p,q) < MinDist(p, e.mbr)$$

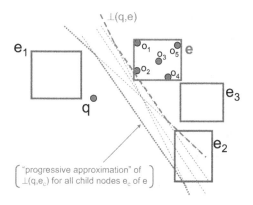

Fig. 7. Illustration of entries that are/are not affected by the refinement of an entry e

Intuitively, those entries are *not* behind the "progressive approximation" of all hyperplanes $\perp(q, e_c)$ of child entries e_c of entry e (cf. Figure 7).

Entries which are affected by the refinement of e are the remaining entries, i.e. those entries e' that fulfill *both* of the following two conditions:

$$\exists p \in e'.mbr : dist(p,q) \leq MaxDist(p, e.mbr)$$

and

$$\forall p \in e'.mbr : dist(p,q) \geq MinDist(p, e).$$

Each entry e' fulfilling the above two conditions, e.g. entry e_2 in our example in Figure 7, has to be checked against the hyperplane representation of each child of e. If the entry e' is behind the hyperplane representation of a child e_c of e, its ranking counter will be increased by $|subtree(e_c)|$.

For the second step, we have to determine the ranking counts of the children of e. For that purpose, we simply have to check all existing entries in $\mathcal{Q}$ and all other children of e whether the current child e_c of e is behind the corresponding hyperplanes. If yes, the ranking count of e_c is increased by the number of objects included in the subtree of the corresponding entry.

Finally, if the top entry e in the queue $\mathcal{Q}$ is a point, i.e. $e \in \mathcal{D}$, the point can be output as a result only if e is *not* beyond any *progressive* approximation of hyperplanes of child nodes of all e' that are currently in the queue, i.e. formally

$$\forall e' \in \mathcal{Q} : dist(e, q) < MinDist(e, e'.mbr).$$

Otherwise, we need to refine any of those entries $e' \in \mathcal{Q}$, for which this condition does not hold. As a consequence, e might get a higher ranking count and might be shifted towards the end of $\mathcal{Q}$ or it may also maintain the top spot of $\mathcal{Q}$.

The pseudocode of the algorithm for the incremental RNN ranking is illustrated in Figure 8 providing the implementation details of the previously discussed steps. First, we initialize an empty result list "result" and the priority queue $\mathcal{Q}$ which stores index entries sorted in ascending order of their ranking count. Ties occuring in the priority

```
ALGORITHM   initializeRanking(root, q)

   input: root = root of index storing D
   input: q = query object

   Q = empty priority queue sorted by RankingCount
   result = ∅
   insert root into Q

METHOD   getNext()

   WHILE Q is not empty DO
     e = dequeued entry from Q
     IF e is a directory entry THEN
       refine(e)
     END-IF
     ELSE // e is a LeafEntry
       e' = refinementRound(e)
       IF e' = NULL THEN RETURN e
       ELSE refine(e')
     END-ELSE
   END-WHILE
```

Fig. 8. Pseudocode of the incremental RNN ranking algorithm

queue are resolved by, first, prefering leaf index entries to directory index entries and, second, by sorting the entries in increasing distance to q.

The priority queue is initialized with the root of the index. For each call of the **get-Next** method, we dequeue the first entry e of Q. If e is a directory node, then it will be refined calling the **refine** routine depicted in Figure 9. During refinement, we first have to find all entries in Q that are candidates for having their ranking count increased due to the refinement of e (see first step above). An entry e' is such a candidate, if there exists a point in the mbr of e' that is closer to q than any point in e (see the predicate in line 3 of the refinement procedure in Figure 9) and if e' has not already been re-ranked by e (see the predicate in line 4 of the refinement procedure in Figure 9). These candidates are stored in a list updateI.

Additionally, we need all entries that are candidates for increasing the ranking count of one of the child entries of e (see the second step above). An entry e'' is such a candidate, if it has not re-ranked e already (first comparison in line 6 of the refinement procedure in Figure 9) and its mbr contains a point that is closer to q than a point in e (second comparison in line 6 of the refinement procedure in Figure 9). These candidates are stored in a list updateII.

Lines 8-12 check for each child node e_c of e and element $e' \in$ updateI, if e re-ranks e' and increases the ranking count of e' if necessary. Analogously, lines 13-17 increase the ranking count of e_c, if an element $e'' \in$ updateII re-ranks e_c.

Then, we increase the ranking count for each child entry e'_c of e that is able to re-rank e_c. Note that e_c and e'_c may be identical, i.e. e_c re-ranks itself. Finally e_c is inserted into the queue Q.

METHOD   refine(e)

　input: e = current directory entry
　updateI= $\{e' \in \mathcal{Q}|$
　　　$\forall p \in e' : MinDist(p,e) \leq MinDist(p,q)\wedge$
　　　$\exists p \in e' : MinDist(p,q) < MaxDist(p,e)\}$
　updateII= $\{e'' \in (queue \cup result)|\exists p \in e :$
　　　$MinDist(p,e'') \leq MinDist(p,q) < MaxDist(p,e'')\}$
　FOR EACH $e_c \in e$ **DO**
　　FOR EACH $e' \in updateI$ **DO**
　　　IF $(\forall p \in e' : MinDist(p,q) \geq MaxDist(p,e_c))$**DO**
　　　　increaseRankingCount(e', e_c.weight);
　　　END-IF
　　END-FOR
　　FOR EACH $e'' \in updateII$ **DO**
　　　IF $(\forall p \in e_c : MinDist(p,q) \geq MaxDist(p,e''))$**DO**
　　　　increaseRankingCount(e_c, e''.weight);
　　　END-IF
　　END-FOR
　　FOR EACH $e'_c \in e$ **DO**
　　　IF $(\forall p \in e_c : MinDist(p,q) \geq MaxDist(p,e'_c))$**DO**
　　　　increaseRankingCount(e_c, e_c'.weight);
　　　END-IF
　　END-FOR
　　queue.insert(e_c)
　END-FOR

Fig. 9. Pseudocode of our refine algorithm

METHOD   refinementRound(e)

　input: e = current leaf entry
　FOR EACH entry $e' \in \mathcal{Q}$ **DO**
　IF$(MinDist(e,e') \leq Dist(e,q) < MaxDist(e,e'))$ **THEN**
　　RETURN e'
　END-IF
　END-FOR
　RETURN NULL

Fig. 10. Pseudocode of the refinement round

If the entry e is a leaf entry, i.e. e is an object, then e obviously cannot be refined. However, we may not yet return e as a result without further checking, because it may be re-ranked due to an entry that has not yet been refined. In that case, we need to scan the queue $\mathcal{Q}$ for an object that is a candidate for re-ranking e by calling the **refinementRound** algorithm which is depicted in Figure 10 and refining (c.f. Figure 9) this object. If no such object exists, e can be returned as the result of the current getNext()-call.

4 Experimental Evaluation

In this section, we present the results of our experiments. We start by explaining in detail the settings of our experiments and those of the competitors. Then, we show the results of our performance evaluation on multi-dimensional data. Finally, we evaluate the effect of the partial spatial pruning on 2D-datasets.

4.1 Test Bed

We compared our novel approach for computing an RkNN ranking, with two adaptions of the TPL [7] approach which is the current state-of-the-art algorithm for RkNN query processing. In fact, we applied two versions of the TPL approach for computing a ranking. The problem of the TPL approach is that we cannot predict the number of getNext()-calls beforehand. Thus, we do not know a suitable value of k to answer all getNext()-calls.

The first variant, called TPL-Lazy, implements a lazy strategy assuming that we have a low number of getNext()-calls. It manages a result list which is initially empty and a counter k_c which stores the current value of k and is initialized with $k_c = 1$. The entries in the result list are ordered by increasing ranking scores. For each call of the getNext() method, this variant checks the result list. If the result list is empty, TPL-Lazy computes a RkNN query with $k = k_c$ using the original TPL approach, adds the result of this query to the result list with a ranking score of k_c, and increments k_c. These three steps are processed iteratively until the result list is no longer empty. Last but not least, the TPL-Lazy method returns the next entry in the result list. Obviously, this variant only issues a new RkNN query if necessary beginning with $k = 1$ and successively incrementing the value of k. The costs for answering l getNext()-calls are the sum of the costs of all queries for $k = 1, \ldots$ necessary to answer the l calls.

The second variant, called TPL-Eager, implements an eager policy assuming a higher but possible fixed maximum number of getNext()-calls. It simply assumes that the maximum number of getNext()-calls will be less than the number of result objects of a RkNN query with a special value of k_{max}, e.g. $k_{max} = 100$. Then, we only need to issue one Rk_{max}NN query using the original TPL approach beforehand and sort the results according to their ranking score. Whenever a getNext()-call is issued (and as long as the assumptions stated above regarding the size of the result and the number of getNext()-calls hold), we can simply return the next object from the result list. The costs for answering l getNext()-calls equal to the costs of answering the Rk_{max}NN query (again, as long as the result contains at least l points). Let us note that there is no direct relationship between the number of getNext()-calls l and the value k_{max}. This makes it even harder for the TPL-Eager approach to guess a proper k_{max} value. In fact, to obtain a fair comparison, we computed the most optimistic scenario for the TPL-Eager variant: we first issued l getNext()-calls with our new ranking method and obtained the ranking count of the resulting point of the last call. This count is the optimal k_{max} value for the TPL-Eager approach and we used this value in all our experiments. Thus, in realistic scenarios, the results of a TPL-Eager approach would be worse than presented here. All experiments are based on an aR*-Tree (aggregate version of R*-Tree) with a page size of 1K. Since all approaches are I/O bound we compared the number of disc pages accessed during the execution of 500 sample RkNN queries and averaged the results.

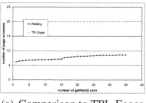
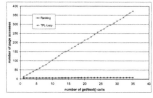

(a) Comparison to TPL-Eager.   (b) Comparison to TPL-Lazy.

Fig. 11. Comparison of our RkNN ranking with the competitors on uniformly distributed data

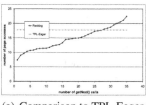
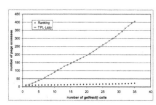

(a) Comparison to TPL-Eager.   (b) Comparison to TPL-Lazy.

Fig. 12. Comparison of our RkNN ranking with the competitors on clustered data

4.2   Performance Evaluation

Synthetic Data. We used two synthetic datasets to compare the performance of our ranking algorithm with the two variants of TPL. The first dataset contains 10,000 uniformly distributed 2D points. Figure 11 displays the performance of the competitors w.r.t. the number of getNext()-calls. As expected, the performance of the TPL-Eager approach (c.f. Figure 11(a))is constant as long as the number of getNext()-calls is smaller than the number of results of the Rk_{max}NN query issued beforehand (which is the case in our scenario – see above). Nevertheless, our ranking algorithm clearly outperforms this TPL variant in terms of query execution times. In fact, the costs of our approach increase only slightly with successive getNext()-calls. In addition, it should be noted that TPL-Eager would need to issue a new RkNN query with a considerably higher value of k if we have more than 35 getNext()-calls because TPL-Eager was optimized for 35 results. Thus, in that case, we would have a jump for the TPL-Eager approach at the 36th getNext()-call while the costs of our ranking algorithm will most likely evolve like in the range of the first 35 getNext()-calls. On the other hand, the costs for the TPL-Lazy variant (cf. Figure 11(b)) increase much faster than the costs of our new ranking algorithm. Again our approach clearly outperforms the competitor in terms of query execution times. Note that the performance of our ranking algorithm is of course the same in both Figures 11(a) and 11(b).

A similar observation can be obtained from Figure 12 which displays the performance of the competitors on a 2D synthetic dataset that contains 10,000 points clustered into four different clusters. The only obvious difference is that here, the TPL-Eager approach performs much better than on the uniform dataset. As illustrated in Figure 14(a), our ranking algorithm outperforms the TPL-Eager variant only for the

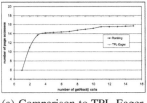
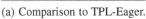
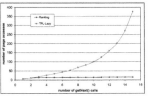

(a) Comparison to TPL-Eager.   (b) Comparison to TPL-Lazy.

Fig. 13. Comparison of our RkNN ranking and the competitors on 5D gene expression data

first 27 getNext()-calls. In this setting, the TPL-Eager slightly outperforms our ranking algorithm for 30 to 35 getNext()-calls. However, please note that, first, the TPL-Eager approach was implemented with the most optimistic assumptions and can be expected to perform considerably worse in a more realistic scenario where the perfect k_{max} value can usually not be determined beforehand. Second, as explained above, TPL-Eager was optimized for 35 results. If we had more than 35 getNext()-calls, then TPL-Eager would be required again to compute a new RkNN query with a considerably higher value of k which would cause significantly higher costs from the 36th getNext()-call on until the next jump limit is reached.

On the other hand, in comparison to the TPL-Lazy variant (cf. Figure 14(b)) our ranking algorithm again performs much better and significantly outperforms the competitor in terms of query execution times. Again, the performance of our ranking algorithm is of course the same in both Figures 14(a) and 14(b).

Real-world Data. We also tested our novel ranking algorithm on real-world data. In Figure 13 the performance of our ranking algorithm is compared with the performances of the TPL-Eager variant (cf. 13(a)) and the TPL-Lazy variant (cf. 13(b)) on a dataset that features the expression level of approx. 6,000 genes under 5 conditions. The result on this 5D dataset is similar to the results on the synthetic datasets reported above. Again, our ranking algorithm clearly outperforms the TPL-Lazy approach. Analogously, the difference to the TPL-Eager approach is less significant but still considerable. It should again be noted that the TPL-Eager variant assumes the most optimistic scenario for its application which is most likely not a realistic setting, and, thus, it can be expected that TPL-Eager performs less accurate in most applications.

4.3 Effect of the Spatial Partial Pruning

We evaluated our spatial partial pruning technique using a uniformly and a clustered 2D-dataset each containing 10,000 datapoints. The results are depicted in Figure 14. Notice that the number of page accesses is reduced by using partial pruning in both experiments. The effect however, is much more significant on the uniformly distributed dataset. Finally, we performed the same experiment on a real world dataset extracted from the Forest Cover Type dataset, retrieved from the UCI KDD repository [15] consisting of 10,000 2D-points. The result in Figure 14(c) shows that the spatial partial pruning technique reduces the number of page accesses by about 25%.

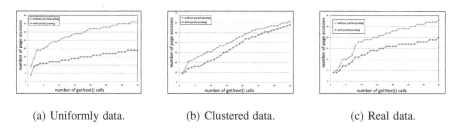

(a) Uniformly data.   (b) Clustered data.   (c) Real data.

Fig. 14. Effect of the spatial partial pruning

4.4   Summary

To summarize the results of our experimental evaluation, our novel ranking algorithm outperforms both adaptions of the existing TPL algorithm to the ranking problem, TPL-Eager and TPL-Lazy, significantly in terms of query execution times. While TPL-Eager seems to be competitive (if at all) only for a higher number of getNext()-calls, TPL-Lazy seems to be competitive (if at all) only for a very low number of getNext()-calls. This result is quite intuitive because TPL-Eager tries to estimate the worst-case by precomputing the maximum number of required results for a maximum number of getNext()-calls by computing one $Rk_{max}NN$ query. Thus, the more the number of getNext()-calls reaches the number of resulting objects of the $Rk_{max}NN$ query, the more the costs for the $Rk_{max}NN$ query pay off. Otherwise, TPL-Eager caused a large portion of unnecessary costs to compute a large number of results that are not needed. On the other hand, TPL-Lazy assumes the best case of very few getNext()-calls and, thus, computes results only if necessary by consecutively issuing a $RkNN$ query with increasing k. Obviously, as long as the number of consecutive $RkNN$ queries, with increasing k necessary to report results, is small, i.e. the number of getNext()-calls is low, this strategy pays off. Otherwise, TPL-Lazy constantly recomputes $RkNN$ queries with the next higher value for k which produces a lot of redundant results w.r.t. the previously computed queries.

Our ranking algorithm obviously performs best because it does not assume worst- or best-cases but focuses on computing the ranking incrementally. Since in a ranking query scenario, it is not known beforehand, how often the method getNext() is called, this is the most efficient solution in the general case but also – as our experiments illustrate – in the borderline cases where either TPL-Eager or TPL-Lazy perform best.

5   Conclusions

In this paper, we formalize a novel ranking problem, the reverse nearest neighbor (RNN) ranking and propose an original solution for it. Our solution extends existing methods for RNN query processing in the following important aspects. First, the mutual-pruning strategy of existing approaches is generalized and adapted so that it can be applied already on higher levels of the index and it can be applied to estimate the ranks of an index entry, rather than just for pruning. Second, we incorporated the idea of self-pruning and explored how this concept can be applied to estimate the ranking of index entries. Third, we explored the concept of partial pruning and derived an efficient solution to integrate the estimation of ranking counts based on this concept for 2D spatial data. Last but not least, we proposed an incremental algorithm for the RNN ranking problem that is based on both introduced ranking estimations. Our experimental evaluation confirms that our new solution outperforms existing methods adapted for the new problem significantly in terms of query execution times.

References

1. Lazaridis, I., Mehrotra, S.: Progressive approximate aggregate queries with a multi-resolution tree structure. In: Proc. SIGMOD (2001)
2. Papadias, D., Kalnis, P., Zhang, J., Tao, Y.: Efficient OLAP operations in spatial data warehouses. In: Jensen, C.S., Schneider, M., Seeger, B., Tsotras, V.J. (eds.) SSTD 2001. LNCS, vol. 2121, p. 443. Springer, Heidelberg (2001)
3. Korn, F., Muthukrishnan, S.: Influenced sets based on reverse nearest neighbor queries. In: Proc. SIGMOD (2000)
4. Yang, C., Lin, K.I.: An index structure for efficient reverse nearest neighbor queries. In: Proc. ICDE (2001)
5. Stanoi, I., Agrawal, D., Abbadi, A.E.: Reverse nearest neighbor queries for dynamic databases. In: Proc. DMKD (2000)
6. Singh, A., Ferhatosmanoglu, H., Tosun, A.S.: High dimensional reverse nearest neighbor queries. In: Proc. CIKM (2003)
7. Tao, Y., Papadias, D., Lian, X.: Reverse kNN search in arbitrary dimensionality. In: Proc. VLDB (2004)
8. Lee, K.C.K., Zheng, B., Lee, W.C.: Ranked reverse nearest neighbor search. IEEE TKDE 20(7), 894–910 (2008)
9. Achtert, E., Böhm, C., Kröger, P., Kunath, P., Pryakhin, A., Renz, M.: Efficient reverse k-nearest neighbor search in arbitrary metric spaces. In: Proc. SIGMOD (2006)
10. Achtert, E., Böhm, C., Kröger, P., Kunath, P., Pryakhin, A., Renz, M.: Approximate reverse k-nearest neighbor search in general metric spaces. In: Proc. CIKM (2006)
11. Tao, Y., Yiu, M.L., Mamoulis, N.: Reverse nearest neighbor search in metric spaces. IEEE TKDE 18(9), 1239–1252 (2006)
12. Guttman, A.: R-Trees: A dynamic index structure for spatial searching. In: Proc. SIGMOD, pp. 47–57 (1984)
13. Beckmann, N., Kriegel, H.P., Schneider, R., Seeger, B.: The R*-Tree: An efficient and robust access method for points and rectangles. In: Proc. SIGMOD, pp. 322–331 (1990)
14. Berchtold, S., Keim, D.A., Kriegel, H.P.: The X-Tree: An index structure for high-dimensional data. In: Proc. VLDB (1996)
15. Hettich, S., Bay, S.D.: The uci kdd archive (1999)

Approximate Evaluation of Range Nearest Neighbor Queries with Quality Guarantee[*]

Chi-Yin Chow[1], Mohamed F. Mokbel[1], Joe Naps[1], and Suman Nath[2]

[1] Department of Computer Science and Engineering, University of Minnesota,
Minneapolis, MN 55455, USA
[2] Microsoft Research, One Microsoft Way, Redmond, WA 98052, USA
{cchow,mokbel,naps}@cs.umn.edu, sumann@microsoft.com

Abstract. The range nearest-neighbor (NN) query is an important query type in location-based services, as it can be applied to the case that an NN query has a spatial region, instead of a location point, as the query location. Examples of the applications of range NN queries include uncertain locations and privacy-preserving queries. Given a set of objects, the range NN answer is a set of objects that includes the nearest object(s) to every point in a given spatial region. The answer set size would significantly increase as the spatial region gets larger. Unfortunately, mobile users in wireless environments suffer from scarce bandwidth and low-quality communication, transmitting a large answer set from a database server to the user would pose very high response time. To this end, we propose an approximate range NN query processing algorithm to balance a performance tradeoff between query response time and the quality of answers. The distinct features of our algorithm are that (1) it allows the user to specify an approximation tolerance level k, so that we guarantee to provide an answer set $\mathcal{A}$ such that each object in $\mathcal{A}$ is one of the k nearest objects to every point in a given query region; and (2) it minimizes the number of objects returned in an answer set, in order to minimize the transmission time of sending the answer set to the user. Extensive experimental results show that our proposed algorithm is scalable and effectively reduces query response time while providing approximate query answers that satisfy the user specified approximation tolerance level.

1 Introduction

Nearest-neighbor (NN) queries have been widely used in location-based services (e.g., see [1, 2, 3, 4, 5]). The problem of traditional NN queries can be defined as follows: "given a set of objects and a query location point p, find the nearest object(s) to p"; and thus, they are referred to as *point* NN queries. *Point* NN queries have been extended to find all NNs for line segments [6] and spatial regions [7, 8, 9] that are referred to as *linear* and *range* NN queries, respectively. A *linear* NN query

---

[*] This work is supported in part by the National Science Foundation under Grants IIS-0811998, IIS-0811935, and CNS-0708604.

N. Mamoulis et al. (Eds.): SSTD 2009, LNCS 5644, pp. 283–301, 2009.

returns an answer set that includes the nearest object(s) to every point in a given line segment. On the other hand, a *range* NN query returns an answer set that includes the nearest object(s) to every point in a given spatial region, where the spatial region can be either a rectangular region [7, 9] or a circular region [8].

Recent research efforts have shown the importance of *range* NN queries in location-based services, as it can be applied to the following realistic scenarios:

- **Uncertain locations.** We have two kinds of location uncertainty, *measurement imprecision* and *sampling imprecision*. The measurement imprecision is due to the limitation of the underlying positioning techniques of network environments, e.g., 2G/3G and Wi-Fi. On the other hand, the sampling imprecision is due to continuous motion, network delays, and location update frequency even with highly accurate positioning devices, e.g., GPS. Thus, we have to use a spatial region where the user is guaranteed to be therein to represent the user location information in order to capture location uncertainty (e.g., see [10, 11, 12, 13, 14]).
- **Privacy-preserving queries.** Mobile users are not willing to reveal their exact location information to location-based service providers, as they want to preserve their location privacy. The most commonly used privacy-enhancing technique is to blur the user's exact location into a spatial region, i.e., spatial cloaking, that satisfies the user's specified privacy requirements (e.g., see [8, 9, 15, 16, 17, 18, 19]).

In these two scenarios, the mobile user sends her NN query along with a spatial region as the query location, i.e., a *range* NN query. Then, a database server returns an answer set that includes the nearest object(s) to every point within the spatial region. The answer set size would substantially increase as the query region gets larger. Unfortunately, the communication bandwidth between the user and the database server is very limited in a mobile environment, i.e., the downlink bandwidth ranges from 128 kbps at vehicular speeds to 2 Mbps at stationary or very slow speeds for 3G mobile subscribers. Transmitting large answer sets to the user would pose very high query response time. Furthermore, as mobile users receive their answer handheld devices with a small screen, it is convenient to return to the users very few answers with high quality.

In this paper, we propose a new *approximate range* NN query processing algorithm that enables the user to tune a trade off between query response time and the quality of query answers. Our proposed algorithm allows the user to specify an *approximation tolerance level* k, where we return an answer set $\mathcal{A}$ such that each object in $\mathcal{A}$ is one of the k nearest objects of every point in the query region. The larger the value of k, the smaller the answer set returned by a database server. Thus, the approximation tolerance level is a tuning parameter that trades off between query response time and the quality of answers. In the case that $k = 1$, we return the exact range NN answer of maximal size to the user. On the other hand, if $k > 1$, we return an approximate answer set, which is smaller than the exact answer, to the user; and thus, the transmission time of sending the answer set to the user is reduced. Since query response time is

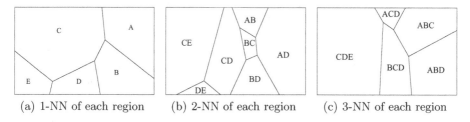

(a) 1-NN of each region   (b) 2-NN of each region   (c) 3-NN of each region

Fig. 1. A motivating example

dominated by the communication overhead between the user and the database server, a larger value of k incurs lower query response time.

Figure 1 depicts a motivating example for our problem where we decompose the query region $Q.R$ of a range NN query Q into disjoint regions and label each region with its k-NN(s). Figure 1a shows the NN of each region. If a user wants to find the nearest object(s) to $Q.R$, i.e., the exact range NN query answer, the required answer set contains the nearest object(s) to every point in $Q.R$, i.e., $\mathcal{A}_1 = \{A, B, C, D, E\}$. Figure 1b shows the 2-NN of each region. If the user is satisfied with the 2-nd nearest object(s) to $Q.R$, the answer set $\mathcal{A}_2$ should contain at least one object among the 2 nearest objects to every point in $Q.R$. For example, if $\mathcal{A}_2 = \{A, B, C, D\}$, regardless of the actual user location within $Q.R$, the user is guaranteed to receive an object among her 2 nearest objects. However, we can still do better. For example, if $\mathcal{A}'_2 = \{A, C, D\}$, the user is still guaranteed to receive an object among her 2 nearest objects, regardless of her actual location within $Q.R$. Thus, the answer set of minimal size is a minimal set of objects such that there is at least one object among the 2 nearest objects of each region. Furthermore, if the user is satisfied with the 3-rd nearest object(s) to $Q.R$, i.e., the required answer set $\mathcal{A}_3$ should contain at least one object among the 3 nearest objects of each region, as depicted in Figure 1c, where the minimal answer set is $\mathcal{A}_3 = \{A, C\}$. Therefore, if a user accepts a higher approximation tolerance in a range NN query answer, the user will receive a smaller answer set and more user convenience.

The main idea of our proposed range NN query processing algorithm is to have an *off-line* process to pre-compute a set of k-order Voronoi diagrams, from order one to a predefined maximum order k_{max}, for a set of stationary data objects, e.g., restaurants, gas stations and hotels. For a k-order Voronoi diagram, each Voronoi cell is associated with a distinct set of k objects that are the k nearest objects to every point in the cell. To efficiently search in a Voronoi diagram, we propose an *incomplete pyramid structure* as an access method to index the Voronoi cells. Given a *range* NN query Q and an approximation tolerance level k, our proposed *on-line* range NN query processing algorithm first determines a set of Voronoi cells $\mathcal{V}$ that intersects the query region by accessing the *incomplete pyramid structure* of the relevant k-order Voronoi diagram. Then, the remaining query processing is reduced to a well-known set-covering problem where we use a greedy approach to select the minimal set of objects, i.e., the answer set $\mathcal{A}$, from the objects associated with the Voronoi cells in $\mathcal{V}$ such that at least one object from each Voronoi cell in $\mathcal{V}$ is selected. As a result, each object in $\mathcal{A}$ is one of the k nearest objects to every Voronoi cell in $\mathcal{V}$, i.e., each object in $\mathcal{A}$ is one of the k nearest objects to every point in the query region. With a larger value of k, there are more common objects associated among the Voronoi cells in $\mathcal{V}$; and hence, we would get smaller answer sets that incur lower query response time. In general, the contributions of this paper can be summarized as follows:

- We introduce a new location-based query type, *approximate range nearest neighbor* (NN) query, that returns an answer set $\mathcal{A}$ such that each object in $\mathcal{A}$ is one of the k nearest object(s) to every point in a given query region. k is a user specified approximation tolerance level that can be used to tune a performance tradeoff between query response time and answer quality.
- We design an *incomplete pyramid structure* as an access method for efficiently retrieving a set of Voronoi cells that intersects a given query region from a Voronoi diagram for our proposed query processing algorithm.
- We propose an approximate range NN query processing algorithm that aims to minimize the number of objects returned in an answer set to improve query response time and user convenience while guaranteeing that the answer set satisfies the user specified approximation tolerance level.
- We provide experimental evidence through a comparison between the state-of-the-art techniques that our proposed query processing algorithm is scalable in terms of query processing time, and it significantly reduces query response time while the returned approximate answer is guaranteed to be satisfied with the user desired approximation tolerance level.

The rest of the paper is organized as follows. Section 2 reviews related works. Section 3 describes our system model. Our proposed approximate range NN query processing algorithm is presented in Section 4. The extensive experimental results are analyzed in Section 5. Finally, Section 6 concludes the paper.

2   Related Works

In location-based services, *point* nearest-neighbor (NN) queries have been extensively studied, e.g., [1, 2, 3, 4, 5]. Existing *point* NN query processing algorithms mainly focus on the scalability and efficiency of finding the nearest object(s) to a given query point. By considering user mobility, the concept of NN searches is extended to line segments [6] (referred to as *linear* NN query). The basic idea of *linear* NN query processing algorithm is to split a line segment into subsegments such that each subsegment has the same nearest object(s). All such nearest objects constitute the answer set of a *linear* NN query. Recently, the concept of NN searches is further extended to rectangular regions [7, 9] (referred to as *range* NN query). A minimal answer set for a *range* NN query includes all objects located in query region and the nearest objects to each edge of the query region [7]. By relaxing the minimality requirement, another existing *range* NN query processing algorithm, *Casper*, computes a candidate answer set that includes the exact answer [9]. The *Casper* algorithm first finds the nearest object to each vertex of

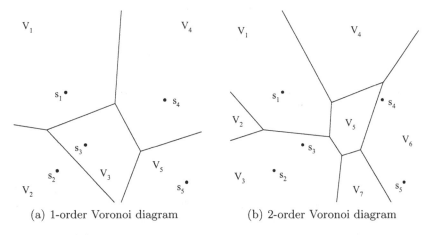

Fig. 2. The 1-order and 2-order Voronoi diagrams for five sites s_1 to s_5

the query region as *filters*, and then extends each edge of the query region to a minimal distance that is computed based on the *filters* to form an extended query region. The candidate answer set is a set of objects that is included in the extended query region. Also, a *range* NN query processing algorithm is proposed for finding the minimal answer set for a circular query region [8].

Our proposed approximate range NN query processing algorithm distinguishes itself from all previous techniques, as (1) it allows the user to specify an approximation tolerance level k for a range NN query, so that each object in the answer set is one of the k nearest objects to every point in a given query region; and (2) it aims to minimize the number of objects returned in the answer set to improve query response time, as the response time is dominated by the transmission time of sending the answer set to the user. In Section 5, we will compare the performance of our proposed algorithm with the state-of-the-art *range* NN query processing algorithms (i.e., [7, 9]).

3   System Model

In this section, we first formally define our problem, and then present the basic concept of Voronoi diagrams and the underlying system architecture.

Problem definition. Our problem is defined as follows: *given a set of objects, a range nearest-neighbor query Q with a query region $Q.R$, and an approximation tolerance level k, find the minimal set of objects $\mathcal{A}$ such that each object in $\mathcal{A}$ is one of the k nearest objects to every point in $Q.R$.*

Voronoi diagrams. Given a set of points $\mathcal{S}$ on the plane, which are the Voronoi *sites*, the Voronoi diagram of $\mathcal{S}$, denoted as $V(\mathcal{S})$, is a decomposition of the space into disjoint regions, *cells*, such that each site s_i is associated with a cell V_j, denoted as $V_j = \{s_i\}$, containing all the points in the plane that are closer to s_i

than any other site in $\mathcal{S}$. In other words, s_i is the nearest site to every point in V_j. Figure 2a depicts a Voronoi diagram of a set of five sites $\mathcal{S} = \{s_1, s_2, s_3, s_4, s_5\}$, $V(\mathcal{S})$. $V(\mathcal{S})$ decomposes the space into five cells V_1, V_2, V_3, V_4, and V_5 that are associated with the sites s_1, s_2, s_3, s_4, and s_5, respectively. For example, given a point p in cell V_1, s_1 is the nearest site to p.

Higher-order Voronoi diagrams. The k-order Voronoi diagram extends the concept of the Voronoi diagram by defining cells based on the k nearest neighbors. The k-order Voronoi diagram of $\mathcal{S}$, where $1 < k \leq |\mathcal{S}| - 1$, denoted as $V_k(\mathcal{S})$, is a decomposition of the space into disjoint cells, such that a distinct set of k sites $S_i = \{s_{i_1}, s_{i_2}, \ldots, s_{i_k}\}$ is associated with a cell V_j, $V_j = \{S_i\}$, containing all the points in the plane that have the sites in S_i as their k nearest sites. In other words, S_i contains k nearest sites to every point in V_j. Figure 2b depicts the 2-order Voronoi diagram of $\mathcal{S}$, $V_2(\mathcal{S})$, that decomposes the space into seven cells, i.e., $V_1 = \{s_1, s_3\}$, $V_2 = \{s_1, s_2\}$, $V_3 = \{s_2, s_3\}$, $V_4 = \{s_1, s_4\}$, $V_5 = \{s_3, s_4\}$, $V_6 = \{s_4, s_5\}$, and $V_7 = \{s_3, s_5\}$. For example, given a point p in cell V_3, the sites s_2 and s_3 are the two nearest sites to p.

System architecture. We consider a mobile environment where mobile users communicate with a location-based database server through a (2G/3G) cellular network. The data/control flow of our system is as follows: The mobile user sends range NN queries to the database server. Our proposed approximate range NN query processing algorithm that is implemented in the database server computes an answer set, and then the server sends the answer set to the user. We use the Euclidean distance as our distance metric.

4   Approximate Range NN Query Processing

In this section, we first describe an *off-line* process to compute Voronoi diagrams, from order one to order k_{max}, where k_{max} is the maximum allowable user specified approximation tolerance level, and present our proposed *incomplete pyramid structure* that is used as an access method for each Voronoi diagram. Then, we present an *on-line* query processing algorithm for approximate range NN queries.

4.1   Building Voronoi Diagrams

We use an off-line process to build k Voronoi diagrams for a set of objects, e.g., restaurants, hotels and gas stations, from order one to order k_{max}, where k_{max} is the maximum user specified approximation tolerance level. Thus, the user can specify her desired approximation tolerance level k from one to k_{max}. Notice that if $k = 1$, our algorithm provides an exact answer set of the minimal size for range NN queries. Given a set of objects $\mathcal{S}$, building a set of Voronoi diagrams, from order one to order k_{max}, i.e., $V_1(\mathcal{S}), V_2(\mathcal{S}), \ldots$, and $V_{k_{max}}(\mathcal{S})$, takes $O(k_{max}^2 N \log N)$ time and $O(\sum_{k=1}^{k_{max}} k^2(N - k))$ space, where N is the number of objects in $\mathcal{S}$ [20]. After we build the k Voronoi diagrams, they are stored for later use in our proposed range NN query processing algorithm. For each Voronoi diagram, we maintain a table to store each Voronoi cell with its associated objects.

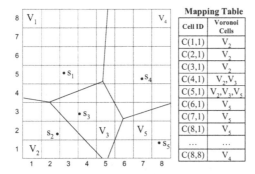

Fig. 3. The base level of an incomplete pyramid structure

4.2   Access Method for Voronoi Diagrams

We will construct an *incomplete pyramid structure* for each Voronoi diagram to support efficient range search among Voronoi cells. The idea of our proposed *incomplete pyramid structure* is that we need to find a set of Voronoi cells $\mathcal{V}$ that intersects a given range query region in order to retrieve their associated objects. Without any index structure, we have to scan every cell in a Voronoi diagram to find $\mathcal{V}$. When the number of objects and/or k is large, scanning all cells in a Voronoi diagram would pose a scalability issue. To this end, we propose an *incomplete pyramid structure* to overcome this issue. The construction of an *incomplete pyramid structure* for each Voronoi diagram includes two main steps.

STEP 1: Base level step. This step decomposes the space into grid cells where each grid cell $C(c, r)$ is uniquely identified by its column number c and row number r. Also, we use a hash table, *mapping table*, that associates each grid cell identity with a list of Voronoi cells that intersects the grid cell. Figure 3 depicts an 8×8 grid structure for the Voronoi diagram given in Figure 2a and the corresponding *mapping table*. For example, given a grid cell identity $C(5, 1)$, we can retrieve a set of Voronoi cells that intersects $C(5, 1)$, i.e., V_2, V_3, and V_5.

STEP 2: Merge step. This step merges quadtree-like neighbor cells to their parent if they intersect the same set of Voronoi cells. The idea of this step is to adaptively determine the height of an *incomplete pyramid structure* for a Voronoi diagram to minimize search time. This is due to the fact that the k Voronoi diagrams we maintain have different structures, e.g., the number of Voronoi cells and Voronoi cell size distribution, with respect to the number of objects, the object distribution and the degree of order (k). Thus, the shape of an *incomplete pyramid structure* would be different for each computed Voronoi diagram. Other than the base level, each cell $C(l, c, r)$ at upper levels is identified by the level of the *incomplete pyramid structure* l, column number c and row number r. This step uses a bottom-up approach to construct the upper levels of an *incomplete pyramid structure*. Starting from the base level, if all quadtree-like sibling cells (i.e., the cells have the same parent) intersect the same set of

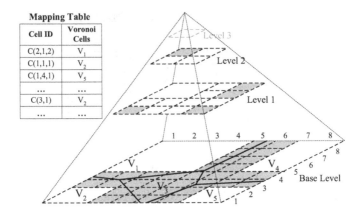

Fig. 4. Incomplete pyramid structure

Voronoi cells, they are merged to their parent. The merge process includes three tasks, (i) adding an entry that associates the parent cell identity with the set of intersected Voronoi cells of its children to the *mapping table*, (2) removing the entries of the merged child cells from the *mapping table*, and (3) annihilating the merged child cells by removing their pointers at their parent.

Figure 4 depicts an *incomplete pyramid structure* with a *mapping table* for the base level given in Figure 3, where the underlying Voronoi diagram is shown at the base level for the sake of illustration. Starting from the base level, we merge the quadtree-like sibling cells to their parent if they intersect the same set of Voronoi cells. For example, the four cells at the left bottom corner ($C(1,1)$, $C(2,1)$, $C(1,2)$, and $C(2,2)$) intersect the same Voronoi cell V_2, these cells are merged to their parent. To complete the merge process, we add an entry with the parent identity $C(1,1,1)$ with the intersected Voronoi cell V_2 to the *mapping table*, remove the entries of the merged child cells from the *mapping table*, and annihilate the merged child cells. We illustrate merged child cells by removing the grid cells. Similarly, we merge the cells at the other corners. At level one, the sibling cells at the left top corner (i.e., $C(1,1,3)$, $C(1,2,3)$, $C(1,1,4)$, and $C(1,2,4)$) intersect the same Voronoi cell V_1, so they are merged to their parent at level two, i.e., $C(2,1,2)$. At level two, since all cells intersect different sets of Voronoi cells, we cannot merge any cells and this step terminates. In this example, the shaded cells depict the lowest maintained cells of the *incomplete pyramid structure*, and there is an entry for each shaded cell in the *mapping table*. The height of the *incomplete pyramid structure* is two.

4.3   Online Query Processing Algorithm

The distinct feature of our proposed approximate range NN query processing algorithm is to enable the user to specify an approximation tolerance k for a range NN query, so that we provide a minimal answer set $\mathcal{A}$ where each object is guaranteed to be one of the k nearest objects to every point in the query region.

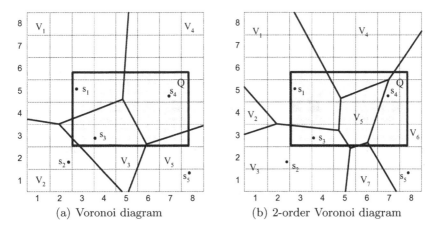

Fig. 5. Range NN query processing using Voronoi diagrams

If $k = 1$, we return an exact answer set with the maximal size to the user. In the case that $k > 1$, we return an approximate answer set with a smaller size than the exact one to the user; and thus, the transmission time of the answer set is reduced. The input of our proposed algorithm is a range NN query Q, a user specified approximation tolerance level k, and a k-order Voronoi diagram that is pre-computed by an off-line process (as described in Section 4.1). Figure 5 depicts a running example for our proposed algorithm where the query region of the input range NN query Q is represented as a bold rectangle and the maximum approximation tolerance level k_{max} is two. The algorithm consists of two key steps, *range search* step (Section 4.3.1) and *query-covering* step (Section 4.3.2).

4.3.1 Range Search Step

In this step, we retrieve a set of Voronoi cells $V_k = \{V_1, V_2, \ldots, V_n\}$ that intersects the query region $Q.R$ and each Voronoi cell in V_k associates with k objects, i.e., $V_i = \{s_{i_1}, \ldots, s_{i_k}\}$ $(1 \le i \le n)$. We use a top-down approach to traverse an *incomplete pyramid structure*. Initially, at the highest maintained level of the *incomplete pyramid structure* of the k-order Voronoi diagram, we find a set of grid cells that intersects $Q.R$, and then recursively search each of these grid cells. During a recursive search, if an encountered grid cell C that intersects $Q.R$ is at the lowest maintained level or base level of the *incomplete pyramid structure*, we retrieve the set of Voronoi cells that intersects C from the *mapping table*, and then return it. Otherwise, we recursively search the four child cells of C.

Figure 6 illustrates the *range search* step for the running example depicted in Figure 5a where $k = 1$. For the sake of illustration, we only show the grid cells intersecting the query region $Q.R$, and the grid cells at the lowest maintained level or base level of the *incomplete pyramid structure* (as depicted in Figure 4) are represented shaded nodes. As shown in Figure 4, the highest maintained level of the *incomplete pyramid structure* is level two where we start the *range search* step. Since all grid cells at level two, i.e., $C(2, 1, 1)$, $C(2, 1, 2)$, $C(2, 2, 1)$

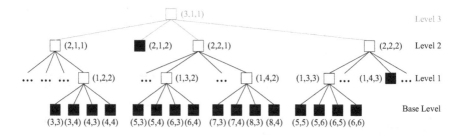

Fig. 6. Example of a range search in the incomplete pyramid structure (Figure 4)

Algorithm 1. Query-Covering Step

1: **function** QueryCovering (RNNQuery Q, ToleranceLevel k, VoronoiCellSet $\mathcal{V}_k$)
2: AnswerSet $\mathcal{A} \leftarrow \{\emptyset\}$
3: Construct an inverted list of $\mathcal{V}_k$, i.e., $\mathcal{L}(\mathcal{V}_k) = \{L(s_1), L(s_2), \ldots, L(s_m)\}$, where $L(s_i) = \{V_{i_1}, V_{i_2}, \ldots, V_{i_{|L(s_i)|}}\}$ and $1 \leq i \leq m$
4: **while** $\mathcal{L}(\mathcal{V}_k) \neq \{\emptyset\}$ **do**
5:     Select the object $s_i \in \mathcal{L}(\mathcal{V}_k)$ with the largest $|L(s_i)|$
6:     **for each** object $s_j \in \mathcal{L}(\mathcal{V}_k)$ $(i \neq j)$ **do**
7:         $L(s_j) \leftarrow L(s_j) - L(s_i)$
8:         **if** $L(s_j) = \{\emptyset\}$ **then**
9:             $\mathcal{L}(\mathcal{V}_k) \leftarrow \mathcal{L}(\mathcal{V}_k) - \{L(s_j)\}$
10:         **end if**
11:     **end for**
12:     $\mathcal{A} \leftarrow \mathcal{A} \cup \{s_i\}$
13:     $\mathcal{L}(\mathcal{V}_k) \leftarrow \mathcal{L}(\mathcal{V}_k) - \{L(s_i)\}$
14: **end while**
15: **return** $\mathcal{A}$

---

and $C(2, 2, 2)$, intersect $Q.R$, we will recursively search these grid cells. For $C(2, 1, 1)$, only one child cell $C(1, 2, 2)$ intersects $Q.R$, so we search their child cells $C(3, 3)$, $C(3, 4)$, $C(4, 3)$, and $C(4, 4)$ at the base level. Since all these child cells intersect $Q.R$, we retrieve the Voronoi cells that intersect them from the *mapping table* and return the set of retrieved Voronoi cells. As $C(2, 1, 2)$ is at the lowest maintained level, we retrieve the Voronoi cells that intersects $C(2, 1, 2)$ from the *mapping table* without further search. Then, we search the grid cell $C(2, 2, 1)$ where it has two child cells $C(1, 3, 2)$ and $C(1, 4, 2)$ intersecting $Q.R$. Since all child cells of $C(1, 3, 2)$ and $C(1, 4, 2)$ at the base level intersect $Q.R$, we retrieve the Voronoi cells that intersect them from the *mapping table* and return the retrieved Voronoi cells. Similarly, we search the grid cell $C(2, 2, 2)$. As a result, the set of Voronoi cells that intersects $Q.R$ is $\mathcal{V}_1 = \{V_1, V_2, V_3, V_4, V_5\}$. For the other running example where $k = 2$, the *range search* step searches the *incomplete pyramid structure* of the 2-order Voronoi diagram and the set of Voronoi cells that intersects $Q.R$ is $\mathcal{V}_2 = \{V_1, V_3, V_4, V_5, V_6, V_7\}$.

4.3.2 Query-Covering Step

Algorithm 1 gives the pseudo code of this step where we aim to compute the minimal set of objects in which each object is one of the k nearest objects to every point in the query region $Q.R$. First, we construct an inverted list of the set of

	1-Order			2-Order	

1-Order
Voronoi Cells   **Inverted List**

$V_1 = \{s_1\}$ $s_1 = \{V_1\}$

$V_2 = \{s_2\}$ $s_2 = \{V_2\}$

$V_3 = \{s_3\}$ ➤ $s_3 = \{V_3\}$

$V_4 = \{s_4\}$ $s_4 = \{V_4\}$

$V_5 = \{s_5\}$ $s_5 = \{V_5\}$

(a) Voronoi cell set $\mathcal{V}_1$

2-Order
Voronoi Cells   **Inverted List**

$V_1 = \{s_1,s_3\}$ $s_1 = \{V_1,V_4\}$

$V_3 = \{s_2,s_3\}$ $s_2 = \{V_3\}$

$V_4 = \{s_1,s_4\}$ ➤ $s_3 = \{V_1,V_3,V_5,V_7\}$

$V_5 = \{s_3,s_4\}$ $s_4 = \{V_4,V_5,V_6\}$

$V_6 = \{s_4,s_5\}$ $s_5 = \{V_6,V_7\}$

$V_7 = \{s_3,s_5\}$

(b) Voronoi cell set $\mathcal{V}_2$

Fig. 7. Inverted lists

Voronoi cells $\mathcal{V}_k$ retrieved from the previous step, $\mathcal{L}(\mathcal{V}_k)$ (Line 3). In the inverted list $\mathcal{L}(\mathcal{V}_k)$, each object s_i has a list of Voronoi cells $L(s_i) = \{V_{i_1}, \ldots, V_{i_m}\}$ $(m \le n)$, where s_i is associated with V_{i_j} $(1 \le j \le m)$.

Figure 7a depicts the inverted list of our running example for $k = 1$, as given in Figure 5a, where the Voronoi cells in $\mathcal{V}_1$ with their associated objects retrieved from the *range search* step are $V_1 = \{s_1\}$, $V_2 = \{s_2\}$, $V_3 = \{s_3\}$, $V_4 = \{s_4\}$, and $V_5 = \{s_5\}$. The inverted list of $\mathcal{V}_1$ is $\mathcal{L}(\mathcal{V}_1) = \{L(s_1) = \{V_1\}, L(s_2) = \{V_2\}, L(s_3) = \{V_3\}, L(s_4) = \{V_4\}, L(s_5) = \{V_5\}\}$. On the other hand, Figure 7b depicts the inverted list of our running example for $k = 2$, as shown in Figure 5b where the Voronoi cells in $\mathcal{V}_2$ with their associated objects retrieved from the *range search* step are $V_1 = \{s_1, s_3\}$, $V_3 = \{s_2, s_3\}$, $V_4 = \{s_1, s_4\}$, $V_5 = \{s_3, s_4\}$, $V_6 = \{s_4, s_5\}$, and $V_7 = \{s_3, s_5\}$. The inverted list of $\mathcal{V}_2$ is $\mathcal{L}(\mathcal{V}_2) = \{L(s_1) = \{V_1, V_4\}, L(s_2) = \{V_3\}, L(s_3) = \{V_1, V_3, V_5, V_7\}, L(s_4) = \{V_4, V_5, V_6\}, L(s_5) = \{V_6, V_7\}\}$.

After constructing the inverted list of $\mathcal{V}_k$, $\mathcal{L}(\mathcal{V}_k)$, our objective is to select the minimal set of objects from $\mathcal{L}(\mathcal{V}_k)$ such that every Voronoi cell in $\mathcal{V}_k$ has at least one associated object selected in the answer set. In other words, we consider the items in the inverted list as sets and the Voronoi cells in $\mathcal{V}_k$ as elements, and then select a minimum number of sets so that the selected sets contain all the elements that are contained in any of the sets in the inverted list. Thus, our problem can be reduced to a well-known *set-covering problem*. Since computing the optimal solution for the set-covering problem is NP-hard [21], we use a greedy approach to compute an answer set. Basically, the greedy approach selects an object with the largest set of Voronoi cells, and then remove the Voronoi cells associated with the selected object from other objects' lists. Then, the selected object and the objects with an empty list are removed from the inverted list. We repeat this procedure until the inverted list is empty (Lines 4 to 14 in Algorithm 1). The set of selected objects is returned as the answer set to the user.

In our running example for $k = 1$, i.e., the user wants to have an exact query answer, the *query-covering* step simply add all objects in the inverted list $\mathcal{L}(\mathcal{V}_1)$ (Figure 7a) to the answer set, i.e., $\mathcal{A}_1 = \{s_1, s_2, s_3, s_4, s_5\}$. On the other hand, Figure 8 depicts the *query-covering* step for our running example for $k = 2$, based on the inverted list of $\mathcal{V}_2$, $\mathcal{L}(\mathcal{V}_2)$, as given in Figure 7b. Since $L(s_3)$ has the largest

Initial Inverted List	Updated Inverted List	Updated Inverted List
$s_1 = \{V_1,V_4\}$	$s_1 = \{V_4\}$	$s_1 = \{\emptyset\}$
$s_2 = \{V_3\}$	$s_2 = \{\emptyset\}$	
$s_3 = \{V_1,V_3,V_5,V_7\}$		
$s_4 = \{V_4,V_5,V_6\}$	$s_4 = \{V_4,V_6\}$	
$s_5 = \{V_6,V_7\}$	$s_5 = \{V_6\}$	$s_5 = \{\emptyset\}$
$A_2 = \{\emptyset\}$	$A_2 = \{s_3\}$	$A_2 = \{s_3,s_4\}$

Fig. 8. Example of the query-covering step, based on Figure 7b

size, we select s_3 and remove the Voronoi cells in $L(s_3)$, i.e., V_1, V_3, V_5, and V_7, from other objects' lists, i.e., $L(s_1)$, $L(s_2)$, $L(s_4)$, and $L(s_5)$. Then, we add s_3 to an answer set A_2 and remove $L(s_3)$ from $\mathcal{L}(\mathcal{V}_2)$. The updated inverted list is $\mathcal{L}(\mathcal{V}_2) = \{L(s_1) = \{V_4\}, L(s_2) = \{\emptyset\}, L(s_4) = \{V_4,V_6\}, L(s_5) = \{V_6\}\}$. After we remove the empty list $L(s_2)$ from $\mathcal{L}(\mathcal{V}_2)$, $L(s_4)$ has the largest size. Thus, we select s_4 and remove the Voronoi cells in $L(s_4)$, i.e., V_4 and V_6, from other objects' lists, i.e., $L(s_1)$ and $L(s_5)$. Then, we add s_4 to A_2 and remove $L(s_4)$ from $\mathcal{L}(\mathcal{V}_2)$. The updated inverted list is $\mathcal{L}(\mathcal{V}_2) = \{L(s_1) = \{\emptyset\}$ and $L(s_5) = \{\emptyset\}\}$. Since all lists are empty, they are removed from the inverted list, and the *query-covering* step terminates and returns the answer set $A_2 = \{s_3, s_4\}$ to the user. From this example, we can see that our proposed approximate range NN query processing algorithm reduces the answer set size by 60%, i.e., from five objects in the exact answer set to two objects in the approximate answer set.

5   Experimental Results

In this section, we evaluate our Approximate Range nearest-neighbor (NN) query processing algorithm (denoted as ARNN) with respect to user specified approximation tolerance levels (k), query region size, the number of objects, downlink bandwidth, and object size. We compare our ARNN algorithm with two state-of-the-art range NN query processing algorithms as baseline algorithms. The first baseline algorithm computes an exact answer set of the minimal size for range NN queries (denoted as Exact) [7], while the other baseline algorithm computes a candidate answer set that contains the exact answer for range NN queries (denoted as Casper) [9].

We have two performance measures: (1) *total processing time* that includes the query processing time at the database server and the transmission time of sending the answer set to the user, and (2) *answer set size* that is the average number of objects returned in the answer sets. The *answer set size* is important as it indicates communication overhead and the power consumed by the user device to receive the answer set and user convenience.

In all experiments, we assume that the user communicates with a database server through a 3G cellular network. The downlink (i.e., from the database server to the user) bandwidth varies with respect to the user mobility speed,

Table 1. Parameter settings

Parameter	Default Value	Range
Approximation tolerance (k)	4	1 to 10
Number of objects	200	100 to 300
Query region size	$(0.05l)^2$	$(0.008l)^2$ to $(0.256l)^2$ (where $l = 1000$)
Downlink bandwidth	384 kbps	128 kbps to 2 Mbps
Object size	10 Kbytes	0.5 Kbytes to 20 Kbytes

i.e., 128 kbps (i.e., kbits per second) at vehicular speeds, 384 kbps at pedestrian speeds, 2 Mbps at stationary or very slow movement speeds. Unless mentioned otherwise, the experiments consider 200 objects in a square space of a length $l = 1000$. The mobile user moving at pedestrian speeds (i.e., the downlink bandwidth is 384 kbps) issues 1,000 range NN queries, and the object size is 10 Kbytes. The default user specified approximation tolerance level (k) is four and the query region size is $0.05l \times 0.05l$. Table 1 summarizes the parameter settings.

5.1 Effect of Approximation Tolerance Levels

Figure 9 depicts the performance of our proposed algorithm (ARNN) with respect to varying the approximation tolerance level (k) from 1 to 10. The performance

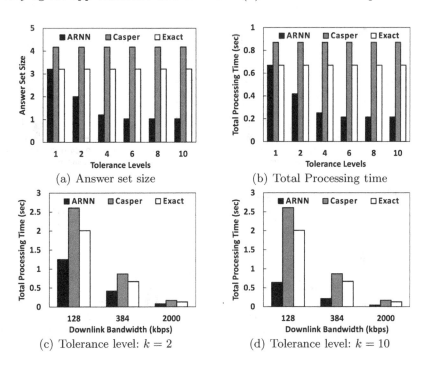

(a) Answer set size   (b) Total Processing time

(c) Tolerance level: $k = 2$ (d) Tolerance level: $k = 10$

Fig. 9. Approximation tolerance levels (k)

of the baseline algorithms (Casper and Exact) is not affected by varying the value
of k. Figure 9a gives the number of objects returned in the answer set, while
Figure 9b indicates the total processing time that includes the query processing
time at the database server and the transmission time of sending the answer set
to the user. Figure 9a shows that ARNN effectively reduces the answer set size
as k gets larger. When $k = 2$ ($k = 10$), ARNN reduces the size of the answer sets
given by Casper and Exact by 34.3% and 66% (79.3% and 89.3%), respectively.
Since the transmission time is much higher than the total processing time, the
total processing time of ARNN decreases as k gets larger (Figure 9b). Figures 9c
and 9d show that ARNN performs better than the baseline algorithms for all
mobility speeds. Since ARNN effectively reduces the answer set size, when the
downlink bandwidth is more limited, ARNN performs much better than Casper
and Exact.

5.2 Effect of Query Region Size

Figure 10 depicts the performance of our proposed algorithm (ARNN) with
respect to increasing the query region size from $(0.008l)^2$ to $(0.256l)^2$, where
$l = 1000$. Figure 10a shows that the answer set of all algorithms gets larger
as the query region size increases. With small query regions, i.e., $(0.008l)^2$, the
answer set size of ARNN is 94.7% and 59.8% smaller than Casper and Exact,
respectively. For large query regions, i.e., $(0.256l)^2$, the answer set size of ARNN

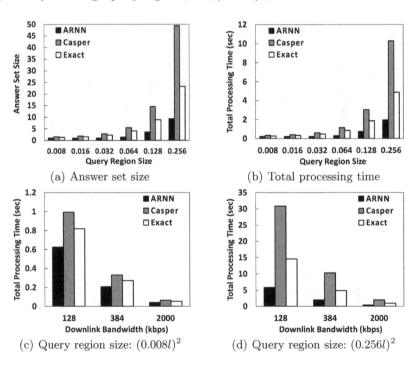

(a) Answer set size   (b) Total processing time

(c) Query region size: $(0.008l)^2$ (d) Query region size: $(0.256l)^2$

Fig. 10. Query region size

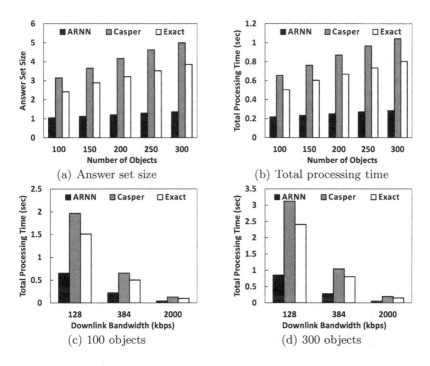

Fig. 11. Number of objects

is 442.1% and 145.3% smaller than Casper and Exact, respectively. Thus, ARNN performs much better than the baseline algorithms for larger query regions. Since the transmission time is much higher than the total processing time, ARNN outperforms the baseline algorithms in terms of query response time (Figure 10b). Figures 10c and 10d depict that ARNN effectively reduces the total processing time of the baseline algorithms regardless of user mobility speeds.

5.3   Effect of Number of Objects

Figure 11 gives the performance of our proposed algorithm (ARNN) with respect to varying the number of objects from 100 to 300. When the number of objects increases, there are more nearest objects to the query region; and thus, the answer set size of all algorithms gets larger (Figures 11a). Similar to the previous experiments, the transmission time is much higher than the total processing time. Since the answer set size of ARNN is smaller than the baseline algorithms Casper and Exact, ARNN incurs the lowest total processing time for any number of objects (Figures 11b). Likewise, ARNN effectively reduces the answer set size, the total processing time of ARNN is better than Casper and Exact for all user mobility speeds, as depicted in Figures 11c and 11d.

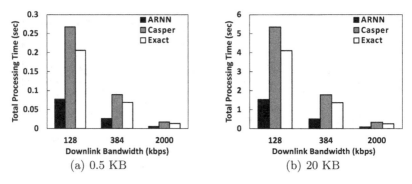

Fig. 12. Object size

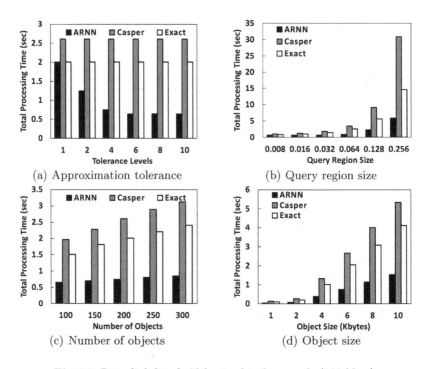

Fig. 13. Downlink bandwidth at vehicular speeds (128 kbps)

5.4   Effect of Object Size

Figure 12 depicts the performance of our proposed algorithm (ARNN) with respect to the object size of 0.5 and 20 Kbytes. Since varying the object size does not affect the answer set size, the answer set size of all algorithms is the same as the case that $k = 4$ in Figure 9a. It is interesting to see that the transmission time is much higher than the total processing time even if the object size is small

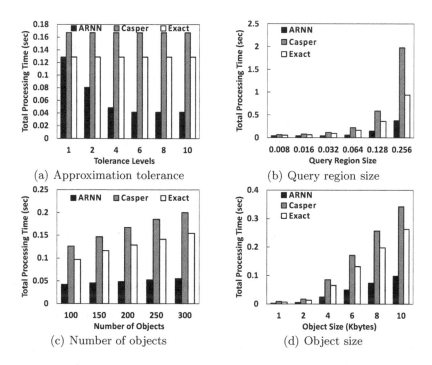

Fig. 14. Downlink bandwidth at stationary or very slow speeds (2 Mbps)

and the answer set is sent to the user through the downlink with the largest possible bandwidth, i.e., 2 Mbps. Therefore, the results indicate that reducing the answer set size is an effective way to improve query response time. This is the motivation of our proposed algorithm ARNN that aims to minimize the answer set size while guaranteeing that the answer set is satisfied with the user specified approximation tolerance level k.

5.5 Effect of Communication Bandwidth

Figures 13 and 14 give the comprehensive evaluation of our proposed algorithm (ARNN) with respect to all parameters for users moving at vehicular speeds and very slow speeds, respectively. Although Casper gives the best query processing time, it suffers from very high transmission time. This is because the candidate answer set provided by Casper is much larger than the answer set of ARNN and Exact. Thus, the total processing time of Casper is always worse than ARNN and Exact. Since ARNN provides approximate answers that satisfy the user specified approximation tolerance level, the answer set size of ARNN is smaller than the exact answer set provided by Exact. As a result, ARNN performs better than Casper and Exact in terms of the total processing time for all parameter settings.

6 Conclusion

In this paper, we propose a new query type, *approximate range nearest-neighbor (NN) query*, for location-based services. The distinct features of this new query type are that (1) It aims to minimize the number of objects returned to the user so as to reduce the transmission time of sending the answer to the user; and (2) It provides quality guarantee for the query answer, i.e., each object in an answer set is one of the k nearest objects to every point in a given query region, where k is a user specified tuning parameter for a tradeoff between query response time (that is dominated by transmission time as shown in all experimental results) and the quality of answers. To achieve these two features, we propose an *approximate range NN query processing* algorithm. The main idea is to have an *off-line* process to compute Voronoi diagrams, from order one to order k_{max}, where k_{max} is the maximum allowable user specified approximation tolerance level, and then build our proposed *incomplete pyramid structure* as an access method for each Voronoi diagram. Given a range NN query and an approximation tolerance level k, our *on-line* query processing algorithm accesses the *incomplete pyramid structure* of the k-order Voronoi diagram to retrieve a set of Voronoi cells that intersects the query region and the k nearest objects to each Voronoi cell. Then, the remaining query processing is reduced to a set-covering problem where we use a greedy approach to find a minimal answer set. Extensive experimental results show that our proposed algorithm is scalable in terms of query processing time, and effective to reduce query response time compared with the state-of-the-art techniques while guaranteeing that the answer set satisfies the user desired approximation tolerance level.

References

[1] Benetis, R., Jensen, C.S., Karciauskas, G., Saltenis, S.: Nearest and reverse nearest neighbor queries for moving objects. VLDB Journal 15(3), 229–249 (2006)

[2] Hu, H., Xu, J., Lee, D.L.: A generic framework for monitoring continuous spatial queries over moving objects. In: SIGMOD (2005)

[3] Mouratidis, K., Papadias, D., Hadjieleftheriou, M.: Conceptual partitioning: An efficient method for continuous nearest neighbor monitoring. In: SIGMOD (2005)

[4] Mokbel, M.F., Xiong, X., Aref, W.G.: Sina: Scalable incremental processing of continuous queries in spatio-temporal databases. In: SIGMOD (2004)

[5] Zheng, B., Xu, J., Lee, W.C., Lee, D.L.: Grid-partition index: A hybrid method for nearest-neighbor queries in wireless location-based services. VLDB Journal 15(1), 21–39 (2006)

[6] Tao, Y., Papadias, D., Shen, Q.: Continuous nearest neighbor search. In: VLDB (2002)

[7] Hu, H., Lee, D.L.: Range nearest-neighbor query. IEEE TKDE 18(1), 78–91 (2006)

[8] Kalnis, P., Ghinita, G., Mouratidis, K., Papadias, D.: Preventing location-based identity inference in anonymous spatial queries. IEEE TKDE 19(12), 1719–1733 (2007)

[9] Mokbel, M.F., Chow, C.Y., Aref, W.G.: The new casper: Query processing for location services without compromising privacy. In: VLDB (2006)

[10] de Almeida, V.T., Güting, R.H.: Supporting uncertainty in moving objects in network databases. In: ACM GIS (2005)

[11] Cheng, R., Kalashnikov, D.V., Prabhakar, S.: Querying imprecise data in moving object environments. IEEE TKDE 16(9), 1112–1127 (2004)

[12] Pfoser, D., Jensen, C.S.: Capturing the uncertainty of moving-object representations. In: Güting, R.H., Papadias, D., Lochovsky, F.H. (eds.) SSD 1999. LNCS, vol. 1651, p. 111. Springer, Heidelberg (1999)

[13] Trajcevski, G., Wolfson, O., Hinrichs, K., Chamberlain, S.: Managing uncertainty in moving objects databases. ACM TODS 29(3), 463–507 (2004)

[14] Yiu, M.L., Mamoulis, N., Dai, X., Tao, Y., Vaitis, M.: Efficient evaluation of probabilistic advanced spatial queries on existentially uncertain data. IEEE TKDE 21(1), 108–122 (2009)

[15] Bamba, B., Liu, L., Pesti, P., Wang, T.: Supporting anonymous location queries in mobile environments with privacygrid. In: WWW (2008)

[16] Cheng, R., Zhang, Y., Bertino, E., Prabhakar, S.: Preserving user location privacy in mobile data management infrastructures. In: Danezis, G., Golle, P. (eds.) PET 2006. LNCS, vol. 4258, pp. 393–412. Springer, Heidelberg (2006)

[17] Ghinita, G., Kalnis, P., Skiadopoulos, S.: Mobihide: A mobile peer-to-peer system for anonymous location-based queries. In: Papadias, D., Zhang, D., Kollios, G. (eds.) SSTD 2007. LNCS, vol. 4605, pp. 221–238. Springer, Heidelberg (2007)

[18] Gedik, B., Liu, L.: Protecting location privacy with personalized k-anonymity: Architecture and algorithms. IEEE Trans. on Mobile Computing 7(1), 1–18 (2008)

[19] Hu, H., Xu, J.: Non-exposure location anonymity. In: ICDE (2009)

[20] Lee, D.T.: On k-nearest neighbor voronoi diagrams in the plane. IEEE Trans. on Computers 31(6), 478–487 (1982)

[21] Cormen, T.H., Leiserson, C.E., Rivest, R.L., Stein, C.: Introduction to Algorithms, 2nd edn. MIT Press, Cambridge (2001)

Time-Aware Similarity Search: A Metric-Temporal Representation for Complex Data

Renato Bueno[1], Daniel S. Kaster[2,*], Agma Juci Machado Traina[1], and Caetano Traina Jr.[1]

[1] Department of Computer Science, University of São Paulo at São Carlos, SP, Brazil
{rbueno,agma,caetano}@icmc.usp.br
[2] Department of Computer Science, University of Londrina, Londrina, PR, Brazil
dskaster@uel.br

Abstract. Recent advances in information technology demand handling complex data types, such as images, video, audio, time series and genetic sequences. Distinctly from traditional data (such as numbers, short strings and dates), complex data do not possess the total ordering property, yielding relational comparison operators useless. Even equality comparisons are of little help, as it is very unlikely to have two complex elements exactly equal. Therefore, the similarity among elements has emerged as the most important property for comparisons in such domains, leading to the growing relevance of metric spaces to data search. Regardless of the data domain properties, the systems need to track evolution of data over time. When handling multidimensional data, temporal information is commonly treated as just one or more dimensions. However, metric data do not have the concept of dimensions, thus adding a plain "temporal dimension" does not make sense. In this paper we propose a novel metric-temporal data representation and exploit its properties to compare elements by similarity taking into account time-related evolution. We also present experimental evaluation, which confirms that our technique effectively takes into account the contributions of both the metric and temporal data components. Moreover, the experiments showed that the temporal information always improves the precision of the answer.

1 Introduction

Recent advances in information technology demand handling complex data types, such as images, video, audio, time series and genetic sequences. Differently from traditional data (such as numbers, short strings and dates), complex data do not possess the total ordering property, yielding relational comparison operators ('<', '≥', '≤' and '>') useless. Even equality comparisons (= and ≠)

---

* On leave at Department of Computer Science, University of São Paulo at São Carlos, SP, Brazil.

N. Mamoulis et al. (Eds.): SSTD 2009, LNCS 5644, pp. 302–319, 2009.

are of little help, as it is very unlikely to have two complex elements exactly equal. Therefore, new query operators are required to compare complex data, and the similarity among elements has been come forth as the most important property of such domains. Similarity relies on a measurement, often given by a distance function called a metric, which quantifies how much two elements are dissimilar. Thus, metric spaces are adequate to represent complex data, as they only require the elements and their pairwise distances [1].

Regardless of the data domain properties, the systems need to track the evolution of data over time. In the literature there are many space models developed to represent time information associated with the data stored in a DBMS. Existing time-aware models are focused either on simple/atomic data (e.g. numbers, dates and small texts) or on dimensional data (e.g. spatial and multidimensional data). However, metric data do not comply to the requirements of either of those spaces, so a new approach is required to represent time together with complex data.

As an example, suppose a health care application aimed at checking patients' treatment evolution based on the results of image-based exams. The temporal information associated with each exam is very important, whether this time data is absolute or relative to an aspect of interest (e.g. the treatment time). Comparing images usually requires extracting a set of features from the images, thereafter comparing the features through a metric. Often, the set of features extracted from each image has different cardinality for each image, as when the features represent shapes of objects in the images. Thus, multidimensional spaces are not adequate to represent such data. Since images and temporal information require integrated treatment, a metric-temporal solution is required to support this application.

Other examples which demand such integration include:

Large buildings monitoring – keeping track of bridges, buildings and towers by monitoring sensors dynamically placed on the structures;
Equipment sensing – supervising sensor data from industrial/scientific apparatus, such as machinery, metallurgical furnaces and ducts in oil refineries;
Temporal series analysis – studying weather, stock exchange and market behavior and tendencies over different periods.

When handling multidimensional data, temporal information is commonly treated as just one or more dimensions. However, metric data do not possess the concept of dimensions, thus including a "temporal dimension" does not make sense. In this paper we propose a novel metric-temporal space representation and exploit its properties to compare elements by similarity taking into account time-related evolution.

The remainder of the paper is structured as follows. Section 2 introduces some basic concepts and surveys related work. The proposed metric-temporal space is described in Section 3. Section 4 shows results of experiments performed to evaluate the properties of the proposed space. Finally, Section 5 presents the conclusions and suggests further work.

2   Background and Survey

In this section we present the main concepts involved in the paper, and review the main related techniques presented in the literature.

2.1   Similarity Search and Metric Spaces

Similarity depends on a function that compares a pair of elements in the data domain and returns a real value quantifying how much they differ. The similarity can be measured as a distance, so smaller values denote more similar elements. There are many kinds of metrics usually employed when indexing complex data in metric access methods, such as the Minkowski family of metrics for multidimensional data, the Metric Histogram Distance (MHD) for images [2] and the Levenshtein distance for character strings.

A similarity query returns the stored elements that satisfy a given similarity criterion. The criterion is usually expressed relatively to one or more reference elements, called the *query center(s)*. The most common similarity operators are the *Range query* (Rq) and the *k-Nearest Neighbors query* (k-NNq) [3].

Although many complex data can be understood as points in a vector or multidimensional space, there are "adimensional" domains, such as words and genetic sequences, for whom no dimensionality can be assigned. If a function that compute distances among the elements is defined, both adimensional and dimensional data can be represented in a metric space. Formally, a metric space is defined as a pair $\mathbb{M} = \langle \mathbb{S}, d \rangle$, where $\mathbb{S}$ is the universe of valid elements and d is a metric, i.e. a distance function $d : \mathbb{S} \times \mathbb{S} \to \mathbb{R}^+$ that satisfies the following properties, $\forall s_1, s_2, s_3 \in \mathbb{S}$: (1) symmetry: $d(s_1, s_2) = d(s_2, s_1)$; (2) non-negativity: $0 < d(s_1, s_2) < \infty$ if $s_1 \neq s_2$ and $d(s_1, s_1) = 0$; and (3) triangular inequality: $d(s_1, s_3) \leq d(s_1, s_2) + d(s_2, s_3)$.

Metric spaces can be aggregated, composing their respective metrics into a new function, provided it also preserves the properties of a metric. Formally, metric aggregations correspond to products of metric spaces, so the resulting metric is usually called the *product metric* [4]. This property allows performing multivariate analysis on metric data composed of several attributes, with varying domains and measurement units.

Several indexing structures that exploit the metric space properties to speed up similarity query answering have been developed, known as Metric Access Methods (MAM). Existing MAM can be classified as: (i) Static, which are constructed in a single operation using the whole dataset and need to be rebuilt upon modifications, such as the BK-Tree [5] and the VP-Tree [6]; and (ii) Dynamic, which allow incremental construction, such as the M-tree [7], the Slim-tree [8] and the BM⁺-tree [9]. However, none of them handle temporal information associated with metric data.

2.2   Time-Aware Databases

The need to handle temporal information is common for many database applications. The data, notwithstanding the domain, can evolve over time and this fact is crucial to several systems. There are many space models described in the literature aimed at representing temporal information in databases. Temporal spaces usually consider one or more additional temporal dimensions over the data, in particular the *transaction time* (the time when the data item is stored, changed or removed) and *valid time* (the period when the data item is valid) [10].

The association of time to complex data has received special attention in geographic applications, leading to the spatio-temporal databases. A spatio-temporal database embodies spatial, temporal, and spatio-temporal concepts, and captures either aspects of data [11]. With the advances in positioning systems and mobile communications, the spatio-temporal research on *moving objects* has become prominent. This area has been impelled by applications that must keep track of geometry changes or movements of the monitored objects over time. These applications demand novel types of queries and database update solutions, since it is usually a problem to maintain the current objects' shape and/or location up to date, especially if the number of objects is large. In this sense, several researchers created spatio-temporal access methods [12,13,14]. Also, new query variations have been suggested, such as joins for moving objects [15,16,17], continuous queries [18,19,20] and clustering [21,22].

The temporal information is commonly treated adding one or more dimensions to the data space. As long as the atemporal data embody the concept of dimensions, the temporal dimensions integrate seamlessly to them. However, this is not the case with metric data, thus the temporal data handling must be kept apart from the treatment of the metric information. To the best of our knowledge, there is no previous work addressing the association of temporal information to metric data, which we propose in this paper.

To do so, we employed the support given by the Fractal theory, which will be briefly explained in the next subsection.

2.3 Fractal Theory Applied to Databases

A fractal is defined as the property of objects being *self-similar*, independently of scale or size, i.e. parts of a fractal object are directly or statistically similar to the whole object [23]. Some experimental evidence has shown that the distribution of distances between pairs of elements, in the majority of real datasets, present self-similarity, thus they can be considered as fractal datasets [24], at least for some distance ranges.

An interesting result of the Fractal theory is that any fractal has intrinsic dimensions, independent from the space where the object is immersed, and they can be measured by its so-called fractal dimensions. One of the most useful fractal dimensions for databases is the *correlation fractal dimension* D_2. Knowing D_2 of a dataset allows predicting its properties as being similar to that of a dimensional dataset with approximately the same embedded dimension [25].

There are two common numerical methods to calculate the intrinsic dimension of a dataset. The *box-counting* method [23] described following, is commonly applied for multidimensional data. Given a dataset immersed in an E-dimensional space, recursively divide the hypercube involving the dataset into cells of side

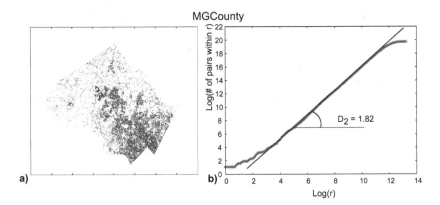

Fig. 1. Crossroads' geographical coordinates of Montgomery county of Maryland, MD, USA.(a) Plot of the dataset. (b) Distance plot graph.

size r, and count the number of elements lying inside each cell until r tends to zero. The plot, in log-log scale of the summation of the squared counts against the side size is called the box-counting plot. The plot of datasets that are perfectly fractals are a straight line, and most of the real datasets results into a curve that can be fitted by a line too. The line slope is a close approximation of the intrinsic dimension of the dataset.

For metric data, the intrinsic dimension can be calculated in the way following [26]. Given a set of elements in a dataset with a metric d, the average number k of neighbors within a given distance r is proportional to r raised to a value $\mathcal{D}$. Therefore, the pair-count $PC(r)$ of elements within distance r follows the power law:

$$PC(r) = K_p \cdot r^{\mathcal{D}} \tag{1}$$

where K_p is a proportionality constant. The graph obtained calculating Equation 1 is called the *distance plot*. When plotted in log-log scales, the graph can be fitted by a line too, and its slope is the exponent $\mathcal{D}$ in Equation 1, called the *distance exponent*. The fundamental property of the distance exponent is that it closely approximates D_2. Figure 1 shows the spatial distribution and the distance plot of a US Bureau of Census dataset whose elements are the crossroads' geographical coordinates of Montgomery county of Maryland, MD, USA, which has intrinsic dimension $D_2 = 1.82$.

The Fractal theory has been successfully applied to many tasks in data management. Particularly, the correlation fractal dimension has a number of useful properties. For example, if the data distribution is kept, it is insensitive to the dataset cardinality, so updates in the dataset do not require its recalculation. Thus, when the dataset is large, it can be calculated based on a small sample. Moreover, the value of the intrinsic dimension reflects the existence of correlations among the attributes of a dataset. Therefore, the fractal dimension provides an estimate of a lower bound for the number of features needed in a similarity search to keep the essential data characteristics and this bound is a function of the intrinsic dimension of the dataset [27].

Following the good results achieved by previous work, in this paper we use the Fractal theory in the definition of a space model to associate temporal information to metric data, as described in the next section.

3   A Metric-temporal Space

In this section we present the main concepts of the proposed metric-temporal space. We also introduce two distance functions to compare time, so both time and metric information can be seamlessly handled, and provide a procedure to assign proper weights for the metric and temporal components of the metric-temporal space that is able to exploit the best contribution of both. Table 1 summarizes the symbols used in this paper.

Table 1. Summary of symbols and definitions

Symbol	Definition
D_2	correlation fractal dimension
$\mathbb{S}$ and $\mathbb{T}$	domains of complex and temporal data
$d_s(s_1, s_2)$ and $d_t(t_1, t_2)$	metrics over complex (s_1, s_2) and temporal (t_1, t_2) elements
$\mathbb{V}$	metric-temporal domain
$d_v(v_1, v_2)$	metric-temporal distance function
$\pi_s(v_i)$ and $\pi_t(v_i)$	metric and temporal projections of the metric-temporal element v_i
w_s and w_t	weights of the metric and temporal components in d_v
p_s and p_t	metric and temporal intrinsic dimensionality
ℓ_s and ℓ_t	metric and temporal component side sizes
δ_{s_max} and δ_{t_max}	largest distances between any two metric projections and any two temporal projections

3.1   Metric-temporal Spaces

Let $\langle \mathbb{S}, d_s \rangle$ be a metric space, where $\mathbb{S}$ is a complex data domain, not necessarily dimensional, and $d_s : \mathbb{S} \times \mathbb{S} \to \mathbb{R}^+$ is the metric employed to compute the similarity between elements of the domain. Let $\langle \mathbb{T}, d_t \rangle$ be another metric space, where $\mathbb{T}$ is a temporal domain that includes every time value that can be associated with an element stored in a database, and $d_t : \mathbb{T} \times \mathbb{T} \to \mathbb{R}^+$ is a metric that calculates the similarity between two time values. Then we define a *metric-temporal space* as follows.

Definition 1 (Metric-temporal space). *A **metric-temporal space** is a pair $\langle \mathbb{V}, d_v \rangle$, where $\mathbb{V} = \mathbb{S} \times \mathbb{T}$ and $d_v : \mathbb{V} \times \mathbb{V} \to \mathbb{R}^+$ is a metric between elements of a metric space with time information associated, called the **metric-temporal distance function**, which aggregates the functions d_s and d_t in a particular way.*

According to Definition 1, an element of a metric-temporal space is basically the association of time information to an element of a metric space. For instance, we can represent a real world object by a set of pairs $\{\langle s_1, t_1 \rangle, \dots, \langle s_n, t_n \rangle\}$, $s_i \in \mathbb{S}$ and $t_i \in \mathbb{T}$, that are the states of the objects in the given time instants.

Definition 2 (Metric and temporal projections). *Given an element $v_i \in \mathbb{V}$, such that $v_i = \langle s_i, t_i \rangle | s_i \in \mathbb{S}, t_i \in \mathbb{T}$, we call s_i as the* **metric projection** *of v_i, denoted as $s_i = \pi_s(v_i)$, and t_i as the* **temporal projection** *of v_i, denoted as $t_i = \pi_t(v_i)$.*

In order to state the identity of elements in a metric-temporal space, we make the definition following.

Definition 3 (Metric-temporal identity). *Two elements $v_1, v_2 \in \mathbb{V}$ are the same iff $\pi_s(v_1) = \pi_s(v_2)$ and $\pi_t(v_1) = \pi_t(v_2)$.*

3.2   Metric-temporal Similarity Functions

A metric-temporal distance function aggregates the metrics of both the metric and temporal components of a metric-temporal space. Here we consider each metric as a "black box" defined by the domain specialist. Existing temporal data models do not treat time information as data in a metric space as we propose in this work. Therefore, we suggest two basic temporal metrics to measure the similarity of the temporal component.

Definition 4 (A metric for instants). *Given a temporal domain $\mathbb{T}_i$ representing instants, such that $\mathbb{T}_i \subseteq \mathbb{R}^+$, the distance between two instants $t_1, t_2 \in \mathbb{T}_i$ is given by $d_{ti}(t_1, t_2) = |t_1 - t_2|$.*

Metric d_{ti} is well suited to compare time instants, stated as the absolute difference between them. This function is essentially an unidimensional Manhattan distance (L_1), so it is a metric.

Definition 5 (A metric for periods). *Let $\mathbb{T}_p$ be a temporal domain representing periods, such that $\forall t_i \in \mathbb{T}_p, t_i = [l_i, u_i] | l_i, u_i \in \mathbb{R}^+$ and $l_i \leq u_i$, where l_i and u_i are respectively the lower and upper instants of the period. The metric to compare two intervals $t_1, t_2 \in \mathbb{T}_p$ is given by:*

$$d_{tp}(t_1, t_2) = |M(t_1) - M(t_2)| + |I(t_1) - I(t_2)| \tag{2}$$

where $M(t_i) = l_i + \frac{(u_i - l_i)}{2}$ is the middle instant and $I(t_i) = u_i - l_i$ is the size of period t_i.

The function d_{tp} clearly satisfies the symmetry and non-negativity properties because it is a summation of absolute values. The triangular inequality property also holds, as shown following, assuming $\forall t_1, t_2, t_3 \in \mathbb{T}_p$:

$$
\begin{aligned}
d_{tp}(t_1, t_2) &= |M(t_1) - M(t_2)| + |I(t_1) - I(t_2)| \\
&= |M(t_1) - M(t_2) + M(t_3) - M(t_3)| + |I(t_1) - I(t_2) + I(t_3) - I(t_3)| \\
&\leq |M(t_1) - M(t_3)| + |M(t_3) - M(t_2)| + |I(t_1) - I(t_3)| + |I(t_3) - I(t_2)| \\
&= d_{tp}(t_1, t_3) + d_{tp}(t_3, t_2)
\end{aligned}
$$

thus, d_{tp} is metric.

In order to develop a similarity measure for a metric-temporal space, it is necessary to adequately compose the two metrics d_s and d_t. As both are metric by definition, the natural way to compose them is defining a metric aggregation as a product metric. We propose here to employ the function following.

Definition 6 (A product metric for metric-temporal spaces). *Given a metric d_s for the metric component and a metric d_t for the temporal component of a metric-temporal space $\mathbb{V}$, the following product metric generates a metric-temporal space, $v_i, v_j \in \mathbb{V}$ and $w_s, w_t \in \mathbb{R}^+$:*

$$d_v(v_i, v_j) = w_s \cdot d_s\big(\pi_s(v_i), \pi_s(v_j)\big) + w_t \cdot d_t\big(\pi_t(v_i), \pi_t(v_j)\big) \tag{3}$$

where w_s is the weight of the metric component and w_t is the weight of the temporal component of the metric-temporal space.

Analyzing the properties of a metric, we see that the non-negativity and symmetry properties follow directly for d_v, because both d_s and d_t are metrics. Furthermore, d_v also satisfies the triangular inequality property for any $v_i, v_j, v_k \in \mathbb{V}$, as follows.

$$
\begin{aligned}
d_v(v_i, v_j) &= w_s \cdot d_s\big(\pi_s(v_i), \pi_s(v_j)\big) + w_t \cdot d_t\big(\pi_t(v_i), \pi_t(v_j)\big) \\
&\leq w_s \cdot \big(d_s\big(\pi_s(v_i), \pi_s(v_k)\big) + d_s\big(\pi_s(v_k), \pi_s(v_j)\big)\big) + \\
&\quad\ w_t \cdot \big(d_t\big(\pi_t(v_i), \pi_t(v_k)\big) + d_t\big(\pi_t(v_k), \pi_t(v_j)\big)\big) \\
&= w_s \cdot d_s\big(\pi_s(v_i), \pi_s(v_k)\big) + w_t \cdot d_t\big(\pi_t(v_i), \pi_t(v_k)\big) + \\
&\quad\ w_s \cdot d_s\big(\pi_s(v_k), \pi_s(v_j)\big) + w_t \cdot d_t\big(\pi_t(v_k), \pi_t(v_j)\big) \\
&= d_v(v_i, v_k) + d_v(v_k, v_j)
\end{aligned}
$$

Thus, the composition function d_v is metric. Notice that temporal metric either for instances or for intervals can be employed as d_t. If more than one time measure exist, for example transaction and valid time, such as multidimensional spaces, all the time-related metrics can be aggregated into a single product metric.

Considering a metric-temporal distance function which follows Definition 6, the challenge now is how to set weights w_s and w_t for the metric and temporal components, in order to achieve good similarity assessment. In other words, we need to answer the following question: *"For each similarity unit obtained by the metric $d_s\big(\pi_s(v_i), \pi_s(v_j)\big)$, what should be the equivalent time unit given by the metric $d_t\big(\pi_t(v_i), \pi_t(v_j)\big)$ to compare v_i and v_j?"* In the next section we propose a solution for this problem.

3.3   A Scale Factor for a Metric-temporal Similarity Function

The main idea to define the weights w_s and w_t is to identify the relative contribution of the metric and temporal components of a metric-temporal space for the final similarity calculation.

A well-known property of the metric spaces theory states that a metric subspace S with cardinality $|S| = s$ always can be mapped into a vector space $\mathbb{R}^{s-1}$ in such a way that the distances in the original space calculated by the original metric are exactly preserved in the mapped space calculated by a Minkowski function of order q. Exact mappings cannot be guaranteed in mapped spaces with less than $s - 1$ dimensions. However, depending on the particular data distribution of the dataset, the errors on the mapped distances can be very small until the number of dimensions drops below a limit related to the intrinsic dimension of the dataset. Here we assume that the correlation fractal dimension provides a close estimate for the intrinsic dimension.

Our approach is to make the weights w_s and w_t proportional to the side sizes of the hypercubes covering the subspaces mapped from the metric and temporal components into corresponding vector spaces. Although we do not need to perform the real mappings, we use these concepts to measure some properties in the dataset to estimate the weights. Therefore, the following definitions are useful.

Definition 7 (Intrinsic component dimensionality). *Let $V \subset \mathbb{V}$ be a dataset such that $S \subset \mathbb{S}$ contains the metric projections of all the elements of V and $T \subset \mathbb{T}$ contains the temporal projections of all the elements of V. Let also the similarity function d_v over $\mathbb{V}$ be a metric-temporal distance function given by Definition 6. Then, we define p_s and p_t respectively as the **intrinsic metric dimensionality** and the **intrinsic temporal dimensionality** of dataset V, calculated as the ceiling of the correlation fractal dimension $\lceil D_2 \rceil$ of the metric and temporal components.*

The intrinsic metric dimensionality p_s can be approximated using the distance plot technique to calculate D_2 over S. In the same way, the intrinsic temporal dimensionality p_t can be approximated using either the box-counting or the distance plot techniques to calculate D_2 over T, depending on the temporal component being multidimensional or not.

Definition 8 (Component side size). *Given a metric-temporal dataset V with the corresponding metric and temporal projections S and T, the **metric side size** ℓ_s of V and **temporal side size** ℓ_t of V are respectively the side sizes of the hypercubes of dimensions p_s and p_t covering the mappings of S and T.*

Although the real side sizes of both hypercubes can only be evaluated performing the mapping of both subspaces, for the sake of this work a close approximation can be evaluated as follows.

The diameter of a hypercube of dimension p_s and side size ℓ_s in a vector space ruled by a Minkowski distance function of order q covering the mapping of the dataset S is $diam(S) = \sqrt[q]{p_s} \cdot \ell_s$. Let δ_{s_max} be the largest distance between any two metric projections in S. The value of δ_{s_max} can be measured directly over the original dataset, and it is commonly referred to as the dataset diameter. Assuming that the diameter δ_{s_max} of the dataset S is nearly equal to the diameter $diam(S)$ of the hypercube covering the mapping of S, we have that the side size of the hypercube can be approximated by $\ell_s = \frac{1}{\sqrt[q]{p_s}} \cdot \delta_{s\_max}$. Figure 2 shows the intuition of this calculation, where the dataset $S \subset \mathbb{S}$ immersed in the metric space

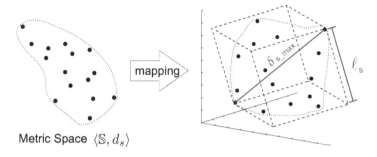

Fig. 2. A metric space mapped to a vector space $\mathbb{R}^3$ ruled by the L_2 distance function. The side size of the cube covering the mapping of the dataset is $\ell_s = \frac{1}{\sqrt{3}} \cdot \delta_{s\_max}$.

$\langle \mathbb{S}, d_s \rangle$ is mapped to a vector space of order $p_s = 3$ ruled by the Euclidean distance function L_2. Thus, the side size of the hypercube covering the mapping of S is $\ell_s = \frac{1}{\sqrt{3}} \cdot \delta_{s\_max}$. The metric space distribution in the figure is merely illustrative, since metric spaces do not need to have a shape in any dimensional space.

Although the value of δ_{s_max} can be measured by computing the distances between every element pair $s_i, s_j \in S$, this operation is quadratic on the number of elements in S. However, cheaper existing techniques can be employed to get close approximations. For example, if a MAM is indexing the data, it is feasible to get a good approximation stating δ_{s_max} as the diameter of the region covered by the MAM's root node.

The value ℓ_t is calculated using in the same procedure as ℓ_s.

We define each component weight as just the inverse of the corresponding component hypercube side size. Therefore, we state one last definition, as follows.

Definition 9 (Component weights). *The **weights** w_s and w_t of the metric and temporal components of metric-temporal distance function following Definition 6 are given respectively by $w_s = \frac{\sqrt[p_s]{p_s}}{\delta_{s\_max}}$ and $w_t = \frac{\sqrt[p_t]{p_t}}{\delta_{t\_max}}$, where p_s, δ_{s_max}, p_t and δ_{t_max} can be measured directly from V.*

In the commonest case where the temporal information in T are single-valued time instants, $\lceil D_2(T) \rceil = 1$. Thus $\sqrt[p_t]{p_t} = 1$ for any q, so $w_t = \frac{1}{\delta_{t\_max}}$.

The experimental evaluation of our technique, shown in the next section, confirms that the proposed way to calculate the weights w_s and w_t gives a suitable representation of the contributions of the metric and temporal components for the metric-temporal similarity function.

4   Experiments

This section presents results of some experiments performed to evaluate the proposed metric-temporal space. The experiments were implemented in C++

using the Slim-tree metric access method available in the *Arboretum library*[1] to index data. The tests were executed in a computer equipped with an AMD Athlon 2.6 GHz processor and 4Gb of RAM. The next subsections describe the dataset setup and discuss the experiments.

4.1   Datasets and Experimental Setup

The experiments explored the scenery where a set of images from evolving subjects need to be analyzed. The images are compared through feature sets extracted by image processing algorithms. We associate to each set the time when the image was obtained, relatively to the time when the subject started to evolve. The image dataset employed is the *Amsterdam Library of Object Images* (ALOI)[2] [28]. It is a collection of color images from one thousand small objects. Figure 3 shows a small sample of those images. Each object was photographed systematically varying an image parameter. For the experiments, we selected a dataset where each object was photographed from 72 different viewing angle, each 5 degree apart from the previous. We assumed that each angle correspond to its relative time stamp. That is, an object rotated by t degrees corresponds to a photography taken after t units of time.

Fig. 3. A sample of the images from the ALOI dataset

Each of the 72,000 images was processed by the three feature extractors summarized in Table 2. The features obtained by each extractor create a dataset. The first dataset is called Zernike, and the features are the first 256 Zernike moments describing the shapes of the images' contents. The second dataset is called Metric Histogram. It is an adimensional dataset containing the *metric histograms* of

---

[1] An open-source software library available at http://gbdi.icmc.usp.br/arboretum, which implements several MAMs.

[2] Available at http://staff.science.uva.nl/~aloi/

Table 2. Datasets used in the experiments

Name	Size	Dim.	Metric	Description
Zernike	72,000	256	L_1	The first 256 Zernike moments of the ALOI images
Metric Histogram	72,000	–	MHD	The metric histograms of the ALOI images
Histogram	36,000	256	L_1	The gray-scale histograms of the ALOI images with rotations from 0 to 175 degrees

the images. The last one is called Histogram, built selecting 36 images of each object (those with rotation angle varying from 0 to 175 degrees) and calculating their gray-scale histograms. The metric to compare the time information, for each dataset, is the absolute time difference. The metrics used to compute the similarity of the metric part were L_1, Metric Histogram Distance (MHD) and L_1, respectively for the datasets Zernike, Metric Histogram and Histogram.

To evaluate the query results quality we employed precision *versus* recall graphs. The recall of the result of a given query is the fraction of the relevant elements which has been retrieved. Thus, $recall = \frac{|Ra|}{|R|}$, where $|Ra|$ is the number of relevant elements retrieved and $|R|$ is the number of relevant elements that should be retrieved. The precision of a query is the fraction of the retrieved elements which are relevant. Thus, $precision = \frac{|Ra|}{|A|}$, where $|Ra|$ is the number of relevant elements retrieved and $|A|$ is the size of the answer set. Precision × recall curves can be summarized by a single numeric value to allow an overall evaluation of different queries at once. There are some strategies to perform this summarization. In this work we employed the *average precision*, which is given by the average of the precision values at all recall levels, as defined in [29]. Intuitively, the average precision is the area below a precision × recall curve.

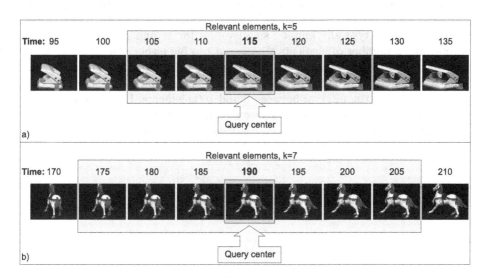

Fig. 4. Elements relevant to the query center shown. Results for a 5-NNq (a) and for a 7-NNq (b).

We assumed the following criterion to define which are the relevant elements for a query: *"The most relevant elements for a query are the images of the query element with the smallest difference between their time stamps."* The support of this decision is the observation that the aspect of an object with similar rotation degrees varies smoothly and the objects recorded are significantly different, which leads to a reasonable assumption that these elements are semantically similar. Figure 4 presents examples that illustrate this idea.

4.2   Inspecting the Ideal Contribution for the Metric and Temporal Components

The goal on the first set of experiments was to identify a near-optimal balance between the components of a metric space in the similarity evaluation. In the experiments we used the Minkowski distance function with $q = 1$ for the mapped space, because it is cheaper to evaluate and we found experimentally that it produces results as good as the others. Thus, we have $w_s = p_s/\delta_{s\_max}$ and $w_t = p_t/\delta_{t\_max}$. We calculated the values of p_s and p_t using the distance plot technique for all datasets. The values δ_{s_max} and δ_{t_max} were also obtained by the distance plot, because this method computes every pairwise distance.

Regarding the three datasets used in the experiments, the temporal component is the same for the three datasets evaluated. The corresponding distance plot can be seen in Figure 5. Thus, the value of the correlation fractal dimension for the metric space $\langle \mathbb{T}, d_t \rangle$ was 0.98, so $w_t = \lceil D_2(T) \rceil = 1$.

The distance plots for the metric component of all datasets are shown in Figure 6. As it can be seen, it is straightforward to fit a line for the Metric Histogram distance plot, but not for the others. The fitting strategy must be suited to the goal for which the value of the fractal dimension will be used. Our interest is on answering queries over a metric-temporal space. In this context, the k-NN queries typically employ small values for k, and range queries aims at returning a number of elements that is not too big either. Thus, we assumed that values of k between 5 and 100 are adequate for most applications. Following this reasoning, we are interested in fitting the line in the part of the distance plot that refers to the interval between 5 and 100 elements. Equation 1 uses the number of pairs $PC(r)$ within a distance r, and we are interested in the number k of elements involved. Therefore we must convert "numbers of elements" into

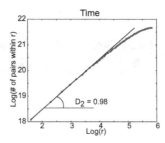

Fig. 5. Distance plot for the temporal component T of the three datasets

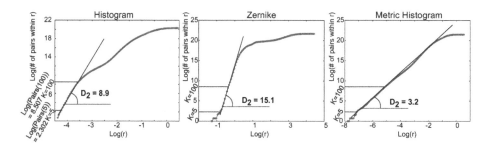

Fig. 6. Distance plot of the metric component S for the datasets: (a) Histogram (b) Zernike (c) Metric Histogram

"numbers of pairs" within a distance. The number of pairs in a subset of k elements (counting each pair once) is $Pairs(k) = k(k-1)/2$. Therefore we fitted the line based on the points that have the ordinate values $\log(Pairs(5)) = 2.302$ and $\log(Pairs(100)) = 8.507$, as indicated in Figure 6. This procedure resulted in correlation fractal dimensions of 8.9, 15.1 and 3.2 respectively for the metric component of the datasets Histogram, Zernike and Metric Histogram, which leads to the metric part power p_s respectively of 9, 16 and 4.

To evaluate if those weights are adequate, we executed a set of queries over each dataset varying the value of p_s. For each query, the query center was randomly selected from the respective dataset and $5 \le k < 10$. Figure 7 shows the average precisions obtained for a range of p_s values for the three datasets, where each point in the average precision plot is the average of evaluating 500 k-NN queries. In this figure, we highlighted the value of $p_s = 9, 16$ and 4 for the respective dataset. As it can be seen, those values of p_s are always the one that lead to the best, or very close to the best precision to answer queries.

Another point to be evaluated is whether using the temporal component helps improving the query answering precision. Therefore, in the next experiment we executed the same set of queries using only the metric component over each dataset. The results are also shown in Figure 7, as the dashed lines (remember that those queries are insensitive to p_s). As it can be seen, using the temporal component is almost always better than not using it, even for bad values of p_s.

As the average precision offers only a rough idea of the precision $\times$ recall plots, we show in Figure 8 the plots for 10-NN queries adjusting the weights to the best values, that is, using $p_s = 9, 16$ and 4 respectively for the Histogram, Zernike and Metric Histogram datasets. This figure also shows the precision $\times$ recall plots obtained executing the same queries using only the metric part of each dataset. The plots show that the metric-temporal space achieved better results at all recall levels in all datasets. In fact, it allows improving the average precision of k-NN queries from 0.88 to 0.92 in the Histogram dataset, from 0.67 to 0.74 in the Zernike dataset, and from 0.82 to 0.92 in the Metric Histogram dataset.

It is worth mentioning that the time required to answer queries using the metric-temporal space was practically identical to answer the same queries using

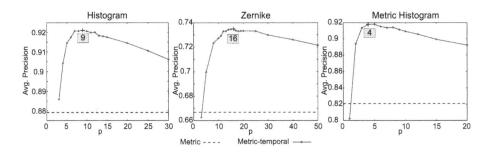

Fig. 7. Average precisions of k-NN queries for varying values of p_s for three datasets. The points show the average precisions regarding the metric-temporal space, and the dashed lines show the average precision considering only the metric components, following Definition 6.

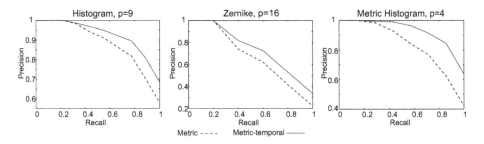

Fig. 8. Precision × recall of 10-NN queries using the proposed metric-temporal strategies endowed with the best weights for the metric and the temporal components, compared with using only the metric components

Table 3. Time measurements to obtain parameters p_s and p_t

Name	Time
Zernike (p_s)	18s
Metric Histogram (p_s)	14s
Histogram (p_s)	5s
Time (p_t)	0.7s

only the metric components, because the additional cost of the temporal component computation was irrelevant when compared to the costs of the original metric. The only significant cost added by the metric-temporal space refers to the calculation of the correlation fractal dimensions, used to define the values of p_s and p_t. However, it can be calculated based on a small sample of the original datasets, since the fractal dimension is invariant to unbiased sampling. We evaluated several sampling ratios, and for those datasets the values of p_s and p_t calculated using samples of up to 1% of the datasets were the same as those calculated using the whole datasets. Table 3 shows the time to determine the values of p_s and p_t using a sampling of 2.5% of the datasets, which were employed in all experiments reported. It is important to emphasize that these values are calculated only once, before starting to answer queries.

5   Conclusions and Future Work

The main contribution of this paper is the introduction of the Metric-Temporal space as an adequate representation for metric data that evolves over time. Distinctly from the multidimensional spaces, where the concepts of dimensions are native, our proposed space does not rely on representing time as dimensions, but as a time-related component associated with the metric data, regardless of the number of distinct time measurements that can be assigned to each metric data element.

Besides the introduction of the metric-temporal space, other related contributions are as follows.

- We proposed a metric approach to compare time measurements using either punctual time events or time periods, leading to a natural way to treat time in metric domains.
- We developed a procedure to assign proper weights for both the metric and the temporal components, which always provided the best answer, and does not rely on any user-provided parameter.

Finally, we performed several experiments showing that our technique effectively takes into account the temporal information associated with the metric data, always helping to improve the precision of similarity queries. The experiments also show that the proposed method to evaluate the contribution of the metric and the temporal components leads to the best weight to be assigned to each component.

As a follow-up of this work, we are now investigating how to perform similarity queries over metric-temporal datasets with missing data, helping to answer questions such as: *"What are the images closest to the image of this particular subject when it was at a given time t, even if there is no image of this subject at that particular time?"*

Acknowledgments

This work has been supported by the Brazilian funding agencies FAPESP, CNPq and CAPES.

References

1. Zezula, P., Amato, G., Dohnal, V., Batko, M.: Similarity Search: The Metric Space Approach. Advances in Database Systems, vol. 32. Springer, Heidelberg (2006)
2. Traina, A.J.M., Traina Jr., C., Bueno, J.M., Marques, P.M.d.A.: The metric histogram: A new and efficient approach for content-based image retrieval. In: VDB, Brisbane, Australia. IFIP Conference Proceedings, vol. 216, pp. 297–311. Kluwer Academic Publishers, Dordrecht (2002)

3. Böhm, C., Berchtold, S., Keim, D.A.: Searching in high-dimensional spaces - index structures for improving the performance of multimedia databases. ACM Comput. Surv. 33(3), 322–373 (2001)
4. Searcóid, M.Ó.: Metric Spaces. Springer Undergraduate Mathematics Series. Springer, Heidelberg (2006)
5. Burkhard, W.A., Keller, R.M.: Some approaches to best-match file searching. Commun. ACM 16(4), 230–236 (1973)
6. Yianilos, P.N.: Data structures and algorithms for nearest neighbor search in general metric spaces. In: SODA, Austin, TX, USA, pp. 311–321. ACM, New York (1993)
7. Ciaccia, P., Patella, M., Zezula, P.: M-tree: An efficient access method for similarity search in metric spaces. In: VLDB, Athens, Greece, pp. 426–435. Morgan Kaufmann, San Francisco (1997)
8. Traina Jr., C., Traina, A.J.M., Faloutsos, C., Seeger, B.: Fast indexing and visualization of metric datasets using Slim-trees. IEEE Trans. on Knowl. and Data Eng. 14(2), 244–260 (2002)
9. Zhou, X., Wang, G., Zhou, X., Yu, G.: BM^+-tree: A hyperplane-based index method for high-dimensional metric spaces. In: Zhou, L.-z., Ooi, B.-C., Meng, X. (eds.) DASFAA 2005. LNCS, vol. 3453, pp. 398–409. Springer, Heidelberg (2005)
10. Snodgrass, R.T.: The TSQL2 Temporal Query Language. Kluwer Academic Publishers, Dordrecht (1995)
11. Sellis, T.K.: Research issues in spatio-temporal database systems. In: Güting, R.H., Papadias, D., Lochovsky, F.H. (eds.) SSD 1999. LNCS, vol. 1651, pp. 5–11. Springer, Heidelberg (1999)
12. Tao, Y., Papadias, D., Sun, J.: The TPR*-tree: An optimized spatio-temporal access method for predictive queries. In: VLDB, Berlin, Germany, pp. 790–801. Morgan Kaufmann, San Francisco (2003)
13. Kollios, G., Papadopoulos, D., Gunopulos, D., Tsotras, V.J.: Indexing mobile objects using dual transformations. VLDB J. 14(2), 238–256 (2005)
14. Chen, S., Ooi, B.C., Tan, K.L., Nascimento, M.A.: St^2B-tree: a self-tunable spatio-temporal B^+-tree index for moving objects. In: SIGMOD, Vancouver, BC, Canada, pp. 29–42. ACM, New York (2008)
15. Zhang, R., Lin, D., Ramamohanarao, K., Bertino, E.: Continuous intersection joins over moving objects. In: ICDE, Cancun, Mexico, pp. 863–872. IEEE, Los Alamitos (2008)
16. Leong Hou, U., Mamoulis, N., Yiu, M.L.: Continuous monitoring of exclusive closest pairs. In: Papadias, D., Zhang, D., Kollios, G. (eds.) SSTD 2007. LNCS, vol. 4605, pp. 1–19. Springer, Heidelberg (2007)
17. Corral, A., Torres, M., Vassilakopoulos, M., Manolopoulos, Y.: Predictive join processing between regions and moving objects. In: Atzeni, P., Caplinskas, A., Jaakkola, H. (eds.) ADBIS 2008. LNCS, vol. 5207, pp. 46–61. Springer, Heidelberg (2008)
18. Mokbel, M.F., Xiong, X., Aref, W.G.: Sina: scalable incremental processing of continuous queries in spatio-temporal databases. In: SIGMOD, Paris, France, pp. 623–634. ACM, New York (2004)
19. Mouratidis, K., Hadjieleftheriou, M., Papadias, D.: Conceptual partitioning: An efficient method for continuous nearest neighbor monitoring. In: SIGMOD, Baltimore, Maryland, USA, pp. 634–645. ACM, New York (2005)
20. Papadopoulos, S., Sacharidis, D., Mouratidis, K.: Continuous medoid queries over moving objects. In: Papadias, D., Zhang, D., Kollios, G. (eds.) SSTD 2007. LNCS, vol. 4605, pp. 38–56. Springer, Heidelberg (2007)

21. Jensen, C.S., Lin, D., Ooi, B.C.: Continuous clustering of moving objects. IEEE Trans. Knowl. Data Eng. 19(9), 1161–1174 (2007)
22. Zhang, Z., Yang, Y., Tung, A.K.H., Papadias, D.: Continuous k-means monitoring over moving objects. IEEE Trans. on Knowl. and Data Eng. 20(9), 1205–1216 (2008)
23. Schroeder, M.: Fractals, Chaos, Power Laws, 6th edn. W.H. Freeman, New York (1991)
24. Faloutsos, C., Kamel, I.: Beyond uniformity and independence: Analysis of R-trees using the concept of fractal dimension. In: PODS, Minneapolis, MN, USA, pp. 4–13. ACM, New York (1994)
25. Belussi, A., Faloutsos, C.: Self-spacial join selectivity estimation using fractal concepts. ACM Trans. on Inf. Systems 16(2), 161–201 (1998)
26. Traina Jr., C., Traina, A.J.M., Faloutsos, C.: Distance exponent: a new concept for selectivity estimation in metric trees. In: ICDE, San Diego, CA, USA, p. 195. IEEE, Los Alamitos (2000)
27. Malcok, M., Aslandogan, Y.A., Yesildirek, A.: Fractal dimension and similarity search in high-dimensional spatial databases. In: IRI, Waikoloa, Hawaii, USA, pp. 380–384. IEEE, Los Alamitos (2006)
28. Geusebroek, J.M., Burghouts, G.J., Smeulders, A.W.M.: The Amsterdam library of object images. Int. J. Comput. Vis. 61(1), 103–112 (2005)
29. Baeza-Yates, R.A., Ribeiro-Neto, B.A.: Modern Information Retrieval. Addison-Wesley, Wokingham (1999)

Adaptive Management of Multigranular Spatio-Temporal Object Attributes*

Elena Camossi[1], Elisa Bertino[2], Giovanna Guerrini[3], and Michela Bertolotto[1]

[1] School of Computer Science and Informatics - University College Dublin,
Belfield, Dublin 4, Ireland
Ph.: +353 (0)1 7162-944/913, Fax: +353 (0)1 2697-262
{elena.camossi,michela.bertolotto}@ucd.ie

[2] CERIAS - Purdue University, 250 N. University Street West Lafayette,
Indiana, USA 47907-2066
Ph.: +1 765 496-2399, Fax: +1 765 494-0739
bertino@cs.purdue.edu

[3] DISI - Università degli Studi di Genova, Via Dodecaneso 35, 16146 Genova, Italy
Ph.: +39 010 353-6701, Fax:+39 010 353-6699
guerrini@disi.unige.it

Abstract. In applications involving spatio-temporal modelling, granularities of data may have to adapt according to the evolving semantics and significance of data. In this paper we define ST^2_ODMGe, a multigranular spatio-temporal model supporting *evolutions*, which encompass the dynamic adaptation of attribute granularities, and the deletion of attribute values. Evolutions are specified as *Event - Condition - Action* rules and are executed at run-time. The event, the condition, and the action may refer to a period of time and a geographical area. The evolution may also be constrained by the attribute values. The ability of dynamically evolving the object attributes results in a more flexible management of multigranular spatio-temporal data but it requires revisiting the notion of object consistency with respect to class definitions and access to multigranular object values. Both issues are formally investigated in the paper.

1 Introduction

The ability of representing datasets with respect to both their spatial layout and their historical evolution is crucial when performing analysis and monitoring changes in the spatial configuration of geographical areas. Moreover, approaches able to present data at different granularities [3] represent an effective solution to facilitate information analysis [1].

The granularity according to which information is represented depends on the the data domain and semantics as well as on the application tasks to be

---

* Research presented in this paper was funded by a Strategic Research Cluster grant (07/SRC/I1168) by Science Foundation Ireland under the National Development Plan. The authors gratefully acknowledge this support. The work of Elena Camossi is supported by the Irish Research Council for Science, Engineering and Technology.

N. Mamoulis et al. (Eds.): SSTD 2009, LNCS 5644, pp. 320–337, 2009.

performed on them. The selection of the appropriate granularity is based on modelling requirements and on a trade-off between application efficiency and data accuracy. A greater detail (i.e., a finer granularity) reduces data indeterminacy and allows to obtain information as accurate as possible. Conversely, storing data at an unnecessary level of detail, causes waste of space and additional costs in aggregating detailed data to the required abstraction level. Thus, the choice of a less detailed representation (i.e., a coarser granularity) makes it possible to store the minimal amount of data, thus reducing storage costs, and could improve application efficiency. Therefore, the selection of attribute granularities is a crucial task, that in existing multigranular systems is done once for all, at schema definition time. To enhance data flexibility, the model at hand must support the ability of dynamically setting and changing the spatio-temporal granularity. A static definition of attribute granularities in the database schema, as supported by current multigranular models, may not be adequate for many important spatio-temporal applications. For instance, in a spatio-temporal database for environmental monitoring, the collection of meteorological parameters such as the amount of rainfall, the strength and direction of the wind, the value of atmospheric pressure must be collected more frequently in the presence of exceptional events like hurricanes and storms. Moreover, such a granularity modification may involve only specific geographical areas (e.g., those affected by the phenomenon), and is required for limited periods of time (e.g., the time when the phenomenon occurs).

In our effort to address these issues we have defined ST^2_ODMGe (*Spatio-(Bi)Temporal ODMG supporting Evolutions*), a spatio-temporal data model that enables the *evolution* of attributes values, that is, the modification of the granularities used in attribute definitions, and the deletion of attribute values at run-time. ST^2_ODMGe evolutions reflect modifications about data significance that arise for several reasons, including: 1) periodic phenomena (e.g., rain and snowfall usually increase during predetermined seasons); 2) modification to the value of an attribute, or its occurrence (e.g., in monitoring systems); 3) the execution of an operation (e.g., in diagnostic systems); 4) data aging (e.g., older data may be aggregated and then maintained at coarser granularities); 5) privacy restrictions (e.g., individual information on user locations, which are collected in traffic analysis, must be aggregated to coarser granularities in order to be made publicly available). Hence, evolutions enhance the flexibility in the management of multigranular spatio-temporal data. They allow one to dynamically adapt the granularities to dynamic events and situations resulting from spatio-temporal attribute value updates and operation executions.

The types of evolutions supported by ST^2_ODMGe include: granularity evolution, granularity acquisition, and value deletion. *Granularity evolution* aggregates existing detailed data at a coarser granularity (e.g., older data that may be stored for future reference), or refines information at a finer granularity (e.g., in data analysis)[1]. By contrast, *granularity acquisition* changes at run-time the

---

[1] The latter operation increases indeterminacy on converted data, as discussed in the paper and in further detail in [8].

granularity used when inserting new values in the database, whenever the domain conditions change (e.g., sales recording during Christmas). Finally, *value deletion* removes attribute values that are no longer useful at a given granularity (e.g., detailed data already aggregated at coarser granularities) from the database.

The ST^2_ODMGe model design extends our previous models ST_ODMG [7], a multigranular spatio-temporal object model that does not support evolutions, and T_ODMGe [6], a multigranular temporal model supporting dynamic objects. It expands on their data definition languages, type systems, and multigranular conversions to support the evolution of spatio-temporal values and the bitemporal domain. Granularity evolutions and value deletions, originally defined for historical data only in T_ODMGe [6], are herein extended to the spatio-temporal domain: the spatio-temporal type system defined in ST_ODMG [7], further extended to include bitemporal support, has been embedded in the new model ST^2_ODMGe. A fundamental difference with our previous evolution model [6], where granularity evolutions only allow one to summarize older data at coarser granularities, is that in ST^2_ODMGe they may be specified also to refine data at finer granularities. Other important differences with T_ODMGe [6] are: 1) evolutions may be specified and executed at run-time, based on the execution model of active databases, instead of being defined statically in the database schema; 2) evolutions of an attribute value may be triggered according to database conditions involving also other attributes, as well as relying on the execution of methods, thus making our evolution approach highly flexible; 3) the introduction of granularity acquisition removes one of the major limiting assumptions of T_ODMGe, where the granularity used for acquiring new data is immutable. The main novelty with respect to ST_ODMG, besides bitemporal support, are the evolution facilities. ST^2_ODMGe enhances the expressive power of ST_ODMG providing a flexible and comprehensive support for run-time modifications of attribute granularities.

Evolutions introduce additional issues that are addressed in the formal design of the model. As a result of the execution of granularity evolutions and acquisitions, the run-time type of an object attribute is a Cartesian product of multigranular types at different granularities. The semantic consistency of the evolved attributes values must therefore be guaranteed and the strategies to access attribute values must be redefined to take advantage of these composed types. Moreover, even if a value is deleted, the access to attribute values may be preserved by considering evolved values at different granularities. In particular, in the paper we formally revise the notion of object consistency, and redefine the strategies to access evolved multigranular attribute values.

The rest of the paper is organized as follows. We first discuss related work (Section 2). In Section 3 we introduce the ST^2_ODMGe type system, objects and classes. Then, in Section 4 we address the definition of evolutions for spatio-temporal data, by means of illustrative examples. In Section 5 we investigate how object consistency is affected by evolutions. We define in Section 6 the access strategies to take advantage of attribute run-time values at multiple granularities, and we show that, under certain assumptions, object access is invariant with respect to the execution of evolutions. Finally, in Section 7 we conclude the paper by outlining future research directions.

2   Related Work

ST^2_ODMGe assumes and extends previous work on efficiently computing historical aggregates for on-line analytical processing (OLAP) of spatio-temporal data streams [16,13,11]. Zhang et al. [16], defined a spatio-temporal extension of the *SB-Tree* [15] structure, that, like our previous work [6], proposes an aggregated indexing approach whereby older data are stored using coarse granularities. Tao and Papadias [13] proposed over the years several indexing structures for the efficient historical aggregation of spatio-temporal data. Recent work focuses on aggregates for trajectories of moving objects [11]. Unlike those approaches, ST^2_ODMGe supports different time granularities and multiple levels of aggregation and refinement, that is, different indexing forms. Moreover the appropriate granularity level can be selected on a per-attribute basis, thus supporting different semantics (i.e., different queries). Furthermore, our notion of evolution refers to the bounds of granules at a given granularity, instead of referring to a given amount of time. Finally, ST^2_ODMGe relies on an widely adopted notion of temporal granularity [5] that considers granularities as data integrity constraints and formalises how different granularities are related to each other.

The approach to deletion we adopt has been inherited from research in the area of temporal databases. In this area data deletion is a crucial issue because answers against historical queries must be preserved [9,14,12]. Garcia-Molina et al. [9] have addressed data deletion in historical databases by proposing an approach whereby data may be removed (i.e., data *expire*) without affecting related views. A similar approach has been proposed by Toman [14] for historical data warehouses, whereby automatic data deletion is supported by preserving answers to a known and fixed set of first-order queries. This approach assumes that conditions for data evolution are inferred from a given set of queries. This approach may complement our work, since conditions for data evolution may be inferred for those attributes for which they are not known at schema definition time. A different approach is proposed by Skyt et al. [12], who address the faithful encoding of data history in temporal databases in the presence of vacuumed data. Such an approach relies on query modification, that is, on accompanying query results with additional information about how the required data may be affected by vacuuming. Unlike that approach, we allow queries be performed on vacuumed data whenever a result at different granularities exists.

3   Preliminaries

In this section we illustrate the main characteristics of ST^2_ODMGe, namely, the spatio-temporal dimensions, the granularities formalization, the multigranular type system, and describe granularity conversions and their properties. We then present ST^2_ODMGe classes and objects.

3.1   Time, Space, and Granularities

The ST^2_ODMGe model is a 4-dimensional multigranular spatio-temporal model that supports two-dimensional space and two temporal dimensions: *valid time* and *transaction time*. In the following, valid time dimension in ST^2_ODMGe (denoted by $\mathcal{VT}$) refers to the time a fact is true in the reality [10]. Transaction time dimension (denoted by $\mathcal{TT}$) represents the time at which database transactions are executed [10]. Moreover, ST^2_ODMGe supports *two-dimensional space*. denoted by $\mathcal{S}$, that refers to the space in which the modelled objects are actually located. Spatio-temporal attributes values refer to valid time and to the space dimensions. By contrast, modification actions, including those triggering the evolutions of attributes, refer to transaction time. Unlike the valid time dimension, transaction time includes references to the current time denoted by variable NOW.

In each ST^2_ODMGe database a set of temporal granularities [5] $\mathcal{G}_T$ and a set of spatial granularities $\mathcal{G}_S$ are defined. We further distinguish between valid time granularities $\mathcal{G}_{\mathcal{VT}}$ and transaction time granularities $\mathcal{G}_{\mathcal{TT}}$. Temporal and spatial granularities are mappings from an index set $\mathcal{IS}$ to the power sets of the temporal [5] and the spatial domains, respectively. For instance, *days* and *weeks* are temporal granularities; *meters*, *yards* and *provinces* are spatial granularities. The temporal dimensions are totally ordered. Temporal and spatial granularities are used to represent ST^2_ODMGe objects attributes and modification actions at different levels of detail.

A *granule* is a subset of a domain corresponding to a single granularity mapping, i.e., given a granularity G and an index $i \in \mathcal{IS}$, $G(i)$ is a granule of G that identifies a subset of the corresponding domain. Granules of the same granularity have disjoint interiors. Moreover, non-empty temporal granules preserve the order of the temporal domains.

Valid time granules bound attribute values, while transaction time granules refers to modification actions. Similarly, spatial granules specify the geographical areas where spatio-temporal attribute values are defined. For instance, consider the value of the daily temperature in Rome the first and the second day of January. To model this value, one can use the labels "01/01", "02/01", and *"Rome"* to denote two temporal granules at granularity *days* and one spatial granule at granularity *municipalities*, respectively.

Granularities differ according to how their granules partition a domain. In this respect, granularities are related by the *finer-than* transitive relationship and its inverse *coarser-than* [5]. For example, granularity *days* is finer-than *months*, which in turn is finer-than *years* (symmetrically, *years* is coarser-than *months*, which is coarser-than *days*). Likewise, *municipalities* is finer-than *countries*. The finer-than relationship is denoted by $\preceq$, while $\prec$ denotes the anti reflexive finer-than.

Given two multigranular values, one at granularity G and one at granularity H such that G and H are not directly related by finer-than, these values may be compared if they are converted, i.e., represented, at the same granularity K, that is finer-than G and H. K is chosen as the granularity that minimizes the number of conversion steps. If K is the coarsest, among the granularities finer-than G

and H, K is referred to as the *greatest lower bound* (GLB) of G and H (denoted as $GLB(G,H)$) [5].

3.2   Multigranular Types and Conversions

Besides conventional database values, a multigranular spatio-temporal database schema includes multigranular spatial, temporal, and spatio-temporal values. Multigranular values are defined as partial functions from the set of granules of the corresponding granularity(ies) to the set of values of a given inner type. Fig. 1 illustrates examples of ST^2_ODMGe multigranular attribute values: `taxpayer_id` is an alphanumeric spatial attribute with type $Spatial_{countries}$(`string`); `address` is an alphanumeric temporal attribute with type $Temporal_{years}$(`string`); finally, `taxes` is a numeric spatio-temporal attribute. The ST^2_ODMGe type of the first value `taxes` in Fig. 1 is $Temporal_{years}(Spatial_{countries}$(`int`)).

In a multigranular database data may be converted at different granularities to increase or reduce their level of detail. In ST^2_ODMGe, the conversion of multigranular geometrical features is obtained through the composition of model-oriented and cartographic map generalisation operators that guarantee topological consistency (e.g., *merge, abstraction*), and refinement operators that perform the inverse functions (e.g., *split, add feature*) [7]. On the other hand, to retrieve, for instance, the annual trend of a phenomenon having a daily representation (e.g., the values of sales in shops located in several countries), also conversion for non geometric attribute values are provided. For example, *average* and *sum* aggregate numeric values; *selection* and its specializations (e.g., *first, main*) coerce alpha-numeric values; *restriction* and *split* refine values.

An interesting property that will be used in the paper to evaluate the correctness of attribute access refers to the *invertibility* of conversions [4]. Indeed, even if when converting a temporal value to a different granularity, and then performing the inverse conversion, we would expect the original value to be returned. However, unfortunately, when converting from a finer to a coarser granularity, we loose some details that we cannot usually recover by applying the inverse conversion to the finer granularity. By contrast, when converting from a coarser to a finer granularity, we introduce some details that we should be able to forget; thus we can recover the original value. Given a pair of conversion functions, we denote them as *quasi-inverse* or *inverse* [4], according to whether the conversions refer to the first

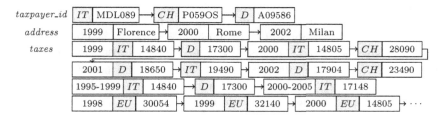

Fig. 1. Example of object state

or the second situation, respectively. In the first case, a measurable indeterminacy is introduced. For example, the pair (average,split) is quasi-inverse, while the pair (split,sum) is inverse.

Granularity conversions may be enriched with user defined functions to suite specific domain requirements (e.g., the RGB spectrum, user credentials). In this case, the user must provide both the conversions to finer and coarser granularities, and he/she has to take care of the invertibiltiy aspect, according to the domain semantics.

3.3 ST^2_ODMGe Classes and Objects

In the following example we illustrate an ST^2_ODMGe class specification.

Example 1. Given an object type for describing *taxpayers*, reporting the value of taxes paid by a person over time in different countries, its definition includes: a spatial attribute taxpayer_id at granularity *countries* to store the fiscal identifiers that a taxpayer holds in different countries; a temporal attribute address at granularity *years* to store the history of his/her fiscal domiciles; and a spatio-temporal attribute taxes, defined at temporal granularity *years* and at spatial granularity *countries*, which stores the amount of taxes the taxpayer pays every year in each country where he/she works (for simplicity we suppose the values are stored according to the same currency in €). □

Given an ST^2_ODMGe class, an ST^2_ODMGe object is defined as follows.

Definition 1. *(ST^2_ODMGe Object). Given a class c, an ST^2_ODMGe object o of c is defined as a 6-tuple $(id, N, v, c, \Upsilon_{VT}^{G_{IT}} \times \Upsilon_{S}^{G_{IS}}, \Upsilon_{TT}^{G_{IT}})$ where: id is the object identifier, unique in the database; N is the set of object names; v is the object state, given as a tuple of attribute values: $(a_1{:}v_1, \ldots, a_n{:}v_n)$, where a_i is an attribute name, and v_i is an attribute value, with $1 \leq i \leq n$; c is the class to which the object belongs; $\Upsilon_{VT}^{G_{IT}} \times \Upsilon_{S}^{G_{IS}}$ is the spatio-temporal object lifespan, represented as set of granules at the temporal chronon and the spatial quantum granularities[2], with respect to valid dimensions; $\Upsilon_{TT}^{G_{IT}}$ is the transactional temporal lifespan of o at the temporal chronon granularity.* ◇

Example 2. Let o be an object of class taxpayer as described in Example 1. According to Definition 1, the values of attributes taxpayer_id, address and the first value of taxes at granularities *years* and *countries* in Fig. 1 define a legal object state v for o. For instance, according to Fig. 1, the contributor in 1999 paid € 14,840 in Italy and € 17,300 in Germany. An example of spatio-temporal lifespan for o is $\{1\ Jan\ 1999\ 00{:}00{:}1, \ldots, 31\ Dec\ 2030\ 23{:}59{:}59\}_{VT}^{seconds} \times \{IT, D, CH\}_{S}^{countries}$, $\{1\ Jan\ 1999\ 00{:}00{:}1, \ldots, NOW\}_{TT}^{seconds}$, where IT is for Italy, D is for Germany, and CH is for Switzerland, assuming *seconds* and *countries* as chronon and spatial quantum granularities. □

---

[2] These are the finest granularities on the spatio-temporal domain.

4   Evolutions

Evolutions are defined and executed at run time on ST^2_ODMGe objects. Our model supports three different types of evolutions: *granularity evolution*, *granularity acquisition*, and *value deletion*.

Granularity evolutions and acquisitions modify the granularity of an attribute. The granularity evolution operation, previously introduced by us [6] in a more limited form and restricted to the valid temporal domain, allows one to define a new portion of an attribute value, specified at a different granularity. The new value, referred to as *target*, is obtained by converting values, referred to as *source*, already stored in the database at a different granularity through the application of granularity conversions. By contrast, granularity acquisitions do not change the database state, but re-define the granularity(ies) that can then be used to insert new attribute values. They have the same effect of a modification of the database schema, as if an SQL ALTER statement were executed. Finally, a value deletion eliminates portions of an attribute value at a given granularity.

Evolutions are performed according to the general execution model of active databases, and have the form: ON *Event* [IF *Condition*] DO *Action*. Given an instance of an ST^2_ODMGe database and a set of evolutions specified for it, the database is continuously monitored. The execution of database transactions modifies the database state and triggers the evolutions whose events refer to such transactions. Then, the corresponding conditions, if present, are evaluated. For the triggered evolutions whose conditions evaluate to TRUE, the corresponding actions are executed. An evolution action is a sequence of operations that may modify attribute granularities and delete attribute values. As a consequence, the database state (or schema, in case of granularity acquisition) may be modified.

The temporal behaviour of ST^2_ODMGe evolutions differs according to their occurrence: ST^2_ODMGe supports *periodic* and *non-periodic* evolutions. These different behaviours may be further characterized with the support of *spatio-temporal bounds*, that may apply to each of the elements of an evolution, thus restricting the occurrence of the evolution event, the evaluation of the condition, and the effects of the action to given temporal periods and geographical areas.

As a consequence of the execution of evolutions, the type of an object state in the ST^2_ODMGe model, that is, of the values of their attributes, changes dynamically. Let a be a multigranular attribute defined in class c. In the general case, the run-time type of a is a Cartesian product of multigranular types, as illustrated by the following example.

Example 3. Let o be the identifier of an object of class taxpayer we described in Example 1. A legal state for o is shown in Fig. 1. The value of attribute taxes in Fig. 1 is a set of spatio-temporal values at different granularities. The first value is the value corresponding to the attribute definition, and is given at temporal granularity *years* and at spatial granularity *countries*. The other two values are obtained from this value through granularity evolutions. They are specified at granularities *5years* and *countries*, and *years* and *ecAlliances* (i.e., economic alliances), respectively. According to the different granularities, they temporally and spatially aggregate the first value (at granularities $years$ and $countries$). The domain of attribute taxes is: $Temporal_{years}(Spatial_{countries}(\texttt{int}))$ $\times\, Temporal_{5years}(Spatial_{countries}(\texttt{int})) \times Temporal_{years}(Spatial_{ecAlliances}(\texttt{int}))$. $\square$

An evolution defined on an attribute a is specified on one of the granular values that compose the value of a. Each value is referred to as a *granularity level*, and is identified by its granularity. More precisely, given an object o of class c, the value of attribute a at a given (either temporal or spatial) granularity G (at temporal granularity G_t and at spatial granularity G_s, respectively) is referred to as the *granularity level* $< G >$ of a ($< G_t, G_s >$ if the attribute is spatio-temporal). Given for instance attribute taxes of Example 3, with the object state depicted in Fig. 1, we have three different granularity levels: $<$ $years, countries >$, $< 5years, countries >$, $< years, ecAlliances >$. In the following example we illustrate the syntax to define evolutions.

Example 4. Given class taxpayer of Example 1, the following evolution summarises the older annual record of taxes at a coarser temporal granularity $5years$.

ON update taxpayer.taxes$< years, countries >$ during $\{1999,\ldots,2014\}^{years}_{VT}$
IF every 5^{years}_{VT}
DO evolve $< years, countries >$ to $< 5years, countries >$ in $\{\texttt{IT}\}^{countries}$
 using average$_{years \to 5years}$, restriction$_{5years \to years}$.

The evolution is defined for attribute taxes, which evolves from granularity level $< years, countries >$ to granularity level $< 5years, countries >$. It is triggered by the updates of the evolution source granularity level, and involves at each execution 5 years of data, as specified by the periodic condition every 5^{years}_{VT}. The evolution involves only the taxes paid between year 1999 and 2014 in Italy, according to the event temporal bound $\{1999, \ldots, 2014\}\,^{years}_{VT}$, and by the action spatial bound in $\{\texttt{IT}\}^{countries}$. Granularity conversion average$_{years \to 5years}$ is applied for creating the target level, while conversion restriction$_{5years \to years}$ may be used to recover the original values whenever these are deleted from the database.

Now suppose that the following evolution is specified from granularity level $< years, countries >$ to $< years, ecAlliance >$, where $ecAlliance$ represents (non-overlapping) economical alliances among different countries:

ON update taxpayer.taxes$< years, countries >$
IF after 1^{years}_{VT}
DO evolve $< years, countries >$ to $< years, ecAlleance >$ in countries($\{\texttt{EC}\}^{ecAlliance}$)
 using sum$_{countries \to ecAlliance}$, split$_{ecAlliance \to countries}$.

The notation $G(\varUpsilon^{G'})$ denotes the conversion of the set of $G'-$granules $\varUpsilon^{G'}$ to granularity G. Note that the spatial bound in countries($\{\texttt{EC}\}^{ecAlliance}$) constraints the action execution, and accordingly the evolution aggregates only the tax logs that refer to European Countries. The evolution is executed periodically, according to the condition after 1^{years}_{VT}. Fig. 1 is an example of object state, after the execution of the evolutions. $\square$

5   Object Consistency

ST^2_ODMGe evolutions affect the conventional notion of object consistency, because at run-time the object state may no longer match their class definitions, as illustrated by Example 4. However, evolved multigranular attribute values are created starting from source granularity levels, which rely in turn on the original granularity level defined for the attribute. The evolved values are semantically consistent with the original ones, to which they are related by a chain of (quasi) inverse granularity conversions. Therefore, the original value may be recomputed from an evolved one when needed (for example when the original data are deleted from the database), with a bounded imprecision. For the same reason, the introduction of granularity acquisition does not pose problems, i.e., the original and the refined granularity levels are related by a pair of (quasi) inverse granularity conversions.

In the following, ST^2_ODMGe object consistency is formalized, taking into account how evolutions modify the object state and relying on the relationships among the granularity levels of object attributes. Such relationships are formalized by the notion of *Granularity Levels Graph* (GLG) of an attribute, which is preliminary to the consistency formalization.

5.1   Attribute Granularity Level Graph

Let a be a multigranular attribute defined in class c. The granularity levels that compose the value of a are pairwise linked by pairs of quasi-inverse granularity conversions, to form a graph that we refer to as the *Granularity Levels Graph (GLG)* of the attribute, which is formalised by the following definition. This structure, that must not be confused with the database granularity lattices we may define for G_{VT}, G_{TT} and G_S relying on finer-than [5], is specific for each multigranular attribute, and relates its granularity levels enabling to navigate among them when accessing the attribute value. In this definition, and in the rest of the paper, for simplicity we consider a multigranular attribute that refers to either the spatial or the temporal domain, whenever the case of spatio-temporal values may be inferred straightforwardly. Whenever needed, we point out the differences of the spatio-temporal case.

Definition 2. (Granularity level graph - GLG) *Given a set of temporal or spatial granularity levels* $< G_i >$ *defined for attribute* a, *where* $\forall i = 1 \ldots n$, $G_i \in \mathcal{G}$ *is either a temporal or a spatial granularity, the granularity level graph of* a, *denoted by* a_{GLG}, *is a graph* (V, E) *such that* $V = \{< G_1 >, \ldots, < G_n >\}$, *and* $E = \{< G_j > \rightleftarrows < G_k >\}$, *if* G_j *and* G_k *are related by the finer-than relationship, and two (quasi)inverse granularity conversions* $f_{G_j \to G_k}$ *and* $g_{G_k \to G_j}$ *have been defined;* $1 \leq j \leq n$, $1 \leq k \leq n$.   ◇

Similarly, given a spatio-temporal attribute a and the set of its spatio-temporal granularity levels $< G_{t_i}, G_{s_i} >$, and given the granularity conversions defined among these granularity levels through evolution specifications, a GLG is defined for a. Given an attribute a, let a_{GLG} denote its GLG. Moreover, given a

GLG a_{GLG} the set of its nodes and edges are denoted by $a_{GLG}.V$ and $a_{GLG}.E$, respectively.

Example 5. Given attribute `taxes` of class `taxpayer` of Example 1, suppose the evolutions of Example 4 have been specified. Hence, $taxes_{GLG} = (V, E)$ is specified as follows:

$taxes_{GLG}.V = \{< years, countries >, < 5years, countries >, < years, ecAlliance >\}$;
$taxes_{GLG}.E = \{< years, countries > \rightleftarrows < 5years, countries >,$
$$< years, countries > \rightleftarrows < years, ecAlliance >\}. \qquad \square$$

The following property ensures that, given two granularity levels in an attribute GLG, it is always possible to compare them even if the granularity levels are not directly linked through granularity conversions in the GLG. Therefore, to solve an attribute access we may navigate among the values defined in the attribute GLG by using the defined granularity conversions, as we will see in the following section.

Property 1. Let $< G_i >$ and $< G_j >$ be two granularity levels in a_{GLG}. Then, one of the following conditions holds:

- $G_i \prec G_j$;
- $G_j \prec G_i$;
- $GLB(G_i, G_j) \in a_{GLG}.V$. $\triangledown$

With the following definition we introduce also the concepts of *bottom* and *top* granularities for a multigranular attribute.

Definition 3. (Bottom and top granularities in an attribute GLG). *Given an object o and a multigranular attribute a defined for o, $G^\perp$ is the set of the (temporal or spatial) bottom granularities of a, that is the finest granularities in a_{GLG}, i.e., no granularity G, with $< G > \in a_{GLG}.V$ exists such that, $\forall G' \in G^\perp$, $G \prec G'$. Symmetrically, $G^\top$ is the set of the (temporal or spatial) top granularities of a, that is, the coarsest granularities in a_{GLG} for which no granularity H, with $< H > \in a_{GLG}.V$ exists such that, $\forall H' \in G^\top$, $H' \prec H$.* $\diamond$

Example 6. Given attribute `taxes` of Example 1 with the GLG of Example 5, $G^\perp_{VT} = \{years\}$ and $G^\top_{VT} = \{5years\}$; $G^\perp_S = \{countries\}$, while $G^\top_S = \{ecAlliance\}$. $\square$

5.2   Consistency Conditions for ST^2_ODMGe Objects

Relying on attribute GLGs we now revisit the consistency of ST^2_ODMGe objects. We define the constraints that are useful to define the access strategies and must therefore be preserved when manipulating object states. Such constraints are expressed with respect to all the dimensions supported by the model.

To guarantee object consistency, every ST^2_ODMGe attribute value must satisfy the following conditions: 1) each attribute value belongs to the set of legal values of the corresponding type; 2) whenever the attribute value is an object

identifier, the referred object exists in the database sometimes during the temporal transactional lifespan of the object; 3) the value of a multigranular attribute
is a tuple of multigranular values, whose spatial and the temporal domains do
not exceed the spatial and temporal lifespan of the object; thus for each defined
value, the corresponding granule intersects the object lifespan[3]; 4) for a multigranular attribute a GLG is defined according to Definition 2, and its edges
preserve the relationships holding among granularities of the granularity levels.
The previous constraints are formalised by Definition 4, which expresses the
notion of run-time consistency for objects in an ST^2_ODMGe database.

Definition 4. (ST^2_ODMGe Consistent Instance). *Let c be a class and attr its
attribute specification $\{(b_1, \tau_1), \ldots, (b_m, \tau_m)\}$, where $\forall j, 1 \le j \le m$, b_j an attribute
name and τ_j an attribute type. Let $\mathcal{LT}$ and $\mathcal{OT}$ be the sets of literal and object
types, respectively, and let T_{geom} be the set of geometric vector types (e.g., point,
line, polygon). Let $[\![\tau]\!]$ be the set of legal values for type τ, and $[\![\tau']\!]_i^{G_{IT}}$ be the
set of legal values defined for τ' in granule $G_{IT}(i)$. Let o be a ST^2_ODMGe object
defined as $(id, N, (a_1 : v_1, \ldots, a_p : v_p), c, \Upsilon_{VT}^{G_{IT}} \times \Upsilon_S^{G_{IS}}, \Upsilon_{TT}^{G_{IT}})$. Then, object o is a
consistent instance of c if all the following conditions hold:*

1. $\forall i, 1 \le i \le p, \exists (b, \tau) \in attr$ *such that* $b = a_i$;
2. $\forall (b, \tau) \in attr, \exists k, 1 \le k \le p$, *such that* $b = a_k$ *and the following conditions
   hold:*
   (a) *if* $\tau \in \mathcal{LT}$, $v_k \in [\![\tau]\!]$ *(Cf. condition 1);*
   (b) *if* $\tau \in \mathcal{OT} \cup T_{geom}$, $v_k \in \bigcup_{\Upsilon_{TT}^{G_{IT}}} \{[\![\tau]\!]_h^{G_{IT}} \mid h \in \mathcal{IS}\}$ *(Cf. condition 2);*
   (c) *if* τ *is a multigranular type at granularity G, all the following conditions
       hold (Cf. conditions 3 and 4):*
       i. $v_k = (v_{k_1}, v_{k_2}, \ldots, v_{k_n})$, *with* $n \ge 1$;
       ii. $\forall j, 1 \le j \le n$, *such that* v_{k_j} *is defined,*
          A. $\exists \tau_j$, *where* τ_j *is a multigranular type at granularity* G_j, *such that*
             $v_{k_j} \in [\![\tau_j]\!]$;
          B. $\forall i \in \mathcal{IS}$ *such that* $v_{k_j}(i)$ *is defined,* $G_j(i) \cap (\bigcup_{\Upsilon_{VT}^{G_{IT}} \times \Upsilon_S^{G_{IS}}} \{G_{IS}(h) \mid
             h \in \mathcal{IS}\}) \ne \emptyset$;
       iii. *a granularity (level) graph* (V_k, E_k) *is defined, such that:*
          A. $V_k = \{< G_1 >, \ldots, < G_n > \mid v_{k_j} \in [\![\tau_j]\!]$ *is defined, with* $1 \le j \le
             n\}$;
          B. $E_k = \{< G_q > \rightleftarrows < G_r >$, *with* $G_q \prec G_r$ *or* $G_r \prec G_q$, $1 \le q \le n, 1 \le
             r \le n\}$ *and two (quasi)inverse granularity conversions* $f_{G_q \to G_r}$
             and $g_{G_r \to G_q}$ *have been defined.*   ◇

Example 7. Given object o of Example 3, we assume the evolutions of Example 4 have been executed on o.taxes, with taxes$_{GLG}$ as defined in Exam-
ple 5. Given $\{1\ January\ 1999\ 00:00:1, \ldots, 31\ December\ 2030\ 23:59:59\}_{VT}^{seconds} \times
\{IT, D, CH\}_S^{countries}$, $\{1\ January\ 1999\ 00:00:1, \ldots, NOW\}_{TT}^{seconds}$, the lifespan of
o, assuming *seconds* and *countries* as chronon and spatial quantum granularities,

---

[3] Border granules may not be completely included in the object lifespan, but their
intersection with it must be non-empty.

and assuming updates on o have been executed after 1998, then object o, with the object state of Fig. 1, is a consistent instance of class taxpayer according to Definition 4. By contrast, it would be inconsistent if its lifespan were $\{1\ January\ 2000\ 00{:}00{:}1, \ldots, 31\ December\ 2002\ 23{:}59{:}59\}_{VT}^{seconds} \times \{IT\}_{S}^{countries}$, $\{1\ January\ 1999\ 00{:}00{:}1, \ldots, NOW\}_{TT}^{seconds}$, because it would intersect neither the values defined before year 2000, nor the countries different from Italy.   □

6   Object Access

In this section we discuss the access to ST^2_ODMGe multigranular attribute values. To simplify the presentation, we introduce a basic access that requires the attribute value defined in a single granule. Access to multiple granular values follows straightforwardly. We further distinguish between two forms of access, *qualified* and *unqualified*, depending on a granularity conversion being specified or not, respectively The strategies to solve them are discussed separately. Therefore, we discuss the invariance of object accesses with respect to evolutions, and characterize unsolvable object accesses.

6.1   Qualified and Unqualified Access

The concept of object access we consider is formalised by the following definition.

Definition 5. (ST^2_ODMGe object access). *Let o be an object identifier, and let a be the name of an attribute defined for o. If a is a multigranular temporal attribute, let G be a temporal granularity. If a is a multigranular spatial attribute, let G be a spatial granularity. Given a granule label l^G, an object access is an expression of the form $o.a \downarrow^{[f]} l^G$, requiring the value of attribute a of object o in granule l^G. If a granularity conversion f is specified, f is applied to compute the access result. In the latter case, the access is referred to as* qualified. *Otherwise it is* unqualified.   ◇

The access to a multigranular spatio-temporal attribute a is expressed as $o.a \downarrow^{[f]} l^{G_t} \downarrow^{[f']} l^{G_s}$, where G_t and G_s are a temporal and a spatial granularity, respectively; l^{G_t}, l^{G_s} are two granule labels for G_t and G_s; f and f' are granularity conversions.

Example 8. Given class taxpayer of Example 1 and object o with the state in Fig. 1, o.taxes $\downarrow \{1998\}_{VT}^{years} \downarrow \{IT\}_{S}^{countries}$ is the unqualified access to the payments made by the contributor during 1998 to the Italian Revenue service. By contrast, object access o.taxes $\downarrow \{1999\}_{VT}^{years} \downarrow^{split[p(x)]} \{IT\}_{S}^{countries}$, where $p(x)$ is the probability distribution: $p(x) = \{(IT, 0.5), (D, 0.5)\}$, is the qualified access to the same payments, requiring the application of the refinement function $split[p(x)]$.   □

6.2   Solving Unqualified Object Access $o.a \downarrow l^G$

To solve the unqualified object access $o.a \downarrow l^G$ we check whether the requested value is available, i.e., if $< G >$ is a granularity level defined for a and if the value of $o.a$ for granule l^G is defined. If so, such value, that we denote as $o.a_G(l^G)$, where $o.a_G$ is the granularity level $< G >$ defined for a, is returned.

Otherwise, the requested value must be computed starting from the values, stored in other granularity levels, that intersect l^G. In this case, two different strategies may be applied for solving the access, depending on whether the user wants to maximize the accuracy of the result or the access efficiency. The *efficiency* maximization strategy minimizes the number of intermediate accesses needed to solve $o.a \downarrow l^G$. According to this strategy, the application of conversion functions from coarser to finer granularities is preferred, because just one value is accessed for each of the granularity levels involved. By contrast, when maximizing *accuracy*, the highest precision is required in computing the result. Therefore the application of granularity conversions from finer to coarser granularities takes precedence, because they minimize the indeterminacy in the returned values (cf. Section. 3).

Fig. 2 reports the algorithm to solve $o.a \downarrow l^G$. The spatio-temporal access $o.a \downarrow l^{G_t} \downarrow l^{G_s}$ follows straightforwardly. We assume that the granularity levels in a_{GLG} are ordered according to the finer-than relationship. Spatio-temporal granularity levels are ordered first according to temporal granularities, and then with respect to spatial granularities. ACCURATE denotes that an accurate answer is preferred, whilst efficiency is the default.

The computational complexity of the algorithm in Fig. 2 is $O(n)$. Indeed, assuming that the set of granularity levels defined for each attribute value is finite, and the time required for the application of granularity conversions is linear, the complexity of the algorithm is mainly given by the sequential access to a given value in a granularity level. If we assume that indexing is applied on granularity levels (e.g., BTree$^+$ for temporal values and R-Tree for spatial values), the complexity may decrease to $O(log(n))$ if the internal nodes of the R-Tree do not overlap. An optimal worst-case complexity is guaranteed also if the indices for spatial data are, for example, PR-Trees [2].

An important result of our work is thus that the introduction of evolutions does not increase the complexity of the access with respect to the conventional multigranular case. Furthermore, complexity may improve whenever the access involves values at granularities among those defined for the attribute, because

$$
\begin{aligned}
&\textbf{if } \exists o.a_G(l^G) \neq \bot \textbf{ then return } o.a_G(l^G)\\
&\textbf{else if } \texttt{ACCURATE} \textbf{ then}\\
&\qquad \textbf{while } \exists o.a_K \text{ s.t. } K \preceq G \text{ and } \forall l_k^K \in K(l^G)\\
&\qquad\qquad \text{s.t. } o.a_K(l_k^K) \neq \bot\\
&\qquad\qquad \textbf{return } f_{K \to G}(o.a_K)(l^G)\\
&\qquad \textbf{return null}\\
&\qquad \textbf{else while } \exists o.a_H \text{ s.t. } G \preceq H\\
&\qquad\qquad \textbf{return } g_{H \to G}(o.a_H)(l^G)\\
&\qquad \textbf{while } \exists o.a_K \text{ s.t. } K \preceq G\\
&\qquad\qquad \textbf{return } f_{K \to G}(o.a_K)(l^G)\\
&\qquad \textbf{return null}
\end{aligned}
$$

Fig. 2. Algorithm for object access $o.a \downarrow l_G^i$

the access result may be already pre-computed in the database. Indeed, in both execution strategies, the access follows an iterative approach, and to solve it we may need to move across several granularity levels. Once a value is found (or a set of values, in the accuracy maximization strategy) that satisfies the access, a sequence of conversions must be performed. If some precomputed value is already available at an intermediate granularity, these values need not be recomputed, thus improving performance.

Example 9. Given access $o.\mathsf{taxes} \downarrow \{1998\}_{VT}^{years} \downarrow \{IT\}_{S}^{countries}$ introduced in Example 8, and object o of class $\mathsf{taxpayer}$ whose state is shown in Fig. 1, the access results in € 14, 840. □

6.3 Solving Qualified Object Access $o.a \downarrow^{f} l^{G}$

If the access is qualified by a granularity conversion f, this function will be used to compute the access result, taking precedence over the functions already specified in granularity evolutions and acquisitions. Differently from unqualified access, if the accuracy maximization strategy is adopted, an existing value for the specified granule is discarded, if it was constructed with a different function. The value would be used instead by the efficiency maximization strategy. If this value is not defined, we distinguish whether f is a conversion to a coarser granularity (CF), or to a finer granularity.

Fig. 3 reports the algorithm for solving a qualified access $o.a \downarrow^{f} l^{G}$. The spatio-temporal object access $o.a \downarrow_{t}^{f} l_{t}^{G} \downarrow^{f'} l_{s}^{G}$ follows straightforwardly. As above, ACCURATE denotes that an accurate answer is preferred.

As in the case of unqualified access, the algorithm for qualified access shown in Fig. 3 has computational complexity $O(n)$, which may reach the optimum if indexing is used on the granularity level values as in the previous case.

Example 10. Given the access $o.\mathsf{taxes} \downarrow \{1998\}_{VT}^{years} \downarrow^{split[p(x)]} \{IT\}_{S}^{countries}$, with $p(x) = \{(IT, 0.5), (D, 0.5)\}$, and object o of class $\mathsf{taxpayer}$ with the object state

```
if ∃o.a_G(l^G) ≠⊥ then
    if ACCURATE then
        if ∃o.a_K s.t. K ⪯ G
            and f_{K→G} is defined between o.a_K and o.a_G
            then return o.a_G(l^G)
        else return o.a_G(l^G)
    if f is a CF then
        while ∃o.a_K s.t. K ⪯ G
            if ACCURATE then
                if ∀l_k^K ∈ K(l^G) s.t. o.a_K(l_k^K) ≠⊥ then
                    return f_{K→G}(o.a_K)(l^G)
                else return null
            else return f_{K→G}(o.a_K)(l^G)
        return null
    else while ∃a_H s.t. G ⪯ H
        return f_{H→G}(o.a_H)(l^G)
```

Fig. 3. Algorithm to solve the qualified object access $o.a \downarrow^{f} l_{G}^{i}$

depicted in Figure 1. When accuracy is required, the access results in € 15,027. This value is computed starting from the aggregate value at granularities $< years, ecAlliances >$.   □

6.4   Evolution Invariant Object Access

In order to preserve the consistency of query answers, evolution execution must not affect access results. In what follows, after a preliminary definition introducing the notion of evolution invariant access, we show that unqualified object access is invariant with respect to the three forms of evolution discussed in this paper, given a bounded approximation introduced by granularity conversions.

Suppose that $< G >$ is one of the granularity levels defined for attribute a, and suppose that from $< G >$ an evolution has been executed involving granule l^G. In the case of acquisitions we consider the insertion of new values in the target granularity level. Suppose the evolution has not been performed yet. Assuming that no updates occurred, if the access $o.a \downarrow l^G$ results in the same value when executed just before and just after the evolution execution, the access is referred to as *evolution invariant*. Considering how we build granularity levels, and the specification of granularity conversions, the following result holds.

Proposition 1. *Given a granularity level $< G >$ defined for a, and provided that a granularity level $< G' >$ exists such that $o.a_{G'}(G'(l^G))$ is defined, every object access $o.a \downarrow l^i_G$ is evolution invariant.*   ◇

Evaluating the access just before and just after the execution of an evolution defined from $< G' >$ to $< G >$, the access results in the same value. By contrast, for granularity acquisitions and deletions the access is evolution invariant but with a bounded imprecision, which is due to the application of granularity conversions. Indeed, if the value defined for granule l^G is deleted, we can recover it if the value has been involved in a granularity evolution to granularity level $< G' >$. In the case of granularity acquisition, the old and the new acquisition levels are related by a pair of (quasi)inverse granularity conversions, which guarantees the value consistency among the two levels, modulo a bounded error.

6.5   Unsolvable Object Access

We may characterize ST^2_ODMGe object accesses that can be statically detected as unsolvable. As usual, null is returned whenever not enough information is available to solve the access. However, we can distinguish between accesses that are statically known to be unsolvable, that is, for which no database state exists such that these accesses will produce a value different from null, and accesses that can produce or not an answer depending on the actual content of the database. Detecting object accesses that are statically unsolvable reduces query execution times, because the system does not need to execute them, but it may return immediately **null**. Given an object access $o.a \downarrow l^G$ (the case of $o.a \downarrow l_t^G \downarrow l_s^G$ follows straightforwardly), the following result holds.

Proposition 2. *Given attribute a defined for an object o, and given value v for a, such that a_{GLG} includes the granularity levels $< G_1 >, \ldots, < G_n >$, the object access $o.a \downarrow l^G$ is unsolvable if one of the following conditions holds:*

- G is not related by $\preceq$ to any of $G_1, \ldots, G_n$;
- $G \prec K$, $K \in G^\perp$;
- $H \prec G$, $H \in G^\top$.   ◇

7 Concluding Remarks

In this paper we have investigated issues related to the evolution of multigranular spatio-temporal objects. The main contribution of this paper is the definition of ST^2_ODMGe, a multigranular spatio-temporal model supporting the adaptive management of multigranular spatio-temporal attributes. Our approach to evolutions allows one to model a large variety of situations. Consistency constraints on attribute values have been relaxed, because the run-time value of a multigranular attribute is a Cartesian product of multigranular values, linked in a connected acyclic graph through the specification of granularity conversions. Relying on such a structure, object accesses may be solved according to different strategies and error tolerances.

The ST^2_ODMGe model may be considered as a basis for future investigations on issues concerning evolutions of multigranular spatio-temporal objects. In particular, the development of a prototype of the model will allow us to investigate the trade off between the flexibility, provided by the model, and the consistency that is guaranteed by the statical specification of evolutions.

Efficient and comprehensive implementations are crucial. Several alternatives can be investigated including 1) implementation of the required features as class libraries on top of an existing DBMS, and 2) extensions to a DBMS engine. Both approaches have shortcomings. The former approach may not be able to support all required features; it may also have performance problems, as it may be impossible to allow the inclusion of specialized indexing techniques or query optimization techniques. The latter approach may require extensive implementation efforts and may also not support all required features, especially the ones depending on the application domain, like specialized spatial conversion operators.

Moreover, since evolution specifications are formulated according to the active database paradigm, it is important that tools for the analysis of evolution triggers be supported to detect non-terminating as well as non-deterministic executions. Note that such issues have been extensively investigated in the area of active DBMS and no general solutions exist. However, for specialized domains, such as, in our case, the evolution of granularities, effective solutions to these issues could be found.

References

1. Andrienko, G., Malerba, D., May, M., Teisseire, M.: Mining spatio-temporal data. J. of Intelligent Information Systems 27(3), 187–190 (2006)
2. Arge, L., de Berg, M., Haverkort, H.J., Yi, K.: The Priority R-Tree: A Practically Efficient and Worst-Case Optimal R-Tree. In: Proc. of SIGMOD Int'l Conf. on Management of Data, pp. 347–358. ACM, New York (2004)
3. Belussi, A., Combi, C., Pozzani, G.: Towards a Formal Framework for Spatio-Temporal Granularities. In: Proc. of 15th Int'l Symp. on Temporal Representation and Reasoning, pp. 49–53. IEEE Computer Society, Los Alamitos (2008)
4. Bertino, E., Camossi, E., Guerrini, G.: Access to Multigranular Temporal Objects. In: Christiansen, H., Hacid, M.-S., Andreasen, T., Larsen, H.L. (eds.) FQAS 2004. LNCS, vol. 3055, pp. 320–333. Springer, Heidelberg (2004)
5. Bettini, C., Jajodia, S., Wang, X.: Time Granularities in Databases, Data Mining, and Temporal Reasoning. Springer, Heidelberg (2000)
6. Camossi, E., Bertino, E., Guerrini, G., Mesiti, M.: Handling Expiration of Multigranular Temporal Objects. J. of Logic and Computation 14(1), 23–50 (2004)
7. Camossi, E., Bertolotto, M., Bertino, E.: A multigranular Object-oriented Framework Supporting Spatio-temporal Granularity Conversions. Int'l J. of Geographical Information Science 20(5), 511–534 (2006)
8. Camossi, E., Bertolotto, M., Bertino, E.: Multigranular spatio-temporal models: Implementation challenges. In: Proc. of 16th SIGSPATIAL Int'l Conf. on Advances in Geographic Information Systems. ACM, New York (2008)
9. Garcia-Molina, H., Labio, W.J., Yang, J.: Expiring Data in a Warehouse. In: Proc. of 24th Int'l Conf. on Very Large Data Bases, pp. 500–511. ACM, New York (1998)
10. Jensen, C.S., Dyreson, C.E., Bohlen, M., Clifford, J., et al.: A Consensus Glossary of Temporal Database Concepts. In: Etzion, O., Jajodia, S., Sripada, S. (eds.) Dagstuhl Seminar 1997. LNCS, vol. 1399, pp. 367–405. Springer, Heidelberg (1998)
11. Orlando, S., Orsini, R., Raffaeta, A., Roncato, A., Silvestri, C.: Spatio-temporal Aggregations in Trajectory Data Warehouses. In: Song, I.-Y., Eder, J., Nguyen, T.M. (eds.) DaWaK 2007. LNCS, vol. 4654, pp. 66–77. Springer, Heidelberg (2007)
12. Skyt, J., Jensen, C.S., Mark, L.: A Foundation for Vacuuming Temporal Databases. Data & Knowledge Engineering 44(1), 1–29 (2003)
13. Tao, Y., Papadias, D.: Historical spatio-temporal aggregation. ACM Transactions on Information Systems 23(1), 61–102 (2003)
14. Toman, D.: Expiration of Historical Databases. In: Proc. of 8th Int'l Symp. on Temporal Representation and Reasoning. IEEE Computer Society, Los Alamitos (2001)
15. Yang, J., Widom, J.: Incremental computation and maintenance of temporal aggregates. The Int'l J. on Very Large Databases 12(3), 262–283 (2003)
16. Zhang, D., Gunopulos, D., Tsotras, V.J., Seeger, B.: Temporal and spatio-temporal aggregation over data streams using multiple time granularities. Information Systems 28(1-2), 61–84 (2003)

TOQL: Temporal Ontology Querying Language

Evdoxios Baratis, Euripides G.M. Petrakis, Sotiris Batsakis, Nikolaos Maris,
and Nikolaos Papadakis

Department of Electronic and Computer Engineering
Technical University of Crete (TUC)
Chania, Greece
{dakis,petrakis,batsakis}@intelligence.tuc.gr,
nickmeet@gmail.com, npapadak@intelligence.tuc.gr

Abstract. We introduce TOQL, a query language for querying time information in ontologies. TOQL is a high level query language that handles ontologies almost like relational databases. Queries are issued as SQL-like statements involving time (i.e., time points or intervals) or high-level ontology concepts that vary in time. Although independent from TOQL, this work suggests a mechanism for representing time evolving concepts in ontologies based on the four-dimensional perdurantist mechanism. However, TOQL prevents users from being familiar with the representation of time in ontologies. To show proof of concept, an application has been developed that supports translation and execution of TOQL queries on temporal ontologies combined with a reasoning mechanism based on event calculus. A real world temporal ontology is also implemented on which several TOQL example queries are processed and discussed.

1 Introduction

Dealing with information that changes in time over the semantic web is a difficult problem to deal with. Recent advances in semantic web technology suggest that this can be achieved by adding the concepts of time and change in a rich semantics ontology representation, allowing time to affect the status of the described concepts [13,6].

Ontologies offer the means for representing high level concepts, their properties and their interrelationships. Dynamic or temporal ontologies will in addition enable representation of time evolving information in ontologies through e.g., versioning [8] or the four-dimensional perdurantist approach [15]. According to this approach, all entities are perdurants: each entity is considered to be an event and has a start and an end point. An entity can be seen as a "space-time worm", with the slices of the worm being temporal parts (time slices) of the entity. A temporal ontology query language is then needed to support searching for temporal concepts and time related information.

The current state of the art of ontology languages requires submitting a textual, description logic (DL) query or SQL-like query [17]. However the logic and syntax of such languages necessitates a tedious effort from users before being

N. Mamoulis et al. (Eds.): SSTD 2009, LNCS 5644, pp. 338–354, 2009.

able to write queries effectively. State-of-the-art ontology query languages such as SeRQL [1] or SPARQL [11] have limited (if not at all) expressive power for handling time in queries (their syntax does not support temporal operators). The present work addresses all these issues.

We introduce TOQL (Temporal Ontology Querying Language), a high-level query language for querying (time) information in ontologies. TOQL handles ontologies almost like relational databases. Queries in TOQL are issued as SQL statements involving time and high-level ontology concepts that (may) vary in time. TOQL maintains the basic structure of an SQL language (SELECT - FROM - WHERE) and treats the classes and the properties of an ontology almost like tables and columns of a database. TOQL supports queries not only on static information in the static part of the ontology (as conventional ontology query languages do) but also supports queries on time evolving information instantiated to the ontology (dynamic part). TOQL supports Allen operators that allow comparisons between time intervals, and the operator AT that allows comparisons between time points or time intervals.

Besides TOQL syntax, this work demonstrates, full query functionality on ontologies in OWL. This includes query translation and execution of temporal queries along with a mechanism for representing time evolving concepts in ontologies inspired by the four-dimensional perdurantist (4D fluent) approach [16]. However, TOQL syntax is independent of any temporal representation and can work with any other mechanism (e.g., versioning). As such, the 4D fluent (perdurantist) mechanism is not part of the language and it is not visible to the user (so the user need not be familiar with peculiarities of the underlying mechanism for time information representation).

In the accompanying implementation, TOQL queries are first translated into equivalent statements in SeRQL which are then executed on the underlying OWL temporal ontology. The query interpreter addresses information in the ontology to generate a projection (in time) of the evolution of the acquired ontology concepts. To show proof of concept, a real world temporal ontology (for enterprise information) is also implemented on which several TOQL example queries are discussed.

Related work in the field of knowledge representation is discussed in Section 2. This includes, discussion on temporal and ontology query languages along with issues related to representing time evolving information in ontologies. The TOQL language is presented in Section 3 (a formal description of the language's syntax in BNF is given in [3]). The implementation of TOQL is discussed in Section 4, followed by conclusions and issues for future work in Section 5. Several query examples are also given and discussed throughout the work.

2   Background and Related Work

Several representation languages are defined for the Semantic Web, the most important of them are referred to as the OWL-family [10] of ontology languages (OWL-Full, OWL-DL and OWL-Lite) for ontology building and knowledge representation. OWL-S [5] is an ontology for describing properties and capabilities of web services. Within OWL-S, a sub-ontology, OWL-Time [6] has been developed that is much simpler and provides a vocabulary for expressing the most needed time-related facts.

Dealing with information that changes over time is a critical problem in Knowledge Representation (KR). Representation languages such as OWL, RDF (which are based on description logics),the same as frame-based and object-oriented languages (F-logic) are all based on binary relations. Binary relations simply connect two instances (e.g., the employee with the company) without any temporal information. Nevertheless, time representation using OWL is feasible, although complicated [16].

2.1   Representation of Time

The OWL-Time temporal ontology describes the temporal content of Web pages and the temporal properties of Web services. Apart from language constructs for the representation of time in ontologies, there is still a need for mechanisms for the representation of the evolution of concepts (events) in time.This is related to the problem of the representation of time in temporal (relational and object oriented) databases [18]. Existing methods are relying mostly on temporal Entity Relation (ER) models [19] taking into account valid time (i.e., time interval during which a relation holds), transaction time (i.e., time at which a database entry is updated) or both. Also time is represented by time points, intervals or finite sets of intervals. However, time representation in OWL differs because (a) OWL semantics are not equivalent to ER model semantics (e.g., OWL adopts the *Open World Assumption* while ER model adopts the *Closed World Assumption*) and (b) relations in OWL are restricted to binary ones. Time representation in Semantic Web can be achieved using *Temporal Description logics, Reification, Versioning or 4D-fluents*.

Temporal Description Logics (TDL) extend Description Logics (DL) with additional time representation operators and semantics such as *"until"* and *"always in the past"*. Many TDLs have been proposed [20,21] with the most expresive of them being undecidable. Contrary to other approaches, temporal description logics offer additional semantics and reasoning mechanisms and they don't suffer from data redundancy. All other approaches except TDLs require temporal semantics to be defined using an additional set of rules combined with a reasoning mechanism, as we did in this work. TDLs disadvantage is that they require extending OWL to represent time (by introducing additional operators and semantics), while the other approaches can be implemented directly using OWL.

Reification is a general puprose technique for perpesenting *n*-ary relations using a language such as OWL that permits only binary relations. Specifically, an *n*-ary relation is represented as a new object that has all the arguments of the *n*-ary relation as attributes. For example if the relation R holds between objects A and B at time t, expressed as $R(A,B,t)$, this is represented in OWL using reification as an object R with attributes A, B and t. Reification suffers from two disadvantages: (a) data redundancy, because a new object is created whenever a temporal relation has to be represented (this is a problem common

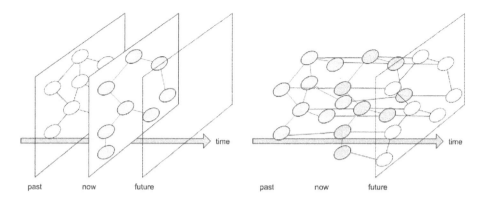

Fig. 1. Schematic representation of the concept of time determined ontology

to all approaches based on non temporal Description Logics such as OWL-DL) and (b) offers limited OWL reasoning capabilities [16].

Versioning [8] suggests that the ontology has different versions (one per instance of time). When a change takes place, a new version is created. Versioning suffers from several disadvantages: (a) changes even on single attributes require that a new version of the ontology be created leading to information redundancy (b) searching for events occurred at time instances or during time intervals requires exhaustive searches in multiple versions of the ontology,(c) it is not clear how the relation between evolving classes is represented. Furthermore, ontology languages such as OWL [10] are based on binary relations (relations connecting two instances) with no time dimension.

The *4D-fluent* (perdurantist) approach [15] shows how temporal information can be represented effectively in OWL. Notice though that it still sufferers from data redundancy. Concepts in time are represented as 4-dimensional objects with the 4th dimension being the time. Time instances and time intervals are represented as instances of a *time interval* class which in turn is related with time concepts varying in time. Changes occur on the properties of the temporal part of the ontology keeping the entities of the static part unchanged.

As illustrated in Figure 1[1] a development in time can only be described by a series of snapshot ontologies each superimposing itself on the previous version of the described reality (left). The 4D-fluent (perdurantist) ontology (on the right) allows the concepts of time and change to become integral parts of the ontology. In TOQL we opt for the later type of representation based on 4D fluents.

Following the approach by Welty and Fikes [15], to add the time dimension to an ontology, classes *TimeSlice* and *TimeInterval* with properties *tsTimeSliceOf* and *tsTimeInterval* respectively are introduced. Class *TimeSlice* is the domain class for entities representing temporal parts (i.e., "time slices") and class *TimeInterval* is the domain class of time intervals. A time interval holds

---

[1] The figure is from "Annex1: Description of Work" document of project TOWL http://www.towl.org

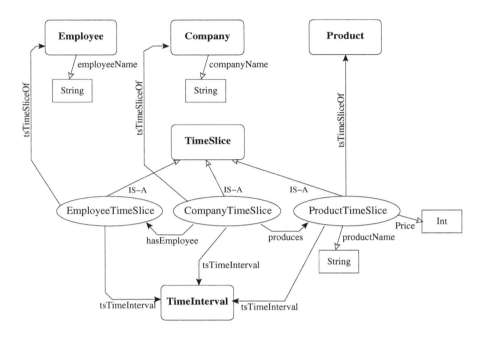

Fig. 2. Dynamic Enterprise Ontology

the temporal information of a time slice. Property *tsTimeSliceOf* connects an instance of class *TimeSlice* with an entity, and property *tsTimeInterval* connects an instance of class *TimeSlice* with an instance of class *TimeInterval*. Properties having a time dimension are called fluent properties and connect instances of class *TimeSlice*.

Figure 2 illustrates the so called "Dynamic Enterprise Ontology" ("*DEn* Ontology") defined in this work: a temporal ontology with classes *Employee* with datatype property *employeeName*, *Company* with datatype property *companyName* and *Product* with datatype properties *price* and *productName*. In this example, *CompanyName* and *EmployeeName* are static properties (their values do not change in time), while properties *produces*, *hasEmployee*, *productName* and *price* are dynamic (fluent) properties whose values may change in time. Because they are fluent properties, their domain (and range) is of class *TimeSlice*. *EmployeeTimeSlice*, *CompanyTimeSlice* and *ProductTimeSlice* are instances of class *TimeSlice* and are provided to denote that the domain of properties *hasEmployee*, *produces*, *productName* and *price* are time slices restricted to be slices of a specific class. For example, the domain of property *productName* is not class *TimeSlice* but it is restricted to instances that are time slices of class *Product*. A knowledge base with the instances of the *DEn* ontology used in the work can be found in [3] or can be downloaded from http://www.intelligence.tuc.gr/~petrakis/downloads/TOQL.zip

2.2   Temporal Query Languages

The main goal of temporal query languages is to maintain simplicity of expression while the time dimension is added. Other desirable features include, temporal upward compatibility (i.e., conventional queries and modifications on temporal relations act on the current state), temporal aggregation (i.e., possibility to request the history of something), point and interval-based views of data, expressive power and ease of implementation.

Examples of temporal query languages for temporal databases include TQuel [14], TSQL2 [9] and ATSQL [4]. Query languages for handling time information in ontologies (e.g., time evolving entities) besides TOQL, are not known to exist. Nevertheless, query languages for RDF and OWL ontological representations are of particular interest as they form the basis for developing the new type of temporal ontology query languages. SeRQL [1] and SPARQL [11] are good representatives of this category of query languages. SPARQL [11] is a W3C recommendation query language. SeRQL is a RDF/RDFS query language combining features of other (query) languages (e.g., RQL [7], RDQL [12], N-Triples, N3). Important features of SeRQL are: Graph transformation, RDF and XML Schema data type support, expressive path expression syntax and optional path matching. SeRQL supports comparison between date times and more query types than SPARQL, which has limitations in the "where" clause, since it doesn't support nested queries.

3   TOQL: Syntax and Semantics

TOQL (Temporal Ontology Query Language) is an SQL-like language for OWL, supporting the basic structure of SQL (SELECT - FROM - WHERE) and treats classes and properties of an ontology almost like tables and columns of a database. The new language takes into account differences in the type of relations in the two representations and also supports time operators: Allen operators (BEFORE, AFTER, EQUALS, MEETS, OVERLAPS, DURING, STARTS, ENDS) and operators AT(time point) and AT(time point, time point). Allen operators [2] compare datatype properties e.g., *A.B like "x" before C.D like "y"*. The language also supports additional functionalities such as LIMIT, OFFSET that limit the number of answers to be returned, and nested queries. A formal description of the language's syntax in BNF can be found in [3] (all keywords are *case insensitive*). TOQL supports most of an SQL language syntax and clauses (see cite[3] for a complete list), the most important of them being:

- **SELECT**: specifies the object property values to be returned.
- **FROM**: declares the class or classes to query from. Always follows SELECT.
- **WHERE**: includes logic operations and comparisons between object property values that restrict the number of answers returned by the query. Always follows FROM.

TOQL supports the following operators:

- **AS**: renames a class (in a FROM clause) or a property (in a SELECT clause). Renaming of a class allows using more than one instances of a class in a query (e.g., FROM Company AS C1, Company AS C2). Renaming of a property allows changing its name in the results (e.g., SELECT Company.companyName AS Name).
- **AND**: connects two expressions involving properties (datatype or object properties) in WHERE, returns objects satisfying both expresssions.
- **OR**: connects two expressions involving properties (datatype or object properties) in WHERE and returns objects satisfying at least one.
- **LIKE**: checks whether a datatype property value matches a specified string in WHERE. Comparison is case sensitive.
- **LIKE "string" IGNORE CASE**: checks whether a datatype property value matches a specified string ignoring case.

Table 1 summarizes TOQL syntax:

Table 1. Generic TOQL syntax

Syntax
SELECT ... AS ...
FROM ... AS ...
WHERE ... LIKE ... AND ... LIKE "string" IGNORE CASE

There are operation clauses connecting two (or more) queries in a nested query:

- **MINUS**: returns query results retrieved by the first operand, excluding results retrieved by the second operand.
- **UNION**: returns the union of results returned by both operands. Duplicate answers are filtered out.
- **UNION ALL**: returns the union of results returned by both operands. Duplicate answers are not filtered out.
- **INTERSECT**:returns the intersection of results retrieved by both operands.
- **EXISTS**: this is a unary operator that has a nested SELECT-query as its operand. The operator is an existential quantifier that succeeds when the nested query has at least one result.
- **ALL**: this is an operator that has a nested SELECT-query as one of its operands. It always follows a comparison operator (i.e., "=", "!=", "<", ">", "<=", ">="). It indicates that for every value of the nested query the comparison must hold.
- **ANY**: has a nested SELECT-query as one of its operands. It always follows a comparison operator (i.e., "=", "!=", "<", ">", "<=", ">="). It indicates for at least one value of the nested query the comparison must hold.
- **IN**: has a nested SELECT-query as one of its operands. Allows set membership checking. The set is defined by the nested SELECT-query.

Table 2 summarizes TOQL syntax with operator clauses:

Table 2. TOQL syntax with operator clauses

Case 1	Case 2	Case 3	Case 4
Query MINUS Query	Query UNION Query	Query UNION ALL Query	Query INTERSECT Query
Case 5	**Case 6**	**Case 7**	**Case 8**
SELECT ... FROM ... WHERE EXISTS (QUERY)	SELECT ... FROM ... WHERE ... CO^2 ALL (Query)	SELECT ... FROM ... WHERE ... CO^2 ANY (Query)	SELECT ... FROM ... WHERE ... IN (Query)

3.1 Dealing with Classes and Properties

In ontologies the basic terms are classes (also named concepts) and properties (object or datatype). Classes represent concepts of the world. Properties represent relations between two concepts or between a concept and a value. Properties relating two classes (concepts) are referred to as object properties, while properties relating a class with a value are referred to as datatype properties. As an example of object property consider the relation between the *Company* and the *Employee*. These two classes are connected with the object property *hasEmployee*. As an example of datatype property consider the name of an *Employee*. Class *Employee* is connected with a name (string value) with datatype property *employeeName*.

TOQL not only uses SQL-like clauses and a similar syntax, but also treats ontologies almost like relational databases. Tables representing concepts correspond to classes and tables representing relations correspond to object properties. Attributes correspond to datatype properties. In addition, 1:1 and 1:N relations correspond to object properties. Table 3 summarizes the mapping between database relations and ontology concepts used by TOQL.

Table 3. Mapping between database relations and ontology concepts

Relational Database	Ontology
Table representing concept	Class
Table representing N:N relation	Object Property
1:N or 1:1 relation	Object Property
Attribute	Datatype Property

In TOQL, classes are declared in FROM clauses just like SQL handles tables. To access a datatype property of a class, the name of the class is followed by a dot (".") and the name of the datatype property, just like SQL handles tables and attributes:

---

[2] CO: comparison operator can be any of "=", "!=", "<", ">", "<=", ">=".

$$ClassName.DatatypePropertyName$$

To access object properties (properties connecting two classes), the name of the domain class is followed by a dot ("."), the name of the object property, double dot (":") and finally the name of range class:

$$DomainClassName.objectPropertyName:RangeClassName$$

The following query can be used to access the names of companies producing products called "x" in the ontology of Figure 2:

> **SELECT** Company.companyName
> **FROM** Company, Product
> **WHERE** Company.produces:Product
> AND Product.productName LIKE "x"

3.2   Dealing with Time

TOQL is a high level language, hiding from the users the implementation of time at the ontology level. A temporal ontology consists of (a) the static part where application classes, properties and their instances are defined and (b) the dynamic part where the additional temporal classes (i.e., classes *TimeSlice*, *TimeInterval*), properties and instances of the above temporal classes and fluent properties are defined (i.e., *tsTimeSliceOf*, *tsTimeInterval*). TOQL automatically determines references to time related information.

To do this, TOQL:

- Retrieves the time slices associated with a class of the static ontology.
- Determines whether a property (object or datatype) in the query is a fluent property (i.e., a property that connects time slices or a time slice with a datatype) or not (i.e., a property that connects "static" classes or a "static" class with a datatype).
- Uses the ontology's dynamic part to answer the query, if a property specified by the query is a fluent one. In case the fluent property is a functional one (i.e. can have only one value at each instance of time) then the reasoner described in Sec. 3.7 is used to answer the query. The rationale behind this choice is that functional properties have unique values, which may change at a later time as the result of events affecting them. For example, if the price of product changes, then the new value substitutes any previously known value, while non functional properties retain both older and newer values.
- Uses the ontology's static part to answer the query, if a property specified by the query is not a fluent one.

TOQL prevents users from being familiar with the representation of time in ontologies. As an example consider the *DEn* Ontology of Figure 2. Typically to retrieve companies that hired employees, one should be familiar with the 4D fluent mechanism and ask for all time slices (instances) of class *Company* and all time slices of class *Employee* and then query on the object property *hasEmployee* that connects those instances. In TOQL (without implementing the high level functionality described above), this is expressed as:

> **SELECT** Company.companyName
> **FROM** Company, Employee, TimeSlice AS T1 ,
> TimeSlice AS T2
> **WHERE** T1.tsTimeSliceOf:Company AND
> T2.tsTimeSliceOf:Employee AND T1.hasEmployee:T2 AND
> Employee.employeeName LIKE "x"

This is a rather complicated expression and requires the user to be familiar with the implementation of time at the level of the ontology (the 4D fluent method in this work). However, this is not necessary in TOQL and the same query can be expressed as:

> **SELECT** Company.companyName
> **FROM** Company, Employee
> **WHERE** Company.hasEmployee:Employee
> AND Employee.employeeName LIKE "x"

The second query is much more easy to write than the first one. Notice that the object property *hasEmployee* is treated like its domain class *Company* and its range class *Employee*.

3.3   Abstract Ontology View

TOQL is a high level language independent from the actual representation of time in an ontology. A user need only be aware of the so called "abstract ontology view". Classes and properties specific to the 4D fluent mechanism are excluded from the abstract view. The fluent properties that connect time slices are considered to connect directly the static classes. Figure 3 illustrates the abstract ontology view corresponding to the *DEn* ontology of Figure 2. Fluent properties are shown in blue color.

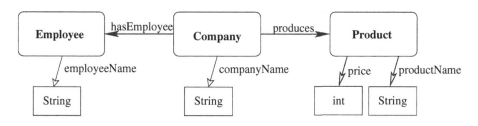

Fig. 3. Abstract ontology view corresponding to the *DEn* ontology of Figure 2

3.4   Allen Operators

In TOQL, the implementation of ALLEN operators correspond to comparisons between fluent properties. Fluent properties connect time slices and time slices are associated with time intervals. Consequently, the implementation of Allen operators correspond to comparisons between time intervals. The following operators are supported in TOQL: BEFORE, AFTER, MEETS, METBY, OVERLAPS, OVERLAPPEDBY, DURING, CONTAINS, STARTS, STARTEDBY, ENDS, ENDEDBY and EQUALS, representing the corresponding relations holding between two time intervals.

The following TOQL query retrieves the name of the company that hired employee "x" and *then* employee "y":

SELECT Company.companyName
FROM Company, Employee AS E1, Employee AS E2
WHERE Company.hasEmployee:E1 BEFORE Company.hasEmployee:E2
AND E1.employeeName like "x" AND E1.employeeName LIKE "y"

3.5   AT, TIME Operators

TOQL also introduces clause "AT" which compares a fluent property (i.e., the time interval in which the property is true) with a time period (time interval) or time point. Notice that the AT clause retrieves data explicitly defined in the knowledge base. As an example, assume the *DEn* Ontology and consider that at time point 5 the price of *Product1* is 10 and that there is no information about its price after time point 5. If a query asks for the price of *Product1* at time point 6 a reasonable answer would be 10 (the last known price in the KB). Answering such queries effectively is achieved by combining TOQL with the reasoner described in Sec. 3.7. In the current implementation:

- **AT(time point)** operation returns true if the time interval holds true at the time specified.
- **AT(start time point, end time point)** operation returns true if the time interval holds true for *all* the time interval.

The following TOQL query retrieves the name of the company employee "x" was working for, from time=3 to time=5:

SELECT Company.companyName
FROM Company, Employee
WHERE Company.hasEmployee:Employee AT(3,5)
AND Employee.employeeName LIKE "x"

Because TOQL is independent of the mechanism implementing time, there is no way to directly access class *TimeInterval* (i.e., the class holding values of time). In order for TOQL to return values of time, the keyword TIME is introduced. It follows datatype or object properties and can be used only in SELECT. It returns the start and end time point (if any) in which the property holds true (the time interval in which the property is true). If no end point exists, it returns only its start point. As an example, the following TOQL query retrieves the time for which a company had employee "x"

SELECT Company.hasEmployee.TIME
FROM Company, Employee
WHERE Company.hasEmployee:Employee AND
Employee.employeeName LIKE "x"

3.6 Special Cases

This section describes TOQL special features. These are related to the way TOQL deals with Class keys, wildcards (*) and Scope.

Dealing with keys. In relational databases each tuple is uniquely characterized by a key. A key can refer to more than one attributes (compound key). Consider a relational database that has the table Company and that this table uses the attribute ID as key. To access this key, in SQL, a user should write:

SELECT Company.ID

In OWL, each class instance and each property have a unique name. This unique name is considered to be equivalent to the unique key of relational databases. The difference is that this unique name is not an ordinary datatype property, and so it can not be accessed by writing the name of the class followed by a dot "." and the datatype property. In TOQL, the (unique) name of a class instance is accessed using the name of the class itself (without reference to a property). For example, to access the unique name of a company we write:

SELECT Company

Dealing with wildcards (*). In TOQL, wildcards can be used only in SELECT. In SQL the presence of wildcard in SELECT implies that all the columns of all the tables declared in clause FROM will be returned. If the wildcard follows a table (*tableName.**), all the columns of the specific table will be returned. In TOQL the presence of wildcard in SELECT implies that all the datatype properties of *all* the classes declared in FROM will be returned. If the wildcard follows a class, the datatype properties of the specific class will be returned. Notice that the class unique name is not returned (only its datatype properties are returned). The following query retrieves companies producing product with unique name "x", as well as the product's name.

SELECT *
FROM Company, Product
WHERE Company.hasProduct:Product
AND Product LIKE "x"

Dealing with scope. TOQL supports set combination operations in queries as well as nested queries. Both set operations and nested queries imply that a TOQL query may be composed of more than one subqueries. Each subquery has its own class declarations, class and property usage and this introduces the need for the handling of scopes.

Queries combined by set operators have different scopes. Classes declared in any of them are local to this query and are not visible to the others. The following query retrieves names of *"Company_1"* and also names of *"Company_2"* from the *DEn* Ontology:

> **SELECT** C1.companyName
> **FROM** Company As C1
> **WHERE** C1 like "Company_1"
> **UNION**
> **SELECT** C1.companyName
> **FROM** Company As C1
> **WHERE** C1 like "Company_2"

This TOQL expression specifies two separate queries combined by the set operator UNION. Each subquery has a different scope: classes declared in the first subquery are not visible to the second one. Even if the same class is used by the second subquery, it must be redeclared.

In TOQL, a nested query inherits all the classes declared in the query it is nested into. A nested query can use these classes, but cannot (re)declare any of them. The following nested TOQL query (a second query follows clause ANY) retrieves products whose price is at least 10 and not smaller than than the price of any other product. Both subqueries use class *Product* but with different names (*P1* and *P2* respectively) otherwise a semantic error will be reported.

> **SELECT** P1
> **FROM** Product As P1
> **WHERE** P1.price >= 10 AND NOT
> P1.price < ANY
> (**SELECT** P2.value **FROM** Product As P2)

3.7   Reasoning in TOQL

TOQL can be used to access temporal information that is explicitly represented in a temporal ontology, but cannot provide answers on information that can be inferred from existing information. For example if the price of a product at time t is p, TOQL should be able to infer that the price of the product remains the same since the last time it was changed. This is exactly the problem the TOQL reasoner is dealing with. The reasoner implements an action theory based on Event Calculus [22]. Event calculus records the events that have taken place. It comprises of events (or actions), fluents and time points. Table 4 illustrates the predicates of Simple Event Calculus. Time points are natural numbers which means that time is ordered, discrete and unbounded. A fluent is a predicate of the form "fluentName1(objectID1)" and the same is an action "actionName1(objectID1,objectID2)".

The definition of the *HoldsAt* and *HoldsBetween* predicates for an arbitrary fluent f is presented below along with rules that state that a fluent retains the same value since the last time it was changed:

Table 4. Predicates of Simple Event Calculus

Predicate	Meaning
Initiates(A, f, x, t)	if action A is executed at time t, then f will have value x at time point t
Terminates(A, f, x, t)	if action A is executed at time t, then f will not have value x after the time point t
HoldsAt(f, x, t)	fluent f has value x at time point t
Initially(f, x)	fluent f has value x in the beginning
HappensAt(A, t)	action A is executed at time point t
t1<t2	time point t1 is before time point t2

$$Started(t1, f, x, t2) \leftarrow \exists a : HappensAt(a, t1) \wedge Initiates(a, f, x, t1) \wedge (t1 < t2)$$

$$Releases(a, f, x, t) \leftarrow \exists a' : HappensAt(a', t) \wedge Initiates(a', f, y, t) \wedge (y \neq x)$$

$$Clipped(t1, f, x, t2) \leftarrow \exists a, t : HappensAt(a, t) \wedge (t1 < t < t2) \wedge Terminates(a, f, x, t)$$

$$HoldsAt(f, x, t) \leftarrow (Initially(f, x) \wedge (0 < t) \wedge \neg Clipped(0, f, x, t))$$

$$\vee (\exists t1 : Started(t1, f, x, t) \wedge \neg Clipped(t1, f, x, t))$$

$$HoldsBetween(f, x, t1, t2) \leftarrow (\exists t : Started(t, f, x, t1) \wedge \neg Clipped(t, f, x, t2))$$

$$\vee (Initially(f, x) \wedge (0 < t1) \wedge \neg Clipped(0, f, x, t2))$$

The reasoner applies when an object property is defined as temporal and functional (e.g. the price of a product, which can have only one value at a time point). For example if the price of the product *"Product4"* is set at 50 euro at time point 2 and 60 euro at time point 4 then the following query:

> **SELECT** Product
> **FROM** Product
> **WHERE** Product.price LIKE "50" AT(9)

will return an empty list as a result, because the reasoner infers that setting the price at 60 euros at time point 4 implies that the price is not 50 euros after that time point. If the reasoner is not used then the query will return *"Product4"* as a result, which is not correct. Thus the AT operator is handled by the reasoner in case of functional fluent properties.

4   TOQL Implementation

To show proof of concept, a TOQL system has been implemented in Java[2]. The system supports query translation and execution of TOQL queries on temporal ontologies in OWL. The input is a query written in TOQL and an ontology in OWL (in RDF/XML or RDF/XML-ABBREV syntax).

---

[2] Available at http://www.intelligence.tuc.gr/~petrakis/downloads/TOQL.zip

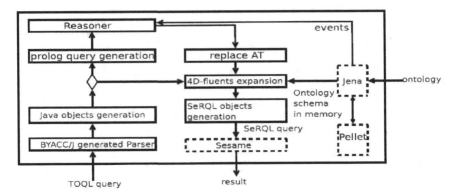

Fig. 4. TOQL system architecture

Figure 4 illustrates the architecture of the proposed system. The TOQL system consists of several modules whose purpose is to translate the TOQL query into an equivalent SeRQL one (which is then executed on the knowledge base).Notice that SeRQL is independent from TOQL. Any other language supporting SQL syntax and comparison between date times (such as SPARQL) would do for this translation. Notice also that executing TOQL statements directly on the ontology is also feasible but the implementation would be more involved. TOQL and SeRQL have different syntax, however, queries are much easier to express in TOQL. SeRQL supports comparison between date times but not the full range of TOQL's time features. Therefore, even simple TOQL queries are translated to complicated SeRQL ones. The complete discussion of the TOQL implementation can be found in [3]. The application loads the ontology schema in memory. TOQL queries are translated into equivalent SeRQL queries which are applied to the knowledge base using SESAME[3]. The TOQL query is parsed and if fluent properties are detected then the query is converted to an equivalent query addressing the underlying 4-D fluent representation, which in turn is translated into a SeRQL query. For example the following TOQL query is translated to the SeRQL query of page 353:

> **SELECT** C1.companyName.TIME as T,
> C1.companyName
> **FROM** Company As C1
> **WHERE** C1 like "Company1"

In case of queries over functional fluent properties fluents are represented as predicates of event calculus and the Prolog reasoner is applied, which transforms the query into an equivalent one that conforms to the event calculus axioms, before the translation to SeRQL occurs. Specifically, at the "java objects generation" phase, if the query uses the AT operator, it is replaced with an equivalent one where every expression that uses the AT operator is replaced with the reasoners answer.

---

[3] http://www.openrdf.org/

The Pellet[4] reasoner applies to the initial ontology schema, thus the schema loaded in memory contains all infered facts using OWL semantics. For example if the class *ComputerCompany* is defined as a subclass of class *Company*, then a query regarding instances of class *Company* will also apply to instances of the class *ComputerCompany*.

SELECT startValue_interval_C1Slice_1,
endValue_interval_C1Slice_1, companyName_C1Slice_1
FROM {interval_C1Slice_1} ex1:startValue {startValue_interval_C1Slice_1},
{interval_C1Slice_1} ex1:endValue {endValue_interval_C1Slice_1},
{C1Slice_1} ex1:companyName {companyName_C1Slice_1},
{C1} rdf:type {ex1:Company},
{C1Slice_1} rdf:type {ex1:TimeSlice},
{interval_C1Slice_1} rdf:type {ex1:TimeInterval},
{C1Slice_1} ex1:tsTimeSliceOf {C1},
{C1Slice_1} ex1:tsTimeInterval {interval_C1Slice_1}
WHERE localName(C1) Like "Company_1"
USING NAMESPACE
ex1= <http://www.owl-ontologies.com/Ontology1197730146.owl#>

5   Conclusions and Future Work

We introduce TOQL (Temporal Ontology Query Language), an ontology query language capable of querying ontologies and temporal information in ontologies. Temporal concepts are assumed to be represented in OWL (or RDF) using the 4D perdurantist approach [15], implementing events occurring at specific time points or time intervals and evolving in time. The language supports a powerful set of operations including Allen operators. An application supporting execution of TOQL queries on OWL temporal (or static) ontologies has been developed and is available on the Web. TOQL is combined with a reasoner based on event calculus to better support queries on temporal ontologies. Query optimization as well as adding new features in TOQL (such as INSERT, UPDATE, DELETE, ORDER BY, GROYP BY operations) are important issues for further research. Extending TOQL's syntax to handle queries on spatial data as well as queries on ontology structure (i.e., sub-classes and super-classes) and improving query performance by applying indexing on ontology information are also directions for further research.

Acknowledgement

This work was supported by project TOWL: "Time-determined ontology based information system for real time stock market analysis" (FP6-STREP, contract number 26896) of the European Union.

---

[4] http://clarkparsia.com/pellet

References

1. Aduna, B.V.: The SeRQL query language. User Guide for Sesame 2.1, Chapter 9, 2002–2008, http://www.openrdf.org/doc/sesame2/2.1.2/users/ch09.html
2. Allen, J.F., Ferguson, G.: Actions and Events in Interval Temporal Logic. Journal of Logic and Computation 4(5), 531–579 (1994)
3. Baratis, E.: TOQL: Querying Temporal Information in Ontologies. Master's thesis, Techn. Univ. of Crete (TUC), Dept. of Electronic and Comp. Engineering (July 2008)
4. Bohlen, M.H., Jensen, C.S.: Seamless Integration of Time into SQL. Technical Report R-96-49, Dept. of Comp. Science, Aalborg University (1996)
5. Martin, D., et al.: OWL-S: Semantic Markup for Web Services. W3C Recommendation (November 2004), http://www.w3.org/Submission/OWL-S
6. Hobbs, J.R., Fang, P.: Time Ontology in OWL. W3C Recommendation (September 2006), http://www.w3.org/TR/owl-time/
7. Karvounarakis, G., Alexaki, S., Christophides, V., Plexousakis, D., Scholl, M.: RQL: A Declarative Query Language for RDF. In: Intern. Conf. on World Wide Web (WWW 2002), Honolulu, Hawaii, USA (May 2002)
8. Klein, M., Fensel, D.: Ontology Versioning for the Semantic Web. In: International Semantic Web Working Symposium (SWWS 2001), California, USA, July-August 2001, pp. 75–92 (2001)
9. Kline, N., Snodgrass, R.T., Cliff Leung, T.Y.: Aggregates. In: The TSQL2 Temporal Query Language, pp. 393–424. Kluwer, Dordrecht (1995)
10. McGuinness, D.L., VanHarmelen, F.: OWL Web Ontology Language Overview. W3C Recommendation (February 2004), http://www.w3.org/TR/owl-features
11. Prud'hommeaux, E., Seaborne, A.: SPARQL Query Language for RDF. W3C Recommendation (January 2008), http://www.w3.org/TR/rdf-sparql-query
12. Seaborne, A.: RDQL - A Query Language for RDF. W3C Recommendation (January 2004), http://www.w3.org/Submission/2004/SUBM-RDQL-20040109
13. Sider, T.: Four-Dimensionalism: An Ontology of Persistence and Time. Oxford University Press, USA (2002)
14. Snodgrass, R.T.: The temporal query language TQuel. ACM Transactions on Database Systems (TODS) 12(2), 247–298 (1987)
15. Welty, C., Fikes, R.: A Reusable Ontology for Fluents in OWL. Fontiers in Artificial Intelligence and Applications 150, 226–236 (2006)
16. Welty, C., Fikes, R., Makarios, S.: A Reusable Ontology for Fluents in OWL. Technical Report RC23755 (Wo510-142), IBM Research Division, T. Watson Research Center, Yorktown Heights, NY (October 2005)
17. Zhang, Z.: Ontology Query Languages: A Performance Evaluation. Master's thesis, The University of Georgia, Comp. Science Dept. (August 2005)
18. Ozsoyglu, G., Snodgrass, R.T.: Temporal and Real-Time Databases: A Survey. Knowledge and Data Engineering 4, 513–532 (1995)
19. Gregersen, H., Jensen, C.S.: Temporal Entity Relationship Models – A Survey. IEEE Transactions on Knowledge and Data Engineering 3, 464–497 (1999)
20. Artale, A., Franconi, E.: A survey of temporal extensions of description logics. Annals of Mathematics and Artificial Intelligence 30(1-4) (2001)
21. Lutz, C., Wolter, F., Zakharyaschev, M.: Temporal description logics: A survey. In: Proc. TIME 2008. IEEE Press, Los Alamitos (2008)
22. Shanahan, M.: The event calculus explained. In: Wooldridge, M., Veloso, M. (eds.) Artificial Intelligence Today. LNCS (LNAI), vol. 1600, pp. 409–430. Springer, Heidelberg (1999)

Supporting Frameworks for the Geospatial Semantic Web

Alia I. Abdelmoty[1], Philip D. Smart[1], Baher A. El-Geresy[2],
and Christopher B. Jones[1]

[1] Cardiff School of Computer Science, Cardiff University, Wales, UK
[2] School of Computing, University of Glamorgan, Wales, UK

Abstract. A lot of information on the web is geographically referenced. Discovering and linking this information poses eminent research challenges to the geospatial semantic web, with regards to the representation and manipulation of geographic data. Towards addressing these challenges, this work explores the potential of the current semantic web languages and tools. In particular, an integrated logical framework of rules and ontologies, using current W3C standards, is assessed for modeling geospatial ontologies of place encoding both symbolic and geometric references to place locations. Spatial reasoning is incorporated in the framework to facilitate the deduction of implicit semantics and for expressing spatial integrity constraints. The logical framework is then extended with geo-computation engines that offer more effective manipulations of geometric information. Example data sets mined from web resources are used to demonstrate and evaluate both frameworks, offering insights to their potentials and limitations.

1 Introduction

Over the past few years, geo-referencing of resources on the web has evolved to become a natural method for organising and linking information with the aim of facilitating its discovery and use. A significant portion of search queries include references to geographic places and spatial relationships [24,9]. In response, geographic information retrieval has emerged as a research domain to address many challenges facing the development of geographically-aware search engines [19] including, geospatial query interpretation, geo-tagging of resources, spatial search and analysis and ranking and presentation of information.

On one hand, many of these challenges are problems that are addressed within the domain of GIS and spatial databases and could benefit from established approaches to their solution. On the other hand, these challenges are also being addressed, at a general level, within the evolving Semantic Web whose aim is to provide common frameworks that allow the sharing and reuse of data and services across applications, enterprise and community boundaries.

This paper studies the following question; Can the current semantic web technologies be "spatially-enabled" to allow the realisation of the geospatial semantic web? Towards answering this question, two frameworks are proposed. The first is

N. Mamoulis et al. (Eds.): SSTD 2009, LNCS 5644, pp. 355–372, 2009.

based entirely on semantic web tools and technologies and is a logical integration of rules and ontologies to provide a platform for expressing and reasoning over symbolic geographic knowledge. The second framework is a hybrid extension of the basic framework with geospatial information processors that are more suited to manipulating the geometrical (location) component of the information. The potential and limitations of the frameworks are explored. Both are implemented using available tools and standards and are tested with some realistic data sets collected from web resources.

The nature of geospatial referencing as used on the web is discussed in section 2, followed by a proposal of a simple place model that encapsulates the dimensions of this data. Section 3 is an evaluation of OWL, as a standard web ontology language, for representing the proposed place model. A discussion of OWL's limitation motivates the use of a rule layer over the ontology. A homogenous approach to the integration of such a rule layer is used in section 4 to form a basic framework for encoding a geospatial ontology and reasoning engine. The framework is evaluated with data sets extracted from Wikipedia. An extension to the framework that incorporates a spatial database system is proposed and evaluated in section 5 and the paper concludes in section 6.

2   Geospatial Referencing on the Web

Place names provide what is probably the most fundamental method of specifying location in natural language and hence also is the the most common form of geo-referencing used in web documents. A name may be a standardised widely recognised name, or informal being locally familiar in certain communities [17]. Further nominal clues are also used to distinguish location, for example, using some address information. If the information described is not exactly associated with the named place, then spatial relationships are used to describe location relative to that place, e.g. "near" and "north of". In addition, the web now offers accessible mapping applications to allow for precise association of resources with a location on a map (e.g. linking photos on Flickr with Google maps). Unless, the resource is geo-located, such as with a GPS, a marker on a map is normally intended as an approximate pointer to the location of the resource.

The same is true when people query geo-referenced information. Typical structure of queries take the form $<subject><relation><somewhere>$ in which the subject specifies the thematic aspect of the web resource, somewhere is the name of a place and the relation stipulates a spatial relationship to the named place [18]. For example, the query "Camp sites in South Wales adjacent to a beach", is a spatial query involving a combination of spatial joins and requires an estimation of the boundary of the region "South Wales". Gazetteers typically only provide a single point (centroid) to approximate the location of geographic regions. In addition, some regions, such as "South Wales" are vernacular and do not have an official recorded boundary. To answer this query, additional knowledge is therefore required.

The web itself acts as a valuable source from which place information can be harvested to complement traditional gazetteers. Research methods (geo-parsing,

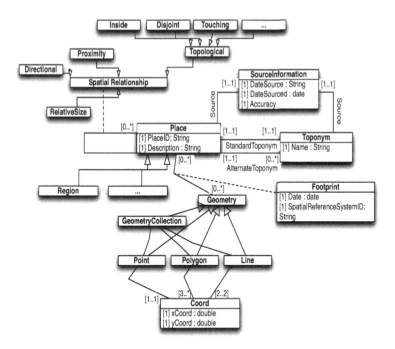

Fig. 1. A Typical Place Ontology Model

coding and tagging) to find and extract this place information are being sought within the field of GIR [25,3]. The task is challenging, involving problems not only with the extraction of information from natural language, but also with reasoning over the extracted data which may be incomplete, fuzzy and in cases contradictory.

Two types of geographic place data can be collected from web resources, qualitative data, in the form of place names and qualitative spatial relationships as well as some geometric information, in the form of mostly point data for the location of some of these places. Similar to processes normally undertaken in GIS and spatial databases, new methods for "cleaning" this geographic information are needed before they can be used as a base for spatial search and analysis. Collecting place information through *Crowdsourcing* (or user collaboration) is emerging and some web databases are already accumulating and serving these geographic data as RDF triples, to facilitate their sharing and integrated use.

In this paper, we use a simple place ontology that captures both types of data above as shown in figure 1. The model captures both qualitative and qualitative spatial description of location through the association of a place concept to a geometric footprint and the recording of different possible types of spatial relationships between places.

To demonstrate and evaluate the frameworks proposed, data sets are mined from the web to populate the place ontology. The following is an example, as RDF(S) triples, of the information mined from Wikipedia articles and stored in the model, where *NS* is the namespace prefix of: http://cf.ac.uk/Place/).

The triples encodes relationships between a set of regions (administrative wards in the city of Cardiff).

```
(<NS:Llanishen> <NS:Inside> <NS:Cyncoed>)
(<NS:Llanishen> <NS:Contains> <NS:Thornhill>)
(<NS:Penylan> <NS:Inside> <NS:Roath>)
(<NS:Penylan> <NS:Inside> <NS:Cathays>)
(<NS:Roath> <NS:Touches> <NS:Penylan>)
(<NS:Llanishen> <http://www.w3.org/1999/02/22-rdf-syntax-ns#type> <NS:Region>)
(<NS:Cyncoed> <http://www.w3.org/1999/02/22-rdf-syntax-ns#type> <NS:Region>)
(<NS:Thornhill> <http://www.w3.org/1999/02/22-rdf-syntax-ns#type> <NS:Region>)
(<NS:Penylan> <http://www.w3.org/1999/02/22-rdf-syntax-ns#type> <NS:Region>)
(<NS:Roath> <http://www.w3.org/1999/02/22-rdf-syntax-ns#type>  <NS:Region>)
(<NS:Region> <http://www.w3.org/1999/02/22-rdf-syntax-ns#type>
                    <http://www.w3.org/1999/02/22-rdf-syntax-ns#Class>)
(<NS:Inside> <http://www.w3.org/1999/02/22-rdf-syntax-ns#type>
                    <http://www.w3.org/1999/02/22-rdf-syntax-ns#Property>)
```

Databases such as Geonames and DBPedia store point coordinates for the places they hold in the form of a latitude-longitude pair. The following is an RDF triple extract from both resources[1]. Interestingly, articles in DBPedia are linked to entries in Geonames using the `owl:sameAs` construct, allowing for possible integration of knowledge from both sources.

```
Geonames - Cardiff University

(<gns:Feature> <http://www.w3.org/1999/02/22-rdf-syntax-ns#about>
                    <http://sws.geonames.org/6697669/>)
(<http://sws.geonames.org/6697669/> <gns:Name> <Cardiff University Queens Buildings>)
(<http://sws.geonames.org/6697669/> <gns:FeatureClass> <http://www.geonames.org/ontology#P.PPL>)
(<http://sws.geonames.org/6697669/> <wgs84_pos:lat> <51.483^^XMLSchema:float>)
(<http://sws.geonames.org/6697669/> <wgs84_pos:long> <-3.16^^XMLSchema:float>)

DPPedia - Cardiff

<http://dbpedia.org/resource/Cardiff> <wgs84_pos:lat>
        <"51.4852777778"^^http://www.w3.org/2001/XMLSchema#float>
<http://dbpedia.org/resource/Cardiff> <wgs84_pos:long>
        <"-3.18666666667"^^http://www.w3.org/2001/XMLSchema#float>
```

Integrating these data resources poses many interesting research problems. The rest of this work focusses primarily on the following two basic problems.

- Are the available web languages and tools able to model this data effectively?
- Can these tools be used to reason effectively with the data to ascertain its consistency?

3   Evaluation of Current Semantic Web Tools

Ontologies are key to the development of the semantic web. They provide platforms for expressing and reasoning over common structures and vocabularies to facilitate sharing as well as machine understanding and reasoning of knowledge [14,13]. Layers of technologies and languages are proposed by the W3C on the semantic web stack to allow for the representation of ontologies, including

---

[1] where $gns = http://www.geonames.org/ontology\#$, $dbns = http://dbpedia.org/resource/\#$ and $wgs84\_pos = http://www.w3.org/2003/01/geo/wgs84\_pos\#$

the resource description framework (RDF), a basic schema definition language RDF(S), and a more expressive web ontology language OWL.

RDF provides a simple knowledge representation model using binary predicates or triples $< subject; predicate; object >$ asserting knowledge described by the predicate about the subject and object. RDF Schema (RDFS)[2] is an extension to RDF that provides base ontological constructs for defining custom vocabularies. RDFS can be considered a simple object-orientated language allowing user defined classes and properties. OWL extends RDFS and provides a richer set of modeling constructs and hence semantics and is considered to be the most complete and expressive web ontology language currently being developed. OWL is based on Description Logics (DL) and allow for the representation of concepts, concept hierarchies, roles and individuals. With its formal logical semantics, description logics support the following key inference tasks:

1. Subsumption reasoning - given concept C and D, determine if C is a subset of D. Checking if the concept D is more general than C.
2. Membership checking - check whether an individual i is a member of the concept C, or find all individuals that are an instance of C (a query).
3. Satisfiability checking - given concept C determine if C is consistent with respect to the knowledge base; checking whether a concept expression does not denote the empty set.

3.1   Using OWL for Representing Geographic Knowledge

The place ontology in figure 1 can be represented using OWL-DL (the description logic subset of full OWL). A sample using XML/RDF syntax is shown below and a range of OWL-DL constructs used in the representation are given in table 3.1.

```
<owl:Class rdf:about="#Place">
    <rdfs:subClassOf>
      <owl:Restriction>
        <owl:onProperty>
          <owl:DatatypeProperty
             rdf:ID="Description"/>
        </owl:onProperty>
        <owl:cardinality rdf:datatype=
             "http://www.w3.org/2001/XMLSchema#int"
        >1</owl:cardinality>
      </owl:Restriction>
      </rdfs:subClassOf>
      ...
  </owl:Class>

<owl:ObjectProperty rdf:ID="Inside">
    <rdfs:domain rdf:resource="#Place"/>
    <rdfs:range rdf:resource="#Place"/>
</owl:ObjectProperty>
```

The expressiveness of OWL makes it a suitable modeling platform for different domains. However, it also has some limitations, as detailed below.

---

[2] http://www.w3.org/TR/rdf-schema/

Table 1. Sample OWL-DL constructs for the Place model

OWL-DL Construct	Description
Place	A Place is a concept
City $\sqsubseteq$ Place	A City is a sub-concept of Place
Ward $\sqsubseteq$ Place	A Ward is a sub-concept of Place
Place $=$ ≥ 1.Name $\sqcap$ $\forall$ partOf.Place	A Place has one or more names, and can be partOf another place
SpatialRelationship	A spatial relationship is a property
Topological $\sqsubseteq$ SpatialRelationship	A topological property is a sub-property of a spatial relationship
Overlap $\sqsubseteq$ Topological	An Overlap property is a sub-property of a spatial relationship
PartOf $\sqsubseteq$ Topological	A PartOf property is a sub-property of a spatial relationship
Equal $\sqsubseteq$ PartOf	An Equal property is a sub-property of a spatial relationship
PartOf$^+$ $\sqsubseteq$ PartOf	PartOf is a transitive property
PartOf $\equiv$ Contains$^-$	PartOf is equivalent to the inverse of the Contains property
City $\equiv$ Stadt	A City concept is equivalent to the concept Stadt (City in German)

1. OWL's first order, open world semantics in combination with the non-unique name assumption makes it unsuitable for constraint checking tasks [5]. For example, qualified cardinality constraints can't be used to constrain and check the possible instantiations of a class.

    Consider the following OWL definition of a Polygon,

$$Polygon > 3.XYCoords$$

    If an individual of type Polygon had two $XYCoords$, the open world assumption would concede that information may exist external to the ontology which can later be added to satisfy the restriction. If an individual had more than three $XYCoords$ then, as OWL does not support the unique name assumption, it will infer that all redundant coordinates are equal.

2. 'Triangular knowledge' is not representable in OWL-DL [15]. In particular, complex property compositions which are inference patterns of the form,

$$\forall x, y, c : R_1(x, y) \wedge R_2(y, c) \rightarrow R_3(x, c)$$

    where R_1, R_2 and R_3 are different relations, can not be handled. OWL v1.1. adds a restricted complex property inclusion axiom that can capture a limited form of an inference rule as follows.

$$R(x, y) \wedge S(y, c) \rightarrow S(x, c)$$

    or

$$R(x, y) \wedge S(y, c) \rightarrow R(x, c)$$

Such axioms only permit the conclusion of a property used in the body of the composition, guaranteeing decidability, but will still not handle the more general form of complex property compositions.

3. Tableaux based reasoners (as used in most DL reasoners) are poor for query answering over individuals [5] and hence will pose a scalability problem for typically large spatial knowledge bases.

4. A further issue, particular to geospatial domains, is related to the representation and manipulation of the geometry. Logic-based paradigms are not suitable for the expression of procedural implementation of spatial operations, nor could they offer efficient storage structures or spatial indexes.

The limitations of OWL has led to proposals for enhancing its expressiveness in particular by exploring approaches for the representation of rules over ontologies. Different methods have been proposed and a rule layer is now part the semantic web stack.

Approaches to the integration can broadly be classified as either hybrid or homogeneous [1], reflecting the degree of interaction between the rule and ontology components. A hybrid approach is a modular approach where both the rule and ontology components are kept distinct. Reasoning is performed separately in both components and entailments from one component are treated as constraints to the other.

A homogenous approach is characterised by the complete translation of one language into the other. Approaches exist based on the expressive union of the two languages, as for example in the standard web ontology rule language (SWRL [16]) . However, the union introduces undecidability in the resultant language [2]. More commonly approaches are built around the common intersection of the language, as for example in the web rule language (WRL)[3] and description logic programs (DLP) [11]. Homogenous approaches offer a better reasoning synergy between ontology and rule components, as they form in effect one language. Furthermore, integrations based on the intersection of rule and ontology component can be used within existing, mature and scalable logic programming engines.

Description Logic Programs (DLP) is an example of a homogenous approach to integration and offers the following useful features.

- A significant number of commonly used constructs of OWL-DL can be captured within DLP[26].
- DLP is considered a sound, practical and extensible paradigm [20] and is the base for the core web rule language WRL.
- DLP can be run by existing forward chaining production systems such as RETE or backward chaining classical logic programs without modification.
- Logic programming engines are better at reasoning with large stores of individuals (as in the case of geospatial knowledge bases) than tableaux-based DL reasoners [21,20].

---

[3] http://www.w3.org/Submission/WRL/

– DLP assumes a more intuitive closed world and unique name assumption and is consequently a suitable language for expressing and implementing integrity constraints, in addition to deductive rules.

In the rest of this paper DLPs are used as a base framework for managing geospatial ontology bases.

4   Description Logic Programs Framework

A Description Logic Program (DLP) framework is proposed here as a base for representing and reasoning over geospatial knowledge base. First, we show how the modeling constructs in OWL can be transformed and expressed in DLP and then how it can be used to represent spatial rule bases for deduction and integrity checking.

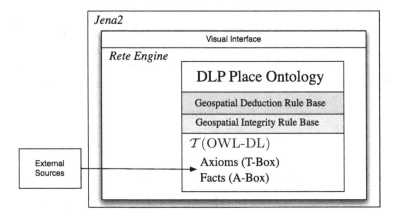

Fig. 2. DLP Place ontology Framework

4.1   Mapping Geospatial Ontologies from OWL to a DLP

A transformation function $\mathcal{T}$, as defined in [12], is used here to map the OWL-DL representation of the place ontology into a DLP as shown in table 2. In practice this transformation can be performed using the KAON2 DLP convert program [21].

Note, that the following constructs of the OWL-DL place ontology could not be represented in a DLP (see [12] for a more in-depth description of features not supported in a DLP):

– Functional properties, for example that each place has a unique ID.
– Cardinality restrictions, for example that each place has only 1 standard name.

In addition to representing the base axioms of the place ontology, a DLP allows for the definition of arbitrary (Horn) rules. Two principle types of rules can be expressed, namely, deduction and integrity, as shown below.

Table 2. Sample DLP Place Ontology using the transformation function $\mathcal{T}$

OWL-DL Syntax	DLP Horn Syntax
Place $\sqsubseteq$ Thing	Place(x) $\rightarrow$ Thing(x)
Region $\sqsubseteq$ Place	Region(x) $\rightarrow$ Place(x)
$\top \sqsubseteq \forall$ PlaceID.xsd:string	PlaceID(x,y) $\rightarrow$ xsd:String(y)
$\top \sqsubseteq \forall$ PlaceID^{-1}.Place	PlaceID(x,y) $\rightarrow$ Place(x)
Topological $\sqsubseteq$ Spatial_Relationship	Topological(x,y) $\rightarrow$ Spatial_Relationship(x,y)
Touches $\sqsubseteq$ Topological	Touches (x,y) $\rightarrow$ Topological(x,y)
...	...

4.2 Deduction Rules

DLP can represent arbitrary deduction rules that can capture certain spatial compositional inferences that result in a definite conclusion (one head predicate) i.e. rules of the form:

$$Inside(A, B) \wedge Disjoint(B, C) \rightarrow Disjoint(A, C)$$

Although not strictly part of a DLP, procedural attachments can be easily added within all logic programming reasoning engines [20]. These are described later in the paper.

4.3 Integrity Rules

The logic programming equivalent of Horn logic used by a DLP assumes a more intuitive closed world and unique name assumption and is consequently a suitable language for expressing and implementing integrity constraints. The bodies of integrity and deduction rules are identical in both specification and functionality. An integrity rule differs from a deduction rule in the use of its head atom. An integrity rule does not assert new information into the ontology[4], instead it asserts errors into an error ontology.

For example, consider the following rule with (where A, B and C are variables).

$$Inside(A, B) \wedge Inside(B, C) \wedge Equal(A, C) \rightarrow error(t_1, ..., t_n) \qquad (1)$$

Here the head predicate is an error predicate that is inferred if the body predicates (relations) exist in the DLP knowledge base. In this rule, if a place bound to the variable A is inside one bound to B, and B is inside a third place bound to C. An invalid state is reached and an error inferred, if a contradictory fact is explicit in the DLP that states that A is Equal to C. A set of integrity rules to

---

[4] As is common in logic programming literature, a rule without head is referred to as an integrity rule.

capture possible invalid states for different types of spatial relations need to represented in the DLP. The resulting inferred error predicates are recorded and can be examined at the end of the inference process to identify the inconsistencies and trace their sources.

5   Framework Implementation

A system has been developed that implements the DLP proposed framework above within the Jena2 Semantic Web toolkit[5]. Jena2's rule engine is based on the Rete pattern matching production system [10] and an XSB [23] logic programming engine.

The system has been tested on real world place information mined from both Wikipedia pages and general web pages. The mined information is stored using the place ontology in OWL and then converted to a DLP program using the KAON2 DLP convert tool, and loaded into Jena2 as a set of logical rules in RDF triple format. A spatial rule base representing the composition of spatial relations has been developed using topological composition table [6,4,22,7,8]. The design and implementation of the spatial reasoning methods are assumed here and are outside the scope of the current paper.

Example. The instantiated place ontology contains 40 regions or neighborhoods within Cardiff, UK and roughly 200 explicit topological spatial relationships between these regions. The following are example of facts.

```
(NS:Penylan rdf:type  NS:Ward)
(NS:Penylan NS:Inside NS:Roath)
(NS:Penylan NS:Inside NS:Cathays)
```

The engine checks the consistency of the ontology and reports the detected problem facts. A visual interface has been designed to allow for the visualisation and editing of the ontologies and rules, as shown in figure 3. The result of the reasoning process is shown on the interface where problem relations (edges) are highlighted. In addition, a trace of the reasoning process can be produced to localise the source of the inconsistency in the data set.

An example of the error detected in this sample data set are the three relationships between the districts Cathays, Roath and Penylan, shown in figure 4(a). In reality, Penylan and Roath are neighbours, as shown in the Google maps view in figure 4(b). To find this inconsistency, the following integrity rules were triggered.

```
[Inside_Meet :  (?x rdf:type NS:Region) (?y rdf:type NS:Region)
Region(?z rdf:type NS:Region) (?x NS:Inside ?y) (?y NS:Meet ?z)
(?x NS:Inside ?z) -> error(?x ?z)]
```

Where Penylan is inside Cathays and Roath meets Cathays implies that Penylan can not be inside Cathays, and hence the rule implies an error.

---

[5] http://dsonline.computer.org/0211/f/wp6jena.htm

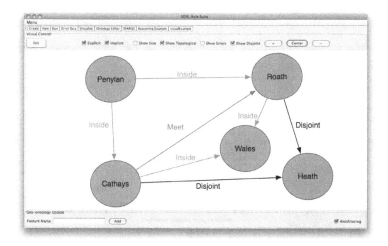

Fig. 3. Place Ontology Visual interface with a sample of the individuals in the ontology

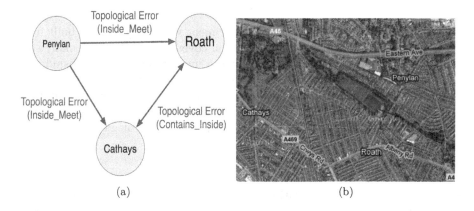

Fig. 4. a) Inconsistencies found between the regions Cathays, Roath and Penylan, b) Google Maps View of the three regions

```
[Contains_Inside: (?x rdf:type NS:Region) (?y rdf:type NS:Region)
Region(?z rdf:type NS:Region)  (?x NS:Contains ?y) (?y NS:Inside ?z)
(?x NS:Meet ?z) -> error(?x ?z)]
```

Where Roath contains Penylan and Roath meets Cathays means that Cathays can not contain Penylan and hence the rule implies an error.

The DLP framework reasons with explicitly stored spatial facts in the ontology base but will not compute the facts if they are stored. Hence, its effectiveness is related to the number and types of spatial relations defined. Figure 5 demonstrates how the number of definite as well as indefinite spatial relations between regions in the ontology varies depending on the number of pre-defined explicit

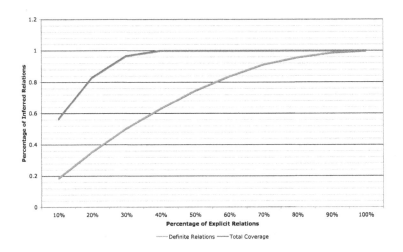

Fig. 5. Percentage of explicit (raw) relations vs. percentage of inferred relations in the sample ontology data set

relations. The figures is based on the experiment with the ontology built from web resources used in the example. The total coverage refers to how many spatial relation in the ontology that are not the universal relation (a disjunction of all possible eight base topological relations). For instance, if the coverage is 100% then every region is connected to every other region by either a definite or indefinite topological relation. The number of definite relations is the percentage of region to region relations that are definite (only one topological relation).

6   The Extended Framework

Information on the object's location, shape and size can be used to directly compute its relationships to other object. A system for managing geo-referenced data need therefore to be able to make effective use of available geometric representations. Logic programming does not naturally support the representation and manipulation of these facts, but it can link up with processors that are more suited to these tasks. In addition, coordinate data representing boundaries of geofeatures can increase the storage (and memory) overhead significantly for an ontology base and stretches the capabilities of current technologies for reasoning with them. A sample geographic ontology base with 10 classes and around 10,000 individuals was created for a data set of European administrative boundaries. Classes were associated with 2 properties and 3 datatype properties. The detailed representation of the boundary data resulted in an OWL ontology that occupied 100MB of persistent storage space and approximately 800MB of memory.

A hybrid extension to the framework is therefore proposed here to integrate an external geometric computation engine, to which the storage and manipulation

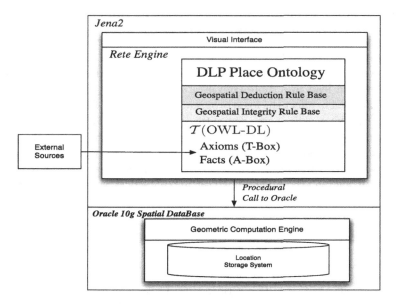

Fig. 6. Extended DLP Framework

of the geometric component of the geospatial ontology bases can be delegated. The extended framework is shown in figure 6. The Location Storage System (LSS) can in practice be a spatial database system (Oracle spatial is used in our case). All geometries are mapped directly in the LSS. An example of the mapping is shown in table 3.

Table 3. Example Geometry Mapping

Place Geometry	(Oracle) Table
District(Roath)→	INSERT INTO locationBase
Geomtry→	VALUES('http://cf.ac.uk/Roath',
polygon→	MDSYS.SDO_GEOMETRY
Coord(3,13)	(2003,8307,null,
Coord(11,13)	MDSYS.SDO_ELEM_INFO_ARRAY
Coord(11,21)	(1,1003,1), MDSYS.SDO_ORDINATE_ARRAY
Coord(3,21)	(3,13,11,13,11,21,3,21,3,13)))
Coord(3,13)	

The unique URI reference to a place instance in the DLP ontology is maintained in the LSS. This allows place instances in the DLP to be linked to their associated geometries in the LSS. In practice, all calls to the LSS take place through procedural attachments from the core DLP.

6.1   Procedural Attachments for Spatial Operators

Many logic engine implementations provide a set of static predefined procedural attachments, denoted *builtins*. *Builtins* commonly revolve around simple arithmetic procedures or comparison procedures. Extending a DLP with procedural attachments can lead to a more complicated semantic treatment if the attachments are allowed to effect the logic program in any way, by for example removing facts from the knoweldge base. Semantically clean builtins are those that only test or compute facts and will not change or remove facts in the DLP.

In addition to standard *builtins*, a set of spatial *builtins* (or spatial operators) needs to be defined to link between the DLP component and the external geo-computation engine. Examples of these procedural attachments are given in table 4.

Table 4. DLP Spatial Procedural Attachments

Procedural Attachment	Arguments	Oracle
exAdjacent	(Ind_1, Ind_2)	SELECT c_b.rdfID, c_d.rdfID, SDO_GEOM.RELATE (c_b.shape, 'TOUCH', c_d.shape, 0.005) FROM \<tableName\> c_b, \<tableName\> c_d WHERE c_b.rdfID = \<ind1\> AND c_d.rdfID = \<ind2\>
Area	(Ind_1, R)	SELECT SDO_GEOM.SDO_AREA (loce.shape, 0.005,'unit= \<unit\>) FROM \<tableName\> loce WHERE loce.rdfID = \<ind1\>
exDisjoint	(Ind_1, Ind_2)	$\cdots$
Distance	(Ind_1, Ind_2, R)	$\cdots$

6.2   Interleaved Reasoning

Typically, all rule body antecedents are matched from existing stored facts (facts derived by rules or explicitly represented). Interleaving forward and backward reasoning modes in a logic program allow for the derivation of facts on the fly if they are not explicitly stored. Consider the following rule:

$$[Region(?x) \wedge Region(?y) \wedge Region(?z) \wedge Inside(?x?z) \wedge Inside(?z?y) \rightarrow Inside(?x?y)]$$

The conclusion of Inside(?x ?y) would only be inferred if both the atoms Inside(?x ?z) and Inside(?z ?y) can be satisfied. These atoms are either satisfied by facts directly stored in the ontology (explicit), or inferred using reasoning rules, or as a last resort satisfied by a rule that calls the external geo-computation engine.

For example, the following is a subset of rules used to derive the inside relationship between two regions. The fifth rule is a call to the external (*exInside* predicate). Hence, `Inside(?x ?y)` will return either true or false, based on whether the relationship exists in the ontology, can be inferred, or whether it can be determined from the geometry.

$Inside(?x\ ?y) \leftarrow Region(?x) \wedge Region(?y) \wedge Region(?c) \wedge Inside(?x\ ?c) \wedge Equal(?c\ ?y)$

$Inside(?x\ ?y) \leftarrow Region(?x) \wedge Region(?y) \wedge Region(?c) \wedge Inside(?x\ ?c) \wedge Inside(?c\ ?y)$

$Inside(?x\ ?y) \leftarrow Region(?x) \wedge Region(?y) \wedge Region(?c) \wedge Inside(?x\ ?c) \wedge CoveredBy(?c\ ?y)$

$Inside(?x\ ?y) \leftarrow Region(?x) \wedge Region(?y) \wedge Region(?c) \wedge CoveredBy(?x\ ?c) \wedge Inside(?c\ ?y)$

$Inside(?x\ ?y) \leftarrow Region(?x) \wedge Region(?y) \wedge Region(?c) \wedge exInside(?c\ ?y)$

Example. The following qualitative relations were mined from Wikipedia related to the region "South Glamorgan"; an administrative subdivision of Wales.

$$contains(\text{Wales,Vale-of-Glamorgan})$$

$$inside(\text{Vale-of-Glamorgan, South-Glamorgan})$$

The spatial deduction rules suggest that South Glamorgan must be connected to Wales through a number of possible relations using the following rule.

$$inside^{-1}(\text{A,B}) \wedge inside(\text{B, C}) \rightarrow Overlap(A,C) \vee Contains(A,C) \vee Inside(A,C)$$
$$\vee\ Equal(A,C) \vee Covers(A,C) \vee CoveredBy(A,C)$$

Consequently, South-Glamorgan can't be disjoint from Wales, as identified by the following integrity rule.

$$inside^{-1}(A,B) \wedge inside(B,C) \wedge disjoint(A,C) \rightarrow error(A,C) \tag{2}$$

Fig. 7. Geonames South Glamorgan Geometric Error

Data are also recorded for the boundary points of Wales as well as point locations for the all the regions concerned (retrieved from Geonames). Firing integrity rule (2) results in interleaved reasoning where each of the predicates (spatial relations) in the rule are determined using the set of spatial composition rules in the system. The relation *disjoint* however, is not stored explicitly. To check this relation, an external call to the geo-computation engine is fired using the builtin `exDisjoint(A,B)`. The call returns "True" indicating the fact that the geometry point location of South-Glamorgan is in fact outside the boundary of Wales. This contradicts with the facts already stored and hence an error is implied. Figure 7 shows the point location for South-Glamorgan, falling in the sea, as recorded in Geonames.

The example demonstrates how the two types of reasoning; qualitative and quantitative, supported by this framework can be complementary to one another. Spatial relations are computed on the fly, when needed, within a logical reasoning framework.

7 Conclusion

In this paper we explore the idea of "spatially-enabling" the semantic web. As geo-referencing of resources on the web becomes more popular, methods to support the search, sharing and linking of these resources are needed. The semantic web offers standard languages and tools to enable the representation and reasoning with the data. This paper demonstrates how these tools can be used for geospatial domains.

In particular, OWL-DL is used to store a basic model of place and spatial relationships. A homogeneous approach to integrating rules with OWL, namely, description logic programs DLPs, was shown to allow the expression of spatial deduction and integrity rules. A framework based on DLPs is proposed and is shown to support, terminological as well as spatial reasoning over geographical ontology bases.

The logical framework will however, not cope well with the demands of the geometric representations of geo-features. An extended framework is proposed to link the DLP with external geometric computation processors. It is shown how this link can be established using procedural attachments. The resultant framework supports both logical and geometric manipulation of geospatial facts and data, thus combining the strengths of both paradigms. Some realistic data sets mined from web sources are used for demonstration and for evaluating the proposed frameworks.

The contribution of the work is in demonstrating possible approaches to geospatial data management on the web and in highlighting the needs of geospatial domains that stretches the current semantic web tools and languages. Future work will consider the issue of scalability and other challenges related to problems of integrating and linking of geospatial data from different sources.

References

1. Antoniou, G., Damásio, C.V., Grosof, B., Horrocks, I., Kifer, M., Maluszyński, J., Patel-Schneider, P.F.: Combining Rules and Ontologies. A survey (2005)
2. Brachman, R.J., Borgida, A., Mcguinness, D.L., Patel-schneider, P.F., Resnick, L.A.: The classic knowledge representation system, or, kl-one: The next generation. In: The Workshop on Formal Aspects of Semantic Networks, Two Harbors, pp. 1036–1043. Morgan Kaufman, San Francisco (1989)
3. Buyukokkten, O., Cho, J., Garcia-Molina, H., Gravano, L., Shivakumar, N.: Exploiting geographical location information of web pages. In: Proceedings of Workshop on Web Databases (WebDB 1999) (June 1999); Held in conjunction with ACM SIGMOD 1999 (1999)
4. Cohn, A., Hazarika, S.: Qualitative spatial representation and reasoning: an overview. Fundamenta Informaticae 45, 1–29 (2001)
5. de Bruijn, J., Lara, R., Polleres, A., Fensel, D.: Owl dl vs. owl flight: conceptual modeling and reasoning for the semantic web. In: WWW 2005: Proceedings of the 14th international conference on World Wide Web, pp. 623–632. ACM Press, New York (2005)
6. Egenhofer, M.: Deriving the composition of Binary Topological Relations. Journal of Visual Languages and Computing 5, 133–149 (1994)
7. El-Geresy, B., Abdelmoty, A.: Towards a general theory for modelling qualitative space. International Journal on Artificial Intelligence Tools, IJAIT 11(3), 347–367 (2002)
8. El-Geresy, B., Abdelmoty, A.: Sparqs: A qualitative spatial reasoning engine. Journal of knowledge-based Systems 17(2-4), 89–102 (2004)
9. Fonseca, F.T., Davis, C.A., Câmara, G.: Bridging ontologies and conceptual schemas in geographic information integration. GeoInformatica 7(4), 355–378 (2003)
10. Forgy, C.: Rete: A fast algorithm for the many patterns/many objects match problem. Artificial Intelligence 19(1), 17–37 (1982)
11. Grosof, B.N., Horrocks, I., Volz, R., Decker, S.: Description logic programs: combining logic programs with description logic. In: WWW, pp. 48–57 (2003)
12. Grosof, B.N., Horrocks, I., Volz, R., Decker, S.: Description logic programs: combining logic programs with description logic. In: Proceedings of the twelfth international conference on World Wide Web, pp. 48–57. ACM Press, New York (2003)
13. Gruber, T.R.: A translation approach to portable ontologies. Knowledge Acquisition 5(2), 199–220 (1993)
14. Guarino, N.: Formal ontology, conceptual analysis and knowledge representation. International Journal of Human-Computer Studies 43(5/6), 625–640 (1995)
15. Horrocks, I.: Owl rules, ok? In: Rule Languages for Interoperability (2005)
16. Horrocks, I., Patel-Schneider, P.F., Tabet, H.B.S., Grosof, B., Dean, M.: Swrl: A semantic web rule language combining owl and ruleml. Internet Report (May 2004), http://www.w3.org/Submission/2004/SUBM-SWRL-20040521/
17. Jones, C.: Geographical Information Systems and Computer Cartography. Longman (1997)
18. Jones, C.B., Abdelmoty, A.I., Fu, G.: Maintaining ontologies for geographical information retrieval on the web. In: Meersman, R., Tari, Z., Schmidt, D.C. (eds.) CoopIS 2003, DOA 2003, and ODBASE 2003. LNCS, vol. 2888, pp. 934–951. Springer, Heidelberg (2003)

19. Jones, C.B., Purves, R., Ruas, A., Sanderson, M., Sester, M., van Kreveld, M., Weibel, R.: Spatial information retrieval and geographical ontologies an overview of the spirit project. In: SIGIR 2002: Proceedings of the 25th annual international ACM SIGIR conference on Research and development in information retrieval, pp. 387–388. ACM, New York (2002)

20. Krötzsch, M., Hitzler, P., Vrandecic, D., Sintek, M.: How to reason with OWL in a logic programming system. In: Eiter, T., Franconi, E., Hodgson, R., Stephens, S. (eds.) RuleML, pp. 17–28. IEEE Computer Society, Los Alamitos (2006)

21. Motik, B., Vrandecic, D., Hitzler, P., Sure, Y., Studer, R.: dlpconvert – converting owl dlp statements to logic programs. In: European Semantic Web Conference 2005 Demos and Posters (2005)

22. Nebel, B., Renz, J.: Efficient methods for qualitative spatial reasoning. Journal of Artificial Intelligence Research, 562–566 (June 19, 1998)

23. Sagonas, K., Swift, T., Warren, D.S.: Xsb: An overview of its use and implementation. Tech. rep., November 2 (1993)

24. Sanderson, M., Kohler, J.: Analyzing geographic queries. In: Workshop on Geographic Information Retrieval SIGIR (August 9, 2004)

25. Silva, M.J., Martins, B., Chaves, M.S., Afonso, A.P., Cardoso, N.: Adding geographic scopes to web resources. Computers, Environment and Urban Systems 30(4), 378–399 (2006)

26. Volz, R.: Web Ontology Reasoning with Logic Databases. PhD thesis, Universität Karlsruhe (TH), Universität Karlsruhe (TH), Institut AIFB, D-76128 Karlsruhe (2004)

Efficient Construction of Safe Regions for Moving kNN Queries over Dynamic Datasets

Mahady Hasan, Muhammad Aamir Cheema, Xuemin Lin, and Ying Zhang

The University of New South Wales, Australia
{mahadyh,macheema,lxue,yingz}@cse.unsw.edu.au

Abstract. The concept of *safe region* has been used to reduce the computation and communication cost for the continuous monitoring of k nearest neighbor (kNN) queries. A safe region is an area such that as long as a query remains in it, the set of its kNNs does not change. In this paper, we present an efficient technique to construct the safe region by using cheap *RangeNN* queries. We also extend our approach for dynamic datasets (the objects may appear or disappear from the dataset). Our proposed algorithm outperforms existing algorithms and scales better with the increase in k.

1 Introduction

With the availability of inexpensive mobile devices, position locators and cheap wireless networks, location based services are gaining increasing popularity. The continuous monitoring of k nearest neighbor (kNN) queries [1,2,3,4] has been widely studied in recent past.

In this paper, we study the problem of moving kNN queries where the query is constantly moving and the objects do not move. Consider the example of a car driver who is interested in five nearest available car parking spaces while driving in a city. Another example is a person looking for the nearest restaurants while walking in a street.

A classical example of the safe region is Voronoi Diagram (VD) [5]. In a VD, each object of the dataset lies within a cell called its voronoi cell. The voronoi cell of an object has a property that any point that lies in it is always closer to that object than any other object in the dataset. For a kNN query, a k order VD can be constructed and k order voronoi cells can be treated as safe regions. The VD based solution has the following major limitations: 1) The VD cannot be precomputed and indexed if the value of k is not known in advance. 2) The VD cannot deal efficiently with update of objects in the underlying dataset.

Our contributions in this paper include: 1) we devise an efficient safe region construction approach that requires cheap *RangeNN*[1] queries; 2) our proposed approach is extended to efficiently update the safe regions of queries for dynamic datasets where the objects may appear or disappear and 3) extensive experiment results show more than an order of magnitude improvement.

---

[1] RangeNN query is to find the nearest object of q from the objects that lie within a given distance from a point p.

N. Mamoulis et al. (Eds.): SSTD 2009, LNCS 5644, pp. 373–379, 2009.

2   Background Information

Continuous k Nearest Neighbor Query. Given a set of objects, a moving query point q , and a positive integer k , the continuous kNN query is to continuously report k closest objects to q at each time stamp.

Definitions and Notations. A *perpendicular bisector $B_{n:o}$* between two points n and o divides the space into two half-spaces. Let $H_{n:o}$ be the half-space containing n and $H_{o:n}$ be the half-space containing o. Every point q in $H_{n:o}$ is always closer to n than it is to o (i.e; $dist(q, n) < dist(q, o)$). Figure 1 shows a bisector $B_{n:o_2}$ between two points n and o_2 and the two half-spaces are also shown.

Safe Region S is a region such that as long as a kNN query q remains in it, the set of its kNNs does not change. If a client (that issued query q) is aware of its safe region, it does not need to contact the server to update its set of kNNs as long as q resides in the safe region. This saves the communication cost as well as computation cost. Now, we formally define the safe region.

Let $N = \{n_1, \cdots, n_k\}$ be the set of kNNs of a query q. The intersection of all half-spaces $H_{n_i:o_j}$ for every $n_i \in N$ and every $o_j \in O - N$ defines a region such that as long as the query resides in it, the set of its kNNs N is unchanged.

Proof. We prove this by contradiction. Assume that q resides in its safe region and $o_j \in O - N$ is an object such that $dist(q, o_j) < dist(q, n_i)$ for any $n_i \in N$. Since safe region is the intersection of all half-spaces $H_{n_i:o_j}$, a query q that resides in it satisfies $dist(q, n_i) < dist(q, o_j)$ which contradicts the assumption.   □

Figure 1 shows an example of the safe region for a NN query. The bisectors between the nearest neighbor n and the objects o_1 to o_4 are drawn and the shaded area is the safe region. Figure 2 shows an example of the safe region for a 2NN query where the two NNs are n_1 and n_2. The bisectors between the NNs and the objects o_1 to o_3 are drawn. For clarity, the bisectors between n_1 and the objects are shown in solid lines and the bisectors between n_2 and the objects are shown in broken lines. The shaded area is the safe region.

Note that not all the bisectors contribute in defining the safe region. A bisector $B_{n_i:o_j}$ that forms an edge of the safe region is called a **representative bisector** (the bisector $B_{n:o_2}$ in Fig. 1). The object o_j that is associated with the representative bisector is called an **influence object** (o_2 in Fig. 1).

A **vertex** is the intersection of two bisectors $B_{n_i:o_j}$ and $B_{n_x:o_y}$. A **confirmed vertex** is the vertex of the safe region (i.e., it is an intersection of two representative bisectors). Vertex v in Fig. 1 is a confirmed vertex whereas the vertex v' is

Table 1. Notations

Notation	Definition
$B_{x:q}$	a perpendicular bisector between point x and q
$H_{x:q}$	a half-space defined by $B_{x:q}$ containing the point x
$H_{q:x}$	a half-space defined by $B_{x:q}$ containing the point q
$dist(x, y)$	the distance between two points x and y
$v \prec B_{n_i:o_j} \cap B_{n_x:o_y} \succ$	a vertex v formed by the intersection of the two bisectors

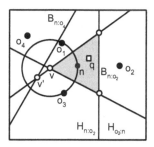
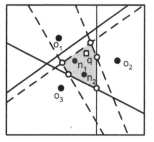
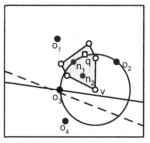

Fig. 1. Safe region for a NN query   **Fig. 2.** Safe region for a 2-NN query   **Fig. 3.** Illustration of Observation 2

not a confirmed vertex. Please note that a confirmed vertex lies at the boundary of the safe region. Table 1 defines the notations used throughout this paper.

The most related work to our technique is proposed in [6]. The authors propose construction of the safe region by using time parameterized kNN queries [7]. Due to space limitations, we omit the details.

3 Technique

Before we present our algorithm, we present observations that can be used to confirm a vertex. First, we present the observation for $k = 1$ and then we extend it for arbitrary value of k.

OBSERVATION 1 :   Let n be the NN of a query q and v be a vertex. The vertex v can be confirmed if no object lies in the circle of radius R centered at v where $R = dist(v, n)$.

Proof. Assume that the circle does not contain any object and o_4 (as shown in Fig. 1) is any object that lies outside the circle. If the vertex v does not lie in the safe region then there must be a half-space $H_{o_4:n}$ that contains v. Any point p that lies in the half-space $H_{o_4:n}$ satisfies $dist(p, o_4) < dist(p, n)$. However, for vertex v, $dist(v, o_4) > dist(v, n)$. Hence there is no such half-space $H_{o_4:n}$ that contains v. So the vertex v lies in the safe region.   □

OBSERVATION 2 :   Let $N = \{n_1, \cdots, n_k\}$ be the set of kNNs of query q and v be any vertex. The vertex v can be confirmed if no object $o \in O - N$ lies in the circle centered at v with radius $R = maxdist(v, N)$ where $maxdist(v, N)$ is $max(dist(v, n_i))$ for every $n_i \in N$.

Proof. Assume that the circle does not contain any object and o_4 is any object that lies outside the circle (as shown in Fig. 3). The vertex v satisfies $dist(v, n_i) < dist(v, o_4)$ for every $n_i \in N$, hence v lies in every $H_{n_i:o_4}$. For this reason, the vertex v lies in the safe region.   □

Algorithm 1 presents the construction of the safe region for a kNN query. The algorithm maintains a set of vertices V (initialized to four vertices of the universal data space). First, the set N containing kNNs of the query q is computed by using BFS [8]. Then, the algorithm randomly selects an unconfirmed vertex v from V and checks whether it can be confirmed or not by using Observation 2. More specifically, the algorithm checks whether there is any object in the circle of range $R = maxdist(v, N)$ centered at v. If there is no object in the circle, the algorithm marks the vertex as confirmed (line 8).

Algorithm 1. Construct Safe Region (q)

1: $V = \{$Vertices of the data space$\}$
2: compute kNNs of q and store in N
3: **while** there is an unconfirmed vertex in V **do**
4:     select any unconfirmed vertex v
5:     $R = maxdist(v, N)$
6:     $o = \text{RangeNN}(q, v, R)$/* `Algorithm 2` */
7:     **if** $o = NULL$ **then**
8:         confirm v
9:     **else**
10:        update V using bisectors between o and each $n_i \in N$

If there are more than one objects in the circle, the algorithm selects the nearest object o to the query q (line 6). The safe region is updated by considering the bisectors between kNNs of q and the object o (line 10). For a given bisector $B_{n_i:o}$, the safe region is updated by removing the vertices from V that lie in $H_{o:n_i}$ and adding the intersection points of $B_{n_i:o}$ and the safe region. The algorithm stops when all the vertices are confirmed. To show the correctness of the algorithm, we need to show that the algorithm finds all the vertices of the safe region and does not include any unconfirmed vertex. The proof of correctness is similar to Lemma 3.1 in [6] and is omitted.

Algorithm 2. RangeNN(q, v, R)

Output: Returns the nearest neighbor of q from the objects that lie within distance
    R from v
1: Initialize a min-heap H with root entry of the tree
2: **while** H is not empty **do**
3:     deheap an entry e
4:     **if** e is an intermediate or leaf node **then**
5:         **for each** of its children c **do**
6:             **if** $mindist(c, v) < R$ **then**
7:                 insert c into H with key $mindist(c, q)$
8:     **else if** e is an object and e is not one of the kNNs of q **then**
9:         **return** e
10: **return** ϕ

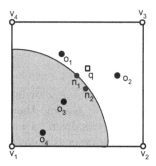
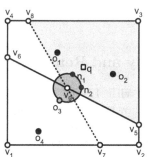
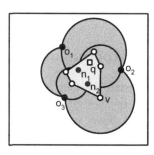

Fig. 4. RangeNN query from v_1

Fig. 5. The safe region after visiting o_3

Fig. 6. Safe Region and impact Region

Algorithm 2 presents the implementation of RangeNN query. This operation can be regarded as finding the nearest object o of q from the objects lying within the range R of a vertex v. Hence, we call it *RangeNN* query.

Example 1. Figure 4 illustrates our algorithm for a 2NN query where n_1 and n_2 are the NNs of q. Initial safe region is the data space bounded by four vertices v_1 to v_4. First, a RangeNN[2] query is issued on vertex v_1 with range $R = dist(v_1, n_1)$ which returns the object o_3. Then, the bisectors between o_3 and the NNs are drawn. In Fig. 5, the bisector between o_3 and n_1 is shown in solid line and the bisector between o_3 and n_2 is shown in broken line. These bisectors update the set of vertices V and the new safe region (the shaded area) now contains vertices v_3, v_5, v_9 and v_8. Then, a RangeNN query is issued on vertex v_9 with range $dist(v_9, n_1)$ and it is marked confirmed because no object is found within the range. The algorithm continues in this way until all the vertices are confirmed. The final safe region is shown in Fig. 6 (light shaded area).

Extension for Dynamic Datasets. First, we define *impact region*. The impact region is an area such that as long as a query remains in its safe region and no object appears or disappears from the impact region, the safe region of the query is unchanged. It is easy to prove that the impact region consists of circles around vertices with radius set to their corresponding nearest neighbors. In Fig. 6, the impact region is shown shaded (both dark and light). Below, we formally define the impact region.

Let V be a set of vertices of a safe region. Let $Circ_v$ be a circle centered at a vertex $v \prec B_{n_i:o_j} \cap B_{n_x:o_y} \succ$ with radius $R_v = dist(v, n_i)$. The impact region is the area covered by all circles $Circ_{v_i}$ for each $v_i \in V$.

We use a grid-based structure and mark all the cells that overlap with the impact region. The results of a query are affected only if an object appears in

---

[2] Note that RangeNN query does not access all the objects within the range. It uses BFS and stops when the NN is found. So the object o_4 is not accessed in the example.

(or disappears from) these marked cells. For such queries, we compute the safe regions again.

4 Experimental Study and Remarks

We compare our algorithm with LBSQ [6]. Other algorithms for moving kNN queries either assume known query trajectory path [7,4] or assume that clients have sufficient computation resources to maintain kNNs from given $(k + x)$ or more NNs [9,10,11]. We use real dataset (http://www.census.gov/geo/www/tiger/) that contains 128,700 unique data points in a data space of 350km×350km. We continuously monitor 500 moving queries created by the spatio-temporal data generator [12].

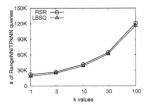
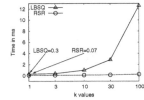
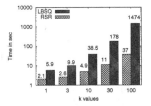

Fig. 7. Total RangeNN / TPkNN queries   **Fig. 8.** Average cost of RangeNN / TPkNN query   **Fig. 9.** The computation time for different k

Figure 7 shows that the number of RangeNN queries is slightly higher than the number of TPkNN queries, but the average cost of a RangeNN query is significantly lower than that of a TPkNN query (Fig. 8).

Figures 9 studies the effect of k on the computation times of both algorithms (shown in log scale). Our algorithm not only outperforms LBSQ but also scales better. We also observed that the number of nodes accessed by our algorithm is lower than that of LBSQ but we do not include the figure due to page limitation.

Previous algorithm uses TPkNN queries to compute the safe region of a kNN query. In this paper, we present an efficient algorithm to construct the safe region by using much cheaper RangeNN queries. Experiment results show an order of magnitude improvement.

References

1. Mouratidis, K., Hadjieleftheriou, M., Papadias, D.: Conceptual partitioning: An efficient method for continuous nearest neighbor monitoring. In: SIGMOD Conference, pp. 634–645 (2005)
2. Yu, X., Pu, K.Q., Koudas, N.: Monitoring k-nearest neighbor queries over moving objects. In: ICDE, pp. 631–642 (2005)
3. Xiong, X., Mokbel, M.F., Aref, W.G.: Sea-cnn: Scalable processing of continuous k-nearest neighbor queries in spatio-temporal databases. In: ICDE, pp. 643–654 (2005)

4. Tao, Y., Papadias, D., Shen, Q.: Continuous nearest neighbor search. In: VLDB, pp. 287–298 (2002)
5. Okabe, A., Boots, B., Sugihara, K.: Spatial tessellations: concepts and applications of Voronoi diagrams. John Wiley and Sons Inc., Chichester (1992)
6. Zhang, J., Zhu, M., Papadias, D., Tao, Y., Lee, D.L.: Location-based spatial queries. In: SIGMOD Conference, pp. 443–454 (2003)
7. Tao, Y., Papadias, D.: Time-parameterized queries in spatio-temporal databases. In: SIGMOD Conference, pp. 334–345 (2002)
8. Hjaltason, G.R., Samet, H.: Ranking in spatial databases. In: SSD, pp. 83–95 (1995)
9. Kulik, L., Tanin, E.: Incremental rank updates for moving query points. In: Raubal, M., Miller, H.J., Frank, A.U., Goodchild, M.F. (eds.) GIScience 2006. LNCS, vol. 4197, pp. 251–268. Springer, Heidelberg (2006)
10. Song, Z., Roussopoulos, N.: K-nearest neighbor search for moving query point. In: Jensen, C.S., Schneider, M., Seeger, B., Tsotras, V.J. (eds.) SSTD 2001. LNCS, vol. 2121, pp. 79–96. Springer, Heidelberg (2001)
11. Nutanong, S., Zhang, R., Tanin, E., Kulik, L.: The v*-diagram: a query-dependent approach to moving knn queries. PVLDB 1(1), 1095–1106 (2008)
12. Brinkhoff, T.: A framework for generating network-based moving objects. GeoInformatica 6(2), 153–180 (2002)

Robust Adaptable Video Copy Detection

Ira Assent[1] and Hardy Kremer[2]

[1] Department of Computer Science, Aalborg University, Denmark
ira@cs.aau.dk
[2] Data management and exploration group, RWTH Aachen University, Germany
kremer@cs.rwth-aachen.de

Abstract. Video copy detection should be capable of identifying video copies subject to alterations e.g. in video contrast or frame rates. We propose a video copy detection scheme that allows for adaptable detection of videos that are altered temporally (e.g. frame rate change) and/or visually (e.g. change in contrast). Our query processing combines filtering and indexing structures for efficient multistep computation of video copies under this model. We show that our model successfully identifies altered video copies and does so more reliably than existing models.

1 Introduction

Video copy detection algorithms aim at automatic identification of video content that is identical to the query or represents an altered version of the original video [6,19,9]. As opposed to content-based similarity search in video databases [7,10,12], the aim is not searching for similar topics or otherwise related content in video material, but to discover videos that have undergone technical or manual changes, such as change in contrast or editing of the order of scenes in the video [9]. Other examples of typical alterations include changes in frame rate due to different video standards or black bars due to varying tv screen aspect ratios.

The changes undergone by video content can be roughly categorized in two groups: first, the video may be altered visually in the image domain, as e.g. in the contrast change example. And, second, the video may have been reordered in the temporal domain, as e.g. in the frame rate change example. The challenge for effective video copy detection therefore lies in both of these domains. Suitable copy detection schemes should be capable of correctly identifying videos altered in one of these domains or in both [9].

In this work, we propose a robust adaptable video copy detection scheme ($RAVC$) that allows effective detection of changes in time and image content or both. Our technique integrates powerful adaptable distance functions for both visual and temporal alterations. Our copy detection scheme does not require prior key frame extraction, but instead works directly on the video frame sequence.

As video features are typically high dimensional, we propose VA-file based indexing. We extend our model to the quantization of features required in VA-file indexing.

N. Mamoulis et al. (Eds.): SSTD 2009, LNCS 5644, pp. 380–385, 2009.

2   Video Copy Detection

Each video is composed of a sequence of frames, i.e. chronologically ordered images. These images are either the result of recording or the result of artifical creation, e.g. in animated movies. Videos can thus be represented as time series of image features. For images, histograms are a popular and simple way of capturing the distribution of properties such as color in the image [14,1,11,2]. An image histogram of resolution d is $h(f) = (h_1, \ldots, h_d)$, and a video histogram of length n is defined as a vector of image histograms: $V = (v_1, \ldots, v_n)$.

Our first goal in video copy detection is to identify videos with visual changes e.g. in contrast. The Earth Mover's Distance (EMD) was introduced in computer vision to mimic the perceived similarity of images [14].

Definition 1. *Earth Mover's Distance (EMD)*
The Earth Mover's Distance between two normalized frame histograms $u = (u_1, \ldots, u_d)$ and $v = (v_1, \ldots, v_d)$ with respect to a ground distance given by a cost matrix $C = [c_{ij}]$ is defined as follows:

$$EMD_C(u, v) = \min_{\mathbf{F}} \left\{ \sum_{i=1}^{d} \sum_{j=1}^{d} c_{ij} f_{ij} \mid Con_{EMD} \right\}$$

with $Con_{EMD} = P \wedge S \wedge T :$ $P : \forall 1 \le i, j \le d : \ f_{ij} \ge 0$

$$S : \forall 1 \le i \le d : \ \sum_{j=1}^{d} f_{ij} = u_i \qquad T : \forall 1 \le j \le d : \ \sum_{i=1}^{d} f_{ij} = v_j$$

where $\mathbf{F}$ denotes the set of possible flow matrices.

Thus, the best possible matching between the two histograms is defined as a minimization over all possible *flow* matrices $f_{ij} \in \mathbf{F}$ from histogram dimension i in u to histogram dimension j in v. The constraints ensure that only non-negative flows are allowed (P), that not more is taken from any histogram dimension i in the *source* histogram u than its value (S), and that all histogram entries j in the *target* histogram are matched (T). In this transportation problem formulation, the EMD can be computed using e.g. the streamlined simplex method from operations research [4,8]. The EMD is robust to small changes in the feature distribution due to its *ground distance*, i.e. the cost c_{ij} associated with any change between feature dimension i and j [14,13].

Our second goal is detecting changes along the time axis of the videos. In speech recognition, Dynamic Time Warping (DTW) was developed to handle nonlinear fluctuations in speech rates [16]. DTW computes the best matching between the time series in an *interdimensional* fashion. Formally,

Definition 2. *Dynamic Time Warping (DTW)*
The DTW distance between two one-dimensional time series $x = (x_1, \ldots, x_n)$

and $y = (y_1, \ldots, y_n)$ *with respect to a band constraint* k *is defined as:*

$$DTW^2(x, y, k) = DTW'(x, y, k) + \min \begin{cases} DTW(start(x), start(y), k) \\ DTW(x, start(y), k) \\ DTW(start(x), y, k) \end{cases} \quad (1)$$

$$DTW(\emptyset, \emptyset, k) = 0, \quad DTW(x, \emptyset, k) = DTW(\emptyset, y, k) = \infty \quad (2)$$

with

$$DTW'((x_1, \ldots, x_u), (y_1, \ldots, y_v), k) = \begin{cases} (x_u - y_v)^2 & |u - v| \leq k \\ \infty & else \end{cases} \quad (3)$$

where $start(x_1, \ldots, x_{n-1}, x_n) = (x_1, \ldots, x_{n-1})$.

This recursive definition of the best alignment subject to a warping band constraint k is the best distance within the path constraint (3) and the minimum of the respectively shorter subproblems (1). (2) ensures that all elements of the time series are compared, starting at the beginning of both time series x_1 and y_1, and ending in x_n and y_n. Note that the band constraint, called *Sakoe-Chiba-Band*, simply ensures that the warping does not degenerate; e.g. matching all elements of one time series to a single element in the second one.

2.1   RAVC

For videos, we use the extension of DTW to multidimensional time series, i.e. each time point is a frame histogram. Frames are compared via Earth Mover's Distance, and entire videos via Dynamic Time Warping on the frame distances.

Definition 3. *Robust Adaptable Video Copy Detection (RAVC)*
The Robust Adaptable Video Copy Detection distance between video histograms $X = (x_1, \ldots, x_n)$ *and* $Y = (y_1, \ldots, y_n)$ *with respect to a ground distance given by a cost matrix* $C = [c_{ij}]$ *and with respect to band constraint* k *is defined as*

$$RAVC(X, Y, k) = DTW_{EMD}(X, Y, k) + \min \begin{cases} RAVC(start(X), start(Y), k) \\ RAVC(X, start(Y), k) \\ RAVC(start(X), Y, k) \end{cases}$$

$$RAVC(\emptyset, \emptyset, k) = 0, \quad RAVC(X, \emptyset, k) = RAVC(\emptyset, Y, k) = \infty$$

with

$$DTW_{EMD}((x_1, \ldots, x_u), (y_1, \ldots, y_v), k) =$$
$$\begin{cases} EMD((x_1, \ldots, x_u), (y_1, \ldots, y_v)) & |u - v| \leq k \\ \infty & else \end{cases}$$

Thus, $RAVC$ is recursively defined just as the DTW distance for univariate time series. The difference in this definition is that EMD is used to find the best matching between the frames in the multivariate video time series.

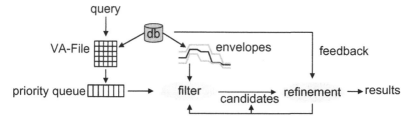

Fig. 1. General multistep indexing scheme

As straightforward calculation of $RAVC$ would be computationally costly (dynamic programming algorithms for DTW are of quadratic complexity, EMD is of worst case exponential complexity, yet in practice quadratic or cubic runtimes are observed [14]), we propose an efficient query processing algorithm in a multistep filter-and-refine architecture [17]. The filter efficiently computes a small set of candidates to reduce the number of videos for which the exact $RAVC$ model has to be calculated. As the filter is lower bounding, filter-and-refine is lossless [5,17]. We use the VA-file, an index structure for high-dimensional data [18], used for music time series in [15]. It quantizes the data space and assigns compact bit codes for quick sequential reading of the compressed data. Details are deferred to an extended version of this paper.

3 Experiments

We evaluate $RAVC$ under different alteration scenarios in the image and time domain that are typically encountered in video copies. Video copies are generated using benchmark scenarios described in [9]: changes in contrast, black bars as a result of screen ratio changes, gauss filters, and, additionally, changes in the temporal order. We used tv news recorded for 33 hours at 30 fps (frames per second) as a real world video data set. The videos have an aspect ratio of 320x200. We measured the accuracy as the recall of the closest match found in the database averaged over 100 queries. The color histograms were computed in extended HLS space [1] and are 20-dimensional.

We first study the effect of temporal changes on the copy detection accuracy. Temporal distortions simulate effects like frame rate change and are achieved through random replication and omission of frames. We vary the maximal temporal distortion, i.e. the maximal number of replicated or omitted frames, to study robustness to frame rate change. We additionally vary the band constraint that determines the degree of temporal change in DTW.

The first setup studies the effect of changes in the temporal domain alone. As we can see in Figure 2, where the change is in the time domain alone, and the images in the original and the potential copies are left unaltered the recall of our method is high for all variations in the band constraint and the Euclidean distance. However, smaller constraints and the Euclidean distance cannot handle larger numbers of frame reorderings and experience a decrease in recall values

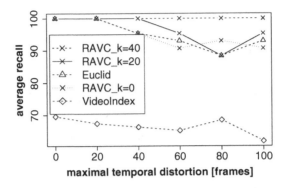

Fig. 2. Temporal change, but no visual change

as the temporal change increases. The *VideoIndex* [3] approach, which ignores temporal change altogether, shows far worse performance.

The next experiment adds the difficulty of changes in the image domain to the previous experiment. The copies under study here have not only been altered in the time domain as before, but additionally three changes that the images were subjected to are applied here as well, i.e. blur by a gaussian filter of radius 2, 25% contrast increase, and an aspect ratio change by black borders of an overall pixel height 16. Figure 3 demonstrates that our approach is clearly better than the Euclidean distance which drops down to only 50% recall for this more complex copy detection problem. We additionally used an even larger band constraint with a value of 80, yet the difference to the previously used value of 40 is negligible.

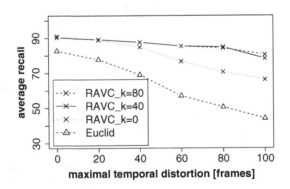

Fig. 3. Temporal change and visual change

4   Conclusion

In this paper, we present a novel technique for video copy detection. Based on the observation that alterations in both the image and time domain matter for reliable identification of copies, we introduce *RAVC* (robust adaptable video copy detection) scheme. Our experiments on real world data validate that our *RAVC* successfully detects copies under typical alterations.

References

1. Assent, I., Wenning, A., Seidl, T.: Approximation techniques for indexing the Earth Mover's Distance in multimedia databases. In: Proc. ICDE (2006)
2. Assent, I., Wichterich, M., Meisen, T., Seidl, T.: Efficient similarity search using the earth mover's distance for large multimedia databases. In: Proc. ICDE, pp. 307–316 (2008)
3. Böhm, C., Kunath, P., Pryakhin, A., Schubert, M.: Effective and efficient indexing for large video databases. In: Proc. BTW, pp. 132–151 (2007)
4. Dantzig, G.: Linear Programming and Extensions. Princeton Univ. Press, Princeton (1998)
5. Faloutsos, C.: Searching Multimedia Databases by Content. Kluwer, Dordrecht (1996)
6. Hampapur, A., Hyun, K., Bolle, R.: Comparison of sequence matching techniques for video copy detection. In: Proc. SPIE, pp. 194–201 (2002)
7. Hanjalic, A.: Content-based Analysis of Digital Video. Kluwer, Dordrecht (2004)
8. Hillier, F.S., Lieberman, G.J.: Introduction to Operations Research. McGraw-Hill, New York (2001)
9. Law-To, J., Chen, L., Joly, A., Laptev, I., Buisson, O., Gouet-Brunet, V., Boujemaa, N., Stentiford, F.: Video copy detection: a comparative study. In: Proc. CIVR, pp. 371–378 (2007)
10. Lee, J., Oh, J., Hwang, S.: STRG-Index: spatio-temporal region graph indexing for large video databases. In: Proc. SIGMOD, pp. 718–729 (2005)
11. Ljosa, V., Bhattacharya, A., Singh, A.K.: Indexing spatially sensitive distance measures using multi-resolution lower bounds. In: Ioannidis, Y., Scholl, M.H., Schmidt, J.W., Matthes, F., Hatzopoulos, M., Böhm, K., Kemper, A., Grust, T., Böhm, C. (eds.) EDBT 2006. LNCS, vol. 3896, pp. 865–883. Springer, Heidelberg (2006)
12. Lu, H., Xue, X., Tan, Y.: Content-Based Image and Video Indexing and Retrieval. In: Lu, R., Siekmann, J.H., Ullrich, C. (eds.) Joint Chinese German Workshops. LNCS, vol. 4429, pp. 118–129. Springer, Heidelberg (2007)
13. Rubner, Y., Puzicha, J., Tomasi, C., Buhmann, J.M.: Empirical evaluation of dissimilarity measures for color and texture. CVIU J. 84(1), 25–43 (2001)
14. Rubner, Y., Tomasi, C.: Perceptual Metrics for Image Database Navigation. Kluwer, Dordrecht (2001)
15. Ruxanda, M.M., Jensen, C.S.: Efficient similarity retrieval in music databases. In: Proc. COMAD, pp. 56–67 (2006)
16. Sakoe, H., Chiba, S.: Dynamic programming algorithm optimization for spoken word recognition. IEEE TAP 26(1), 43–49 (1978)
17. Seidl, T., Kriegel, H.-P.: Optimal multi-step k-nearest neighbor search. In: Proc. SIGMOD, pp. 154–165 (1998)
18. Weber, R., Schek, H.J., Blott, S.: A quantitative analysis and performance study for similarity-search methods in high-dimensional spaces. In: Proc. VLDB, pp. 194–205 (1998)
19. Yang, X., Sun, Q., Tian, Q.: Content-based video identification: a survey. In: Proc. ITRE, pp. 50–54 (2003)

Efficient Evaluation of Static and Dynamic Optimal Route Queries

Edward P.F. Chan and Jie Zhang

David R. Cheriton School of Computer Science
University of Waterloo
Waterloo, Ontario
Canada, N2L 3G1
epfchan@uwaterloo.ca, janezhangj@hotmail.com

Abstract. We investigate the problem of how to evaluate efficiently, with a general algorithm, static or dynamic optimal route queries on a massive graph. A graph is said to be *dynamic* if its edge weights are changed (increased or decreased) over time. Otherwise, it is *static*. A route query is *static* (*dynamic*) if the underlying graph is static (dynamic, respectively). The answer to an optimal route query is a shortest path that satisfies certain constraints imposed on paths in a graph. Under such a setting, a general and efficient algorithm called *DiskOP_{HBR}* is proposed to evaluate classes of static or dynamic optimal route queries. The classes of queries that can be evaluated by the algorithm are exactly those the constraints of which can be expressed as a set of edge weight changes. Experiments are conducted on this algorithm to show its desirability.

1 Introduction

In a route information system, like Yahoo!Map or Google!Map, or a moving object database [2] in which a user may issue queries to find optimal routes from sources to destinations. It is imperative that the queries allowed are not restrictive, and the answers can be generated fast. In such a system, a network is represented as a labeled graph G. The answer to an optimal route query, which involves a source s, a destination d, and a constraint or predicate θ on paths in G [1], is an optimal s-d path in G.

Recently, a general and efficient disk-based algorithm named *DiskCP* is derived to evaluate classes of (static) optimal route queries. The classes of optimal route queries are called *constraint preserving* (*CP*) [1]. CP query classes are static and encompass, among others, SP, forbidden nodes and edges and α-autonomy [1]. Instead of finding an efficient evaluating algorithm for individual query class, the approach taken in [1] is to find a unified algorithm to evaluate as many different query classes as possible. However, in order to have a fast evaluation, a pre-processing is required on the network to generate some materialized data for the specific query class before queries can be posted to the system.

We observe that in some applications, the optimal route query classes may not be known in advance, or the network on which a route query is posted is dynamic. In these situations, due to the size of a graph, it is impractical to re-compute all materialized data, as is required by *DiskCP*, for fast query evaluation. On the other hand,

N. Mamoulis et al. (Eds.): SSTD 2009, LNCS 5644, pp. 386–391, 2009.

without some pre-computed information on paths in a graph, it is impossible to speed up the search process. In this paper, we focus on the problem of fast evaluation, with a general algorithm, of an optimal route query, in both the static and dynamic environments, and without knowing the query or query classes in advance.

In Section 2, we define some basic notation. In Section 3, we briefly discuss the optimal route query evaluation algorithm. In Section 4, experimental results are presented. Finally, a summary is given in Section 5.

2 Definition and Notation

2.1 Optimal Route Queries

An optimal route query returns an SP in a graph G that satisfies certain constraint [1]. Let θ be a constraint imposed on paths in a graph G. If θ is *null* (Λ), then any path in G is satisfying wrt θ. An optimal route query, denoted as $Q(G,\theta,s,d)$, where s and d are two distinct nodes in G. The answer to $Q(G,\theta,s,d)$ is a satisfying s-d path in G, wrt θ, and no other satisfying s-d path in G with a shorter length. A graph is said to be *dynamic* if its edge weights are changed (increased or decreased) over time. If a graph is dynamic, then the answer to a query is wrt the graph at a time t. An *optimal route query class* $Q(G,\theta)$ is the set of optimal route queries $\{Q(G,\theta,s,d) \mid s$ and d are distinct nodes in $G\}$.

It has been shown empirically that certain static optimal route queries (named *constraint preserving* (*CP*)) can be evaluated very fast with distance materialization [1]. CP query classes include, among others, SP, forbidden nodes/edges, α-autonomous, 2-consecutive nodes and CP hypothetical weight changes. The following is an important property about CP queries.

Corollary 1. Let $Q(G,\theta)$ be a CP optimal query class, and p be a path in G. The path p is satisfying wrt θ iff every edge in p is satisfying wrt θ.

Given a CP query $Q(G,\theta,s,d)$, the *answer* to Q can be computed by finding an SP in graph G', which is obtained from G by eliminating all edges not satisfying θ [1]. In this work, we are interested in both static and dynamic optimal route queries. We define a more general optimal route query, for both static and dynamic settings, as follow.

Let Σ be a set of edge weight changes on a graph $G = (V,E,w)$, where $\Sigma=\{<e_i, \tau_i> \mid e_i \in E$ and $\tau_i \geq 0$ or $\tau_i=+\infty\}$. Σ is a *modified edge set*. Semantically, each element in Σ assigns a new weight τ_i to an edge e_i. Syntactically, an *optimal route query* is denoted as $Q(G,\Sigma,s,d)$, where G is a graph, s and d are two distinct nodes in G, and Σ is a modified edge set defined on edges in G. The answer to $Q(G,\Sigma,s,d)$ is a shortest s-d path in $G' = (V,E,w')$, where $\forall e \in E$, $w'(e)$ is τ if $<e, \tau> \in \Sigma$, and is $w(e)$ otherwise. The graph G' is said to be obtained from (G, Σ), or equivalently $G'=(G, \Sigma)$. Suppose $G'=(G, \Sigma)$. Then computing a path in G wrt Σ is the same as computing a path in G'. In other words, in the remainder of this work, an optimal route query is modeled as a set of edge weight changes on a graph G, which is called a *based* graph.

It is worth noting that in this work, when computing the answer to a query, G is *hypothetically* changed, according to Σ, to generate a graph G' from which the answer is computed. We call G' a *modified graph*. Given a CP route query, it can be expressed as an optimal query $Q(G,\Sigma,s,d)$. If the graph G is dynamic, the edge weight changes can be incorporated easily into Σ. Consequently, all static and dynamic CP queries can be expressed as $Q(G,\Sigma,s,d)$. Thus, the optimal route queries investigated in this work are quite general and including many real-life route queries.

The answer to an optimal route query is called an *optimal path* (*OP*). If Σ is the empty set Λ, then $Q(G,\Sigma,s,d)$ is an SP from s to d in G. Unless confusion arises, $Q(G,\Sigma,s,d)$ denotes an optimal route query as well as its answer. The distance of an optimal u-v path in G, denoted as $SD(G,\Sigma,u,v)$, is defined as its length, if it exists, and $+\infty$ otherwise.

2.2 Graph Partitioning and Fragments

A *fragment* is a connected sub-graph such that an edge connects two nodes in a fragment precisely when the two nodes are connected by the same edge in the original graph G. A node is a *boundary* node if it belongs to more than one fragments, otherwise it is an *interior* node. A *partition* $P(G)$ of $G = (V, E, w)$ is a collection of fragments $\{F_1 = (V_1, E_1, w_1),\ldots, F_n = (V_n, E_n, w_n)\}$ such that $\bigcup_i V_i = V$, $\bigcup_i E_i = E$, and $\forall f$ $\forall e \in E_f$, $w_f(e) = w(e)$. The resulting partition called *fragment database* is stored in a disk-based structure. Conceptually, once a graph is partitioned, one can apply a route query evaluation algorithm to it, by reading in fragments and their auxiliary data structures from the disk whenever they are needed, and swapping them out when their usefulness expires. However, this brute-force method may not be effective [1], especially if the search space is huge. For some classes of route queries, query evaluation can be sped up by pre-computing some optimal distances. For each fragment F in a partition $P(G)$, a *distance matrix* is created to record the distance of an OP from one boundary node to the other. That is, for each pair of boundary nodes u and v, the distance $SD(F, \Lambda, v, u)$ is recorded in a distance matrix. The edge $<u, v>$ is called a *super edge*. All these matrices collectively are called a *distance (matrix) database (DMDB)* and are stored on some secondary storage device.

3 Algorithm *DiskOP_{HBR}*

DiskOP_{HBR} assumes the existence of DMDB for a based graph G. A super edge is said to be *affected* if it is embedded in an affected fragment. A fragment F is said to be *affected* if there is some modified edge in Σ that belongs to F. Otherwise it is *unaffected*. In Dijkstra's, when a vertex is closed, its adjacent edges are relaxed. However, for boundary vertices in *DiskOP_{HBR}*, the relaxation performed may be different, depending on whether the fragment involved is affected or not.

In *DiskOP_{HBR}*, instead of relaxing affected super edges, its adjacent edges in the modified fragment (F,Σ) are relaxed. That is, when a boundary node of a fragment F is closed, if F is unaffected, we relax its adjacent super edges using the DMDB, otherwise, we relax its adjacent edges in the modified fragment (F,Σ). Since the relaxation of adjacent edges of a closed boundary node could be either its super edges or its modified graph edges, $DiskOP_{HBR}$ is an algorithm based on the novel concept of *hybrid relaxation*.

4 Experiments

This section focuses on the experiment conducted on a road network. Since there is no similar algorithm, $DiskOP_{HBR}$ is compared with a brute-force disk-based algorithm without any data materialization, named $DiskOP_{BF}$. To our best knowledge, the most general route query evaluation algorithm comparable to $DiskOP_{HBR}$ is the algorithm $DiskCP$ [1]. We shall also compare it with $DiskOP_{HBR}$ in this work.

4.1 Experimental Setup

4.1.1 Factors Evaluated
The road systems of Connecticut, Massachusetts, New Jersey, New York, and Pennsylvania extracted from Tiger/Line file, are chosen as our test case. This data set is called *East5* in the rest of the discussion. Since different applications may have distinct characteristics, they could have different influences on the algorithms evaluated. In order to draw a meaningful conclusion, we extract some factors from the general situations, and examine how they affect these algorithms. Table 1 lists these factors with sample values used in the experiment.

Table 1. Samples of Evaluated Factors

Factor	Samples for *East5*
paf (all cases) %	1, 5, 10, 30, 50, 70, 90
pce (all cases) %	0.005, 0.1, 1, 2, 5
pcw (increase cases only) %	101, 200, 10000
(decrease cases only) %	10, 50, 90
pie (mixed cases only) %	10, 50, 90

Percentage of Affected Fragments (***paf***). It is the percentage of affected fragments.

Percentage of Changed Edges (***pce***). It is the percentage of edges with modified weights. The changed edges are evenly distributed among all the affected fragments.

Percentage of Changed Weight (***pcw***). It is the new edge's weight expressed as the percentage of its original weight. There are two groups of samples: weight increases and weight decreases. Note that these weight changes do not apply to the mixed cases.

Percentage of Increased Edges (***pie***). The mixed cases correspond to the situation in which some edge weights are increased while others are decreased. After selecting a group of modified edges, we vary the ratio between the number of the increased edges and the number of the decreased edges in this group. The new edge weight is $y \times wt$, where wt is the original edge weight, and y is randomly chosen, with $1 < y \leq 100$ and $0 \leq y < 1$ for increase and decrease cases, respectively.

4.2 Algorithms Evaluation

In this section, we show that $DiskOP_{HBR}$ outperforms, in terms of execution time, $DiskOP_{BF}$. To evaluate these algorithms, we measure the execution time (*seconds*) and the amount of I/O (*MBs*) accessed. Twelve queries of various lengths are used in these tests.

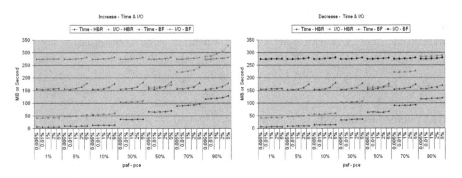

Fig. 1.   **Fig. 2.**

4.2.1 DiskOP_{HBR} vs. DiskOP_{BF}

In Section 4.1.1, we identify several factors that may be important in determining the applicability or desirability of an algorithm. We investigate if and how these factors influence the performance of algorithms studied. For the rest of this section, *HBR* and *BF* in a plot denote the algorithms $DiskOP_{HBR}$ and $DiskOP_{BF}$, respectively.

Effects of Weight Changes

It turns out that the factor *pce* has little or no effect on the performance of these algorithms.

Effects of Percentage of Affected Fragments

Figures 1 and 2 show the results, for the increase case and decrease case, of the effects of affected fragments on the performance of the two algorithms. These graphs record the average query evaluation time and I/O accesses. The corresponding plot for the mixed case is almost identical, and thus not included here.

This result suggests that *paf* has a significant effect on the performance of $DiskOP_{HBR}$, while it has little or no influence on $DiskOP_{BF}$. The performance of $DiskOP_{BF}$ does not vary much over all values of *paf*; the evaluation time, I/O accesses and the number of queue operations remain relatively flat, since the processing done by $DiskOP_{BF}$ is not influenced by the number of affected fragments. On the other hand, $DiskOP_{HBR}$ makes use of the DMDB of the based graph as much as possible in finding a skeleton path. Thus, the amount of processing and I/O accesses increases with the increase of *paf*. This results in staircase-like plots for $DiskOP_{HBR}$. Thus, the evaluation time increases with the increase of *paf*.

4.2.2 DiskOP$_{HBR}$ vs. DiskCP

There are three classes of optimal route query (*SP*, *cluster*, and *forbidden edge*) used in the experiment in [1]. We use them in this section for comparison. For the SP and cluster query classes, the two algorithms have a very comparable performance, except *DiskOP$_{HBR}$* requires a little more I/O accesses. The reason for this phenomenon is that the percentage of affected fragments in these two cases either is zero or is very small. This shows that *DiskOP$_{HBR}$* is very effective when there is little or no affected fragment. However, for forbidden edge query class, *DiskCP* outperforms *DiskOP$_{HBR}$*, over all query types, and by a wide margin. It is not surprising since with 89% affected fragments, the performance of *DiskOP$_{HBR}$* deteriorates rapidly.

5 Conclusion

We have studied the problem of how to evaluate efficiently, with a general algorithm, classes of static and dynamic optimal route queries on a massive graph. We observe that, many static and dynamic queries can be expressed as queries with a set of modified edge weights. Under such a setting, we found an efficient algorithm *DiskOP$_{HBR}$* to evaluate classes of static and dynamic optimal route queries. The main idea behind *DiskOP$_{HBR}$* is to make use of the pre-computed distance database as much as possible during a query evaluation. To achieve this objective, it employs the novel idea of *hybrid relaxation*. Experiments are conducted on this algorithm to demonstrate its desirability.

Acknowledgement

The authors wish to thank the financial support of Natural Sciences and Engineering Research Council of Canada.

References

1. Chan, E.P.F., Zhang, J.: A Fast Unified Optimal Route Query Evaluation Algorithm. In: Proceedings of ACM 16th Conference on Information and Knowledge Management (CIKM 2007), Lisboa, Portugal, November 2007, pp. 371–380 (2007)
2. Vazirgiannis, M., Wolfson, O.: A Spatial temporal Model and Language for Moving Objects on Road Networks. In: Jensen, C.S., Schneider, M., Seeger, B., Tsotras, V.J. (eds.) SSTD 2001. LNCS, vol. 2121, pp. 20–35. Springer, Heidelberg (2001)

Trajectory Compression under Network Constraints

Georgios Kellaris, Nikos Pelekis, and Yannis Theodoridis

Department of Informatics, University of Piraeus, Greece
{gkellar,npelekis,ytheod}@unipi.gr
http://infolab.cs.unipi.gr

Abstract. The wide usage of location aware devices, such as GPS-enabled cell-phones or PDAs, generates vast volumes of spatiotemporal streams modeling objects movements, raising management challenges, such as efficient storage and querying. Therefore, compression techniques are inevitable also in the field of moving object databases. Moreover, due to erroneous measurements from GPS devices, the problem of matching the location recordings with the underlying traffic network has recently gained the attention of the research community. So far, the proposed compression techniques are not designed for network constrained moving objects, while map matching algorithms do not consider compression issues. In this paper, we propose solutions tackling the combined, map matched trajectory compression problem, the efficiency of which is demonstrated through an experimental evaluation using a real trajectory dataset.

Keywords: Trajectory Compression, Road Network, Map-Matching.

1   Introduction

A Moving Object Database (MOD) is a collection of objects whose location changes over time. The preservation of vast volumes of moving objects' trajectories for future reference raises compression aspects, i.e., the *trajectory compression* problem (hereafter, called *TC*). Locations are often recorded by GPS receivers which embed an error of some meters. Thus, there arises the problem of matching these data points onto a network, also known as the *map-matching problem* (hereafter, called *MM*).

In this work, we study the combined problem of the compression of a moving object's trajectory keeping it at the same time matched on the underlying road network (hereafter, called map matched trajectory compression problem - MMTC).

To the best of our knowledge, there is no related work directly addressing the MMTC problem. Regarding its two components, TC and MM, the state-of-the-art is [6] and [1], respectively. In particular, Meratnia and de By [6] propose a compression technique that uses the Douglas-Peucker method and, moreover, takes the parameter of time into account. In particular, it replaces the Euclidean distance used in Douglas-Peucker by a time-aware one, called *Synchronous Euclidean Distance* (SED) [6]. The time complexity of the algorithm proposed in [6] is $O(n^2)$, where n is the number of points composing the trajectory. Regarding MM, Brakatsoulas et al. [1] propose the following methodology: for every point P_i, given that point P_{i-1} has already been

N. Mamoulis et al. (Eds.): SSTD 2009, LNCS 5644, pp. 392–398, 2009.

matched, the adjacent edges to this edge are the candidate edges to be matched to P_i and they are evaluated. In order to choose among the candidate edges two measures are used that take into consideration the distance and the orientation of the edges. The higher the sum of these measures is, the better the match to this edge is. The quality of the result is improved by using a "look ahead" policy. That is, the total score of each candidate edge is calculated by adding the scores of a fixed number of edges, which are ahead of the current position, to the initial one. The time complexity of the algorithm proposed in [1] is $O(n)$.

Also related to ours is the work by Tiakas et al. [7], where a method for trajectory similarity search under network constraints is proposed. In particular, the cost to travel from one node to another (which could be travel distance, average travel time, etc.) is used to calculate the *network distance* between two trajectories as the average of the equivalent node distances, and, on top of that, the total similarity D_{total} between two trajectories is expressed as a weighted average of their network and *time* distances.

The most related to our work is the one by Cao and Wolfson [2], which explored the combination of the map-matching with the storage-space problem by proposing a solution that uses the a priori knowledge of the road network. However, our approach is different as our goal is to reduce the size of the MOD keeping the trajectory data without altering its infrastructure, but only by removing and/or altering certain records. Of course, the proposed solution is not a lossless compression technique, but it preserves the structure of the original data providing at the same time data ready to be used without any preprocessing.

The contribution of the paper is three-fold:

1. Taking into consideration off-the-shelf TC and MM algorithms, we propose two working solutions for the combined MMTC problem.
2. Formulating MMTC as a cost-optimization problem, we present a theoretical analysis for finding the optimal solution accompanied, due to its high computational cost, by an approximate algorithm, called MMTC-App.
3. Performing an experimental study conducted over a real trajectory dataset, we demonstrate the effectiveness and the efficiency of the proposed solution.

The paper outline is as follows. In Section 2 we formalize the MMTC problem. Section 3 presents two solutions based on existing TC and MM algorithms as well as our novel approximate solution (MMTC-App) followed by their experimental evaluation in Section 4. Section 5 concludes the paper.

2   MMTC Problem Formalization

Before we present our solutions for the MMTC problem, we formalize the MMTC problem. Due to space limitations, the definitions of the key concepts of our work (*Network, Trajectory, Map-matched trajectory, Map-matched counterpart of a trajectory*, and *Compressed version of a trajectory*) are presented in detail in [5]. On top of those definitions, the MMTC problem is formalized as follows:

Definition 1 (MMTC Problem). Given a road network $G(V, E)$ consisting of graph vertices V and edges E and a trajectory T consisting of time-stamped point locations $<x, y, t>$, the *map-matched trajectory compression* (MMTC) problem asks to find a network-constrained trajectory, T_{MMTC}, which is (a) a compressed version of the map-matched counterpart of T, called T', and (b) as similar to T' as possible.

The degree of compression $Comp(T_{MMTC}, T')$ between T_{MMTC} and T' is measured as follows:

$$Comp(T_{MMTC}, T') = \begin{cases} 1 - \dfrac{|T_{MMTC}|}{|T'|}, if \ |T_{MMTC}| < |T'| \\ 0, otherwise \end{cases} \quad (1)$$

while the degree of similarity $Sim(T_{MMTC}, T')$ between T_{MMTC} and T' is measured as follows:

$$Sim(T_{MMTC}, T') = 1 - D_{total}(T_{MMTC}, T') \quad (2)$$

where $D_{total}(T_{MMTC}, T')$ is calculated as in [0] using the network distance defined in [5].

Since it is expected that optimizing both $Comp()$ and $Sim()$ is contradicting, an overall quality measure $0 \leq Q(T_{MMTC}, T') \leq 1$ is defined as an aggregate of $Comp()$ and $Sim()$. The trajectory T_{MMTC} that maximizes Q among all possible network-constrained trajectories is the solution to the MMTC problem. ∎

In the rest of the paper, we adopt

$$Q(T_{MMTC}, T') = Comp(T_{MMTC}, T') \cdot Sim(T_{MMTC}, T') \quad (3)$$

which requires the values of both components to be as high as possible in order for its value to be high. Of course, other quality functions could be equally taken into consideration.

In the following section, we investigate the MMTC problem and propose solutions either exploiting off-the-shelf TC and MM techniques or designing from scratch.

3   Solutions to the MMTC Problem

By combining the algorithms of data compression and map-matching we devise two naïve solutions:

- 1st approach (called, TC+MM): the original trajectory T is compressed to T_{TC} using TC, which afterwards is map-matched to T_{MMTC} using MM.
- 2nd approach (called, MM+TC+MM): the original trajectory T is map-matched to T_{MM} using MM, which afterwards is compressed to T'_{MMTC} using TC, which afterwards is map-matched to T_{MMTC} using MM (since applying a general TC algorithm on a map-matched trajectory destroys its map-matched properties).

If we adopt the algorithms [6] and [1], with $O(n^2)$ and $O(n)$ time complexity, respectively, it turns out that the complexity of both TC+MM and MM+TC+MM is $O(n^2)$.

Alternatively, in this paper we propose a novel compression method. The motivating idea is that the compressed trajectory is built by replacing some paths of a given map-matched trajectory with shorter ones. This could be done by executing a shortest path algorithm (hereafter, called *SP*) on appropriately selected points of the trajectory without considering the weights of the edges. We ignore the weights in order to get the result with the minimum number of nodes and, hence, achieve a high compression. Moreover, we need to choose among all the possible shorter paths, the one that maximize the value of the quality measure Q defined in Eq. (3).

In particular, we adopt the *Minimum Description Length* (MDL) principle. There are two components that comprise MDL, namely $L(H)$ and $L(D|H)$, where H denotes the hypothesis and D denotes the data, as presented in [3]. The best hypothesis H to explain D is the one that minimizes the sum of $L(H)$ and $L(D|H)$. Mapping the above discussion to our problem, $L(H)$ represents the compression achieved by a compressed trajectory and $L(D|H)$ represents the difference between the compressed trajectory and the original one as explained in [5].

Recalling Definition 1, the solution we envisage for the MMTC problem is the path that minimizes the sum $L(H) + L(D|H)$ adopting the MDL principle. Unfortunately, the cost of finding this path is prohibitive since we need to consider every combination of shortest paths between the points of the original trajectory. Actually, it is similar to the problem of finding all possible acyclic paths between two graph nodes. Since we cannot expect to find the optimal trajectory in reasonable time, we propose an algorithm that would approximate it.

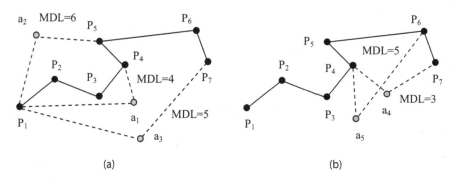

(a)   (b)

Fig. 1. Example of the MMTC-App algorithm: shortest paths and their MDLs (a) from the first node and (b) from the fourth node

The main idea of our approach is illustrated in Fig. 1 whereas the pseudocode of the proposed approximate algorithm, called MMTC-App, is listed in [5] (due to space limitations). First, we calculate a map-matched counterpart T_{MM} of the original trajectory T simply by matching every point of the trajectory to a network edge.

In particular, we choose the edge that is closest to the examined point and it is adjacent to the previous selected. Following the example illustrated in Fig. 1, given T_{MM} $\{P_1, P_2, P_3, P_4, P_5, P_6, P_7\}$, the algorithm calculates the MDL of every SP as illustrated in Fig. 1(a), and chooses the one with the minimum MDL value. Let us suppose that this value is given by the SP from P_1 to P_4 via node a_1. Then we can replace the sub-path $\{P_1, P_2, P_3, P_4\}$ of the temporary result by $\{P_1, a_1, P_4\}$. We also need to calculate the time the object is located at node a_1. This can be estimated by using the temporal information on nodes P_1 and P_4 and considering the object is moving at constant speed. The algorithm stores this result and continues by considering as first point the last node of the already found SP and by checking the SPs from this node to its next ones. In our running example, P_4 is checked against its next points. Finally, the SP with the best score is the one from P_4 to P_7 via a_4, as illustrated in Fig. 1(b). Since P_7 is the end point of the trajectory, the algorithm terminates by returning the compressed trajectory $\{P_1, a_1, P_4, a_4, P_7\}$. In case no SP is found in a sub-path which is under evaluation, the algorithm adds the remaining points to the output and, then, terminates.

The MMTC-App algorithm requires the SPs from the original nodes to all others to be pre-calculated. This is an offline procedure of general purpose. Theoretically, the time complexity for running an all-pair shortest path algorithm on a network $G(V, E)$ is $O(|V|^2 \log|V|)$ [4]. The complexity of MMTC-App algorithm is $O(n^2 \log n)$ on average, where n is the number of points composing the trajectory, excluding the cost of shortest path calculations, with the proof found in [5].

4   Experimental Study

We evaluated the proposed techniques over a real dataset consisting of trajectories of vehicle movements in the city of Milano (described in [5]). For each trajectory, a compressed and map-matched version of it was constructed, following one of the above techniques. The resulting trajectories were compared against their map-matched counterparts with respect to (a) a quality criterion (Eq.(3)) (b) the compression achieved (Eq.(1)) and (c) the execution time.

In our experiments, we included MMTC-App together with TC+MM(low), TC+MM(high), MM+TC+MM(low), MM+TC+MM(high), where low and high indicate different threshold values of TC, thus, the rate of compression. The details of the experimentation are discussed in [5].

The first set of experiments concentrates on the effectiveness of the proposed methods, i.e., as high compression (according to Eq.(1)) and overall quality (according to Eq.(3)) as possible. Fig. 2 illustrates the compression achieved (labeled 'Comp' in the chart) together with the resulting quality ('Q'). It is clear that MMTC-App significantly compresses the original trajectory keeping about 60% of its size while, at the same time, the overall quality is at least twice the quality of any of the naïve approaches. It is also worth to be mentioned that high speed and agility has offered slightly higher compression for methods MM+TC+MM and TC+MM with respect to their behavior in low speed and agility.

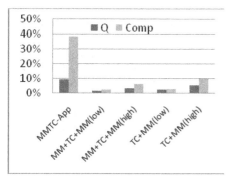
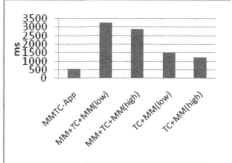

Fig. 2. Compression and overall quality achieved

Fig. 3. Execution time

The second set of experiments concentrates on the efficiency of the proposed techniques. Fig. 3 illustrates the execution time required to run the proposed techniques over the Milano dataset. The proposed MMTC-App algorithm completes running in half second while the naïve approaches required a few seconds to perform their task (depending on the compression threshold set for the TC component).

5   Conclusion

MOD literature offers solutions to the problems of trajectory compression and map-matching succinctly, but none satisfies the combined problem of trajectory compression under network constrains.

In this paper, apart from straightforward solutions to the so-called MMTC problem, we have proposed a method for trajectory compression under network constraints. According to our results, our approximate solution turns out to be both efficient and effective, offering successful compression while at the same time retaining the quality of the output.

References

1. Brakatsoulas, S., Pfoser, D., Salas, R., Wenk, C.: On Map-Matching Vehicle Tracking Data. In: Proc. 31st International Conference on Very Large Data Bases (VLDB) (2005)
2. Cao, H., Wolfson, O.: Nonmaterialized Motion Information in Transport Networks. In: Eiter, T., Libkin, L. (eds.) ICDT 2005. LNCS, vol. 3363, pp. 173–188. Springer, Heidelberg (2004)
3. Grünwald, P., Myung, I.J., Pitt, M.: Advances in Minimum Description Length: Theory and Applications. MIT Press, Cambridge (2005)
4. Johnson, D.B.: Efficient algorithms for shortest paths in sparse networks. Journal of the ACM 24(1), 1–13 (1977)

5. Kellaris, G., Pelekis, N., Theodoridis, Y.: Trajectory Compression under Network Constraints, UNIPI-INFOLAB-TR-2009-01, Technical Report Series, InfoLab, Univ. Piraeus (April 2009), http://infolab.cs.unipi.gr
6. Meratnia, N., de By, R.A.: Spatiotemporal Compression Techniques for Moving Point Objects. In: Bertino, E., Christodoulakis, S., Plexousakis, D., Christophides, V., Koubarakis, M., Böhm, K., Ferrari, E. (eds.) EDBT 2004. LNCS, vol. 2992, pp. 765–782. Springer, Heidelberg (2004)
7. Tiakas, E., Papadopoulos, A.N., Nanopoulos, A., Manolopoulos, Y.: Trajectory Similarity Search in Spatial Networks. In: Proc. 10th International Database Engineering and Applications Symposium (IDEAS) (2006)

Exploring Spatio-Temporal Features for Traffic Estimation on Road Networks

Ling-Yin Wei, Wen-Chih Peng, Chun-Shuo Lin, and Chen-Hen Jung

Institute of Computer Science and Engineering
National Chiao Tung University
Hsinchu, Taiwan, ROC
{lywei.cs95g,wcpeng,zvn.cs97g,clare.csie94}@nctu.edu.tw

Abstract. In this paper, given a query that indicates a query road segment and a query time, we intend to accurately estimate the traffic status (i.e., the driving speed) on the query road segment at the query time from traffic databases. Note that a traffic behavior in the same time usually reflects similar patterns (referring to the temporal feature), and nearby road segments have the similar traffic behaviors (referring to the spatial feature). By exploring the temporal and spatial features, more GPS data points are retrieved. In light of these GPS data retrieved, we exploit the weighted moving average approach to estimate traffic status on road networks. Experimental results show the effectiveness of our proposed algorithm.

Keywords: Traffic Patterns, Data Mining, Trajectory Data.

1 Introduction

In recent years, the global position system (GPS) is widely used in sensor networks and technical products, such as navigation devices, GPS loggers, PDAs and mobile phones. At the same time, with the explosion of map services and local search devices, many GPS-related Web services are built. Many research efforts have implemented GPS data collection platforms, which are based on client-server architectures [5,2,3]. Recent studies in [5,2,1,4] utilized GPS data to estimate the traffic status. However, the challenge issue is that the GPS data reported along with a road segment required may contain less amount of GPS data for traffic estimation. As a result, the estimated driving speed cannot closely reflect the real traffic status.

In this paper, we explore spatio-temporal features to obtain more GPS data points from historical data. We can consider historical data (referring to temporal features) since traffic on road segments usually follows a certain pattern. Furthermore, we also consider traffic information on nearby road segments (referring to spatial features) for predicting traffic information on a given road segment at a given time slot. Consequently, in this paper, by exploring the temporal features of road networks, we are able to collect more GPS data for traffic estimation. Note that GPS data points spread over different day times. Thus, we exploit the

N. Mamoulis et al. (Eds.): SSTD 2009, LNCS 5644, pp. 399–404, 2009.

weighted moving average approach to estimate traffic information on road networks. Experimental results show the effectiveness of our proposed algorithm.

The rest of the paper is organized as follows. In Section 2, some assumptions and the problem statement are described. In Section 3, we propose a traffic estimation algorithm. Performance study is presented in Section 4. Finally, Section 5 concludes with this paper.

2   Preliminary

Same as other research in road networks, a road network is represented as a directed graph $G = (V, E)$. In this paper, a road segment is sensitive to its own driving direction. Furthermore, each road segment belongs to one road type, such as freeway, urban roads, and street, to name a few. To facilitate the presentation of this paper, a set of road types is denoted as $C_G = < c_1, c_2, \ldots, c_m >$, where each road type $c_i = i \in \mathbb{N}$ and the number of road types is m. Therefore, each road segment in a road network G has its corresponding road type. Without loss of generality, each road type (e.g., c_i) has its own speed limit (represented as $|c_i|$). In our paper, the road type in C_G are sorted in strictly decreasing order according to speed limit of road types. Based on the road network and the traffic database TDB, the goal of our paper is to estimate the traffic information on road segments. In this paper, the traffic information on a road segment is represented as a speed value aggregated from a set of GPS data points [3].

3   Traffic Estimation Algorithm

3.1   Retrieving GPS Data Points by Temporal Feature

The traffic is usually heavy in rush hours. Furthermore, the traffic on a given road segment (e.g., r_q) has correlations with prior traffic on r_q (i.e., the traffic on r_q at previous time slots). Above observations are referred to the temporal feature of road networks. Thus, based on the temporal feature, we are able to extract more GPS data points from traffic databases. Explicitly, we not only extract GPS data points of the same road segment whose time is close to the query time on the same day but also collect GPS data points at the same query time from previous days. To facilitate the presentation of our paper, we define Neighboring Time Slot (abbreviated as NTS).

Definition 1. *(Neighboring time slot) Given a query $Q = (r_q, t_q)$, a time interval $[t_q - \Delta, t_q + \Delta]$, where Δ is a window size, is called a neighboring time slot (NTS).*

According to the above definition, we should retrieve those GPS data points whose time is within time interval NTS. For NTS, we should further specify their date time. As such, LD_k denotes a set of k days considered for retrieving GPS data points.

3.2   Aggregating GPS Data Points by Temporal Feature

Given a query time t_q, one road segment r_e, a parameter Δ, and LD_k, we can extract GPS data points of road segment r_e. Specifically, we extract GPS data points within NTS on each day in LD_k, and the set of these extracted GPS data points is expressed by P_{r_e}. Furthermore, according to the date information, $P_{r_e|D_j}$ is a subset of P_{r_e} and $P_{r_e|D_j}$ contains those GPS data points whose date time is D_j. For GPS data points in the same day, we assign different weights in accordance with the GPS time with respect to the query time t_q.

In general, if the time of GPS data points are near or close to the query time, larger weights are assigned. We assign the weight of a GPS data point by considering both the relative days and the relative time with respect to the query time (i.e., the time associated with date information).

Definition 2. *(d-weight function) Given a GPS data point (denoted as p) of some road segment r_e with its date time as $D_j \in LD_k$ (i.e., $p \in P_{r_e|D_j}$), the d-weight function is formulated as*

$$w_d(p) = \frac{k-(j-i+1)+1}{\sum_{l=1}^{k}(l+1)}.$$

Definition 3. *(t-weight function) Given a GPS data point (denoted as p) of some road segment r_e with its date time as D_j and the time slot as p.t, the t-weight function is defined by*

$$w_t(p) = \frac{\frac{1}{|p.t-t_q|+1}}{\sum_{l=1}^{|P_{r_e|D_j}|}\frac{1}{|p_l.t-t_q|+1}}.$$

Clearly, we have $\sum_{p \in P_{r_e|D_j}} w_t(p) = 1$ for each $P_{r_e|D_j} \subseteq P_{r_e}$. Then, we utilize these two weight functions to formally derive *the time weighted function* as below.

Definition 4. *(Time weighted function) Given a set of extracted GPS data points (i.e., P_{r_e}) for a road segment r_e, let $T : P_{r_e} \to (0,1] \subset \mathbb{R}$ be the time weighted function with its formula as follows:*

$$T(p) = w_d(p) \cdot w_t(p).$$

Therefore, given a query time $Q = (r_q, t_q)$ on D_i and a road segment r_e, we can derive a set of extracted GPS data points (i.e., P_{r_e}). For the road segment r_e, an aggregated speed V_{r_e} is formulated as follows:

$$V_{r_e} = \sum_{p \in P_{r_e}} (T(p) \cdot p.v), \text{ where } p.v \text{ is the speed of a GPS data point p.}$$

3.3   Retrieving GPS Data Points by Spatial Feature

Given a road network G, its corresponding road category C_G, and two road segments r_i and r_j in G, the road type distance between r_i and r_j is formulated as $TD(r_i, r_j) = |c_i - c_j|$, where the types of r_i and r_j are c_i and c_j, respectively. In addition, the spatial distance between r_i and r_j, denoted as $SD(r_i, r_j)$, is the minimal number of nodes appearing in connected paths between r_i and r_j.

Accordingly, we could further define the set of *kth*-connected spatial relations among road segments as follows:

Definition 5. *(kth-connected spatial relation between two road segments) Given a road network G and two road segments r_i and r_j in G, r_j is a kth-connected road segment of r_i if $SD(r_i, r_j) = k$ and r_i and r_j have the same driving direction. The set of kth-connected road segments of r_i is denoted as $R_k(r_i)$.*

The set of extracted road segments is regarded as the spatial feature of road networks. The set is called the restricted set, denoted as $rSet$.

Definition 6. *(rSet) Given a road network G, a query $Q = (r_q, t_q)$, and two thresholds d_t and d_s, each road segment r_e in $rSet$ satisfies the both conditions: (1) $0 \leq TD(r_e, r_q) \leq d_t$, and (2) $1 \leq SD(r_e, r_q) \leq d_s$.*

With the above definition of $rSet$, we could formulate $rSet$ as follows:

$$rSet = \cup_{1 \leq i \leq d_s} \cup_{c_l \in C_G} E_i^{c_l}.$$

The size of $rSet$, denoted as $|rSet|$, is the number of subsets in $rSet$. Without loss of generality, we suppose that $|rSet| = d_s$, and $E_i^{c_l} \neq \phi$ for each $E_i^{c_l}$. Note that $E_i^{c_l}$ is a set of edges satisfying road type c_l and ith-connected.

To determine the traffic statuses of neighboring road segments, we could utilize the temporal feature mentioned above.

3.4   Aggregating GPS Data Points from Spatial Feature

For each road segment in $rSet$, we can derive an aggregated speed by the time weighted function. In light of aggregated speed of nearby road segments, we could derive the aggregated speed of a given road segment via the weighted moving average approach. Similar to the weight functions in Section 3.2, we will derive two weight functions for the nearby road segments.

Given a query $Q = (r_q, r_t)$ and its corresponding $rSet$ with some thresholds, we adopt *the spatial weighted function* to integrate speeds of nearby road segments, and then derive an estimated speed of a query road segment r_q.

Aggregated speeds of road segments should be set to different weights in accordance with the spatial distance (referring to the s-weight function) and the type distance (referring to the c-weight function) of road segments. Assume that the query road segment is r_q and the nearby road segment is r_e. Then, we define s-weight function as follows:

Definition 7. *(s-weight function) let $w_s : E \rightarrow (0, 1] \subset \mathbb{R}$ be the s-weight function formulated as follows:*

$$w_s(r_e) = \frac{d_s - SD(r_e, r_q) + 1}{\sum_{1 \leq l \leq d_s} l}.$$

Definition 8. *(c-weight function) let $w_c : E \rightarrow (0, 1] \subset \mathbb{R}$ be the c-weight function formulated as follows:*

$$w_c(r_e) = \frac{1}{|E_{SD(r_e, r_q)}^{c_l}|}$$

where c_l is the road type of the road segment r_e.

Definition 9. *(Spatial weighted function) Given a query* $Q = (r_q, t_q)$ *and a road segment* r_e, *let* $S : E \rightarrow (0, 1] \subset \mathbb{R}$ *be the spatial weighted function defined by*

$$S(r_e) = w_s(r_e) \cdot w_c(r_e).$$

3.5   Velocity Estimation

Especially, to adaptively adjust the influence of the spatial feature of road networks, we set a parameter α and $0 \leq \alpha \leq 1$. If α is set to 1, we only consider the temporal feature of the query road segment. Thus, the aggregated speed of a query road segment r_q, denoted as $STV(r_q)$, is formulated as below:

$$STV(r_q) = \alpha \cdot V_{r_q} + (1 - \alpha) \cdot \left(\sum_{r_e \in rSet} S(r_e) \cdot V_{r_e} \right).$$

4   Experimental Evaluation

In this section, we evaluate our proposed algorithm STW by using the real dataset from CarWeb [3]. The query road segments include the four major road segments. The parameters setting are $\Delta = 1.5$ hours, a spatial distance threshold $d_s = 4$, and a road type distance d_t is the total number of types in the road category. For each road segment, we query different time about eight times, and then average the error between the real speed of the query road segment and the estimated speed of the query road segment to derive the average accuracy.

First, we examine the effect of α on the temporal and spatial features. In Figure 1(a), in most cases, the error decreases and then increases while α gradually increases. Figure 1(a) also illustrates that the average accuracy is somewhat better as the value of α is between 0.6 and 0.8. Moreover, with a smaller value of k (e.g., $k = 1$), the average error is smaller by setting a smaller value of α, showing the advantage of the spatial feature. However, in Figure 1(a), as $\alpha \geq 0.8$, the average error slightly increases while k increases. Since more GPS data points extracted for traffic estimation, some GPS data points are noise data, which may increase the average error. Also, GPS data points with their date time close to the query

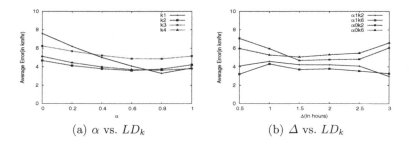

(a) α vs. LD_k (b) Δ vs. LD_k

Fig. 1. Accuracy of the spatio-temporal weighted algorithm

date time are more important for the traffic estimation even though the amount of GPS data points increases. Therefore, we do not need to extract GPS data points on many dates if we can extract GPS data points on the query date.

In Figure 1(b), if $\alpha = 1$, the average error gradually decreases while Δ increases. However, if $\alpha = 0$, the average error gradually decreases and then increases while Δ increases. In addition, the average error of $\Delta = 1.5$ is smaller than that of other values of Δ while $\alpha = 0$. The reason is that when α is set to 0, we do not utilize the GPS data points on the query road segment, and only retrieve GPS data points on the other road segments. Even though we extract more GPS data points from a lager NTS on the other road segments, those lower relevant GPS data points would increase the average error.

5   Conclusion

In this paper, we fully utilized the spatio-temporal feature to obtain the historical GPS data and neighboring traffic information. In light of the GPS data that includes the real-time GPS data and the historical GPS data, we are able to derive an aggregated speed of the query road segment at the query time specified by users. Experimental results showed the effectiveness of our proposed algorithm and by exploiting the spatio-temporal feature, the traffic estimation is very close to the true real-time traffic status.

References

1. de Fabritiis, C., Ragona, R., Valenti, G.: Traffic estimation and prediction based on real time floating car data. In: Proc. of the 11th International IEEE Conference on Intelligent Transportation Systems, pp. 197–203 (2008)
2. Kriegel, H.-P., Renz, M., Schubert, M., Zuefle, A.: Statistical density prediction in traffic networks. In: Proc. of the 8th SIAM Conference on Data Mining, pp. 692–703 (2008)
3. Lo, C.-H., Peng, W.-C., Chen, C.-W., Lin, T.-Y., Lin, C.-S.: Carweb: A traffic data collection platform. In: Proc. of the 9th International Conference on Mobile Data Management, pp. 221–222 (2008)
4. Sananmongkhonchai, S., Tangamchit, P., Pongpaibool, P.: Road traffic estimation from multiple gps data using incremental weighted update. In: Proc. of the 8th International Conference on ITS Telecommunications, pp. 62–66 (2008)
5. Yoon, J., Noble, B., Liu, M.: Surface street traffic estimation. In: Proc. of the 5th International Conference on Mobile Systems, Applications and Services, pp. 220–232 (2007)

A Location Privacy Aware Friend Locator

Laurynas Šikšnys, Jeppe R. Thomsen, Simonas Šaltenis, Man Lung Yiu,
and Ove Andersen

Department of Computer Science, Aalborg University
DK-9220 Aalborg, Denmark

Abstract. A location-based service called friend-locator notifies a user if the user is geographically close to any of the user's friends. Services of this kind are getting increasingly popular due to the penetration of GPS in mobile phones, but existing commercial friend-locator services require users to trade their location privacy for quality of service, limiting the attractiveness of the services. The challenge is to develop a communication-efficient solution such that (i) it detects proximity between a user and the user's friends, (ii) any other party is not allowed to infer the location of the user, and (iii) users have flexible choices of their proximity detection distances. To address this challenge, we develop a client-server solution for proximity detection based on an encrypted, grid-based mapping of locations. Experimental results show that our solution is indeed efficient and scalable to a large number of users.

1 Introduction

Mobile devices with geo-positioning capabilities are becoming cheaper and more popular. Consequently users start using *friend-locator* services (e.g., Google Latitude, FireEagle) for seeing their friends' locations on a map and identifying nearby friends.

In existing services, the detection of nearby friends is performed manually by the user, e.g., by periodically examining a map on the mobile device. This works only if the user's friends agree to share either exact or obfuscated location. However, LBS users usually demand certain level of privacy and may even feel insecure if it is not provided [5]. Due to the poor support for location privacy in existing friend-locator products, it is sometimes not possible to detect nearby friends if location privacy is desired. The challenge is to design a communication-efficient friend-locator LBS that preserves the user's location privacy and yet enables automatic detection of nearby friends.

To address the challenge, we develop a client-server, location-privacy aware friend-locator LBS, called the `FriendLocator`. It first employs a grid structure for cloaking the user's location into a grid cell and then converts it into an encrypted tuple before it is sent to the server. Having received the encrypted tuples from the users, the server can only detect proximity among them, but it is unable to deduce their actual locations. In addition, users are prevented from knowing the exact locations of their friends. To optimize the communication cost, the `FriendLocator` employs a flexible region-based location-update policy where regions shrink or expand depending on the distance of a user from his or her closest friend.

The rest of the paper is organized as follows. We briefly review related work in Section 2 and then define our problem setting in Section 3. The `FriendLocator`

N. Mamoulis et al. (Eds.): SSTD 2009, LNCS 5644, pp. 405–410, 2009.

is presented in Section 4. Section 5 presents experimental results of our proposal and Section 6 concludes the paper.

2   Related Work

In this section, we review relevant work on location privacy and proximity detection.

Location privacy. Most of the existing location privacy solutions employ the *spatial cloaking* technique, which generalizes the user's exact location q into a region Q' used for querying the server [4]. Alternative approaches [6,11,3] have also been studied recently. However, all these solutions focus on range/kNN queries and assume that the dataset is public (e.g., shops, cinemas). In contrast, in the proximity detection problem, the users' locations are both queries and data points that must be kept secret.

Proximity detection. Given a set of mobile users and a distance threshold ϵ, the problem of proximity detection is to continuously report all events of mobile users being within the distance ϵ of each other. Most existing solutions (e.g., [1]) focus on optimizing the communication and computation costs, rather than location privacy.

Recent solutions were proposed [10,8] to address location privacy in proximity detection. Ruppel et al. [10] develop a centralized solution that applies a *distance-preserving mapping* (i.e., a rotation followed by a translation) to convert the user's location q into a transformed location q'. Unfortunately, Liu et al. [7] point out that distance-preserving mapping can be easily attacked. Mascetti et al. [8] employ a server and apply the filter-and-refine paradigm in their secure two-party computation solution. However, it lacks distance guarantees for the proximity events detected by the server, and leads to low accuracy when strong privacy is required. Unlike our approach, the central server in their proposal knows that a user is always located within his or her cloaked region.

Our solution is fundamentally different from the previous solutions [10,8] because we employ encrypted coordinates to achieve strong privacy and yet the server can blindly detect proximity among the encrypted coordinates.

3   Problem Definition

In this section we introduce relevant notations and formally define the problems of proximity detection and its privacy-aware version.

In our setting, a large number of mobile-device users form a social network. These mobile devices (MD) have positioning capabilities and they can communicate with a central location server (LS). We use the terms *mobile devices* and *users* interchangeably and denote the set of all MDs (and their users) in the system by $\mathbf{M} \subset \mathbb{N}$.

The friend-locator LBS notifies two users $u, v \in \mathbf{M}|u \neq v$ if u and v are friends and the proximity between u and v is detected. Given the distance thresholds ϵ and λ, the proximity and separation of two users u and v are defined as follows [1]:

1. If $dist(u, v) \leq \epsilon$, then the users u and v are in proximity;
2. If $dist(u, v) \geq \epsilon + \lambda$, then the users u and v are in separation;
3. If $\epsilon < dist(u, v) < \epsilon + \lambda$, then the service can freely choose to classify users u and v as being either in proximity or in separation.

Here, $dist(u, v)$ denotes the Euclidean distance between the users u and v. The parameter ϵ is called the *proximity distance*, and it is agreed/selected by u and v. The parameter $\lambda \geq 0$ is a service precision parameter and it introduces a degree of freedom in the service. As different pairs of friends may want to choose different proximity distances, we use $\epsilon(u, v)$ to denote the proximity distance for the pair of users $u, v \in \mathbf{M}$. For simplicity we assume mutual friendships, i.e., if v is a friend of u, then u is a friend of v, and we let the proximity distance to be symmetric, i.e., $\epsilon(u, v) = \epsilon(v, u)$ for all friends $u, v \in \mathbf{M}$.

A proximity notification must be delivered to MDs when proximity is detected. Any subsequent proximity notification is only sent after separation have been detected.

The friend-locator LBS must be efficient in terms of mobile client communication and provide the following privacy guarantees for each user $u \in \mathbf{M}$: (i) The exact location of u is never disclosed to other users or the central server. (ii) User u only permits friends to detect proximity with him.

4   Proposed Solution

In this section we propose a novel, incremental proximity detection solution based on encrypted grids. It is designed for the client-server architecture, it is efficient in terms of communication, and it satisfies user location-privacy requirements (see Sec. 3).

Grid-based encryption. Let us consider three parties: two friends, u_1 and $u_2 \in \mathbf{M}$, and the location server (LS). Both users can send and receive messages to and from LS. User u_1 is interested in being informed by LS when user u_2 is within proximity and vice versa.

Assume that users u_1 and u_2 share a *list of grids*, where a grid index within the list is termed *level*. Grids at all levels are coordinate-axis aligned and their cell sizes, i.e., width and height, at levels $l = 0, 1, 2, \ldots$ are fixed and equal to $L(l)$. We let $L(l) = g \cdot 2^{-l}$, where g is some level zero cell size. Then sizes of cells gradually decrease going from lower to higher levels, level zero cells being the largest.

Each column (row) of each of these grids is assigned a unique *encryption number*. A grid within the list, together with encryption numbers, constitutes a Location Mapping Grid (LMG). Each user generates such a list of LMGs utilizing two shared private functions L and ψ, where $\Psi : \mathbb{N} \mapsto \mathbb{N}$ is a one-to-one encryption function (e.g., AES) mapping a column/row number to an encryption number.

Incremental proximity detection. Assume that users u_1 and u_2 use an LMG of some level l. Whenever a user moves into a new cell of LMG, the following steps are taken:

(i) The user maps the current location (x, y) into an LMG cell $(k,m)=(\lfloor x/L(l) \rfloor, \lfloor y/L(l) \rfloor)$.
(ii) The user computes an encrypted tuple $e = (l,\alpha^-,\alpha^+,\beta^-,\beta^+)$ by applying $E_\Psi(l, k, m) = (l, \Psi(k), \Psi(k+1), \Psi(m), \Psi(m+1))$, where (α^-,α^+) and (β^-,β^+) are encrypted values of adjacent columns k and $k+1$ and adjacent rows m and $m+1$ respectively.
(iii) The user sends the encrypted tuple e to LS.

Since u_1 and u_2 use the same list of LMG, with the same encryption-number assignments for each column and row, the LS can detect proximity between them by checking if the following function is true:

$$\Gamma(e_1, e_2) = (e_1.l = e_2.l) \wedge ((e_1.\alpha^- = e_2.\alpha^-) \vee (e_1.\alpha^- = e_2.\alpha^+) \vee (e_1.\alpha^+ = e_2.\alpha^-))$$
$$\wedge ((e_1.\beta^- = e_2.\beta^-) \vee (e_1.\beta^- = e_2.\beta^+) \vee (e_1.\beta^+ = e_2.\beta^-)).$$

Parameters e_1 and e_2 are encrypted tuples delivered from users u_1 and u_2 respectively. Note that since Ψ is a one-to-one mapping, Γ is evaluated to *true* if and only if k_{u_1} or $k_{u_1} + 1$ matches k_{u_2} or $k_{u_2} + 1$ and m_{u_1} or $m_{u_1} + 1$ matches m_{u_2} or $m_{u_2} + 1$, where (k_{u_1}, m_{u_1}) and (k_{u_2}, m_{u_2}) are LMG cells of users u_1 and u_2 respectively.

In the extended version of this paper we prove that an LMG at level l can be used to detect proximity with the following settings $\epsilon = L(l)$, $\lambda = L(l) \cdot (2\sqrt{2} - 1)$, i.e., Γ is always *true* when $dist(u_1, u_2) \leq L(l)$ and always *false* when $dist(u_1, u_2) \geq L(l) \cdot 2\sqrt{2}$. Every two friends $u_1, u_2 \in M$ choose an LMG level, called *proximity level* $L_\epsilon(u_1, u_2)$ that corresponds best to their proximity detection settings. Then our approach forces every user to stay at the lowest-possible level such that few grid-cell updates are necessary. Only when proximity between friends $u_1, u_2 \in M$ is detected at a low level, are they asked to switch to a higher level. This repeats until required level $L_\epsilon(u_1, u_2)$ is reached or it is determined that users are not in proximity.

Figure 1 illustrates the approach. It shows the geographical locations of two friends u_1 and u_2, and their mappings into LMGs at 4 snapshots in time. Note that lower level grids are on top in the figure. Assume that u_1 and u_2 have agreed on $L_\epsilon(u_1, u_2) = 2$ and have already sent their encrypted tuples, for levels 0 and 1 to LS. Figure 1a visualizes when LS detects a proximity at level 0, but not at level 1. As $L_\epsilon(u_1, u_2) > 0$, nothing happens until a location change. In Figure 1b both users have changed their geographical location. User u_2 did not go from one cell to another at his current level 1, thus he did not report a new encrypted tuple. User u_1 however, changed cells at both level 1 and level 0, he therefore sends a new encrypted tuple for level 0. The LS detects a proximity between u_1 and u_2 at level 0 and asks u_1 to switch to level 1, because $L_\epsilon(u_1, u_2) > 0$. Figure 1c shows user LMG mapping when u_1 has delivered new encrypted tuple for level 1. Again, LS detects proximity at level 1 and commands both users u_1 and u_2 to switch to level 2. When both encrypted tuples for level 2 are delivered to LS, it detects the proximity at this level (see Figure 1d) and, because $2 = L_\epsilon$, proximity notifications are sent to u_1 and u_2.

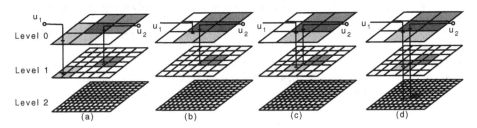

Fig. 1. Two-user proximity detection in the `FriendLocator`

Note that the presented algorithms implement an adaptive region-based update policy. If a user is far away from his friends, then he stays at a low-level grid with large cells, resulting in few updates for the user's future movement. Only when the user approaches one of his friends, he is asked to switch to higher levels with smaller grid cells. Thus, at a given time moment, the user's current communication cost is not affected by the total number of his friends, but by the distance to his closest friend.

5   Experimental Study

The proposed `FriendLocator` and a competitor solution, called `Baseline`, were implemented in C#. In this section, we study their communication cost in terms of messages received by the clients and the server. The network-based generator [2] is used to generate a workload of users moving on the road network of the German city Oldenburg. A location record is generated for each user at each timestamp.

Competitor Solution. The `Baseline` employs the filter-and-refine paradigm for proximity detection among friend pairs. Each user cloaks its location by using a uniform grid, and sends its cell to the server. Filtering is performed at the LS, which calculates the *min* and *max* distances [9] between the cells c_i and c_j of the users u_i and u_j. The LS then checks the following conditions:

1. If $maxdist(c_i, c_j) \leq \epsilon$, then LS detects a proximity.
2. If $mindist(c_i, c_j) > \epsilon$, then LS detects no proximity.
3. If $mindist(c_i, c_j) \leq \epsilon < maxdist(c_i, c_j)$, then users u_i and u_j invoke the peer-to-peer Strips algorithm [1] for the refinement step.

The resulting communication cost is lower than Strips due to the use of a centralized (untrusted) server. Observe that, the `Baseline` does not use encrypted tuples as in our `FriendLocator` solution, so it offers a weaker notion of privacy.

Experiments. We first study the impact of the proximity detection distance ϵ on the cost per user per timestamp (Fig. 2a). Both `Baseline` and `FriendLocator` have similar

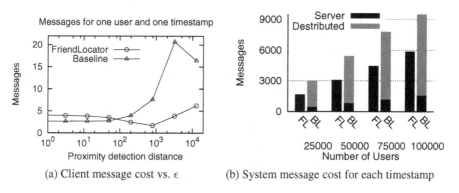

(a) Client message cost vs. ϵ (b) System message cost for each timestamp

Fig. 2. Effect of various parameters on the communication cost

performance at small ϵ (below 10). As ϵ increases, Baseline invokes the refinement step frequently so its cost rises rapidly. At extreme ϵ values (above 10000), most of the pairs are within proximity so the frequency and cost of executing the refinement step in Baseline are reduced. Observe that the cost of FriendLocator is robust to different values of ϵ, and its cost rises slowly when ϵ increases. Figure 2b shows the total number of messages during 40 timestamps as a function of the total number of users in the system. Clearly, FriendLocator incurs substantially lower total cost than Baseline. In Fig. 2b the distributed messages represent peer-to-peer messages.

6   Conclusion

In this paper we develop the FriendLocator, a client-server solution for detecting proximity among friend pairs while offering them location privacy. The client maps a user's location into a grid cell, converts it into an encrypted tuple, and sends it to the server. Based on the encrypted tuples received from the users, the server determines the proximity between them blindly, without knowing their actual locations. Experimental results suggest that FriendLocator incurs low communication cost and it is scalable to a large number of users.

In the future, we plan to extend the proposed solution for privacy-aware proximity detection among moving users on a road network, in which the distance between two users is constrained by the shortest path distance between them.

References

1. Amir, A., Efrat, A., Myllymaki, J., Palaniappan, L., Wampler, K.: Buddy Tracking - Efficient Proximity Detection Among Mobile Friends. In: INFOCOM (2004)
2. Brinkhoff, T.: A Framework for Generating Network-Based Moving Objects. GeoInformatica 6(2), 153–180 (2002)
3. Ghinita, G., Kalnis, P., Khoshgozaran, A., Shahabi, C., Tan, K.-L.: Private Queries in Location Based Services: Anonymizers are not Necessary. In: SIGMOD (2008)
4. Gruteser, M., Grunwald, D.: Anonymous Usage of Location-Based Services Through Spatial and Temporal Cloaking. In: USENIX MobiSys. (2003)
5. Heining, A.: Stalk your friends with google (February 2009)
6. Khoshgozaran, A., Shahabi, C.: Blind Evaluation of Nearest Neighbor Queries Using Space Transformation to Preserve Location Privacy. In: Papadias, D., Zhang, D., Kollios, G. (eds.) SSTD 2007. LNCS, vol. 4605, pp. 239–257. Springer, Heidelberg (2007)
7. Liu, K., Giannella, C., Kargupta, H.: An Attacker's View of Distance Preserving Maps for Privacy Preserving Data Mining. In: Fürnkranz, J., Scheffer, T., Spiliopoulou, M. (eds.) PKDD 2006. LNCS, vol. 4213, pp. 297–308. Springer, Heidelberg (2006)
8. Mascetti, S., Bettini, C., Freni, D., Wang, X., Jajodia, S.: Privacy-Aware Proximity Based Services. In: MDM (to appear, 2009)
9. Roussopoulos, N., Kelley, S., Vincent, F.: Nearest Neighbor Queries. In: SIGMOD (1995)
10. Ruppel, P., Treu, G., Küpper, A., Linnhoff-Popien, C.: Anonymous User Tracking for Location-Based Community Services. In: Hazas, M., Krumm, J., Strang, T. (eds.) LoCA 2006. LNCS, vol. 3987, pp. 116–133. Springer, Heidelberg (2006)
11. Yiu, M.L., Jensen, C.S., Huang, X., Lu, H.: SpaceTwist: Managing the Trade-offs Among Location Privacy, Query Performance, and Query Accuracy in Mobile Services. In: ICDE (2008)

Semantic Trajectory Compression

Falko Schmid[1], Kai-Florian Richter[1], and Patrick Laube[2]

[1] Transregional Collaborative Research Center SFB/TR 8 Spatial Cognition,
University of Bremen, P.O. Box 330 440, 28334 Bremen, Germany
{schmid,richter}@sfbtr8.uni-bremen.de
[2] Department of Geomatics, The University of Melbourne, VIC 3010, Australia
plaube@unimelb.edu.au

Abstract. In the light of rapidly growing repositories capturing the movement trajectories of people in spacetime, the need for trajectory compression becomes obvious. This paper argues for *semantic trajectory compression* (STC) as a means of substantially compressing the movement trajectories in an urban environment with acceptable information loss. STC exploits that human urban movement and its large–scale use (LBS, navigation) is embedded in some geographic context, typically defined by transportation networks. STC achieves its compression rate by replacing raw, highly redundant position information from, for example, GPS sensors with a semantic representation of the trajectory consisting of a sequence of *events*. The paper explains the underlying principles of STC and presents an example use case.

Keywords: Trajectories, Moving Objects, Semantic Description, Data Compression.

1 Motivation

Trajectories, the representation of movement by means of positioning fixes, usually contain data which can be considered as highly redundant information. Movement often happens along network infrastructure, such as streets or railway tracks, and the significant behavior patterns, such as stops, are performed along it as well. Especially in dense urban environments there are not many alternatives for reaching a certain destination other than to move along available network links. The representation and storage of trajectories by means of lists of fixes also pose questions about knowledge gain and further processing; trajectories are only meaningful when their spatial context is considered. Relating trajectories to their spatial context at an early stage will lead to improved means of analyzing them with standardized methods in a later stage.

Figure 1b is a visualization of a stream of raw positional data (Figure 1a) produced by a tracking system (e.g. GPS). Figure 1c depicts the same movement embedded in its geographical context. An object moved through a system of streets to reach its destination. The actual information contained in trajectories is the sequence of implicitly encoded spatio-temporal events, i.e., a single datum is usually not of interest, but rather the *significant* information with respect to

N. Mamoulis et al. (Eds.): SSTD 2009, LNCS 5644, pp. 411–416, 2009.

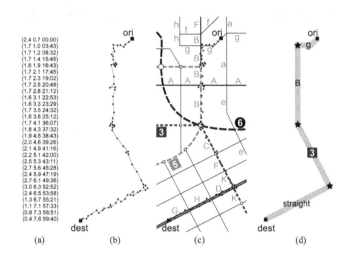

Fig. 1. Problem overview: (a) Raw positional data; (b) trajectory in two-dimensional space, moving from origin to destination; (c) trajectory embedded in geographic context, the semantically annotated *map* features a train line #6 with two stations; tram lines #3 and #5 with several stops; major streets (Upper case) and minor streets (lower case); (d) minimalist representation of the same trajectory, used for the semantic compression: origin, street *g*, street *B*, tram line #3, straight, destination

the movement. Significance depends on the application context, but is always based on the spatial course and the determination of events (usually stops).

Having a look at the trajectory, it becomes obvious that its course can be described by referring to elements of the network without loosing relevant information. The course can be expressed by the streets and tram tracks it moves along (street g, street B, tram line #3, and straight along the streets D, G, H). This is a minimalist representation of the movement (Figure 1d). Instead of using a large number of point coordinates, the movement can be described with elements of the transportation network, annotated with the behaviorally significant elements (in this case origin, stops, and destination).

Networks as a constraining basis for movement reduce the dimensionality of space and, thus, allow efficient indexing structures for moving objects [1]. Most work focused on the geometry of the underlying networks. However, the database community has acknowledged semantics—the meaningful annotation of moves with labels from the embedding environment —as being paramount for the interpretation and analysis of raw trajectory data [2, 3]. Whereas exploiting semantics is a young branch in spatial database research, in spatial cognition the semantics of movement has long been exploited for designing better wayfinding instructions [4, 5]. Going a step beyond utilizing spatial infrastructure as a suitable representation, a semantic representation of the trajectory can be implemented that focuses on qualitative change in course and events without loosing the conceptual information of the movement data.

2   Semantics in Trajectories

The majority of systems tracking the movement of individuals produce lists of time-stamped position samples, so-called *fixes* in the form of tuples (x, y, t). Even though this is a discrete approximation of the movement behavior, it is widely accepted to model the respective movement as a sequence of fixes, connected with straight line segments. In its most simple form a *trajectory* is a 2-dimensional polygonal line connecting the fixes of a moving individual. For example, in Figure1, an individual has moved from location origin ori to destination dest, starting at 00:00 arriving at 59:40. Figure 1a illustrates a trajectory's raw data, Figure 1b the respective trajectory in two-dimensional space. Note the raw data column only illustrates a subset of the plotted fixes.

A *map* is a semantically annotated network of edges and nodes. A map represents the transport network of an urban environment, featuring streets, bus, tram and train lines (see Figure 1c). In a map vertices are unambiguously defined, either by IDs or by (x, y) coordinate tuples. Further, edges may have a label (street name, or bus, tram, train line). This is a $n : 1$-relation as several edges can have the same label. Vertices of bus, tram, and train lines are stops and stations; these may be labeled with the stops' names. The labeling of edges and vertices can extend several levels. An edge may at the same time have a local street name (e.g., "Ostertorsteinweg"), be part of a national highway system (e.g., "A7"), and be part of a bus or tram line (e.g., "Tram #3").

3   Semantic Trajectory Compression

If movement happens in a transport network, as is usually the case in urban environments, trajectories can be mapped to a map representing this environment. The mapping of fixes to vertices and edges of the transport network then allows for exploiting this structure to restricting a trajectory's representation to the significant events. A network reduces the dimensionality of a two-dimensional movement space. It allows for concise positioning of a moving object through time-stamping along edges and at vertices, which both have unique identifiers. In a semantically annotated map, edges and vertices can be aggregated according to shared labels, for example their street names or the train lines. Often, several consecutive edges represent the same street and, thus, share the same label. Tram and bus lines may extend over large sections of an urban transport network. Thus, the semantic annotation of the network offers a high-level reference system for urban spaces, which is exploited in STC.

Taking this perspective, streets and tram, bus or train lines are viewed as *mobility channels* that moving objects hop on, ride for a while, and hop off again to catch another channel that brings them closer to their destination. In terms of trajectory compression, this perspective has the advantage that only little information needs to be retained for describing the movement of an individual in terms of riding such channels. For most kinds of movement storing a sequence of the identifiers of the specific channels and hop-on and hop-off times results in a sufficient approximation of the individual's movement through the network. At the same time, this drops a large amount of fixes, which are highly correlated and, hence, redundant. Semantic compression of trajectories makes use of principles and methods that have been previously implemented for the generation of cognitively motivated route descriptions (the GUARD process, cf. [6]). Broadly, it is based on three steps:

1. Identify the relevant events along the trajectory. Relevant events are origin and destination, as well as street intersections and public transport stops (see Figure 1c).
2. For each event, determine all possible descriptions of how movement continues from here. These descriptions are egocentric direction relations (straight, left, right, etc.; in Figure 1c straight from edge D to edge G) or changes in labels of network elements (in Figure 1c change from label street B to tram line #3) for capturing the motion continuation of an event.
3. Based on the descriptions, combine consecutive events into sequences of events. These sequences are termed (*spatial*) *chunks* [5]. The compressed trajectory consists of sequences of such spatial chunks (Figure 1d).

In decompression, the aim is to reconstruct movement through an environment. In the chosen semantic approach, decompression does not restore the original trajectory, but rather the path through the network along with inferred timestamps. The path contains all information on changes of direction as well as places along the way; each such event point is coupled with a time-stamp stating when in the travel behavior it occurred. Note that for all reconstructed points in the decompressed trajectory, i.e., those that are not original points retained in the compressed trajectory, the time-stamp is calculated based on an assumed linear movement behavior between start and end point of a chunk. While this time estimation is a simplification resulting in information loss, it provides no limitation for the targeted applications (see Section 5).

In a nutshell, the decompression algorithm iterates through the sequence of chunks stored in the compressed trajectory. It returns a sequence of vertices that are a geometric representation of the travelled path through the network. In more detail, beginning with the start vertex of a chunk the algorithm adds geometric edges to the reconstructed path until the end vertex is reached. To this end, it uses different strategies to determine which edge is to be added; these strategies depend on the description used for chunking. Each added vertex is linked to a time-stamp, which is calculated assuming constant movement speed, i.e., representing a fraction of time corresponding to the fraction of the distance travelled between start and end vertex.

4   Example Use Case

Figure 2 shows an example use case of applying semantic trajectory compression. The geometric representation of the path contains 115 points in space-time (115 tuples of (x, y, t)). It further comprises 52 events, i.e., 52 intersections and stops along the way. Performing compression yields the following 6 elements as result:

((3490254.00 5882064.00 00:00) (3490057.00 5882110.00 01:12) "Bei den Drei Pfählen")
((3490057.00 5882110.00 01:12) (3489534.00 5882241.50 04:47) "Am Hulsberg")
((3489534.00 5882241.50 04:47) (3488929.50 5882100.00 08:21) "Am Schwarzen Meer")
((3488929.50 5882100.00 08:21) (3488222.50 5882314.50 13:09) "Vor dem Steintor")
((3488222.50 5882314.50 13:09) (3487688.75 5882291.00 16:17) "Ostertorsteinweg")
((3487688.75 5882291.00 16:17) (3487544.75 5882351.00 17:21) "Am Wall")

As can be seen, STC achieves a high compression rate. Instead of the 115 original points, it ends up with only 6 items, which corresponds to a compression
rate of 94.78%. Considering that each item in the compressed trajectory consists
of three elements, the ratio is still 18 to 115 elements or 84.35%. Decompressing
the compressed trajectory reconstructs the original path. It also keeps the timestamps explicitly stated in the compressed trajectory. There are some differences
in the geometric representation—in this case the reconstructed path contains 3
coordinates more than the original path. This can be explained with ambiguities
in the underlying geographic data set that for some streets has individual representations of different lanes, resulting in different geometric representations.
However, there is no visual or semantic difference between the original and the
reconstructed path; all events of the original path are correctly reconstructed.
Regarding time, the original time-stamps stored in the compressed trajectory
are retained; all other reconstructed events are annotated with estimated timestamps assuming linear movement within a chunk.

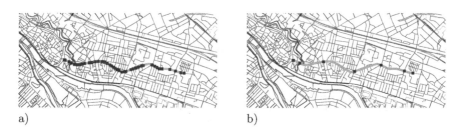

a)   b)

Fig. 2. The map shows part of the inner-city region of Bremen, Germany. a) The
displayed path (the bold line) runs from right to left. The dots on the path mark all
event points along the way. b) The events stored in the compressed trajectory.

5   Conclusions and Outlook

This paper presents a novel approach for compressing large volumes of trajectory data by exploiting the semantic embedding of movement in a geographical
context. Inspired by network-constrained object indexing and techniques used
in spatial cognition and wayfinding, the paper presents *semantic trajectory compression* (STC). STC matches the movement to the underlying map and aggregates chunks based on identical semantic descriptions. Initial experiments with a
set of use case trajectories captured with volunteers in the city of Bremen serve

as a proof of concept, deliver promising results for future experiments and help to identify limitations and a road map for future work. After implementing an STC prototype, future work will focus on evaluating the STC algorithm with large and diverse trajectory data. Extensive experiments with real, recorded trajectory data shall identify possible conceptual shortcomings and reveal the runtime characteristics of the STC algorithm for various scenarios.

As a main contribution, the paper illustrates that the embedding of human movement in the geographic context of an urban street network can successfully be exploited for compressing large volumes of raw trajectory data with acceptable information loss. The reconstructed information is suited for a number of applications based on individual spatial profiles which are not built upon a fine-grained analysis of movement dynamics (e.g., ascending and descending velocity). Specifically, this holds for prior-knowledge based navigation support [7] which relies on previously visited places and traveled paths. Also, most applications within the field of Location Based Services that rather rely on a clean *model* of movement than its detailed dynamics will benefit from semantically compressed trajectories.

Acknowledgments

Support by the German Science Foundation (DFG) and Group of Eight / DAAD Australia Germany Joint Research Co-operation Scheme is acknowledged.

References

[1] Li, X., Lin, H.: Indexing network-constrained trajectories for connectivity-based queries. Int. Journal of Geographical Information Science 20(3), 303–328 (2006)

[2] Alvares, L.O., Bogorny, V., Kuijpers, B., Fernandes de Macedo, J.A., Moelans, B., Vaisman, A.: A model for enriching trajectories with semantic geographical information. In: GIS 2007: Proc. of the 15th annual ACM international symposium on Advances in GIS, pp. 1–8. ACM, New York (2007)

[3] Spaccapietra, S., Parent, C., Damiani, M.L., de Macedo, J.A., Portoa, F., Vangenot, C.: A conceptual view on trajectories. Data and Knowledge Engineering 65(1), 126–146 (2008)

[4] Tversky, B., Lee, P.U.: How space structures language. In: Freksa, C., Habel, C., Wender, K.F. (eds.) Spatial Cognition 1998. LNCS (LNAI), vol. 1404, pp. 157–175. Springer, Heidelberg (1998)

[5] Klippel, A., Hansen, S., Richter, K.F., Winter, S.: Urban granularities – a data structure for cognitively ergonomic route directions. GeoInformatica 13(2), 223–247 (2009)

[6] Richter, K.F.: Context-Specific Route Directions - Generation of Cognitively Motivated Wayfinding Instructions. DisKI, vol. 314. IOS Press, Amsterdam (2008); also appeared as SFB/TR 8 Monographs Volume 3

[7] Schmid, F.: Knowledge based wayfinding maps for small display cartography. Journal of Location Based Services 2(1), 57–83 (2008)

Pretty Easy Pervasive Positioning

René Hansen, Rico Wind, Christian S. Jensen, and Bent Thomsen

Center for Data-Intensive Systems, Department of Computer Science, Aalborg University
Selma Lagerlöfs Vej 300, DK-9220 Aalborg Ø, Denmark
{rhansen,rw,csj,bt}@cs.aau.dk

Abstract. With the increasing availability of positioning based on GPS, Wi-Fi, and cellular technologies and the proliferation of mobile devices with GPS, Wi-Fi and cellular connectivity, ubiquitous positioning is becoming a reality. While offerings by companies such as Google, Skyhook, and Spotigo render positioning possible in outdoor settings, including urban environments with limited GPS coverage, they remain unable to offer accurate indoor positioning.

We will demonstrate a software infrastructure that makes it easy for anybody to build support for accurate Wi-Fi based positioning in buildings. All that is needed is a building with Wi-Fi coverage, access to the building, a floor plan of the building, and a Wi-Fi enabled device. Specifically, we will explain the software infrastructure and the steps that must be completed to obtain support for positioning. And we will demonstrate the positioning obtained, including how it interoperates with outdoor GPS positioning.

1 Introduction

Positioning is a key requirement for a number of useful mobile services including personal navigation, personalized shopping assistance, tourist guidance, and friend finder services. Unfortunately, many services are constrained to working exclusively in either only outdoor or only indoor settings. By offering ubiquitous positioning, the utility of many services can be improved greatly. For example, a ubiquitous navigation system can guide a user from his home to the airport and also to the relevant check-in counter and departure gate inside the airport. And it can inform the user about available shops, restaurants, restrooms, etc. A personal shopping assistant application can display relevant nearby stores. Once the user is inside a store, the application might provide information about special offers or where to find a particular item of clothing. Similarly, a ubiquitous tourist guide can offer a high-level overview of tourist attractions found in an area, as well as provide background information on a piece of art in a museum that a user visits. Finally, friend finder applications are of limited value if they can only locate a person's friends in areas with GPS coverage *or* in restricted indoor environments.

Companies such as Google [1], Skyhook [2], and Spotigo [3] have developed solutions that offer GPS-less positioning. They use so-called lateration techniques to infer locations based on knowledge of the placement of 802.11 (Wi-Fi) base stations and cell towers. This knowledge is gathered through a process called war-driving where drivers travel the streets, scanning for nearby base stations. Lateration-based techniques generally infer a user's position at the granularity of a building or even a particular region of

N. Mamoulis et al. (Eds.): SSTD 2009, LNCS 5644, pp. 417–421, 2009.

a building. However, they are unable to position users at the granularity of individual rooms. In fact, these systems have no notion of a room. Moreover, positions are delivered as two-dimensional latitude/longitude coordinates, which renders it difficult to meaningfully position a user or point of interest inside a multi-floor building.

The Streamspin [4,5] system seeks to overcome these limitations by using accurate location fingerprinting to position users indoor. Location fingerprinting works by building a database, called a *radio map*, of Wi-Fi base station signal strengths observed at different locations; each pair of a location and a set of signal strengths is called a fingerprint. Users are subsequently positioned by reversing the process, i.e., given a set of measured signal strengths, a position is inferred by finding the best match between the measured signal strengths and the fingerprints already stored in the radio map. The accuracy of the technique is typically within a few meters, corresponding to the size of many office building rooms [6,7,8,9,10].

The most limiting aspect of using location fingerprinting is that of constructing and calibrating the needed radio maps. Signal strengths need to be collected with relatively fine-grained spacing, i.e., typically every two to three meters, to ensure the high accuracy. Moreover, collection of signal strengths generally has to be repeated at regular intervals to account for the environmental dynamics of the wireless channel [6]. This naturally incurs a much larger overhead than for war driving. Streamspin enables the distribution of this administrative burden by letting the users themselves contribute to calibrating the radio maps. This is done by letting the users upload signal strength information to a central server. In addition, Streamspin lets users upload floor plans and symbolic information for locations in the buildings for which positioning is being provided. Once a radio map for a building has been added to the Streamspin server, the system is able to provide indoor positioning in the building, with the same ease as using GPS, i.e., transparently to the user. We are not aware of any other system, commercial or academic, that offers support for ubiquitous positioning by automatic sharing of user-generated radio maps.

Section 2 describes the process of building indoor positioning for a given building. Section 3 explains how Streamspin provides a seamless and transparent handover between outdoor and indoor positioning, by identifying automatically the radio map that applies to a building when a user enters the building. Section 4 summarizes and identifies promising research directions. Finally, Section 5 gives an overview of the content of the proposed demonstration.

2   Enabling Indoor Positioning in Streamspin

Enabling positioning in a building entails the construction of a radio map that supports fine-grained positioning; but it also entails the provisioning of a floor plan and symbolic indoor location information that provide users with understandable location information. More specifically, the following steps are involved:

1. Upload a floor plan to the Streamspin server.
2. Geo-position the floor plan globally.
3. Calibrate a radio map using the Streamspin client.
4. Supply symbolic indoor location information via the Streamspin client.

Steps 1 and 2. The system architecture assumed in the following is shown in Figure 1. The users must upload a floor plan of the building, in JPEG format, to the Streamspin server through the system's web interface. In case of several floors, several JPEG images are to be uploaded. When a floor plan is initially uploaded, the user is asked to position the image on the surface of the Earth using Google Maps. This is done in order to position the building globally. For a building with multiple floors, this is done only for the first image; subsequent images are snapped to the region already defined for the first image. When uploading images for multiple floors, the user must specify the floor level of each image. A unique building identifier is generated at the server for later use.

Step 3. Using the mobile Streamspin client, a user can download the images that were previously uploaded to the server. The downloaded images include decorations that let the user see locations that have already been fingerprinted, i.e., locations that were previously added to the building's radio map. This allows the user to see the current coverage of the indoor positioning system. The user can then contribute to the calibration of the radio map by choosing either an already fingerprinted location or by selecting a new location that has not yet been fingerprinted. In either case, the user moves to the chosen location, selects "Start measuring signal strengths," and later (at the user's discretion) selects "Stop measuring signal strengths." For each Wi-Fi access point detected during that period, a histogram of frequencies of the recorded signal strength values is saved together with the user's location, and this is uploaded to the radio map of the building.

The Streamspin server maintains a list of the access points visible in each building known to the system. Each time a signal strength measurement is contributed, any new access points are added to the list. The access point list is later used to identify the building a user enters, so that the appropriate radio map can be downloaded automatically.

Step 4. Users can supply symbolic location information about places in a building by simply selecting a coordinate or dragging a region and then entering textual information, e.g., "Dr. X's Office" or "The Gym."

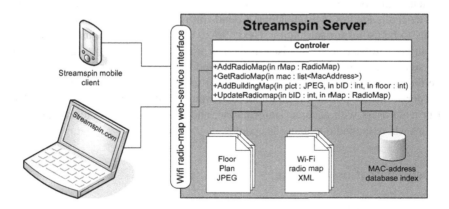

Fig. 1. Architecture of the Wi-Fi Radio Map Subsystem of Streamspin

3   Performing Indoor Positioning in Streamspin

Once a radio map for a building is in place, users can be positioned within the building. Specifically, users can be positioned at any fingerprinted location (these locations represent the possible state space). The system handles the handover between outdoor GPS positioning and indoor positioning completely without user intervention.

When a Streamspin client looses the GPS signals, it initiates Wi-Fi scanning to determine whether indoor positioning is available. Specifically, the client records any access points that can be detected and sends their MAC addresses to the server. The server then searches for an appropriate radio map by matching the MAC addresses against its collection of access point lists. If a match is found, the server returns the identifier of the appropriate building. The client uses the identifier to determine whether the corresponding radio map is already stored in its local cache. If not, the client downloads the radio map. As soon as a radio map is available, the client initiates Wi-Fi based positioning. Eventually, when the client detects that the user has again left the building, the system reverts back to GPS. The specific handover policies are covered elsewhere [11,12].

4   Summary and Research Directions

This paper presents an open, easy-to-use approach to enabling user-generated indoor positioning. This approach, implemented as part of the Streamspin system, enables users to build and deploy indoor Wi-Fi-based positioning, which in turn enables ubiquitous indoor/outdoor location-based services. Users contribute by calibrating radio maps, and they provide floor plans and symbolic indoor location information. Once built and plugged into Streamspin, the radio maps offer the users automatic, fine-grained indoor positioning. Specifically, the usage is transparent and the appropriate radio map is found automatically when a user enters a building.

Several interesting directions for future work exist. It is worth studying whether the trajectories that capture user movements can be used for radio map calibration without explicit user involvement, thus serving as a beneficial supplement to active user calibration. The challenge lies in establishing ground-truth user positions that are sufficiently accurate to be useful for calibration. Next, due to the infrastructure's inherently openness, an issue related to trust arises. Effective measures are needed to guard against faulty signal strength information being uploaded by malicious users, as this may degrade the performance of the radio maps. Finally, the ability to report on the quality of a reported indoor position is an important feature planned for implementation. This will provide users with accuracy information, and it will enable the identification of locations where the radio map needs updates.

5   Demonstration Content

This section describes the content envisioned for the demonstration.

The demonstration will contain a short introduction to the overall Streamspin system, including how users can easily create, share, and use location-based services.

Next, the indoor positioning technology is described and the general architecture for creating, uploading, and distributing the radio maps is presented. We will adopt a multi-step approach, introducing first the Streamspin web site from where the creation of the radio map is initiated. Then the floor plan image in JPEG format is uploaded to the web site and spherically positioned using a map, to allow positions inside the building to be reported in latitude/longitude coordinates.

This is followed by a demonstration of the Streamspin client. This includes the functionality for recording a radio map using the just uploaded JPEG image. The process is quite simple. The user clicks on the current location on the displayed floor plan, and the system will record the signal strengths until requested to stop by the user. By doing this, we demonstrate the ease with which a radio map can be built.

Although recording a radio map is straightforward, it is a relatively time-consuming process, and only a few position recordings are demonstrated. A pre-made radio map is used for demonstrating the actual positioning, including moving outside to illustrate the handover between Wi-Fi based and GPS positioning.

The positioning demonstration will include a service where the user actually receives content based on the locations reported by the system. Possible services include a small tour of the perimeter, involving both outdoor and indoor points of interest, or simply a service showing the user's track.

References

1. Google Latitude, http://www.google.com/latitude
2. Skyhook Wireless, http://www.skyhookwireless.com
3. Spotigo, http://www.spotigo.com
4. Jensen, C.S., Vicente, C.R., Wind, R.: User-Generated Content—The Case for Mobile Services. IEEE Computer 41(12), 116–118 (2008)
5. Wind, R., Jensen, C.S., Pedersen, K.H., Torp, K.: A testbed for the exploration of novel concepts in mobile service delivery. In: Proc. MDM, pp. 218–220 (2007)
6. Bahl, P., Padmanabhan, V.N.: RADAR: An In-Building RF-Based User Location and Tracking System. In: Proc. INFOCOM, pp. 775–784 (2000)
7. Krumm, J., Horvitz, E.: Locadio: Inferring motion and location from Wi-Fi signal strengths. In: Proc. MobiQuitous, pp. 4–13 (2004)
8. Ladd, A.M., Bekris, K.E., Rudys, A., Marceau, G., Kavraki, L.E., Wallach, D.S.: Robotics-based location sensing using wireless ethernet. In: Proc. MOBICOM, pp. 227–238 (2002)
9. Saha, S., Chaudhuri, K., Sanghi, D., Bhagwat, P.: Location determination of a mobile device using IEEE 802.11b access point signals. In: Proc. IEEE Wireless Communications and Networking Conf., pp. 1987–1992 (2003)
10. Hansen, R., Thomsen, B.: Accurate and Efficient WLAN Positioning With Weighted Graphs. In: Proc. Intl. Conf. on Mobile Lightweight Wireless Systems, 15 pages (to appear, 2009)
11. Hansen, R., Wind, R., Jensen, C.S., Thomsen, B.: Seamless Indoor/Outdoor Tracking Handover for Location Based Services in Streamspin. In: Proc. MDM, 6 pages (to appear, 2009)
12. Hansen, R., Jensen, C.S., Thomsen, B., Wind, R.: Seamless Indoor/Outdoor Positioning with Streamspin. In: Proc. MobiQuitous, 2 pages (2008)

Spatiotemporal Pattern Queries in SECONDO

Mahmoud Attia Sakr and Ralf Hartmut Güting

Databases for New Applications, Fernuniversität Hagen, Germany
{mahmoud.sakr,rhg}@fernuni-hagen.de

Abstract. We describe an initial implementation for spatiotemporal pattern queries in SECONDO. That is, one can specify for example temporal order constraints on the fulfillment of predicates on moving objects. It is shown how the query optimizer is extended to support spatiotemporal pattern queries by suitable indexes.

Keywords: Spatiotemporal pattern queries, SECONDO, Query optimizer.

1 Introduction

Moving objects are objects that change their position and/or extent with time. Having the trajectories of these objects stored in a suitable database system allows for issuing spatiotemporal queries. One can query for example for animals which crossed a certain lake during a certain time.

Spatiotemporal pattern queries (STPQ) provide a more complex query framework for moving objects. They are used to query for trajectories which fulfill a certain sequence of predicates during their movement. For example, suppose predicates P, Q and R that can hold over a time interval or a single instant. We would like to be able to express spatiotemporal pattern conditions like the following:

- P then (later) Q then R.
- P ending before 8:30 then Q for no more than 1 hour.
- (Q then R) during P.

The predicates P, Q, R, etc. might be of the form

- Vehicle X is inside the gas station S.
- Vehicle X passed close to the landmark L.
- The speed of air plane Z is between 400 and 500 km/h.

STPQs are the intuitive way to query the moving objects by their movement profile. We demonstrate a novel approach for STPQs along with its implementation in SECONDO. Some other approaches can be found in [3], [4], [5] and [6].

2 SECONDO Platform

SECONDO [1] is an extensible DBMS platform that doesn't presume a specific database model. Rather it is open for new database model implementations. It consists of three loosely coupled major components: the kernel, the GUI and the query optimizer.

N. Mamoulis et al. (Eds.): SSTD 2009, LNCS 5644, pp. 422–426, 2009.

It supports two syntaxes: an SQL-like syntax and a special syntax called *SECONDO executable language*.

All three SECONDO components can be extended. The kernel, for example, is extended by algebra modules. In an algebra module one can define new data types and/or new operations. In this work, we added the *spatiotemporal pattern algebra*.

Due to its clean architecture, SECONDO is suitable for experimenting with new data types, database models, query types, optimization techniques and database operations. The source code and documentation for SECONDO are available for download [1].

3 Spatiotemporal Pattern Queries

A main idea underlying our approach to spatiotemporal pattern queries is to reuse the concept of *lifted predicates* [2]. Basically by this concept we define time dependent versions of static predicates. A static binary predicate has the signature

$$\sigma_1 \times \sigma_2 \rightarrow \underline{bool}$$

where σ_1 and σ_2 are of static types (i.e: *point*, *int* or *region*). Lifted counterparts of this predicate are obtained by replacing one or both parameters by their moving versions. The signature of a lifted predicate is one of the following:

$$\underline{moving}(\sigma_1) \times \sigma_2 \rightarrow \underline{moving}(\underline{bool})$$

$$\sigma_1 \times \underline{moving}(\sigma_2) \rightarrow \underline{moving}(\underline{bool})$$

$$\underline{moving}(\sigma_1) \times \underline{moving}(\sigma_2) \rightarrow \underline{moving}(\underline{bool})$$

Since the arguments are time dependent, the result is time dependent as well.

SECONDO uses the sliced representation for moving data types. For a *moving(bool)*, for example, the sliced representation could be thought of as an array of units with every unit consisting of a time interval and a boolean value. Therefore, each unit is a mapping between a time interval/ slice and a value that holds during this time. Units are not allowed to overlap in time.

For many static predicates lifted counterparts are already implemented in SECONDO. We use lifted predicates in formulating patterns. Hence we can easily leverage a considerable part of the available infrastructure.

A simple STPQ may be

```
Find all cars that entered a gas station, parked close to the
bank, and then drove away fast (candidate bank robbers).
```

This can be expressed in SQL as follows:

```
SELECT * FROM car AS c, landmark As l WHERE l.type= "gas
station" and pattern(c.trip inside l.region then
distance(c.trip, bank)< 0.2 then speed(c.trip)> 100.0)
```

where *c.trip* is a *moving(point)* storing the trajectory of the car, the distance is measured in km and the speed in km/h. This pattern consists of three predicates connected to each other by the *then* temporal connector. In our context, *then* is a binary temporal connector with the meaning that the second parameter must start to be true any time greater than or equal to the time when the first parameter starts to be true. In the following illustration we call the three predicates *P*, *Q* and *R* in order. They are lifted predicates that return a *moving*(*bool*) since one of the parameters to these predicates is of a moving type.

At the executable level of SECONDO we introduce a new predicate (operator) *stpattern*. Its arguments are a tuple and a sequence of lifted predicates (i.e., functions mapping the tuple into a *moving*(*bool*)).

Fig. 1 shows how the *stpattern* operator works applied to the sequence <*P*, *Q*, *R*>. The time is indicated on the horizontal axis and the results of evaluating the three predicates are shown on the vertical axis. The top most result *P*, for example, is a *moving*(*bool*) consisting of three units: false before t_1, true between t_1 and t_2 and false again after t_2. Note that we draw only the true intervals.

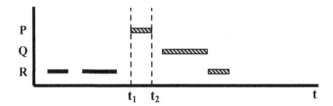

Fig. 1. Evaluating the result of a STPQ for one tuple

The pattern is fulfilled if and only if all the results of the predicates in the sequence have true units and the temporal order of these units meets the order in the sequence. In this example, the order of the sequence is *P* then *Q* then *R* and it is fulfilled by the three hatched intervals in Fig. 1.

The *stpattern* operator is a lazy operator. It evaluates the predicates in order. Once it is sure that the sequence of lifted predicates is not fulfilled (e.g., one predicate is always false) it returns *false*. This makes the operator more efficient by skipping the evaluation of unnecessary predicates. For example, if *P* is always false then *Q* and *R* are not evaluated.

More complex pattern descriptors may include temporal connectors like *later*, *meanwhile* or *immediately* or temporal constraints in the form of *P for 10 min then Q for no more than an hour meanwhile R*. We have ongoing research which considers such ideas but these are not yet included in this first implementation.

4 Query Optimization for Spatiotemporal Pattern Queries

SECONDO is a complete system in which even the query optimizer is accessible and, in fact, relatively easily extensible. Hence in contrast to previous work we are able to actually integrate pattern queries into the overall optimization framework. Obviously for an efficient execution of pattern queries on large databases the use of indexes is mandatory. We now consider how pattern predicates can be mapped to efficient index accesses.

The basic idea is to add each of the lifted predicates in a modified form as an extra "standard predicate" to the query, that is, a predicate returning a boolean value. The optimizer then should have for the rewritten predicate already some optimization rule available ("translation rules" in SECONDO) to map it into efficient index access.

For example, consider the lifted predicate "*c.trip inside l.region*" with *moving*(*point*) *c.trip*. We can rewrite it to a standard predicate "*c.trip passes l.region*" returning a *bool* value. The optimizer then already has rules to find trajectories passing the given region if their units are appropriately indexed.

A general way to convert a lifted predicate into a standard predicate is to apply the operation *sometimes*. It is a SECONDO predicate that takes a *moving*(*bool*) and returns a *bool*. It returns *true* if its parameter ever assumes *true* during its lifetime, otherwise *false*. So the strategy we implemented is to rewrite each lifted predicate P into a predicate *sometimes*(P) and to add translation rules that were missing.

Here is a brief description of how the optimizer processes STPQs along with the required extensions:

1. The parser parses the SQL-like query. For that we extended the parser to accept the syntax of the pattern operator.
2. The *query rewriting* rules are invoked. We added rules that for every predicate P in the pattern descriptor add a condition *sometimes*(P) to the where-clause of the query. Clearly the rewritten query is equivalent to the original one. We keep a list of these additional conditions for further processing.
3. We added translation rules for *sometimes*(Q) terms that were not yet available.
4. The optimizer continues its cost based optimization procedure trying to utilize available indexes.
5. After the best plan is chosen, the optimizer invokes the rules for translating the best plan into the *SECONDO executable language* and passes the query to the kernel for execution. A rule is added here to remove the additional *sometimes* predicates (with the help of the maintained list described in step 2) before passing the query to the kernel.

5 What Will Be Demonstrated

In this demo we will run several spatiotemporal pattern queries in SECONDO. The database used in the demonstration is the *berlintest* database coming with the SECONDO distribution containing geodata and objects of types *moving*(*point*) and *moving*(*region*). We will also demonstrate the query optimizer and the process of optimizing the STPQs. SECONDO will be invoked in its complete configuration where the three modules (kernel, GUI, and optimizer) are running together.

In Fig. 2, the GUI displays the result of an STPQ. The query is first written in SQL-like syntax. The GUI passes the SQL query to the optimizer. The optimizer generates a query plan which is passed back to the GUI (note that the plan accesses an index called Trains_Trip_sptuni). The GUI passes the executable query to the kernel, the kernel executes the query, the kernel passes the results to the GUI. Finally, the GUI tries to find the best viewer available and renders the results (in this case the Hoese- Viewer is chosen).

In Fig. 2, the trains fulfilling the STPQ are displayed as white circles with black outlines. The rest of the trains are black circles. "trip" is a *moving*(*point*) and "msnow" is a *moving*(*region*). They change their locations with time (see the time meter). The train with id 419 fulfills the pattern. It moves downwards and its path is indicated by a dashed line.

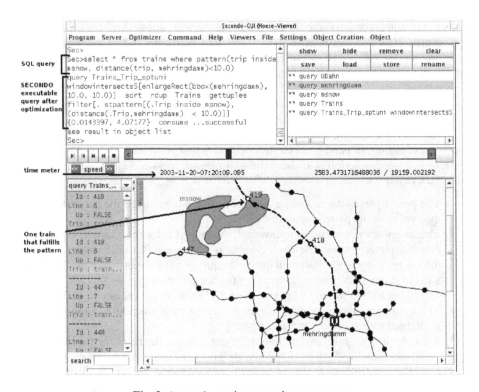

Fig. 2. A sample spatiotemporal pattern query

References

1. SECONDO home page, http://dna.fernuni-hagen.de/Secondo.html/index.html
2. Güting, R.H., Böhlen, M.H., Erwig, M., Jensen, S.J., Lorentzos, N.A., Schneider, M., Vazirgiannis, M.: A foundation for representing and quering moving objects. ACM Transactions Database Systems 25(1), 1–42 (2000)
3. Du Mouza, C., Rigaux, P.: Mobility Patterns. In: 2nd Workshop on Spatio-Temporal Database Management, STDBM 2004 (2004)
4. Erwig, M.: Toward Spatiotemporal Patterns. In: De Caluwe, et al. (eds.) Spatio-Temporal Databases, pp. 29–54. Springer, Heidelberg (2004)
5. Schneider, M.: Evaluation of Spatio-Temporal Predicates on Moving Objects. In: 21st Int. Conf. on Data Engineering, ICDE, pp. 516–517 (2005)
6. Hadjieleftheriou, M., Kollios, G., Bakalov, P., Tsotras, V.J.: Complex Spatio-Temporal Pattern Queries. In: VLDB 2005, pp. 877–888 (2005)

Nearest Neighbor Search on Moving Object Trajectories in SECONDO

Ralf Hartmut Güting, Angelika Braese, Thomas Behr, and Jianqiu Xu

LG Datenbanksysteme für neue Anwendungen
Fakultät für Mathematik und Informatik, Fernuniversität in Hagen
D-58084 Hagen, Germany

Abstract. In the context of databases storing histories of movement (also called trajectories), we present two query processing operators to compute the k nearest neighbors of a moving query point within a set of moving points. Data moving points are represented as collections of point units (i.e., a time interval together with a linear movement function). The first operator, *knearest*, processes a stream of units arriving ordered by start time and returns the set of units representing the k nearest neighbors over time. It can be used to process a set of moving point candidates selected by other conditions. The second operator, *knearestfilter*, operates on a set of units indexed in an R-tree and uses some novel pruning techniques. It returns a set of candidates that can be further processed by *knearest* to obtain the final answer. These nearest neighbor algorithms are presented within SECONDO, a complete DBMS environment for handling moving object histories. For example, candidates and final results can be visualized and animated at the user interface.

1 Introduction

Moving objects databases allow one to represent and query time dependent geometries, in particular continuously changing geometries. For moving point objects their behaviour over time is conceptually a function from time into 2D space. This can be represented in a moving point data type (*mpoint* for short), also called a trajectory. In this demo we address the problem of computing within a large database of stored moving points the k nearest neighbors to another moving point, the query point. Note that the solution is time dependent as well, hence, a set of moving points. The problem can be stated precisely as follows [1]. Let $d(p, q)$ denote the Euclidean distance between points p and q. Let $mp(i)$ denote the position of moving point mp at instant i.

Definition 1 (k-NN Query). *A spatiotemporal k-nearest neighbor query is defined as follows: Given a query* mpoint *mq and a relation R with an attribute* mloc *of type* mpoint, *return a subset R' of R where each tuple has an additional attribute mloc' such that the three conditions hold:*

1. *For each tuple $t \in R'$, there exists an instant of time i such that $d(t.mloc(i), mq(i))$ is among the k smallest distances from the set $\{d(u.mloc(i), mq(i)) | u \in R\}$.*

N. Mamoulis et al. (Eds.): SSTD 2009, LNCS 5644, pp. 427–431, 2009.

2. *mloc' is defined only at the times condition (1) holds.*

3. *$mloc'(i) = mloc(i)$ whenever it is defined.*

In other words, the query selects a subset of tuples whose moving point belongs at some time to the k closest to the query point and it extends these by a restriction of the moving point to the times when it was one of the k closest.

Whereas a lot of work exists for various kinds of nearest neighbor queries including continuous online maintenance of nearest neighbors, NN queries on historical trajectory databases have been addressed only more recently [2,3]. In this demo, we present a new, efficient solution for this problem within a full-fledged historical MO database system, SECONDO [6,7], which outperforms those earlier approaches by a large margin.

2   Representing Moving Objects in SECONDO

Within SECONDO, a large part of the data model proposed in [5] has been implemented which defines a set of data types for moving objects together with a comprehensive set of operations.

A moving point object can be represented in SECONDO in two ways: (i) In the compact representation, it is represented by a single tuple of a relation with an *mpoint* attribute. An *mpoint* value is represented as a sequence of so-called *units* where each unit consists of a time interval and a linear movement function. Time intervals of units are pairwise disjoint and ordered temporally within the *mpoint* representation. One can think of the storage scheme for an *mpoint* as an array of units. (ii) Units are also available as an independent data type *upoint*, and in the unit representation, a moving object is represented by a set of tuples with a *upoint* attribute, each tuple containing one of the units of the whole object. Operations are available to freely convert in query processing between the two representations.

For indexing moving points, standard R-trees are available. Indexes can be built in many ways. For example, for the compact representation, one can index the complete *mpoint*, or each of its units by its spatial or temporal projection or its spatiotemporal value. In the same way one can index each unit in the unit representation.

3   Small and Large Test Database

The demo will run on a small and a large test database. The SECONDO distribution [7] is equipped with a database called *berlintest*. Besides many other relations with spatial objects in the city of Berlin, there is a relation

    Trains (Id: int, Line: int, Up: bool, Trip: mpoint)

containing trains of the underground network (called U-Bahn") travelling according to schedule on a particular day (Nov. 20, 2003) between about 6 am and 10 am. There are 562 trains with about 100 units each (51544 units in total).

After a startup phase, usually about 90 trains are present at any instant of time. This is the *small database*. The *large database* at scale n^2 is created from the small one by making n^2 copies of each train, translating the geometry n times in x and n times in y-direction (see Figure 2 for a database at scale 25).

4   Operator *knearest*

A first query processing operator, *knearest*, has the following signature (with syntax indicated to the right, # denoting the operator and _ an argument):

```
knearest: stream(Tuple) × ai × mpoint × int
→ stream(Tuple)                          - #[-, -, -]
```

It receives (i) a stream of tuples where each tuple contains an attribute a_i of type *upoint*, (ii) the attribute name a_i, (iii) a query moving point, and (iv) the number k of nearest neighbors to be found. It returns a stream of tuples with the same structure as the input stream. However, the output units in attribute a_i belong to the k closest moving points represented in the input stream and are restricted to the times when they are among the k closest. It is required that the first argument stream arrives ordered by unit start times.

This operator is implemented essentially by plane sweep over the distance curves of all units (relative to the query *mpoint*), finding intersections of distance curves and reporting always the units belonging to the k lowest curves. Distance functions are in general quadratic polynomials. A query using *knearest* is:

```
query UnitTrainsOrdered feed filter[.Line = 7]
      knearest[UTrip, train5, 5] consume
```

which finds the five continuous nearest neighbors among trains of line 7 of *train5*, an *mpoint* value. Here *UnitTrainsOrdered* is a version of relation *Trains* that has been converted into unit representation. Additionally units are ordered by start time. This query can be posed as shown at the SECONDO user interface. Operator *feed* produces from a relation a stream of tuples, *filter* evaluates a predicate on each tuple. The output stream from *knearest* is collected into a relation by *consume* which is then displayed at the user interface.

5   Operator *knearestfilter*

Operator *knearestfilter* is designed to work on a 3D R-tree indexing the set of units of a given set of moving objects. Hence the arguments are basically the index, the query *mpoint*, and the number k. The exact signature is shown below. The operator returns a stream of unit tuples ordered by start time that are candidates to belong to the k nearest neighbors. This stream can be fed into the *knearest* operator to obtain the final result.

This algorithm is more complex and uses some sophisticated data structures and algorithms. Hence we cannot explain it in the limited space of this demo paper; a detailed description will be presented elsewhere [4]. One of the central

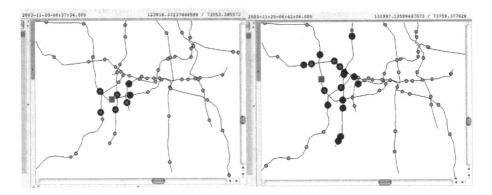

Fig. 1. Two instants during animation of k-nearest neighbor query

ideas is to compute for each node p of the index in preprocessing its *coverage*, a time dependent integer, which represents the number of units represented in the subtree and present at a given time during the node's bounding temporal interval. From the coverage curve, minimal coverages for temporal subintervals are determined which in turn are used for pruning during the R-tree traversal. The R-tree is carefully custom-built to obtain good coverage curves (using a bulk-loading technique).

In principle one could store these coverage values within the nodes of the R-tree. However, in a system context it does not make sense to modify the general index structure for each specific query algorithm. Therefore instead a separate relation is computed that is indexed by node identifiers of the R-tree. During the tree traversal, minimal coverage numbers for a given node are retrieved from this structure. Hence the signature of this operator is:

```
knearestfilter: rtree(Tuple) × rel(Tuple) ×
    btree(CoverTuple) × rel(CoverTuple) × mpoint × int
    → stream(Tuple)          _ _ _ _ #[_, _]
```

An example query on a large database at scale 100 is

```
query UnitTrains100_UTrip UnitTrains100
    UnitTrains100Cover_RecId UnitTrains100Cover
    knearestfilter[train742, 6]
    knearest[UTrip, train742, 6] consume
```

After *knearestfilter*, there appear 2241 candidate units out of the 5154400 units in the relation. The result has 941 tuples. Note that in this database on the average at any instant of time 9000 units are present.

An experimental comparison on a database at scale 25 and this query for $k = 5$ yields CPU times of 7.3 seconds for this approach, 52 seconds for the algorithm of [2], 33 seconds for the algorithm of [3], taking the average of 10 query trains. A systematic comparison will be provided in [4].

6   What Will Be Demonstrated

The demo will cover the following aspects:

- Live demonstration of queries on the small and large database as illustrated in Figures 2 and 1. Figure 2 shows the SECONDO GUI with an overview of the large database at scale 25. The query *mpoint* is shown as a large square, candidates from *knearestfilter* as large circles, solutions as small squares.

  Figure 1 shows two snapshots of the animation; here additionally data *mpoints* from the area are shown as small circles.
- On the small database we will demonstrate commands for the construction of the custom-built R-tree, visualizing also its node layout in space.
- We show the computation of coverage curves for nodes of the R-tree and of derived minimum coverages. Such curves can also be visualized.

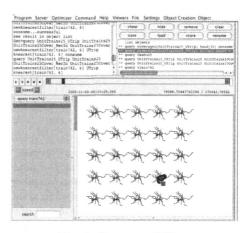

Fig. 2. SECONDO GUI

References

1. Düntgen, C., Behr, T., Güting, R.H.: BerlinMOD: A benchmark for moving object databases. The VLDB Journal (online first, 2009)
2. Frentzos, E., Gratsias, K., Pelekis, N., Theodoridis, Y.: Algorithms for nearest neighbor search on moving object trajectories. GeoInformatica 11(2), 159–193 (2007)
3. Gao, Y., Li, C., Chen, G., Chen, L., Jiang, X., Chen, C.: Efficient *k*-nearest-neighbor search algorithms for historical moving object trajectories. J. Comput. Sci. Technol. 22(2), 232–244 (2007)
4. Güting, R.H., Behr, T., Xu, J.: Efficient *k*-nearest neighbor search on moving object trajectories (manuscript in preparation, 2009)
5. Güting, R.H., Böhlen, M.H., Erwig, M., Jensen, C.S., Lorentzos, N.A., Schneider, M., Vazirgiannis, M.: A foundation for representing and quering moving objects. ACM Trans. Database Syst. 25(1), 1–42 (2000)
6. Güting, R.H., de Almeida, V.T., Ansorge, D., Behr, T., Ding, Z., Höse, T., Hoffmann, F., Spiekermann, M., Telle, U.: SECONDO: An extensible DBMS platform for research prototyping and teaching. In: ICDE, pp. 1115–1116. IEEE Computer Society, Los Alamitos (2005)
7. Secondo Web Site (2009),
   http://dna.fernuni-hagen.de/Secondo.html/index.html

A Visual Analytics Toolkit for Cluster-Based Classification of Mobility Data

Gennady Andrienko[1], Natalia Andrienko[1], Salvatore Rinzivillo[2], Mirco Nanni[2], and Dino Pedreschi[3]

[1] Fraunhofer IAIS, Sankt Augustin, Germany
[2] ISTI - CNR, Pisa, Italy
[3] Università di Pisa, Pisa, Italy

Abstract. In this paper we propose a demo of a Visual Analytics Toolkit to cope with the complexity of analysing a large dataset of moving objects, in a step wise manner. We allow the user to sample a small subset of objects, that can be handled in main memory, and to perform the analysis on this small group by means of a density based clustering algorithm. The GUI is designed in order to exploit and facilitate the human interaction during this phase of the analysis, to select interesting clusters among the candidates. The selected groups are used to build a classifier that can be used to label other objects from the original dataset. The classifier can then be used to efficiently associate all objects in the database to clusters. The tool has been tested using a large set of GPS tracked cars.

1 Introduction

The technologies of mobile communications and ubiquitous computing pervade our society, and wireless networks sense the movement of people and vehicles through their location-aware devices, generating large volumes of mobility data of unprecedented quantity, quality and timeliness at a very low cost. However, raw mobility data, such as collections of GPS tracks, are very complex, as they represent rough approximations of complex human activities, and at the same time semantically poor. Therefore it is extremely challenging to develop analysis techniques capable of mastering the complexity of the data and extracting meaningful abstractions, in particular, by discovering and interpreting groups of people or vehicles that exhibit similar mobility behavior. The mentioned problem can be formulated as a trajectory clustering problem: find, for the spatial area and the time interval under analysis, the natural clusters of similar trajectories, together with an intuitive way of presenting the discovered clusters to a human analyst for interpretation, i.e. attaching semantics. A few trajectory clustering techniques have been proposed, but their direct application to realistic mobility datasets, such as the object of study in this paper, is simply unfeasible. Real mobility data are both computationally and analytically complex, and require the involvement of a human analyst with her background knowledge and understanding of the properties of space and time. We present here a visual analytics environment, which enables to progressively find and refine trajectory clusters

N. Mamoulis et al. (Eds.): SSTD 2009, LNCS 5644, pp. 432–435, 2009.

and associated representative prototypes; our experiments demonstrate that natural clusters are found in the data, which characterize movement behaviors at a suitable abstraction level, understandable by mobility managers.

2   Analysis Process Description

The tool we present here is designed to support a stepwise analysis process for a large trajectory dataset. The analyst can consider a small portion of the dataset, extracting all the interesting clusters from this sample, by means of a density based clustering algorithm, i.e. OPTICS [1]. The visual environment allows the user to validate, refine and revise the found clusters. For each cluster, the system computes and proposes a set of specimens (i.e. representants) that serves as a classifier of the whole cluster: any other new object belongs to that cluster (i.e. has a similar behavior) if it is close to one of these specimens.

Given a trajectory dataset D, the analytic process can be formalized as follow:

- Extract a sample D' of trajectories from the database D
- Apply OPTICS with a suitable distance function d [2] and get a set of density-based clusters $\{C_1, C_2, \ldots, C_m\}$
- For each cluster C_i
    • Select s specimens in C_i, with $1 \leq s < |C_i|$, namely $\{c_{i1}^{\epsilon_1}, c_{i2}^{\epsilon_2}, \ldots, c_{is}^{\epsilon_s}\}$, such that the cluster C_i may be described as the set of objects in D' whose distance from one of the objects $c_{ij}^{\epsilon_j}$ is less than the threshold ϵ_j, i.e. $C_i = \{c \in D' | \exists\, j\ s.t.\ d(c, c_{ij}^{\epsilon_j}) < \epsilon_j, j = 1, 2, \ldots, s\}$
- Visually inspect and refine the selected specimens. The set of the specimens for all clusters forms a classifier
- Apply the classifier to the remaining trajectories, attaching each new trajectory to the closest specimens. The trajectories with no close specimen remain unclassified
- Possibly, restart the whole process again for the unclassified trajectories.

3   Presentation of the Tool

The user interface of the tools consists of an operational window (Figure 1(b)) and a map window (Figure 1(a)). The two windows are linked, only the selection of the operational window is showed in the map window. The application automatically selects a set of specimens for each cluster and partitions the objects in each group according to the closest specimen. After that, the analyst can manipulate the specimens and, in parallel, her actions are reflected on the corresponding trajectories. The refinement actions available for each cluster are: *(1)* merging two or more sub-clusters, *(2)* removing a sub-cluster, *(3)* splitting a sub-cluster. To show the functionalities of the tool, consider the cluster in Figure 1(a). On the map the specimens and the trajectories are represented with different colors. The analyst can select one subset of specimens as the most representative for the cluster

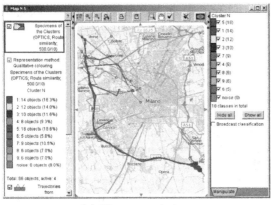 

(a) The map window   (b) The operational window

Fig. 1. The user interface of the application

(a) The main represen- (b) Two specimens of (c) The specimens are
tative specimen of the Cluster 1   extracted to create a
Cluster 1   new cluster (Cluster 10)

Fig. 2. Selection and merging of two specimens. Split of the cluster.

(Figure 2(a)). If other subclusters are visually inspected and valued as too dissimilar, it is possible to split the cluster. For example, the two specimens in Figure 2(b)
are moved in a new cluster (Figure 2(c)) since they describe shorter paths. It is also
possible to discard a sub-cluster from the analysis by tagging it as noise.

When the analyst concludes the refinement phase, she can use the resulting
classifier to classify all the other trajectories of the original dataset. Figure 3(a)
shows 745 trajectories from the database that have been attached to Cluster
1. Figure 3(b) represents these trajectories in a summarized form. For comparison, a summarized representation of Cluster 10 (133 trajectories) is shown in
Figure 3(c).

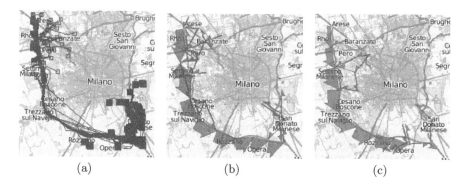

(a)   (b)   (c)

Fig. 3. Here we show complete clusters based on select specimens: (a)745 trajectories of the 1st cluster, (b) their summarized representation, (c) and summarized representation of 133 trajectories in the 2nd cluster

4   Conclusions

We have presented a tool to tackle the problem of analyzing a large dataset of trajectories, where the size and the complexity of the data prevent the effective application of traditional data mining methods. The visual tool is based on a step wise user-driven approach, based on the reduction of the problem complexity through sampling and filtering. The proposed methodology is able to find natural clusters in a complex and large dataset, hence it is able to underline meaningful mobility behaviors. In addition to clusters, the method also produces a classifier that can be used for many purposes such as movement prediction, detection of abnormal behaviors etc. The scalability of the tool have been tested on a real dataset of GPS tracked cars (around 200.000 trajectories from 17.000 cars equipped with an on-board GPS receiver, collected during 7 days in the city of Milan, Italy, resulting in more than 2,000,000 irregularly sampled positions).

5   Nature of the Demonstration

The demo will present the functionalities provided by the tool. Depending on the availability of newtwork connection, it is possible to test the application both on the online dataset and on a local dump of the data. We will show the steps followed to analyze a real dataset of GPS tracked cars using the visual analytic environment.

References

1. Rinzivillo, S., Pedreschi, D., Nanni, M., Giannotti, F., Andrienko, N., Andrienko, G.: Visually driven analysis of movement data by progressive clustering. Information Visualization 7(3-4), 225–239 (2008)
2. Pelekis, N., Kopanakis, I., Marketos, G., Ntoutsi, I., Andrienko, G.L., Theodoridis, Y.: Similarity search in trajectory databases. In: TIME, pp. 129–140 (2007)

ELKI in Time: ELKI 0.2 for the Performance Evaluation of Distance Measures for Time Series

Elke Achtert, Thomas Bernecker, Hans-Peter Kriegel, Erich Schubert, and Arthur Zimek

Ludwig-Maximilians-Universität München
Oettingenstr. 67, 80538 München, Germany
{achtert,bernecker,kriegel,schube,zimek}@dbs.ifi.lmu.de
http://www.dbs.ifi.lmu.de

Abstract. ELKI is a unified software framework, designed as a tool suitable for evaluation of different algorithms on high dimensional real-valued feature-vectors. A special case of high dimensional real-valued feature-vectors are time series data where traditional distance measures like L_p-distances can be applied. However, also a broad range of specialized distance measures like, e.g., dynamic time-warping, or generalized distance measures like second order distances, e.g., shared-nearest-neighbor distances, have been proposed. The new version ELKI 0.2 now is extended to time series data and offers a selection of these distance measures. It can serve as a visualization- and evaluation-tool for the behavior of different distance measures on time series data.

1 Introduction

In high dimensional data, and especially in time series data, the choice of a distance measure suitable and meaningful w.r.t. the data in question is essential. In many implementations of algorithms, either provided by authors or implemented in general frameworks, the Euclidean distance is invariably used as a standard distance measure.

In the software system described in this paper, we facilitate the use of a wide range of different algorithms along with a wide choice of distance measures. The framework provides the data management independently of the tested algorithms. So all algorithms and distance measures are comparable on equal conditions. But even more important is an intuitive and easy-to-understand programming style to invite additional contributions. This way, the interested users can easily provide a new algorithm or a new customized distance measure and compare their performance with existing solutions.

2 An Overview on the Software System

Focus and strength of Weka [1] and YALE [2] as popular environments for data mining algorithms is mainly in the area of classification, while clustering approaches are somewhat under-represented. However, first steps towards incorporating subspace clustering into Weka have been presented recently [3]. Although

N. Mamoulis et al. (Eds.): SSTD 2009, LNCS 5644, pp. 436–440, 2009.

both, Weka and YALE, support the connection to external database sources, they are based on a flat internal data representation. Thus, experiments assessing the impact of an index structure on the performance of a data mining application are not possible using these frameworks. Furthermore, in both frameworks, the user often cannot select the distance measure to be used by a certain algorithm but the distance computation is coded deeply in the implementation of an algorithm. Especially, neither Weka nor YALE do support special requirements like specialized distance measures for time series data. On the other hand, frameworks for index structures, such as GiST [4], do not provide any precast connection to data mining applications. Finally, data mining tools specialized for time series data, like T-Time [5], do support specialized distance measures for time series but usually provide only a small selection of algorithms for clustering or classification of time series data that can be used in combination with suitable distance measures.

To combine these different aspects in one solution, we built the Java Software Framework ELKI (**E**nvironment for Deve**L**oping **K**DD-Applications Supported by **I**ndex Structures). ELKI version 0.1 [6] already comprised a profound and easily extensible collection of algorithms for data mining applications, such as item-set mining, clustering, classification, and outlier-detection. On the other hand, ELKI incorporates and supports arbitrary index structures to support even large, high dimensional data sets. But ELKI does also support the use of arbitrary data types and respective distance functions. Thus, it is a framework suitable to support the development and evaluation of new algorithms at the cutting edge of data mining as well as to incorporate experimental index structures or to develop and evaluate new distance measures, e.g., to support complex data types.

ELKI intends to ease the development of new algorithms and new distance measures by providing a wealth of helper classes and methods for algebraic and analytic computations, and simulated database support for arbitrary data types using an index structure at will.

2.1   The Environment: A Flexible Framework

As a framework, our software system is flexible in a sense, that it allows to read arbitrary data types (provided there is a suitable parser for your data file or adapter for your database), and supports the use of any distance or similarity measure appropriate for the given data type. Generally, an algorithm needs to be provided with a distance function of some sort. Thus, distance functions connect arbitrary data types to arbitrary algorithms.

The architecture of the software system separates data types, data type-specific distance measures, data management, and data mining applications. So, different tasks can be implemented independently. A new data type can be implemented and used by many algorithms, given a suitable distance function is defined. An algorithm will perform its routine irrespectively of the data handling which is encapsulated in the database. A database may facilitate efficient data management via incorporated index structures. Index structures are encapsulated in database objects. These database objects facilitate range queries using arbitrary distance functions. Algorithms operate on database objects irrespective of the underlying index structure. So the implementation of an algorithm, as pointed out above, is not concerned with the details of handling the data which can be supported by arbitrary efficient procedures.

2.2   Arbitrary Distance Measures

Often, the main difference between clustering algorithms is the way to assess the distance or similarity between objects or clusters. So, while other data mining systems usually predefine the Euclidean distance as the only possible distance between different objects, ELKI allows to flexibly define and use any distance measure. This way, for example, subspace clustering approaches that differ mainly in the definition of distance between points (like e.g. COPAC [7] and ERiC [8]) can use the same algorithmic routine and become, thus, highly comparable in their performance.

Distance functions are used to perform range queries on a database object. Any implementation of an algorithm can rely on the database object to perform range queries with an arbitrary distance function and needs only to ask for k nearest neighbors not being concerned with the details of data handling.

A new data type is supposed to implement the interface `DatabaseObject`. A new algorithm class suitable to certain data types O needs to implement the Interface `Algorithm<O extends DatabaseObject>`. The central routine to implement the algorithmic behavior is `Result run(Database<O> database)`. Here, the algorithm is applied to an arbitrary database consisting of objects of a suitable data type. The database supports operations like

`<D extends Distance<D>> List<DistanceResultPair<D>> kNNQueryForObject(`
`    O queryObject, int k, DistanceFunction<O,D> distanceFunction)`

performing a k-nearest neighbor query for a given object of a suitable data type O using a distance function that is suitable for this data type O and provides a distance of a certain type D. Such a query method returns a list of `DistanceResultPair<D>` objects encapsulating the database IDs of the collected objects and their distance to the query object in terms of the specified distance function. A `Distance` object (here of the type D) in most cases just encapsulates a `double` value but could also be a more complex type, e.g. consisting of a pair of values as often used in subspace or correlation clustering algorithms like DiSH [9] or ERiC [8]. This list is sorted in ascending order w.r.t. the distance from the query object. As such, this method or related methods for epsilon-range queries are not only used in any clustering algorithm but also for comparing different distance measures. In ELKI 0.2, the performance of different distance measures can be directly assessed and visualized to enable the researcher to get a feeling for the meaning, benefits and drawbacks of a specific distance measure.

For that purpose, in a data set of time series, a specific time series can be picked and a k-NN query can be performed for this time series within the data

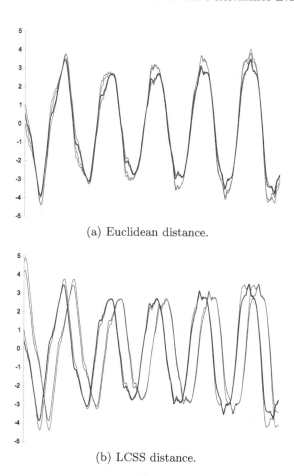

(a) Euclidean distance.

(b) LCSS distance.

Fig. 1. ELKI 0.2: Visualization of k-NN query results for different distance measures

set for any k and any distance function. The result of the query is e.g. visualized by assigning colors of degrading similarity to the time series in the query result according to the decreasing similarity w.r.t. the given distance measure. An example is shown in Figure 1: the query time series (blue) and its 3 nearest neighbors (color blending from blue to red with increasing distance). In this case, the different behavior of Euclidean distance (Figure 1(a)) and LCSS distance (Figure 1(b)) is demonstrated.

For time series data, in ELKI 0.2 especially the following exemplary distance measures are incorporated: the *Dynamic Time Warping (DTW)* distance [10], the *Longest Common Subsequence (LCSS)* distance [11], the *Edit Distance on Real sequence (EDR)* [12] and the *Edit distance with Real Penalty (ERP)* [13]. Any clustering or classification algorithm may therefore use a specialized distance function and implement a certain routine using this distance function on an arbitrary database.

2.3 Availability and Documentation

Via `http://www.dbs.ifi.lmu.de/research/KDD/ELKI/` the framework ELKI, documentation of the implementation and usage as well as examples to illustrate how to expand the framework by integrating new algorithms are available.

3 Conclusion

The software system ELKI presents a large collection of data mining algorithms which can be supported by arbitrary index structures and work on arbitrary data types given supporting data classes and distance functions. ELKI 0.2 is also able to visualize the behavior and the possibly different partialities of different distance measures for time series data. We therefore expect ELKI 0.2 to facilitate broad experimental evaluations of algorithms and distance measures – existing and newly developed ones alike.

References

1. Witten, I.H., Frank, E.: Data Mining: Practical machine learning tools and techniques, 2nd edn. Morgan Kaufmann, San Francisco (2005)
2. Mierswa, I., Wurst, M., Klinkenberg, R., Scholz, M., Euler, T.: YALE: Rapid prototyping for complex data mining tasks. In: Proc. KDD (2006)
3. Müller, E., Assent, I., Günnemann, S., Jansen, T., Seidl, T.: OpenSubspace: an open source framework for evaluation and exploration of subspace clustering algorithms in WEKA. In: Proc. OSDM@PAKDD (2009)
4. Hellerstein, J.M., Naughton, J.F., Pfeffer, A.: Generalized search trees for database systems. In: Proc. VLDB (1995)
5. Aßfalg, J., Kriegel, H.P., Kröger, P., Kunath, P., Pryakhin, A., Renz, M.: T-Time: threshold-baed data mining on time series. In: Proc. ICDE (2008)
6. Achtert, E., Kriegel, H.P., Zimek, A.: ELKI: a software system for evaluation of subspace clustering algorithms. In: Ludäscher, B., Mamoulis, N. (eds.) SSDBM 2008. LNCS, vol. 5069, pp. 580–585. Springer, Heidelberg (2008)
7. Achtert, E., Böhm, C., Kriegel, H.P., Kröger, P., Zimek, A.: Robust, complete, and efficient correlation clustering. In: Proc. SDM (2007)
8. Achtert, E., Böhm, C., Kriegel, H.P., Kröger, P., Zimek, A.: On exploring complex relationships of correlation clusters. In: Proc. SSDBM (2007)
9. Achtert, E., Böhm, C., Kriegel, H.P., Kröger, P., Müller-Gorman, I., Zimek, A.: Detection and visualization of subspace cluster hierarchies. In: Kotagiri, R., Radha Krishna, P., Mohania, M., Nantajeewarawat, E. (eds.) DASFAA 2007. LNCS, vol. 4443, pp. 152–163. Springer, Heidelberg (2007)
10. Berndt, D., Clifford, J.: Using dynamic time warping to find patterns in time series. In: KDD Workshop (1994)
11. Vlachos, M., Kollios, G., Gunopulos, D.: Discovering similar multidimensional trajectories. In: Proc. ICDE (2002)
12. Chen, L., Özsu, M., Oria, V.: Robust and fast similarity search for moving object trajectories. In: Proc. SIGMOD (2005)
13. Chen, L., Ng, R.: On the marriage of Lp-norms and edit distance. In: Proc. VLDB (2004)

Hide&Crypt: Protecting Privacy in Proximity-Based Services

Dario Freni, Sergio Mascetti, and Claudio Bettini

DICo, Università di Milano

Abstract. A particular class of location based services, called *friend finder*, is becoming very popular. Using this kind of service, a subscriber obtains to know location information about other participants (called buddies). Current commercial applications of this service imply the acquisition by the service provider of the user location at the best possible precision. This can discourage many users to use this service, since the precise location is a private information. We present a privacy-aware friend finder system called *Hide&Crypt* that notifies a user whether her buddies are within a user-specified threshold distance. The user can specify her location privacy requirements both with respect to the service provider and to the other buddies. The system includes a server application, and clients for mobile devices and desktop computers.

1 Introduction

Friend finders are one of the most popular location based applications. Using a friend finder, it is possible to obtain location information about the so-called *buddies*, which are other users participating in the service. The buddies associated to a user can be predetermined, like with a contact list, or chosen dynamically, e.g. as a result of a query about users' interests. Many applications available in the market allow users to see the exact location of the buddies on the geographical map, others let users only know the exact distance between their position and their buddies. Most of these services require the client applications to periodically update their location to the service provider (SP) at the most precise resolution. This may arise a privacy concern to those users that consider their exact location as a sensitive information. For the same reason, some users would prefer to use a friend finder service, while being able to tune the precision of the location information revealed to the other buddies.

One of the functionalities of a friend finder application allows a user to discover which of her buddies are in proximity, i.e., which buddies are located within a *distance threshold* specified by that user. Technically, this is equivalent to compute a range query over a database of moving entities (the buddies). In this demo we show a fully functional friend finder system, which implements privacy preserving techniques that allows a user to choose the minimum privacy requirements with respect to the service provider and each of her buddies. The system implements the protocol *Hide&Crypt*, designed to privately compute range queries [2].

N. Mamoulis et al. (Eds.): SSTD 2009, LNCS 5644, pp. 441–444, 2009.

Several friend finder applications already exists. Among the others, two of the most famous are Google Latitude [5], that allows the user to view the location of her buddies on a map, and Loopt [6], which also has social network functionalities. Unfortunately, none of the existing friend finders provides a sophisticate privacy preserving technique. For example, Google Latitude let users choose whether the location information sent to one buddy is provided at the maximum available precision or generalized to the city level. One could argue that this solution lacks of flexibility. In addition, in any case, the protocol requires the user to send her exact location to the SP. To the best of our knowledge, the only friend finder application that has been specifically designed to guarantee users' privacy is NearbyFriend, which implements one of the solutions presented in [3]. Technically, the main difference with our approach is that NearbyFriend uses a peer-to-peer protocol. This approach prevents the release of any location information to the SP. However, it requires a user to contact all of her buddies each time a range query is issued. This can be prohibitively expensive in terms of communication costs when the number of buddies is large. Vice versa, our solution takes advantage of a centralized SP that enhances the computation of range queries. This improvement is achieved at the cost of revealing some location information to the SP. However, users can choose how much location information they reveal to the SP: the more precise location information is provided, the most efficient the service is.

2   The *Hide&Crypt* System

Location privacy can be expressed as the uncertainty that an external entity has about the position of a user. Hence, a user can specify a *minimal uncertainty region* from where an adversary cannot exclude any point as a possible position. For example, Alice specifies that Bob should never be able to find out the specific building where Alice is within the campus. To formally model the totality of these uncertainty regions, the notion of *spatial granularity* can be used. Analogously to time granularities [1], a spatial granularity is a subdivision of the spatial domain into a discrete number of non-overlapping regions, called *granules*. In our approach, users can specify their minimum privacy requirements through a granularity[1] G^{SP} for the SP and a granularity G^U for each of her buddies.

The *Hide&Crypt* protocol is composed by two steps. In the former, called *SP-Filtering*, a user sends to the SP her generalized location. While the formula to compute this generalization considers G^{SP} as well as all the G^U granularities, the intuition is that the generalized location is a region including the granule of G^{SP} that contains the position of the user. When two users A and B sends their generalized location, the SP computes the minimum and maximum distance between the two regions. If the minimum distance is larger than the distance threshold δ_A defined by A, the SP communicates to A that B is not in proximity. Indeed, independently from the precise location of the two users, the distance between A and B is larger than δ_A. Analogously, if the maximum distance is

---

[1] In the following we use "granularity" to mean "spatial granularity".

smaller than δ_A, the SP communicates to A that B is in proximity. Finally, if δ_A is between the minimum and the maximum distance, the SP is not able to compute the proximity of B with respect to A and communicates to A that the second step of the protocol should be initiated. The same computations on the SP are run considering the distance threshold defined by B.

In the second step of the protocol a user A initiates a two-parties secure computation with a user B. First, A retrieves from the SP the granularity G^U defined by B as the minimum privacy requirement with respect to A^2. Then, A computes the set of granules of G^U that intersects the circle C centered in the exact location of A, having radius equal to the distance threshold of A. Running a secure computation for the set-inclusion problem, A discovers whether B is located in one of the granules in that set. If this is the case, A derives that B is in proximity, otherwise she derives that B is not in proximity. During this step, B does not acquire any new knowledge about the location of A.

The *Hide&Crypt* protocol has been proved to guarantee the minimum privacy requirements defined by the users. In addition, it has also been shown that it is able to achieve a level of privacy significantly larger than the minimum required [2]. However, it should be observed that, due to the use of granularities, a form of approximation is introduced. Indeed, it can happen that a user B is located in a granule g that intersects with C, while the distance between the exact locations of A and B is above the distance threshold. It has been experimentally shown that false positives are very rare and with practically useful settings the protocol has precision close to 1.

As required by the filtering step of the *Hide&Crypt* protocol, the architecture is centralized. In principle, the two-parties communication would require a direct connection between two buddies. However, in practice, it is not always possible to establish a communication between two peers (e.g. due to the presence of NAT or firewalls). Consequently, in our solution, the SP also acts as a gateway for the (encrypted) communications between users.

As observed in the introduction, the set of buddies associated to a user can be predetermined or chosen dynamically. Our application supports the computation of proximity in both cases. In the former, the computation of proximity is performed among the buddies that are in the contact list of a user. In the latter, the proximity query also includes as parameter some profile preferences. The SP is then able to compute the proximity among the participants matching that user's preferences.

3   Implementation

For what concerns the implementation, the server component is developed using Java and it is running on a Windows 2003 Server machine. Currently, the server offers the functionalities to compute the proximity only. However, it is designed to support extra functionalities we are planning to develop soon. These functionalities

---

^{2} We assume that the minimum privacy requirements are public knowledge.

includes: instant messaging, microblogging and the possibility to show the location of the buddies on a map if the proper authorization is provided.

We developed two client applications, one for desktop computers and one for mobile devices. The desktop software is a web application using Google Gears API [4] to obtain a user's location. The application for the mobile devices is implemented using Java and runs on the Android platform. This makes it possible to use the platform's API to acquire the location.

For testing purpose, we also developed some additional software that will be shown during the demo. One software simulates the behavior of the participants in the service, including their movements and their requests for proximity. Another software graphically shows the impact of the privacy preferences on the privacy that is provided to users, the system performance and the service precision.

4   Demonstration Outline

- We introduce the user to the privacy problem by showing the information transmitted by existing friend finder applications.
- We show the main functionalities of the *Hide&Crypt* client application. In particular we present the GUI that shows which buddies are in proximity. We also illustrate how it is possible to set the privacy requirements with respect to the SP and to the buddies.
- We run the application using both real buddies and simulated ones.
- We introduce our testing software. In particular we present the application that shows on the map the location data acquired by the other buddies, that measures the communication costs, and that detects false positives results of the proximity computation. Using this tool we explain how the privacy preferences affect the three main parameters characterizing the service: user's privacy, system costs and service precision.

Acknowledgments

This work was partially supported by National Science Foundation under grant CT-0716567 and by Italian MIUR under grant PRIN-2007F9437X.

References

1. Bettini, C., Wang, X.S., Jajodia, S.: Time Granularities in Databases, Temporal Reasoning, and Data Mining. Springer, Heidelberg (2000)
2. Mascetti, S., Bettini, C., Freni, D., Wang, X.S., Jajodia, S.: Privacy-aware Proximity Based Services. In: Proc. of the 10th International Conference on Mobile Data Management. IEEE Computer Society, Los Alamitos (2009)
3. Zhong, G., Goldberg, I., Hengartner, U.: Louis, Lester and Pierre: Three protocols for location privacy. In: Borisov, N., Golle, P. (eds.) PET 2007. LNCS, vol. 4776, pp. 62–76. Springer, Heidelberg (2007)
4. http://gears.google.com/
5. http://www.google.com/latitude/
6. http://www.loopt.com/

ROOTS, The ROving Objects Trip Simulator

Wegdan Abdelsalam[1], Siu-Cheung Chau[2], David Chiu[1], Maher Ahmed[2],
and Yasser Ebrahim[3]

[1] University of Guelph, Canada
[2] Wilfrid Laurier University, Canada
[3] Prince Sultan University, Saudi Arabia

Abstract. This paper introduces a new trip simulator, ROOTS. ROOTS creates moving objects with distinct characteristics in terms of driving style and route preference. It also creates a road network and associates each road with some characteristics. The route taken and the moving object speeds during the trip are determined based on both the characteristics of the moving object, those of the road being travelled, and other contextual data such as weather conditions and time of day.

1 Introduction

In previous work we introduced the idea of using moving objects modelling as a means of lowering uncertainty in location tracking applications [3]. Our focus is on human moving object travelling over a network. We refer to this class of moving objects as Roving objects (ROs). The location of the RO at a certain point in time is determined based on his/her reported location, estimated route, and estimated speed(s). Causality relationships between contextual variables such as road speed limit, road type, day of week, time of day, and weather conditions; and RO speed and route are used to build the RO model. The RO model is what we use to estimate his/her speed, route, and other factors that may affect his/her location (e.g., stopping pattern).

In this paper we present a new trip simulator, the ROving Objects Trip Simulator (ROOTS) we developed to test our modelling-based approach. The primary motivation behind developing ROOTS is the need for a trip simulator that creates ROs with distinct characteristics and preferences that are taken in account when trips are generated. ROOTS focuses on the RO characteristics that affect the speed and route. In ROOTS, ROs differ in the way they respond to a number of contextual variables, such as the time of day, day of week, weather conditions, and road type.

Although a number of network-based trip generators is available [2], none of them take in account moving-object-specific information such as object's preferences and habits in account. As a matter of fact, many such tools dispose of the moving object at the end of the trip. Since context-aware and location-aware applications require objects to be persistent and information about RO behaviour to be collected for modelling purposes, none of the currently available generators can be used. ROOTS is the first trip simulator to fill this void.

N. Mamoulis et al. (Eds.): SSTD 2009, LNCS 5644, pp. 445–449, 2009.

2   ROOTS: ROving Object's Trip Simulator

ROOTS simulates a *city road network* by creating a number of *roads* that form a grid. Each road is divided into a number of *road segments*. These roads are travelled by a number of ROs. Each RO goes on a number of *trips* each of which is associated with a *route* between the trip's source and destination. During the trip, the RO generates a number of *location reports*.

ROOTS generates several ASCII files representing the entities shown in italics in the previous paragraph. Two sets of files are generated, one is in a comma delimited format, the other is in Spatial Data Loader (SDL) format. The SDL format is used with AutoDesk's MapGuide and other World Wide Web map authoring tools.

2.1   Roving Objects

Each RO is associated with an ID , Aggressiveness level (AGGRESSIVE, AVERAGE, SAFE), Route preference (FASTEST, SHORTEST), and Typical trips' source/destination, time of day, and day of week. The aggressiveness level of the RO is reflected in his/her speed. The aggressiveness level is assigned randomly to each RO at creation time. Because each aggressiveness level is equally probable, about one third of the ROs belong to each of the three aggressiveness level classes.

An initial RO speed is randomly generated based on the road segment's speed limit and the RO's aggressiveness level. For each RO aggressiveness level, there is a probability distribution that determines the probability of the RO being above, within, or below the speed limit. Going above the speed limit, means going on a speed anywhere from the speed limit up to $x\%$ above the speed limit. Going within the speed limit, means deviating no more than $y\%$ of the speed limit either way. Going below the speed limit, means going anywhere from the speed limit down to $z\%$ below the speed limit. x, y, and z are determined by the user.

The initial RO speed can be negatively affected by the driving conditions and the road level of service. The time-of-day, day-of-week, and the area information is used to calculate a Level of Service (LoS) based on a formula determined by the user. There are 6 levels of service ranging from A to F going from best to worst. The LoS negatively affects the previously calculated speed according to a user-defined percentage for each level of service.

The RO speed is also affected by driving conditions which is in turn affected by the weather conditions and the road type. There are three driving conditions: GOOD, FAIR, and BAD. The effect of weather conditions on speed differs based on the RO's aggressiveness level. Note that these reductions are applied to the speed after being adjusted by the LoS. The amount of reduction in speed is determined by the user. Fig. 1 depicts the relationships among the variables affecting the RO speed.

The route preference can either be that of distance (i.e., the shortest route) or speed (i.e., the fastest route). The route speed is determined based on the speed limits on its constituent road segments. At the time of determining the route of a trip, the RO's route preference is used to choose between the two possible

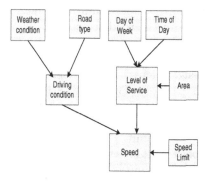

Fig. 1. RO speed model

routes for the trip (i.e., the shortest and the fastest routes). Each RO has an equal chance of having a distance or speed route preference.

Each RO has one or more *typical trip(s)* that he/she makes frequently around the same time. Typical trips my be the trip to/from work, supermarket, gym, ...etc. The source, destination, day-of-week, and time-of-day of each typical trip are all chosen randomly. The typical trip(s) info is stored to be used at the time of creating trips for the RO. The typical trips percentage of all the trips made by the RO is a user specified constant.

Each trip (typical or not) is associated with an ID , Source and destination address (the coordinates of a road segment start/end point), Day of week, Time, Weather condition (DRY, RAINY, or SNOWY), and RO. All these attributes are determined randomly except for the trip ID wich is determined sequentially. Each trip is also associated with a route. Based on the RO preference, either the fastest or the shortest route between the source and destination is associated with the trip.

2.2 Roads and Routes

Each road is associated with an ID, Direction (East-West or North-South), Type (EXPRESSWAY, PRIMARY HIGHWAY, SECONDARY HIGHWAY, MAJOR ROAD, LOCAL ROAD, or TRAIL), and a Road number which determines the road's location on the grid. Road numbers are multiples of 100 with the first road having the number 0.

There are two sets of roads, those running North-South and those running East-West. The roads are spaced equally to form a uniform grid. There are no diagonal or winding roads. The road type is assigned randomly using a distribution that is determined by the user.

Each road is divided into a set of road segments. A road segment is the road stretch between two consecutive intersections. All roads have the same number of road segments and all road segments have the same length. The road segment length and the number of road segments within each road are user defined constants.

Each road segment has an ID, Speed limit (a numeric value representing the maximum allowable speed on the road segment in kilometres per hour K/h), Area (a numeric value that represent the area of city the road segment is part of such as downtown), Start and end coordinates (x,y coordinates of either ends of the road segment), Length (length of the road segment in meters),and Road (the road the segment is part of). The length attribute is added here to allow future versions of the simulator to handle variable road segment lengths.

For road types EXPRESSWAY and PRIMARY-HIGHWAY, all segments of the road have the same speed limit. For other road types, when the first road segment is created, its speed limit is determined by its road type. For subsequent segments of the road, the speed limit is set to that of the preceding road segment $k\%$ of the time. In $100 - k\%$ of the time, the speed limit varies by m kph from the preceding road segment. The direction of the change (up or down) is determined randomly. As a result, the same road will have stretches that have different speed limits in simulation of real roads. The speed limit change is allowed as long as it does not result in a value that exceeds the speed limit of the road type, or goes below a certain minimum speed. k, m, the road type speed limit, and the minimum speed limit are all user defined constants.

The road segments are grouped into rectangular shaped areas. A set of attributes are associated with each area, ID , Lower left coordinates, Upper right coordinates. The number of areas, the coordinates for each area, and the effect each area has on the level of service is determined by the user.

A route is a set of road segments between the source and destination of a trip. Each route has an ID, a unique string that uniquely identifies the route. Routes are optimized for either time or distance. The road network is represented as a graph and Dijkstra's shortest path algorithm is used to determine the shortest distance and the shortest time routes between the source and destination points. In only $l\%$ of the time the route selection will conform to the RO's preference. This is to reflect the possibility that the RO may choose a different route in exceptional situations. l is determined by the user.

3   ROOTS Demonstration

The demonstration presents an overview of ROOTS's design and approach to produce synthetic datasets for moving object databases, including a demonstration of ROOTS current capabilities and a look into the ROOTS's development goals for its 1.0 release.

ROOTS version 1.0 was implemented in Java version 1.6.0. It consists of three parts, ROOTS configure, ROOTS server, and ROOTS viewer (i.e., client). ROOTS configure is used to customize the ROOTS variables and generate the datasets. ROOTS server is a small scale database management system that loads a specific dataset, opens a connection, and waits for the ROOTS client queries. ROOTS viewer is a graphical user interface connected to the server to view the dataset. ROOTS viewer allows the user to pan, zoom in and out, and use elastic band zooming. Since ROOTS adopts a client/server architecture, many instants

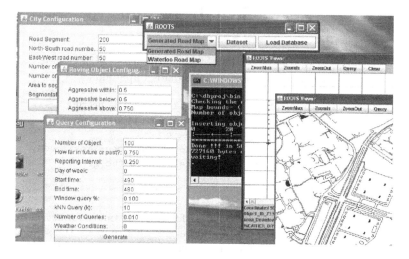

Fig. 2. Screenshot of ROOTS

of the ROOTS viewer may be loaded simultaniously. Fig. 2 shows a screenshot of the ROOTS demo under Microsoft Windows environment.

4   Conclusions and Future Work

This paper presents ROOTS, a trip simulator that simulates moving objects' characteristics and preferences. The data provided by ROOTS can be used to build moving object models that can be used to better estimate the location of roving objects.

ROOTS is an ongoing project. Currently we are working on turning it into an extensible platform that allows the user to create new characteristics and behaviours of the ROs. This highly customizable environment will allow ROOTS to be tailored to the needs of specific applications and environments and to generate trips on any arbitrary network (i.e., road map) given by the user.

References

1. Brakatsoulas, S., Pfoser, D., Tryfona, N.: Modeling, storing and mining moving object databases. In: Proceedings of International Database Engineering and Applications Symposium, pp. 68–77. IEEE, Los Alamitos (2004)
2. Brinkhoff, T.: Generating traffic data. Bulletin of the Technical Committee on Data Engineering 26, 19–25 (2003)
3. Abdelsalam, W., Siu-Cheung, C., Ahmed, M., Ebrahim, Y.: A roving user modeling framework for location tracking applications. In: 9th International IEEE Conference on Intelligent Transportation Systems, IEEE ITSC 2006, pp. 169–174 (2006)

The TOQL System

Evdoxios Baratis, Nikolaos Maris, Euripides G.M. Petrakis,
Sotiris Batsakis, and Nikolaos Papadakis

Department of Electronic and Computer Engineering
Technical University of Crete (TUC)
Chania, Greece
{dakis,petrakis}@intelligence.tuc.gr, nickmeet@gmail.com,
{batsakis,npapadak}@intelligence.tuc.gr

Abstract. TOQL, is a query language for querying time information in
ontologies. An application has been developed that supports translation
and execution of TOQL queries on temporal ontologies. A Graphical User
Interface (GUI) has been also developed to facilitate user interaction and
supports operations such as syntax highlighting, code autosuggestion,
loading of the ontology into the main memory, results and error display.

1   Introduction

TOQL (Temporal Ontology Querying Language), is a high-level query language
for querying (time) information in ontologies. TOQL handles ontologies almost
like relational databases. TOQL maintains the basic structure of an SQL language (SELECT - FROM - WHERE) and treats the classes and the properties
of an ontology almost like tables and columns of a database. The following table
summarizes TOQL syntax:

Table 1. Generic TOQL syntax

Syntax
SELECT ... AS ...
FROM ... AS ...
WHERE ... LIKE ... AND ... LIKE "string" IGNORE CASE ... AT...

The TQQL system supports query translation and execution of temporal
queries (i.e. queries that contain the AT and Allen temporal operators) along
with a mechanism for representing time evolving concepts in ontologies inspired
by the four-dimensional perdurantist approach [4]. The 4D perdurantist mechanism is not part of the language and it is not visible to the user (so the user
need not be familiar with peculiarities of the underlying mechanism for time
information representation). A graphical user interface has also been developed
to facilate user interaction with both TOQL and with the knowledge base (the
termporal ontology).

N. Mamoulis et al. (Eds.): SSTD 2009, LNCS 5644, pp. 450–454, 2009.

2   System Architecture

The TOQL system has been implemented in Java. The system supports query translation and execution of TOQL queries on OWL ontologies containing temporal information.

Figure 1 illustrates the architecture of the TOQL system. The TOQL system consists of several modules, the most important of them being the TOQL interpreter whose purpose is to translate the TOQL query into a SeRQL [1] query, which is executed on the knowledge base. TOQL and SeRQL have different syntax and SeRQL does not support the full range of TOQL's time features. Part of the interpreter is a reasoner implemented in Prolog. The reasoner handles queries concerning properties that conform to the event calculus axioms [5]. The complete discussion of the TOQL implementation can be found in [2].

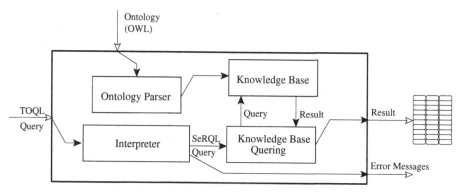

Fig. 1. TOQL system architecture

2.1   Ontology Parser

The input is a query written in TOQL and an ontology in OWL. The ontology is parsed using JENA[1] and SESAME[2] and is loaded into the main memory. The ontology is checked for consistency with the 4D fluent mechanism [2] (error messages are reported to the output). The SeRQL query addresses both the ontology structure (TBOX) and the knowledge base instances (ABOX). Inferred facts are asserted in the KB using the Pellet[3] OWL reasoner. A query language alone can be used to access temporal information that is explicitly represented in a temporal ontology, but cannot provide answers on information that can be inferred from existing information (e.g., if the price of a product at time t is p, TOQL should be able to infer that the price of the product is the same since the last time it was changed). TOQL is combined with a reasoner based on event calculus [5] to better support queries on temporal ontologies. The output to a query is a table with the results. If errors have been encountered during interpretation, the output is one or more error messages.

---

[1] http://sourceforge.net/projects/jena
[2] http://www.openrdf.org
[3] http://clarkparsia.com/pellet

2.2   Interpreter

A TOQL query is initially lexically, syntactically and semantically analyzed. Lexical analysis converts a sequence of characters to tokens (i.e., meaningful labels). The lexical analysed is implemented using JFlex[4]. For example, the token SELECT is given the meaning SELECT, while the token "Linux" is given the meaning NAME. The next step is syntax analysis (parsing) whose purpose is to analyze a sequence of tokens to determine grammatical structure (i.e., allowable expressions) with respect to a given formal grammar [2]. The parser transforms the query to a syntax tree, a form suitable for further processing. The parser is implemented using Byacc/J[5]. Syntactical errors are reported in the output. If the query is lexically and syntactically correct, query translation proceeds with semantic analysis. Semantic analysis adds semantic information to the parse tree and builds the symbol table. This phase performs two types of semantic checks. The first type needs no external knowledge. Semantic errors reported in this case include, use of a class in a SELECT or WHERE clause without having it declared in the FROM clause, use of a property in a SELECT or WHERE clause without a class preceding it,and use of more than one properties in the SELECT clause of a nested query.

   Detection of the second type of semantic errors needs external knowledge (i.e., the ontology). This requires that the ontology is first loaded into the main memory. The ontology is parsed using JENA and SESAME libraries. The semantic analyzer checks if a class or property used in a query exists in the ontology, if a property is a property of a specific class and finally, if a property is a fluent property (so that keyword TIME can be applied to it). A complete list of error messages that can be reported by the semantic checker can be found in [2].

Code generation: The last phase of query processing is the actual translation of a TOQL into an equivalent SeRQL query. Code generation performs the following steps:

- Intermediate code generation.
- Intermediate code parsing and instantiation to Java objects (representing the TOQL query).
- Processing of Java objects and expansion with 4D fluent elements.
- Processing of Java objects and mapping to Java objects (representing the SeRQL query).
- SeRQL query generation.

Finally the SeRQL query is applied to the Knowledge Base using SESAME and the result is presented to the user.

3   Graphical User Interface

A Graphical User Interface (GUI) has been also developed to facilitate user interaction with TOQL. It supports operations such as syntax highlighting and

---

[4] http://jflex.de
[5] http://byaccj.sourceforge.net

loading of the ontology into the main memory. The toolbar panel provides buttons for query editing (undo, redo, copy, cut, paste), for displaying query results as well as for displaying the SeRQL equivalent query. Syntactic and semantic errors are also displayed. The interface, provides options for ontology loading (the "Load Ontology" button loads a new ontology into the memory) and ontology viewing (i.e., "View Ontology" button displays the Abstract Ontology View referred to [2], which hide the temporal mechanism representation from the user). The query editing panel contains the query editor and a toolbar panel that contains buttons useful for querying editing (save, save as, load query, run).

Query formulation is supported by *TOQL syntax highlighting* (recognizes TOQL clauses keywords and classes-properties) and by *Code autosuggestion* (each time the user writes a class name followed by ".", a list with the class properties is displayed to choose from).

The results panel has two tabs. The first one displays the results returned by the query, while the second one displays the errors returned as the result of query parsing. These errors can be either due to inconsistencies with the 4D fluent representation or due to errors in TOQL syntax.

4   Conclusions

The TOQL system provides a high level user interface to TOQL, an SQL-like language for querying temporal information in ontologies. TOQL is currently being extended to handle queries on ontology structure (i.e., sub- classes and super-classes) as well as open-schema query functionality by allowing variables in the class position (e.g. using rdf:type as a property in a query triple).

References

1. Aduna, B.V.: The SeRQL query language. User Guide for Sesame 2.1, Chapter 9 (2002–2008), http://www.openrdf.org/doc/sesame2/2.1.2/users/ch09.html
2. Baratis, E.: TOQL: Querying Temporal Information in Ontologies. Master's thesis, Techn. Univ. of Crete (TUC), Dept. of Electronic and Comp. Engineering (July 2008)
3. McGuinness, D.L., VanHarmelen, F.: OWL Web Ontology Language Overview. W3C Recommendation (February 2004), http://www.w3.org/TR/owl-features
4. Welty, C., Fikes, R., Makarios, S.: A Reusable Ontology for Fluents in OWL. Technical Report RC23755 (Wo510-142), IBM Research Division, T. Watson Research Center, Yorktown Heights, NY (October 2005)
5. Shanahan, M.: The event calculus explained. In: Veloso, M.M., Wooldridge, M.J. (eds.) Artificial Intelligence Today. LNCS (LNAI), vol. 1600, pp. 409–430. Springer, Heidelberg (1999)

The TOQL System- Demonstration Outline

The TOQL sytem demonstration will include:

- Presentation of TOQL query examples.
- Presentation of the TOQL system interface and of its functionality (Figure 2).

Fig. 2. Graphical user interface: TOQL query with answer

- Comparison between TOQL and SeRQL syntax demonstrating the rich expressive power and simpler syntax of TOQL.

PDA: A Flexible and Efficient Personal Decision Assistant*

Jing Yang, Xiyao Kong, Cuiping Li, Hong Chen, Guoming He, and Jinghua Tian

School of Information, Renmin University of China, Beijing 100872, China
{jingyang,licuiping,chong,kongxiyao,hegm,jinghuatian}@ruc.edu.cn

Abstract. Despite a rich set of techniques designed to process specific types of optimization queries such as top-k, skyline, NN/kNN and dominant relationship analysis queries, a unified system for a general process of such kinds of queries has not been addressed in the literature. In this paper, we propose PDA, a interactive personal decision assistant system, for people to get the desired optimal decision easily. In addition to supporting the basic queries mentioned above, PDA can also support some sophisticated queries. Several novel technologies are employed to improve the flexibility and efficiency of the system and a visualization interface like google map is provided for people to view the query result interactively.

1 Introduction

In many business applications such as customer decision support, market data analysis, e-commerce, and personal recommend, users often need to optimize their selections of entities. While databases have been applied predominately in business settings with well-defined query logic, they are now frequently used in retrieving or analyzing data. In these scenarios, the desired decisions are described with some qualifying constraints, which specify what subsets of data should be considered valid, and some qualifying functions, which measure their degrees of matching. Such queries are often referred as constrained optimization query.

Many data exploration queries can be cast as constrained optimization queries, such as top-k queries [3,5,6], skyline queries [2,4,10], NN queries and its variants [1,9], dominant relationship analysis queries [8]. These queries can all be considered to be a type of optimization query with different constrains and object functions.

Despite a rich set of techniques designed to process specific types of queries, a unified system for a general process of such kinds of constrained optimization queries has not been addressed in the literature. In the real life, users often resort to different tools for specific query types.

In this demonstration, we present a web-based personal decision assistant system called PDA as a general processing platform for a variety of query types. It seamlessly integrates efficient query evaluation and versatile processing model for large amounts

---

* Supported by the National Science Foundation of China (60673138, 60603046), Program for New Century Excellent Talents in University.

N. Mamoulis et al. (Eds.): SSTD 2009, LNCS 5644, pp. 455–459, 2009.

of data and provides powerful optimization decision functionality. In addition to supporting basic queries mentioned above, PDA can support more sophisticated queries. For example, it can process those queries whose constraint is a dynamic range and the object function is an ad hoc aggregation function over the tuples within the range. The presence of such kind of constraints and functions presents significant challenges for optimization query processing since in this case it is hard to compute tight bounds and determine an efficient early stopping condition.

To address the above challenges and to improve the query flexibility and efficiency, PDA employs several practical techniques: a special index structure for sophisticated query, progressive approximate result retrieving, and multi-threaded query processing. In addition, PDA provides users with a web-based map exploration method to assist them to browse the query result easily and interactively by clicking and dragging the mouse.

2 System Architecture

The architecture of PDA is shown in Figure 1. The system is implemented in Java and follows B/S architecture. There are four main components in this system: Data Loading Module, User Interface, Query Execution Engine, and Results Displaying Module. Data Loading Module loads external data and converts it into the fixed format for PDA. User Interface provides a friendly interface which gives users the freedom to input their requirements. Query Execution Engine is the core part of our system. It processes certain kind of query according to the user's action in the User Interface. After analyzing the query constraints and object functions that the user imputed, it chooses the optimal algorithm to deal with the data set and returns the desired result. At last, Results Displaying Module shows the result set retrieved by the Query Execution Engine. Results Displaying Module is another key component of PDA. It provides a visual interface which is as intuitive and expressive as current web mapping services like Google maps.

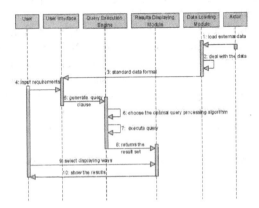

Fig. 1. System Overview

3   Core Query Execution Engine

In order to process various kinds of query types mentioned in the above specification, PDA adopts and substantially extends the versatile optimization query processing model proposed by [5]. As discussed earlier, the presence of blocking operators (Group by and order by) in the query plan makes the query evaluation wasteful. In order to use the general query processing framework of [5], a possible remedy would be that we first guess a value k'(k' > k) and ask the dynamic source to compute its top-k' results upfront, with k' being sufficiently large so that the global operator never needs any scores of items that are not in the local top-k'. PDA treats such situations by employing the following practical technologies.

Index Structures Organization. PDA uses a multi-resolution R-tree proposed in [7] instead of a regular R-tree to process sophisticated queries. The multi-resolution R-tree augments to each non-leaf entry of the R-tree an aggregate measure of all data points in the subtree pointed by it. Thus, algorithms for searching using the aggregate R-tree are identical to those of the corresponding plain tree because search does not affect the aggregate values stored. Figure 2 shows an example of a multi-resolution R-tree whose aggregate functions are min, max, count and sum respectively.

Progressive Approximate Result Retrieving. Since the user's goal behind a query usually is not to find exactly the best data objects with regard to some ranking model, but rather to explore and identify one or a few relevant pieces of information, it is often acceptable to provide approximate answers to users at a significantly lower computational cost by providing a good estimate without accessing large portions of the database or index. PDA provides an approximate ranking range aggregation algorithm based on the multi-resolution R-tree. It produces monotonically improving quality answers when the algorithm iteratively searching the multi-resolution R-tree level by level.

Multi-threaded Query Processing. For queries that cover a large part of the data space, the above progressive approximate query processing method works well since it need not explore deep nodes of the aggregate R-tree to retrieve the high quality answers. But if the query range is relatively small, and on the other hand, the number of group by is relatively large, the performance of the progressive method will decrease dramatically in this case since the algorithm will have to explore the very low nodes of the multi-resolution R-tree. Considering the existence of the natural parallelism among the multiple group-bys, we adopt a multi-threads query processing policies to do the sequential data scan and aggregate computation simultaneously when the required

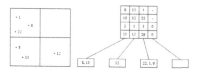

Fig. 2. An Example of Multi-Resolution R-tree

aggregation range below a pre-defined threshold. One thread will be activated for each group-by to scan the source and do the aggregation on the fly. After getting all the aggregations, it selects an appropriate top-k algorithm to get the final result.

4   Visual Result Display

When the query is executed, users can get the results from the Result Display Module. PDA's Result Display Module shows two kinds of information: the final result computed by the Query Execution Engine and the dynamic query execute procedure on how the results are produced.

The final query results can be displayed in various ways such as report forms, graphs, and so on. The report forms provide the basic information to users. The graphic interface offers a visual and interactive way for people to explore the result information.

The screenshot in Figure 3 shows a typical scenario when a user wants to find top-10 most popular hotels between July and September in Beijing. After the query is executed by the Query Execution Engine, our system not only provides a top-10 most popular hotels list including their basic information such as the hotel name, price and occupancy rate, but also locates the result hotels on the map in the right of the web page. This map is a true Beijing map, and when users click on the hotel's icon, the full information about it will be further provided.

To develop this map, we embed Google maps in PDA's web pages using Google Maps API2.0. The Google Maps API is now integrated with the Google AJAX API loader, which creates a common framework for loading and using multiple Google AJAX APIs. By employing the functions that the Google Maps API provides, we can easily develop a map and locate those result hotels on it.

Generalizing this idea, for any multi-dimensional data set, we can select its initial layout on two typical dimensions as a starting point and then if users click on some object, the full information of the object can be provided further. We implemented this function by using JavaScript to create a coordinate plane whose axes are the two dimensions initially selected by users. And then we can locate those result objects on this

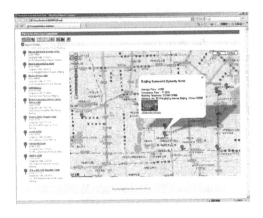

Fig. 3. Visual Result Display

coordinate plane according to these two dimensions. For each object on the coordinate plane, there is an Action Listener. And when users click on the object, full information of the object can be provided in the pop window.

5   User Interface Overview

The current design of the user interface of PDA is shown in Figure 4. It consists of several components. We discuss them below. Please note that this is a universal interface for all types of queries, therefore only some components are valid for certain type of query.

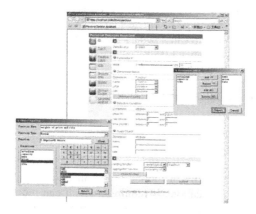

Fig. 4. User Interface

References

1. Corral, A., Manolopoulos, Y., Theodoridis, Y., Vassilakopoulos, M.: Algorithms for processing k-closest-pair queries in spatial databases. Data Knowl. Eng. 49(1), 67–104 (2004)
2. Fu, G., Papadias, D., Tao, Y., Seeger, B.: An optimal and progressive algorithm for skyline queries. In: SIGMOD (2003)
3. Han, J., Xin, D., Chang, K.C.-C.: Progressive and selective merge: Computing top-k with ad-hoc ranking functions. In: SIGMOD Conference, pp. 103–114 (2007)
4. Tan, K., et al.: Efficient progressive skyline computation. In: VLDB (2001)
5. Gibas, M., Zheng, N., Ferhatosmanoglu, H.: A general framework for modeling and processing optimization queries. In: VLDB Conference (2007)
6. Hristidis, V., Koudas, N., Papakonstantinou, Y.: Prefer: A system for the efficient execution of multi-parametric ranked queries. In: SIGMOD Conference (2001)
7. Lazaridis, I., Mehrotra, S.: Progressive approximate aggregate queries with a multi-resolution tree structure. In: SIGMOD Conference (2001)
8. Li, C., Ooi, B.C., Tung, A.K.H., Wang, S.: Dada: a data cube for dominant relationship analysis. In: SIGMOD Conference, pp. 659–670 (2006)
9. Roussopoulos, N., Kelley, S., Vincent, F.: Nearest neighbor queries. In: SIGMOD Conference, pp. 71–79 (1995)
10. Kossmann, D., Borzsonyi, S., Stocker, K.: The skyline operator. In: ICDE (2001)

A Refined Mobile Map Format and Its Application

Yingwei Luo, Xiaolin Wang, and Xiao Pang

Dept. of Computer Science and Technology, Peking University, Beijing, China, 100871
lyw@pku.edu.cn

Abstract. Byte-Map is a kind of vector format with different blocks through different levels. The basic cell of Byte-Map is a block, which is fixed in size of 255 units*255 units according to different coordinates systems and thus the co-ordinates of all the features in a certain block can be encoded with only two bytes. Based on Byte-Map, a LBS supporting platform LBS-p is built to illustrate mobile online map service.

Keywords: Mobile Online Map Service, Byte-Map, LBS-p.

1 Introduction

In order to achieve high quality of mobile online map services, an outstanding map data format for mobile applications is required, which can reduce map data volume and simplify map data processing complexity while keeping a reasonable display result in mobile device, so as to satisfy the situation of limited bandwidth of wireless network, low storage and CPU capability of mobile devices.

Mobile SVG [1] is now the most widely used data format in mobile map services [2] [3]. cGML [4] is a compressed version for GML. By using small tags, server side pre-projected and pre-scaled coordinates, cGML allows development and deployment of map-based software for mobile facilities with strong constraints on connections, CPU and memory. However, both Mobile SVG and cGML are based on XML, so it still needs lots of tags to identify the attributes, and the parsing time is not ignorable.

This paper presents Byte-Map, which is a novel data format for representing vector map. Also, LBS-p will use Byte-Map to provide mobile online map service.

2 Byte-Map Specification

2.1 The Design Metrics of Byte-Map

In order to minimize the data volume to satisfy the mobile applications, Byte-Map data is organized in different blocks through different levels.

(1) Level. LOD (Level of Detail) is a common model to organize massive data for visualization. Due to the limited screen size of mobile devices, mobile map data should adopt a same organization mode. The content of map to display should depend on the current scale. With a small scale, it's sufficient to show only some important features with their rough profiles. So different levels map data should be

N. Mamoulis et al. (Eds.): SSTD 2009, LNCS 5644, pp. 460–464, 2009.

pre-generated from the original data. Different Levels have different contents in different details to meet the demand of displaying in different map scales.

Take 3 levels for example, as shown in Figure 1. The highest displaying level "Level 0", which is supposed to meet a smallest scale, contains the fewest features going with the roughest profiles. The lowest displaying level "Level 2" contains map data in details to meet a big scale. Accordingly, the map content in "Level 1" is between in "Level 0" and in "Level 2".

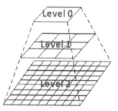 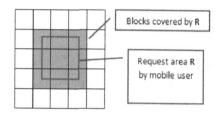

Fig. 1. Different Levels of Map Data   **Fig. 2.** Map Divided into Blocks

(2) Block. The area browsed by mobile users is only a small part of the whole map due to the limited screen size of mobile devices. Hence, it's reasonable to divide the whole map area into blocks, and each block has the same size. Server can only provide the blocks covered by the user's requested area, as shown in Figure 2.

Here, the choice of block size is a critical issue. To minimize the data volume, we set the block size as 255 units * 255 units according to different coordinates systems. Let the left-bottom point of a block be the base point, so in a certain block, the value of coordinates x or y is ranged from 0 to 255 according to the base point, which means that we can use only one byte to present x or y.

When showing map data of high level (e.g. Level 0), if the scope which one coordinate unit presents is the same as that in low level, the performance will decrease due to the large number of blocks. To avoid this, we can increase the coordinate unit of block. Different coordinate units will be adopted for different levels. For example, as shown in Figure 1, we can set "1cmeter" as the coordinate unit of blocks in Level 2, "10 meters" in Level 1 and "100 meters" in Level 0.

2.2 The Basic Cell of Byte-Map: Block

2.2.1 Structure of Block

The geographic range of each block in Byte-Map is 255 units*255 units. The left-bottom point is set to be the base point, which is the identifier of each block.

Figure 3 shows the structure of block. The data in block is stored as different features. The features with different attributes belong to different layers and features in the same layer are displayed in the same style. Besides geometric attribute, each feature only contains the most basic information, such as name, identifier and to which layer it belongs. Any other attribute information can be added when needed.

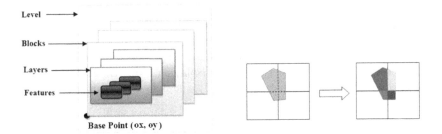

Fig. 3. Structure of Block   **Fig. 4.** Large Feature's Incision

2.2.2 Block Coordinates

For each block, the identifier is the base point (ox, oy), which is a real geographic coordinates. The coordinates of all features in a certain block is relative coordinates. Each coordinates (x, y) in a block is the real geographic coordinates minus that of the base point. Hence the range of x or y is 0~255, and can be encoded by one byte.

2.2.3 Feature Incision

Each block only includes the data exactly within its range. Hence large features may need to be incised because they may cross different blocks. The feature, as shown in Figure 4, crosses 4 blocks and becomes 4 parts after incision. Each part of the feature becomes an individual, but sharing the same name, identifier and the layer ID with other parts. Mobile terminal retrieves the original large feature by recognizing the incised parts with the same identifier.

2.3 Structure of Byte-Map

Map data is organized in blocks in Byte-Map. Figure 5 shows a slice of Byte-Map data while answering a data request from mobile terminals.

3 A LBS Supporting Platform LBS-p Based on Byte-Map

Based on Byte-Map, we have developed LBS-p, a platform supporting location-based services. Conformed to client-server architecture, LBS-p consists of two parts: LBS-p

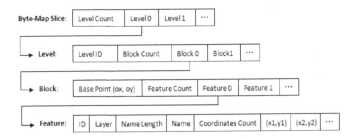

Fig. 5. Structure of Byte-Map

Mobile and LBS-p Server. LBS-p Mobile (as shown in Figure 6) can display map by providing some basic map functions, such as moving, zooming, etc. Besides, LBS-p Mobile can send request to LBS-p Server with specified range and levels.

Once the LBS-p Server (as shown in Figure 7) receives the message, it generates a spatial data set in the form of blocks covered by the requested range, encapsulates them in a Byte-Map slice, and then transmits back to LBS-p Mobile.

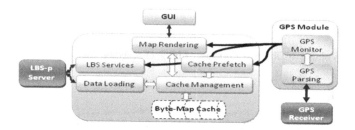

Fig. 6. LBS-p Mobile

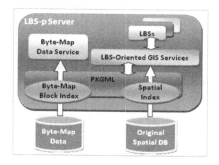

Fig. 7. LBS-p Server

4 Performance Evaluation

The critical issue of mobile online map service is to try to transmit much less map data and display map with much less response time in mobile terminal. In this section, we designed several experiments to compare the data volume in different format: Byte-Map, Mobile SVG Tiny (SVGT) and PNG. Also, we compared the handling complexity of Byte-Map data in mobile terminal with that of SVGT. The original map data used in the following experiments is about Beijing in the format of GML.

Figure 8 shows the amount of data encapsulated in different formats. Figure 9 shows the displaying time of SVGT (with Tinyline) and Byte-Map (Nokia N73, Symbian OS v9.1, CPU: 220MHz, memory: 64MB).

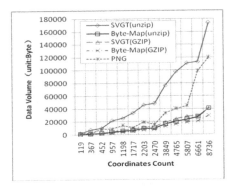 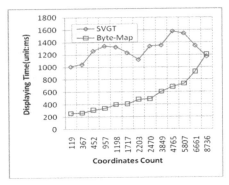

Fig. 8. Data Volume of Different Formats   **Fig. 9.** Displaying Time of SVGT and Byte-Map

5 Conclusions

Byte-Map is a novel kind of vector format with different blocks through different levels. The basic cell of Byte-Map is block, and the coordinates of all the features in a certain block can be encoded with only two bytes.

Acknowledgments

This work was supported by the National Grand Fundamental Research 973 Program of China under Grant No.2006CB701306; the National Science Foundation of China under Grant No. 40730527.

References

[1] W3C, Mobile SVG Profiles: SVG Tiny and SVG Basic (2003),
    http:// www.w3.org/TR/SVGMobile/
[2] Binzhuo, W., Bin, X.: Mobile Phone GIS Based on Mobile SVG. In: Proceedings of 2005 IEEE International Conference on Geoscience and Remote Sensing Symposium (IGARSS 2005)(2005)
[3] Li, D., Zhang, Y., Yu, B., Gu, N., Peng, Y.: Research on Mobile SVG Map Service Based on Java Mobile Phone. In: The 2nd IEEE Asia-Pacific Service Computing Conference (2007)
[4] De Vita, E., Piras, A., Sanna, S.: Using compact GML to Deploy Interactive Maps on Mobile Devices, WWW 2003 (2003), http://www2003.org/cdrom/papers/poster/p051/p51-devita.html

Author Index